Effects of Polyphenol-Rich Foods on Chronic Diseases

Effects of Polyphenol-Rich Foods on Chronic Diseases

Editors

Sonia de Pascual-Teresa
Luis Goya

MDPI

Basel • Beijing • Wuhan • Barcelona • Belgrade • Novi Sad • Cluj • Manchester

Editors
Sonia de Pascual-Teresa
Institute of Food Science,
Food Technology and
Nutrition (ICTAN)
Spanish Research Council
(CSIC)
Madrid, Spain

Luis Goya
Institute of Food Science,
Food Technology and
Nutrition (ICTAN)
Spanish Research Council
(CSIC)
Madrid, Spain

Editorial Office
MDPI
St. Alban-Anlage 66
4052 Basel, Switzerland

This is a reprint of articles from the Special Issue published online in the open access journal *Nutrients* (ISSN 2072-6643) (available at: https://www.mdpi.com/journal/nutrients/special_issues/polyphenol_foods_nutrients).

For citation purposes, cite each article independently as indicated on the article page online and as indicated below:

Lastname, A.A.; Lastname, B.B. Article Title. *Journal Name* **Year**, *Volume Number*, Page Range.

ISBN 978-3-0365-9262-6 (Hbk)
ISBN 978-3-0365-9263-3 (PDF)
doi.org/10.3390/books978-3-0365-9263-3

Contents

About the Editors

Sonia de Pascual-Teresa

Sonia de Pascual-Teresa holds a degree in Pharmacy and a Ph.D. from the University of Salamanca (1999). Since 2003, she has been working at ICTAN–CSIC (formerly Instituto del Frio), first thanks to a Ramón y Cajal contract, and then, since 2007, as a tenured scientist and latterly as a research scientist (2023). Previously, she worked at the Quadram Institute in Norwich (formerly IFR), UK, thanks to a Marie Curie Individual Fellowship, and at the University of Reading, UK, under a BBSRC grant. From the beginning, her entire career has been devoted to the field of dietary polyphenols, first from an analytical point of view and later focusing on their biological activity, moving from biochemical tests to cell culture models and human trials. In recent years, she has focused on demonstrating that dietary nutrients and bioactive compounds are capable of preventing the neurocognitive and cardiometabolic deterioration associated with aging.

Luis Goya

Luis Goya: Ph.D. with an extraordinary award from the Pharmacy School of the University Complutense, Madrid, in 1987. After a four-year postdoctoral research period at the University of California at Berkeley, he returned to Spain in 1993 to the Institute of Biochemistry (CSIC–UCM), where he studied the regulation of the insulin-like growth factor (IGF) systems by nutrients. For these and previous studies, the research group received the Reina Sofía National Award in 1994 for research focused on the prevention of deficiencies. He is currently a CSIC-affiliated research investigator in the Department of Metabolism and Nutrition at the Institute of Food Science and Technology and Nutrition (ICTAN), within the group of Metabolism and Bioactivity of Phytochemicals (acronym BIOCELL), where he is studying the biochemical and molecular mechanisms of action of bioactive compounds from diet. He has published more than 150 SCI-indexed articles, with most of them included in the first quartile of the research fields of biomedicine, molecular biology, pharmacology, toxicology, and nutritional biochemistry. He has been a member of the scientific committee of the Observatory of Cocoa along with Doctors Ramón Estruch and María Izquierdo Pulido since 2015 and a member of the editorial board of the scientific journal *Food Research International* of the Canadian Institute of Food Science and Technology since 2016, and his h index is 45.

Preface

Polyphenols comprise an extended number of compounds belonging to groups such as different flavonoids and hydroxycinnamic acids found mainly in plant foods and derived products, but they have also been found in trace amounts in animal products after they have consumed such polyphenol-rich plant products.

Polyphenols have been gaining attention since the 1990s, and although not considered nutrients, their study in relation to their health implications has produced thousands of scientific publications. However, it is still difficult to define their implications concerning health maintenance and chronic disease prevention or even treatment. There are many difficulties in their study, with the first being the fact that this group of compounds includes such a diverse array of chemical structures and effects as part of the food products in which they coexist with many other nutrients and non-nutrients. However, today, it is generally accepted that they exert a vasoprotective effect, that they may modulate blood pressure and exert a regulatory effect both in lipid and glucidic metabolism, and that they have an immunomodulatory and gut microbial effect. It is through all these means that polyphenols might be of use in the prevention and treatment of chronic diseases as prevalent as obesity and cardiovascular disease or neurocognitive decline.

Sonia de Pascual-Teresa and Luis Goya
Editors

nutrients

MDPI

Editorial

Effects of Polyphenol-Rich Foods on Chronic Diseases

Luis Goya and Sonia de Pascual-Teresa *

Departamento de Metabolismo y Nutrición, Instituto de Ciencia y Tecnología de Alimentos Nutrición (ICTAN-CSIC), C/José Antonio Novais, 10, 28040 Madrid, Spain; luisgoya@ictan.csic.es
* Correspondence: s.depascualteresa@csic.es

Citation: Goya, L.; de Pascual-Teresa, S. Effects of Polyphenol-Rich Foods on Chronic Diseases. *Nutrients* **2023**, *15*, 4134. https://doi.org/10.3390/nu15194134

Received: 14 September 2023
Revised: 15 September 2023
Accepted: 18 September 2023
Published: 25 September 2023

Ever since the French paradox raised the research interest pertaining to the high potential of certain phytochemicals—until then regarded as anti-nutrients—as positive bioactive compounds for health, research on the biological and molecular effects of polyphenols has subsequently been continuously increasing. Epidemiological studies, clinical trials and interventions, animal experimentation and cell culture models have more recently benefited from new research tools that have significantly increased information and led to vast possibilities for future approaches. All these new data are necessary in order to establish nutritional claims for these compounds that might be validated by international regulating organizations and help set new healthier dietary patterns for the population as well as produce more functional foods and nutraceuticals. The present Special Issue of the journal *Nutrients* contains 16 of such new studies and reviews, which mostly report that the beneficial effects of phytochemicals validate their inclusion in a healthy human diet. A brief summary of each of the studies is offered below.

In their study on the influence of the APOE genotype, docosahexaenoic acid (DHA) and flavanol intervention on brain DHA and the lipidomics profile in aged transgenic mice, Martinsen and coworkers [1] establish a correlation between dietary flavanol intake and brain omega-3 polyunsaturated fatty acid (n3-PUFA) levels in humanized apolipoprotein E3 (APOE3) and APOE4 targeted replacement transgenic mouse models. Although only a modest—and interestingly limited to APOE3—effect on brain DHA values is observed after flavanol administration, the study by the University of East Anglia group points out the promising connection between the dietary intake of flavanols, lipid profile and brain availability of n3-PUFA [1].

Jeong-Yeon On and colleagues [2], in their study concerning the effects of fermented *Artemisia annua* L. and *Salicornia herbacea* L. on the inhibition of obesity in vitro in 3T3-L1 adipocytes and in C57BL/6 mice fed a high-fat diet, report the anti-obesity effects of the metabolites of both plant extracts after fermentation. In addition, the authors describe the inhibitory effect on adipocyte differentiation and fat accumulation. This study by the Korean group concluded that biotransformation in vitro displayed significant metabolite changes that increased the anti-obesity effects of plant extracts in mice [2].

Similarly related to the effect of food fermentation on its biological activity, Lüersen and colleagues [3], in their German–Japanese collaborative research on the anti-diabetic properties of a soy extract rich in hydroxylated isoflavones in vitro and in *Drosophila melanogaster* in vivo, show that fermentation with *Aspergillus evoked* produced an enrichment of hydroxy-isoflavones that is accompanied by an enhanced free radical scavenging activity. Both pre- and post-fermented extracts significantly showed anti-diabetic properties on oxidative stress and inflammatory markers, and the supplementation of a high-starch diet with post-fermented hydroxyl-isoflavones-rich extract decreased the triacylglyceride levels in female *D. melanogaster*, confirming its anti-diabetic properties in an in vivo model [3].

In their observational study on the dietary sources of anthocyanins and their association with consumption biomarkers and cardiometabolic risk factors, Mostafa and colleagues [4] reveal the associations between dietary intake, microbial metabolism and the cardiometabolic health benefits of anthocyanins. A targeted metabolomic analysis of a subsample of the Diet, Cancer, and Health-Next Generations (DCH-NG MAX) study with 1351 samples from 624 participants reveals two metabolites, salsolinol sulfate and 4-methylcatechol sulfate, associated with the intake of anthocyanins from berries, and inversely related to visceral adipose tissue. The group from the University of Barcelona confirms the strong dependence of plasma metabolome biomarkers of dietary anthocyanins on the dietary source in order to correlate dietary intake with cardiometabolic health benefits [4].

On the other hand, Navarro-Masip and coworkers [5] reveal changes in body weight gain and lipolysis in adipose tissue in their research on the modulation by the grape-seed proanthocyanidins of adipose tissue adaptation to obesity in rats submitted to different photoperiods and fed a cafeteria diet for five weeks. Supplementation with grape-seed proanthocyanidins for four weeks prevented excessive body weight gain under a long photoperiod, which could be explained by an increased lipolysis in the adipose tissue. The authors conclude that the impact on obesity in the adipose tissue of flavonoids is photoperiod dependent [5].

In two intervention studies—one acute and one three weeks long—on the effectiveness of a combination of green tea catechins and coffee chlorogenic acid on postprandial glycemic responses in healthy men, Yanagimoto and colleagues [6,7] report that a combined ingestion of the two phenolic compounds significantly altered the incretin response and reduced glucose and insulin levels, suggesting an effective minimum dose of 540 mg of green tea catechins and 150 mg of coffee chlorogenic acid [6]. In addition, the randomized, double-blinded, placebo-controlled crossover trial shows that the consumption of the combined phenolics enhanced insulin sensitivity and increased postprandial GLP-1 as compared to the placebo group [7].

In their research on the modulatory effect of chlorogenic acid and coffee extracts on Wnt/β-catenin pathway in colorectal cancer cells, Villota and coworkers [8] report that polyphenol-rich coffee extracts and chlorogenic acid regulate the Wnt pathway on colorectal cancer SW480 cells, with a reduction in the transcriptional activity of β-catenin. The Colombian group concludes that the results establish a starting point for the discovery of a mechanism of action of chlorogenic acid on Wnt pathway and confirm the anti-colorectal cancer potential of polyphenols present in coffee [8].

In their report on the chemical characterization, antioxidant capacity and anti-oxidative stress potential of the South American medicinal plant *Desmodium tortuosum*, Rodríguez and coworkers [9] report the relevant amount of antioxidant phytochemicals in the plant and their antioxidant capacity in vitro. Furthermore, the study shows the cytoprotective capacity of *Desmodium tortuosum* extract both in endothelial and neuronal-like cell lines subjected to oxidative stress. The Peruvian–Spanish team suggests that this chemoprotective effect must be due to the high content of phenolic compounds such as phenolic acids, flavonoids, carotenoids and other antioxidant compounds, and confirm the medicinal use of the plant [9].

In their study on the effect of hesperidin supplementation on rat mucosal immunity after an intensive chronic training and an exhausting exercise, Ruiz-Iglesias and colleagues [10] focused on fecal microbiota and composition and the function of mesenteric lymph node lymphocytes and mucosal immunoglobulin A. The authors conclude that, although hesperidin supplementation did not prevent exercise-induced changes in the distribution and function of lymphocytes, it was able to enhance immunoglobulin A synthesis in the intestinal compartment. This effect could be important in enhancing the immune intestinal barrier in this stressful situation [10].

In a metabolomics approach on the protective role of cocoa in Zucker Diabetic Rats via 1H-NMR-based approach, Fernández-Millán and colleagues [11] identified 14 differential urinary metabolites in Zucker diabetic fatty rats fed with a 10% cocoa-rich diet for 10 weeks. A correlation analysis of pathways indicated major associations between some of the urine metabolites (mainly valine, leucine and isoleucine) and body weight, glycaemia, insulin sensitivity and glycated hemoglobin levels. The authors conclude that an untargeted metabolomics approach provides a clear metabolic fingerprint associated with chronic cocoa intake that can be used as a marker for the improvement of glucose homeostasis in a diabetic context [11].

In a randomized, blinded, cross-over, controlled clinical trial carried out in normocholesterolemic and hypercholesterolemic subjects on the effect of olive pomace oil (OPO) on cardiovascular health and associated pathologies, González-Rámila and colleagues [12] report a lack of effect on any of the markers related to lipid profile, blood pressure and endothelial function in both groups. However, a significant decrease in visceral fat in both groups of subjects was observed after OPO intake, accompanied by an increment of leptin only in the hypercholesterolemic group. The authors conclude that reducing visceral fat after prolonged OPO intake might contribute to improving cardiometabolic status, with a potentially positive effect on the vascular tone [12].

In another intervention trial from the same research group, Seguido and his colleagues [13] report that the sustained consumption of a decaffeinated green coffee nutraceutical (300 mg hydroxycinnamates twice daily for two months) has limited effects on phenolic metabolism and bioavailability in overweight/obese subjects. The study of plasma and urinary pharmacokinetics, and the fecal excretion of phenolic metabolites via LC-MS-QToF, showed a significant increase in reduced forms of caffeic, ferulic and coumaric acids, or 3-(3′-hydroxypenyl)propanoic, and 3,4-dihydroxybenzoic acids in feces, and a decrease in coumaroylquinic and dihydrocoumaroylquinic acids in urine. The authors conclude that the nutraceutical product shows a small overall effect on the bioavailability of polyphenols [13].

In a comprehensive overview of in vitro and in vivo animal and human trials of the anti-diabetic potential of fruits of the rosaceae Maleae tribe, covering 131 articles published this century, Rutkowska and Olszewska [14] review the anti-diabetic effects and potential mechanisms of the action of fruits from 46 species. The first part of this review focuses on the effects on tissue-specific glucose transport and the expression or activity of proteins in the insulin signaling pathway; whereas the second part covers the phytocompounds responsible for the activity of particular fruits—primarily polyphenols (e.g., flavonols, dihydrochalcones, proanthocyanidins, anthocyanins, phenolic acids), but also polysaccharides, triterpenes and their additive and synergistic effects. The authors conclude that fruits from the Maleae tribe seem promising as functional foods for diabetes management [14].

In another overview on the polyphenol avenanthramide, a group of polyphenolic compounds found abundantly in oat (*Avena sativa* Linn.), as regulators of PI3K signaling in the management of neurodegenerative diseases, Wankhede and colleagues [15] report that these compounds have been shown to modulate PI3K/AKT signaling. Since the dysregulation of PI3K signaling has been implicated in the pathogenesis of various neurodegenerative diseases such as Alzheimer's and Parkinson's, the recovery of regular activity might lead to increased neuronal survival, reduced oxidative stress, and improved cognitive function. The authors conclude that Avenanthramides have emerged as promising candidates for neuroprotection due to their immense antioxidant, anti-inflammatory and anti-apoptotic properties [15].

Finally, in a review proposing the regular inclusion of pistachio in our diet, Mateos and coworkers [16] report the healthy nutritional profile of pistachio that, together with its rich composition in phytochemicals, such as tocopherols, carotenoids and, importantly, phenolic compounds, make these seeds a powerful food ingredient to explore its involvement in the prevention of prevalent pathologies. The review gathers recent data regarding the most beneficial effects of pistachio on lipid and glucose homeostasis, endothelial function, oxidative stress and inflammation that essentially result in a protective/preventive effect on the onset of pathological conditions, such as obesity, type 2 diabetes, cardiovascular disease and cancer [16].

Author Contributions: All authors contributed equally to the writing of this editorial. All authors have read and agreed to the published version of the manuscript.

Funding: Spanish MINECO grant BFU2017-82565-C2-2-R and MCIN/AEI/10.13039/501100011033/. Grant number PID2019-107009RB-100.

Conflicts of Interest: The authors declare no conflict of interest.

References

1. Martinsen, A.; Saleh, R.N.M.; Chouinard-Watkins, R.; Bazinet, R.; Harden, G.; Dick, J.; Tejera, N.; Pontifex, M.G.; Vauzour, D.; Minihane, A.-M. The Influence of *APOE* Genotype, DHA, and Flavanol Intervention on Brain DHA and Lipidomics Profile in Aged Transgenic Mice. *Nutrients* **2023**, *15*, 2032. [CrossRef] [PubMed]
2. On, J.-Y.; Kim, S.-H.; Kim, J.-M.; Park, S.; Kim, K.-H.; Lee, C.-H.; Kim, S.-K. Effects of Fermented *Artemisia annua* L. and *Salicornia herbacea* L. on Inhibition of Obesity In Vitro and in Mice. *Nutrients* **2023**, *15*, 2022. [CrossRef]
3. Lüersen, K.; Fischer, A.; Bauer, I.; Huebbe, P.; Uekaji, Y.; Chikamoto, K.; Nakata, D.; Hiramatsu, N.; Terao, K.; Rimbach, G. Soy Extract, Rich in Hydroxylated Isoflavones, Exhibits Antidiabetic Properties In Vitro and in *Drosophila melanogaster* In Vivo. *Nutrients* **2023**, *15*, 1392. [CrossRef] [PubMed]
4. Mostafa, H.; Meroño, T.; Miñarro, A.; Sánchez-Pla, A.; Lanuza, F.; Zamora-Ros, R.; Rostgaard-Hansen, A.L.; Estanyol-Torres, N.; Cubedo-Culleré, M.; Tjønneland, A.; et al. Dietary Sources of Anthocyanins and Their Association with Metabolome Biomarkers and Cardiometabolic Risk Factors in an Observational Study. *Nutrients* **2023**, *15*, 1208. [CrossRef]
5. Navarro-Masip, È.; Colom-Pellicer, M.; Manocchio, F.; Arola-Arnal, A.; Bravo, F.I.; Muguerza, B.; Aragonès, G. Grape-Seed Proanthocyanidins Modulate Adipose Tissue Adaptations to Obesity in a Photoperiod-Dependent Manner in Fischer 344 Rats. *Nutrients* **2023**, *15*, 1037. [CrossRef]
6. Yanagimoto, A.; Matsui, Y.; Yamaguchi, T.; Saito, S.; Hanada, R.; Hibi, M. Acute Dose–Response Effectiveness of Combined Catechins and Chlorogenic Acids on Postprandial Glycemic Responses in Healthy Men: Results from Two Randomized Studies. *Nutrients* **2023**, *15*, 777. [CrossRef] [PubMed]
7. Yanagimoto, A.; Matsui, Y.; Yamaguchi, T.; Hibi, M.; Kobayashi, S.; Osaki, N. Effects of Ingesting Both Catechins and Chlorogenic Acids on Glucose, Incretin, and Insulin Sensitivity in Healthy Men: A Randomized, Double-Blinded, Placebo-Controlled Crossover Trial. *Nutrients* **2022**, *14*, 5063. [CrossRef] [PubMed]
8. Villota, H.; Santa-González, G.A.; Uribe, D.; Henao, I.C.; Arroyave-Ospina, J.C.; Barrera-Causil, C.J.; Pedroza-Díaz, J. Modulatory Effect of Chlorogenic Acid and Coffee Extracts on Wnt/β-Catenin Pathway in Colorectal Cancer Cells. *Nutrients* **2022**, *14*, 4880. [CrossRef] [PubMed]
9. Rodríguez, J.-L.; Berrios, P.; Clavo, Z.-M.; Marin-Bravo, M.; Inostroza-Ruiz, L.; Ramos-Gonzalez, M.; Quispe-Solano, M.; Fernández-Alfonso, M.S.; Palomino, O.; Goya, L. Chemical Characterization, Antioxidant Capacity and Anti-Oxidative Stress Potential of South American Fabaceae *Desmodium tortuosum*. *Nutrients* **2023**, *15*, 746. [CrossRef] [PubMed]
10. Ruiz-Iglesias, P.; Estruel-Amades, S.; Massot-Cladera, M.; Franch, À.; Pérez-Cano, F.J.; Castell, M. Rat Mucosal Immunity following an Intensive Chronic Training and an Exhausting Exercise: Effect of Hesperidin Supplementation. *Nutrients* **2023**, *15*, 133. [CrossRef] [PubMed]
11. Fernández-Millán, E.; Ramos, S.; Álvarez-Cilleros, D.; Samino, S.; Amigó, N.; Correig, X.; Chagoyen, M.; Álvarez, C.; Martín, M.Á. Urinary Metabolomics Study on the Protective Role of Cocoa in Zucker Diabetic Rats via ^{1}H-NMR-Based Approach. *Nutrients* **2022**, *14*, 4127. [CrossRef] [PubMed]
12. González-Rámila, S.; Sarriá, B.; Seguido, M.Á.; García-Cordero, J.; Bravo-Clemente, L.; Mateos, R. Effect of Olive Pomace Oil on Cardiovascular Health and Associated Pathologies. *Nutrients* **2022**, *14*, 3927. [CrossRef] [PubMed]
13. Seguido, M.Á.; Tarradas, R.M.; González-Rámila, S.; García-Cordero, J.; Sarriá, B.; Bravo-Clemente, L.; Mateos, R. Sustained Consumption of a Decaffeinated Green Coffee Nutraceutical Has Limited Effects on Phenolic Metabolism and Bioavailability in Overweight/Obese Subjects. *Nutrients* **2022**, *14*, 2445. [CrossRef]
14. Rutkowska, M.; Olszewska, M.A. Anti-Diabetic Potential of Polyphenol-Rich Fruits from the Maleae Tribe—A Review of In Vitro and In Vivo Animal and Human Trials. *Nutrients* **2023**, *15*, 3756. [CrossRef]
15. Wankhede, N.L.; Kale, M.B.; Bawankule, A.K.; Aglawe, M.M.; Taksande, B.G.; Trivedi, R.V.; Umekar, M.J.; Jamadagni, A.; Walse, P.; Koppula, S.; et al. Overview on the Polyphenol Avenanthramide in Oats (*Avena sativa* Linn.) as Regulators of PI3K Signaling in the Management of Neurodegenerative Diseases. *Nutrients* **2023**, *15*, 3751. [CrossRef]
16. Mateos, R.; Salvador, M.D.; Fregapane, G.; Goya, L. Why Should Pistachio Be a Regular Food in Our Diet? *Nutrients* **2022**, *14*, 3207. [CrossRef]

nutrients

MDPI

Article

Lonicera Caerulea Juice Alleviates Alcoholic Liver Disease by Regulating Intestinal Flora and the FXR-FGF15 Signaling Pathway

Baixi Zhang *,†, Lijuan Niu † and Xinwen Huang

National Engineering Research Center for Functional Food, School of Food Science and Technology, Jiangnan University, Wuxi 214122, China; 6200113070@stu.jiangnan.edu.cn (L.N.); 6200112030@stu.jiangnan.edu.cn (X.H.)
* Correspondence: zbx@jiangnan.edu.cn
† These authors contributed equally to this work.

Abstract: Alcoholic liver disease (ALD) is a growing public health issue with high financial, social, and medical costs. *Lonicera caerulea*, which is rich in polyphenolic compounds, has been shown to exert anti-oxidative and anti–inflammatory effects. This study aimed to explore the effects and mechanisms of concentrated *Lonicera caerulea* juice (LCJ) on ALD in mice. ALD was established in mice via gradient alcohol feeding for 30 days. The mice in the experimental group were given LCJ by gavage. The reduction of aspartate transaminase (AST) and alanine transaminase (ALT) in the serum of mice indicated that LCJ has a liver-protective effect. LCJ improved the expression of AMPK, PPARα, and CPT1b in ALD mice to reduce the liver lipid content. Additionally, LCJ increased the expression of farnesoid X receptor (FXR), fibroblast growth factor 15 (FGF15), and fibroblast growth factor receptor 4 (FGFR4), which lowers the expression of cytochrome P450 7A1 (CYP7A1) and lessens bile acid deposition in the liver. In mice, LCJ improved the intestinal barrier by upregulating the expression of mucins and tight junction proteins in the small intestine. Moreover, it accelerated the restoration of microbial homeostasis in both the large and small intestines and increased short–chain fatty acids in the cecum. In conclusion, LCJ alleviates ALD by reducing liver and serum lipid accumulation and modulating the FXR–FGF15 signaling pathway mediated by gut microbes.

Keywords: *Lonicera caerulea;* alcoholic liver damage; gut microbiota; FXR–FGF15 signaling pathways; short-chain fatty acid

Citation: Zhang, B.; Niu, L.; Huang, X. Lonicera Caerulea Juice Alleviates Alcoholic Liver Disease by Regulating Intestinal Flora and the FXR-FGF15 Signaling Pathway. *Nutrients* **2023**, *15*, 4025. https://doi.org/10.3390/nu15184025

Academic Editor: Antonio Colecchia

Received: 2 August 2023
Revised: 11 September 2023
Accepted: 14 September 2023
Published: 17 September 2023

1. Introduction

The chronic daily intake of a specified quantity of alcohol can lead to the development of alcoholic liver disease (ALD), which comprises a spectrum of ailments ranging from reversible hepatic steatosis to acute alcoholic hepatitis, chronic fibrosis, cirrhosis, and hepatocellular carcinoma [1]. ALD is a leading cause of morbidity and mortality worldwide [2,3]. Upon entering the body, the majority of alcohol is promptly transported to the intestines, where over 90% of it is absorbed into the bloodstream [4]. The first-pass metabolism of alcohol occurs in the stomach; however, compared with the liver, the stomach has a lower amount of alcohol-metabolizing enzymes. Thus, the liver serves as the primary organ responsible for alcohol metabolism and breakdown within the body [5]. The intestines and liver are, thus, the main organs that are damaged by alcohol. The pathogenesis of ALD is complex, and the existing treatment methods of abstinence, diet, and drug therapy for ALD have some drawbacks [6–8]. Therefore, exploring the pathogenesis of ALD to identify a suitable treatment for the disease is essential.

Prior research has demonstrated that edible plants and their bioactive compounds, such as polyphenols, can safeguard against ALD by modulating the gut microbiota [9,10]. By inhibiting the expression of genes associated with lipid production and encouraging the expression of genes linked to lipid breakdown in ALD mice, 200 mg/kg of blueberry polyphenol extract can help to alleviate liver damage [11]. Grape–leaf extract—which is rich in phenolic compounds (250–500 mg/kg)—was orally given to ethanol-induced rats for 12 days, which reduced liver injury by improving antioxidant activity and inhibiting nuclear factor kappa B p65 and pro–inflammatory cytokines (tumor necrosis factor-alpha) [12]. The honeysuckle plant *Lonicera caerulea* (LC) is classified as a "homology of medicine and food" fruit due to its high concentration of bioactive compounds [13]. High levels of phenolic chemicals were found in LC, including catechin, procyanidin, chlorogenic acid, anthocyanin 3–glucoside (C3G), etc. [14]. The total phenolic compound content of LC reached 1140.06 mg/100 g FW, which is much higher than that of blueberries (94.60–137.74 mg/100 g FW) [15,16]. In addition, LC polyphenol extract alleviates non-alcoholic fatty liver by inhibiting proinflammatory cytokine production and lipid peroxidation. [17]. LC polyphenols can regulate the intestinal epithelial barrier function and microbiome of Sprague Dawley rats on a high-fat diet to inhibit fat absorption [18].

Herein, we hypothesized that LC—which is high in polyphenols—can alleviate ALD. We developed a mouse model of ALD to determine if *Lonicera caerulea* juice (LCJ) can alleviate the disease and examine potential mechanisms. In addition to promoting the thorough development and utilization of LC resources, our results may also provide a new reference for the treatment of ALD.

2. Materials and Methods

2.1. Preparation of LCJ

Pure water and *Lonicera caerulea* were combined in a colloid mill at a ratio of 1:1 (*w*/*v*), and the pulp was processed repeatedly. The juice was collected after the pulp had been centrifuged at 3000 rpm for 30 min. Finally, the juice was evaporated by rotation into two types of LCJ, each with a total phenol content of 9 mg/mL and 18 mg/mL. The *Lonicera caerulea* was provided by the Heilongjiang Fengran Agricultural Group Company. The company certified that the *Lonicera caerulea* L. was the "Beilei" variety, belonging to the *Lonicerae* genus of the *Caprifoliaceae* family.

2.2. Determination of Phenolic Components in LCJ by Liquid Chromatography–Mass Spectrometry

LCJ with a total phenolic content of 18 mg/mL was diluted by a factor of four, and then about 100 μL of the sample was transferred to an Eppendorf tube. After the addition of 300 μL of extract solution (methanol), samples were vortexed for 30 s, sonicated for 10 min in an ice–water bath, and incubated for 2 h at −20 °C to precipitate the proteins. Then, the sample was centrifuged at 12,000 rpm for 15 min at 4 °C. The resulting supernatant was transferred to a fresh glass vial for analysis. A triple quadrupole compound linear ion trap liquid-mass coupling instrument (Applied Biosystems, Inc, CA, USA) was used to detect phenols in LCJ [19]. Chromatographic separations were performed in a BEH C18 column (2.1 × 100 mm i.d., 1.7 μm particle diameter). Solvent A was water (0.1% formic acid aqueous solution), and solvent B was acetonitrile. The column temperature was kept at 40 °C. The program was performed as follows: 0–10 min, 98% A and 2% B; 10–12 min, 2% A and 98% B; and 12–15 min, 98% A and 2% B at a flow rate of 0.30 mL/min. And the injection volume was 1 μL.

2.3. Animal Experiments

It was reported that the chronic plus—binge ethanol challenge induced a greater degree of adipose tissue inflammation and liver injury in female mice than in male mice [20,21]. Therefore, the choice of female mice as the research subjects in our experiment was advantageous for successful modeling of alcoholic liver disease in mice as well as for future population studies on liver damage in the female alcoholic population. Thirty–six female mice of the C57BL/6N background (10 weeks old, weighing 20 ± 2 g) were purchased from Beijing Vital River Laboratory Animal Technology (Beijing, China) and housed at a temperature of 22 ± 2 °C and a humidity of 50 ± 10% under a 12 h light/dark cycle in a specific pathogen-free animal facility. All animal experiments were examined and authorized by the Laboratory Animal Ethics Committee of Jiangnan University (Ethical approval number: JN. No20220615c0450803223). Mice had access to water and a normal diet for five days of the adaptation period.

It has been proven that 200 mg/kg of blueberry polyphenol and grape leaf extract containing 250 mg/kg of polyphenol can effectively relieve ALD [11,12]. Since phenolic compounds are the main active substances in LC, the test groups were divided into two groups according to their total phenol intake. After the adaptation period, mice were randomly divided into 4 groups with 9 mice in each group: the control group (CG), the alcohol group (AG), the low-dose LCJ group (LG) (total phenol intake: 150 mg/kg), and the high-dose LCJ group (HG) (total phenol intake: 300 mg/kg). The dosage of concentrated fruit juice given to mice was 300 mg/kg and 150 mg/kg. This dose was realistic, as it is equivalent to 100 mL of LCJ (the total phenol concentration is 9 mg/mL or 18 mg/mL) per day for humans.

To establish a model of chronic intake and single binge drinking, all mice were given the Lieber-DeCarli control liquid diet for the first 5 days. The model group and the test group were fed with gradient alcohol adaptation mode for 9 days, which was 1.7% (*w*/*v*) alcohol for 3 days, 2.5% (*w*/*v*) alcohol for 3 days, and 3.3% (*w*/*v*) alcohol for 3 days. Then, they were given a Lieber-DeCarli diet containing 5% (*w*/*v*) alcohol for 16 days. On the last day, the experimental group was given a concentration of 31.5% (vol/vol) ethanol solution by gavage, and the volume of the gavage (μL) = body weight (g) $\times$ 20. The control group was administered with an isocaloric maltodextrin solution. Mice were sacrificed nine hours after intragastric administration.

2.4. Histopathological Analysis

Specimens of the live large intestine and small intestine were collected for histological analysis. They were fixed in 4% paraformaldehyde/phosphate-buffered saline (PBS) (*v*/*v*) and embedded in paraffin. Samples were cut into small sections for staining with hematoxylin and eosin (HE) and oil red O; sections were then examined using a microscope. According to Osho et al. [22], new large intestine and small intestine tissues were stained with Alcian blue and periodic acid–Schiff reagent for 30 min for goblet cells. Afterward, samples were rinsed in tap water for 10 min, exposed to periodic acid (5 g/L) for 5 min, rinsed in lukewarm water for 10 min, and then counter–stained with Coleman's Schiff reagent.

2.5. Biochemical Assays of Serum and Tissue

Serum was extracted from blood samples by centrifugation at 4000 g for 15 min at 4 °C after being maintained at 4 °C for 2 h. Subsequently, serum levels of cholesterol (TC), alanine transaminase (ALT), aspartate transaminase (AST), triglycerides (TG), and total bile acid (TBA) were evaluated using a BK–400 automatic biochemical analyzer. In addition, the kits of serum ALT, AST, TG, TC, and TBA were purchased from Nanjing Jiancheng Biological Engineering Research Institute (Nanjing, China). Lipopolysaccharides (LPS) ELISA kits were purchased from SenBeiJia Biological Technology Co., Ltd. (Nanjing, China).

A liver sample weighing 0.1 g was homogenized using a high-speed homogenizer in 0.9 mL of physiological saline to produce a liver homogenate. The supernatants were collected by centrifugation at 3000 g for 15 min at 4 °C. The levels of TC, TG, and TBA in the liver were determined using commercial kits obtained from Nanjing Jiancheng Biological Engineering Research Institute (Nanjing, China). Fecal bile acid levels were detected according to a method previously described by Guo et al. [23].

2.6. Gut Microbiota Analysis

All the contents of the large and small intestines were scraped and stirred well. Samples were then divided into 6 groups: SCG, small intestine in the control group; SAG, small intestine in the alcohol group; SHG, small intestine in the high–dose LCJ group;

LCG, large intestine in the control group; LAG, large intestine in the alcohol group; LHG, large intestine in the high–dose LCJ group. The total DNA of microorganisms in intestine contents was extracted using a fecal DNA extraction kit (MP Biomedicals, Santa Ana, CA, USA). DNA was successfully separated by agarose gel electrophoresis.

The highly variable region (V3–V4) of bacterial 16S ribosomal DNA was used for polymerase chain reaction (PCR) amplification. The primers were 341F (5′-ACTCCTACGG GAGGCAGCAG-3′) and 806R (5′-GGACTACHVGGGTWTCTAAT-3′). According to the instructions of Nanjing Vazyme Biotech Co., Ltd. (Nanjing, China) reagent kit, a PCR reaction system was prepared by taking a qualified genomic DNA sample of 30 ng and corresponding fusion primers. The PCR products were purified according to the instructions of the AxyPrep DNA Gel Extraction Kit (Axygen company, CA, USA) reagent kit. The Agencourt AMPure XP (Beckman CoulterTM, Brea, CA, USA) magnetic beads were used to dissolve and purify the PCR amplification products before labeling them to finish the library construction. An Agilent 2100 bioanalyzer (Agilent, Santa Clara, CA, USA) was used to analyze libraries for fragment range and concentration. Libraries that passed the test were selected for sequencing on a HiSeq platform (BGI Genomics Co., Ltd., Shenzhen, China) according to insert sizes. The data were filtered, and the remaining high-quality, clean data were used in post-analyses. Sequence splicing was performed using FLASH software (Fast Length Adjustment of Short reads, v1.2.11), which assembles pairs of reads from double-end sequences into a single sequence using overlap relationships to obtain tags for high–variation regions. Amplicon sequence variants (ASVs) were obtained, compared with the database, and annotated with the species using the DADA2 (divisive amplicon denoising algorithm) method in QIIME2. Sample species complexity analysis, intergroup species variation analysis, association analysis, and functional prediction were carried out based on the results of OTU and annotation. The BGI platform made the bioinformatics analytic techniques available for free online.

2.7. Determination of Short Chain Fatty Acids (SCFA) in Cecal Contents

About 0.09 g of fecal content homogenate was added to 0.5 mL of saturated sodium chloride solution precooled to 4 °C. Then, 10 μL of a 10% (v/v) sulfuric acid solution precooled at 4 °C was added to extract the fatty acids, followed by the addition of 1 mL of anhydrous ether (containing 1 mmol/L of internal standard 2–ethylbutyric acid) precooled at 4 °C. The whirlpool was then shaken for 30 s. Samples were centrifuged at 3000 rpm for 15 min at 4 °C. Finally, SCFAs in the supernatant were analyzed using gas chromatography–mass spectrometry. [24].

2.8. Real-Time Quantitative PCR

From the liver, total RNA was extracted, and 1 μg of this RNA underwent reverse transcription to produce cDNA. Using SYBR Green qPCR Master Mix, quantitative RT-PCR was carried out on a CFX96 Touch real–time PCR machine. These analyses were carried out with the corresponding commercial kits from Nanjing Vazyme Biotech Co., Ltd. (Nanjing, China). The expression level of the liver gene was normalized to the β-actin gene, while the expression level of the small intestine gene was normalized to the GAPDH gene and calculated utilizing the $2^{-\Delta\Delta Ct}$ method. Primer sequences are presented in Supplementary Table S1.

2.9. Western Blotting

Three mice in each group were selected for the Western blot assay. The tissues were lysed with RIPA buffer (1 mL) containing proteinase inhibitors (Beyotime, Shanghai, China). The homogenate was centrifuged at 3000 rpm for 5 min at 4 °C. The protein content was determined by the BCA protein concentration assay kit (Beyotime, Shanghai, China) and diluted to a suitable concentration for the experiment.

The protein sample from each mouse was separated by electrophoresis in SDS-PAGE (Amresco, Solon, OH, USA) and transferred onto a polyvinylidene difluoride (PVDF) membrane (Millipore, Schwalbach, Germany). The membrane was incubated with only one primary antibody every time, then incubated with the secondary antibody and detected. Then, all the detection reagents from the blot on the same membrane are removed and reprobed with the other primary antibody for the detection of the second target protein. Specific operations are as follows: Equal quantities of protein were separated by electrophoresis in SDS–PAGE and then deposited onto PVDF membranes. After the membranes had been blocked with 5% nonfat dry milk, the corresponding primary antibodies were incubated on the membranes overnight at 4 °C. After being washed with Tris–buffered saline with Tween, followed by incubation with secondary antibodies for 1 h at room temperature. Both a luminescence imaging workstation (Bio–Rad Laboratories, Inc., CA, USA) and an ECL chemiluminescence kit (Beyotime, Beijing, China) were utilized to create the target protein. When multiple exposures of the same PVDF membrane were required, Western blot stripping buffer was shaken at room temperature for 45 min and washed with TBST 3 times. The subsequent steps were the same as before. Image Pro Plus 6.0 software was used to analyze the optical density value (Tanon Science & Technology Co., Ltd., Shanghai, China). The following antibodies were used: MUC4 (1:100, Santa Cruz), CYP7A1 (1:5000, Santa Cruz), FXR (1:1000, Cell Signaling Technology), MUC2 (1:1000, Abcam), FGFR4 (1:1000, Abcam), ZO-1 (1:1000, Abcam), Claudin-1 (1:1000, Abcam), FGF15 (1:1000, Abcam), GAPDH (1:1000, Abcam), β-actin (1:1000, Abcam), AMPK (1:5000, Abcam), CPT1b (1:1000, Abcam), PPARα (1:1000, Abcam).

2.10. Statistical Analysis

Statistical analysis was performed using GraphPad Prism 8.0 software and SPSS 25.0 software. All the data in this study were expressed as the mean ± standard deviation. The statistical significance of the results was analyzed by one-way analysis of variance (ANOVA) based on Duncan's multiple range test using SPSS. The significance level in the analyses was considered $p < 0.05$.

3. Results

3.1. The Content of the Phenolic Components in LCJ

The content of 16 phenolic substances in LCJ was determined (Supplementary Table S2 and Figure S1), which was consistent with the literature reports on the identification of LC components [14]. The highest content in LCJ was cyanidin-3-O-glucoside (C3G), whose content was 394.57 ± 0.06 mg/100 mL. Among the detected chemicals, many have not been quantitatively reported in LC, such as cyanidin–3–galactoside, phlorizin, neodiosmin, naringenin, astragalin, and 7-hydroxycoumarin. Since cyanidin–3–galactoside is an isomer of C3G, its quantitative analysis may have been neglected in previous studies. The content of cyanidin–3–galactoside in LCJ was 275.48 ± 0.22 mg/100 mL, which is the same order of magnitude as that of C3G. In addition, the contents of chlorogenic acid, quercetin, and isochlorogenic acid C were relatively high: 96.44 ± 0.02 mg/100 mL and 16.77 ± 0.31 mg/100 mL, respectively.

3.2. Effects of LCJ Intervention on Body Growth Performance and Hepatocyte Damage in ALD Mice

Figure 1a shows the changes in the BW. The BW of the AG group was lower than that of the CG group from the beginning, suggesting that prolonged alcohol intake can inhibit normal growth in mice. The BW of mice in the LG and HG groups was positively correlated with the dose of LCJ intragastric administration, which was higher than that in the AG group, but there was no significant correlation ($p > 0.05$). These results indicated that LCJ could increase the BW of ALD mice. Furthermore, alcohol consumption did not induce any notable alteration in liver weight, and there was no significant difference in liver weight observed among any of the groups (Figure 1b). The liver index of mice treated with ethanol was much higher than that of control mice, and a low dose of LCJ did not significantly lower this proportion, which was significantly lowered by a high dose of LCJ

(Figure 1c). These findings support the notion that drinking in this manner may negatively impact the BW and liver index, but high doses of LCJ can alleviate this impact. In addition, alcohol increased serum AST, ALT, and AST/ALT ratios (Figure 1d–f), which is consistent with the results of Wu et al. [25], suggesting that this type of alcohol intake can lead to liver cell damage in ALD mice. Nevertheless, the administration of high–dose LCJ orally resulted in a substantial reduction in the atypical increases of ALT and AST serum activity as well as the AST/ALT ratio induced by alcohol, indicating the hepatoprotective potential of LCJ in ALD mice.

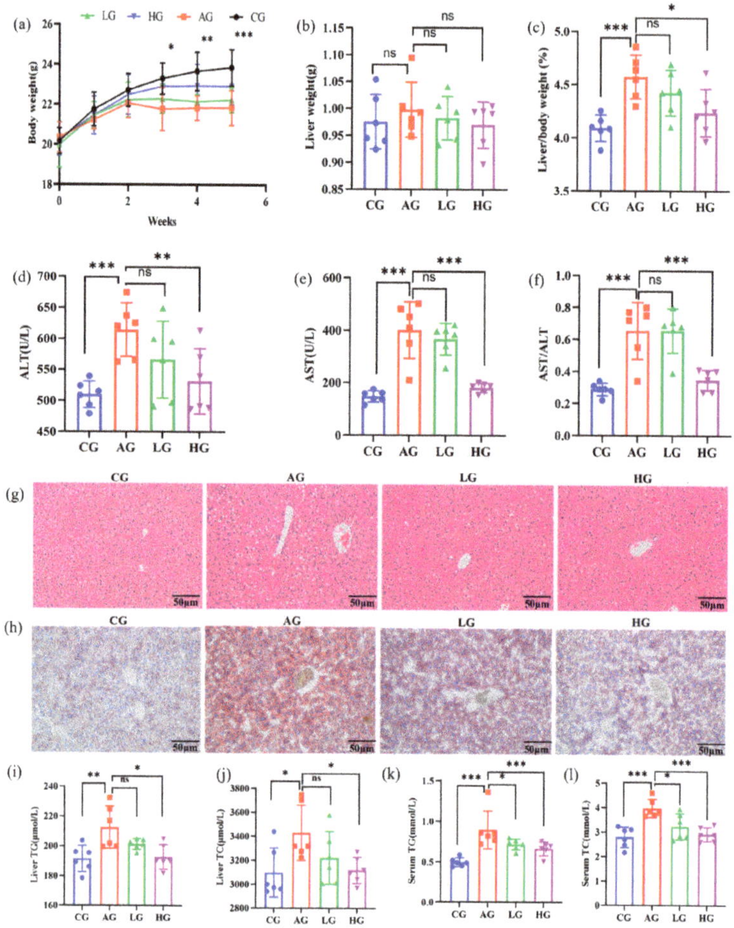

Figure 1. Effects of LCJ intervention on body weight, liver weight, liver index, and lipid accumulation in mice. (**a**) Changes in body weight from week 0 to week 5; (**b**) liver weight; (**c**) liver to body weight ratio; (**d**) serum alanine aminotransferase (ALT); (**e**) serum aspartate aminotransferase (AST); (**f**) ratio of AST/ALT; (**g**) HE staining (200 primary magnification); (**h**) oil red O staining (200 primary magnification); (**i**) liver triglycerides (TG); (**j**) liver cholesterol (TC); (**k**) serum TG; (**l**) serum TC; $n = 6$. ^{ns} $p > 0.05$, * $p < 0.05$, ** $p < 0.01$, and *** $p < 0.001$ vs. the model group (ANOVA).

3.3. LCJ Improved Liver Fat Accumulation in ALD Mice

The HE staining study revealed that chronic alcohol use resulted in hepatocyte destruction and liver cell steatosis in AG groups (Figure 1g). However, both the LG and HG groups had decreased hepatic steatosis and hepatocyte injury when compared to the model group, and the impact was favorably correlated with dose. LCJ can lower hepatic lipid buildup in ALD mice, according to oil–red O staining (Figure 1h). The increase of serum TC and TG as well as liver TG and TC in AG group mice was brought on by continuous alcohol feeding, which was consistent with the results obtained by Zhou et al. in the alcoholic fatty liver model group [26]. High–dose LCJ could reduce the accumulation of blood lipids and liver lipids in ALD mice (Figure 1i–l). According to the biochemical indicators of lipids in the serum and liver, only high–dose LCJ significantly alleviates ALD in mice. As a result, the subsequent investigation of the effects of LCJ on the expression of substances involved in lipid synthesis and metabolism pathways using qPCR and Western blot methods was limited to the CG group, AG group, and HG group. We found that high–dose LCJ can promote the β oxidation of fatty acids by increasing the protein expression of AMPK, PPARα, and CPT1b in the liver of ALD mice (Figure 2a–c), thereby reducing the accumulation of TG and TC in the liver of ALD mice.

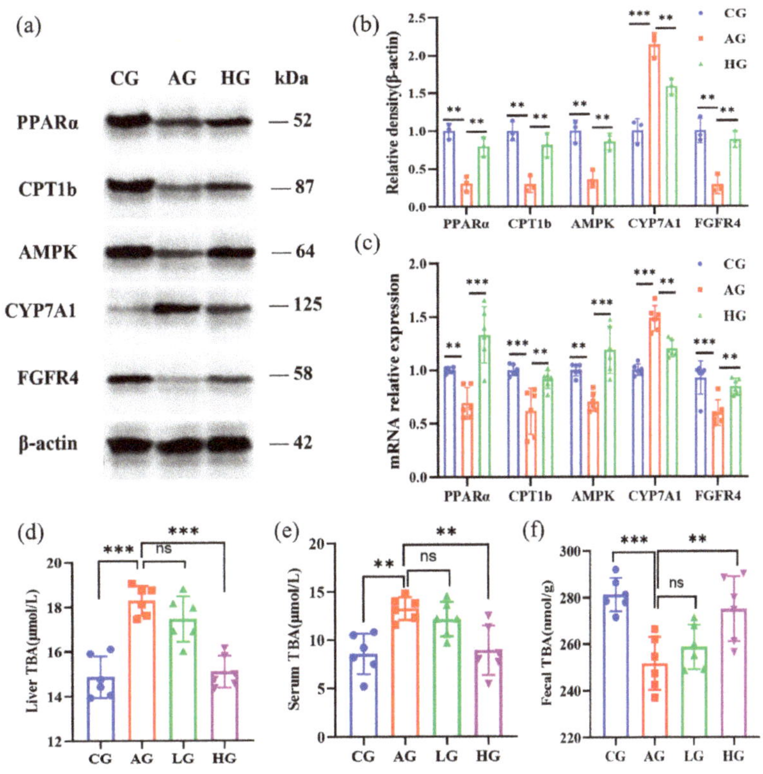

Figure 2. Effects of LCJ on bile acid content, proteins involved in bile acid metabolism, and lipid metabolism in the liver. (**a**,**b**) Protein expression of PPARα, CPT1b, AMPK, CYP7A1, and FGFR4 in the liver of mice (n = 3); (**c**) mRNA levels of PPARα, CPT1b, AMPK, CYP7A1, and FGFR4; (**d**) liver TBA; (**e**) serum TBA; (**f**) feces TBA (n = 6). ^{ns} $p > 0.05$, ** $p < 0.01$, and *** $p < 0.001$ vs. the model group (ANOVA).

3.4. LCJ Alleviates Hepatic BA Deposition in ALD Mice by Regulating the FXR-FGF15 Axis

Compared with the CG group, TBA levels in the liver and serum of the AG group were significantly higher, whereas the TBA levels in the feces were significantly lower. However, high–dose LCJ reduced TBA in the liver and serum and increased TBA in the feces of ALD mice, while low–dose LCJ had no significant effect on TBA in the liver, blood, or feces (Figure 2d–f). Therefore, we determined the content of BA metabolization-related genes in the liver and small intestine of the CG, AG, and HG groups by RT–qPCR and Western blotting. Alcohol reduced the levels of fibroblast growth factor receptor 4 (FGFR4) in the liver and fibroblast growth factor 15 (FGF15) in the small intestine while raising the levels of cytochrome P450 7A1 (CYP7A1) in the liver of the AG group. High doses of LCJ, however, restored these proteins to levels comparable to those in the CG group in ALD mice (Figures 2a and 3c).

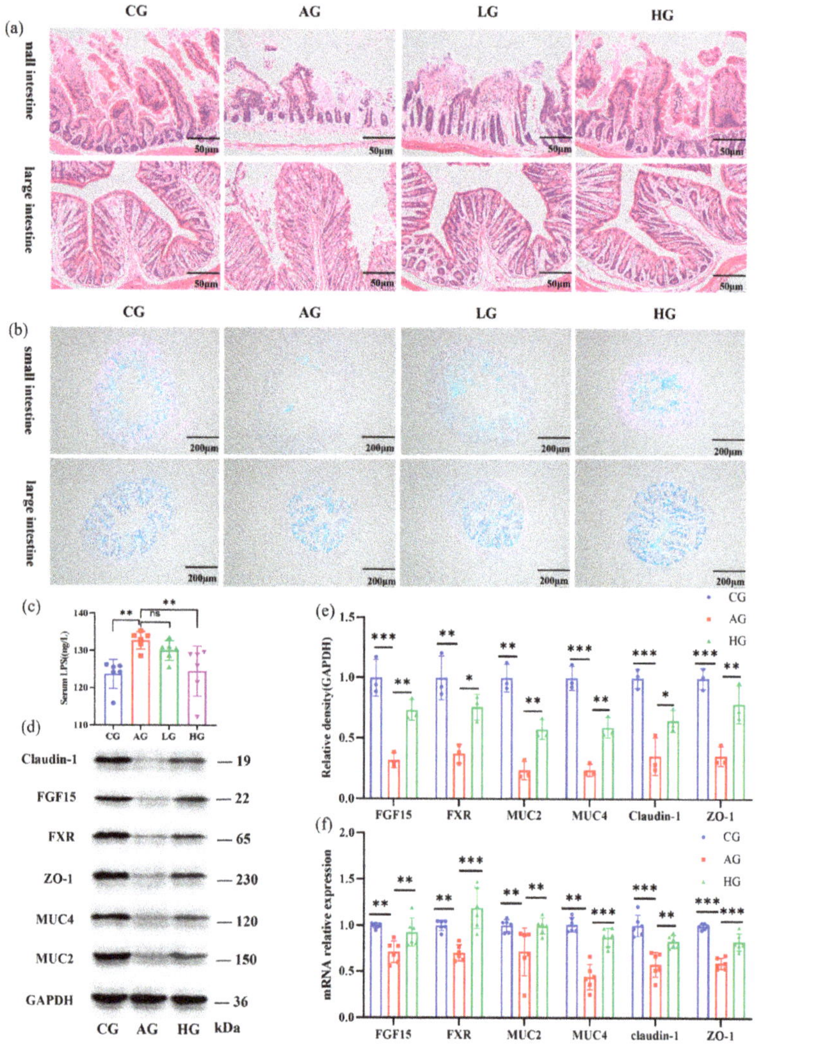

Figure 3. Effects of LCJ on intestinal histological characteristics and the intestinal barrier in mice. (**a**) Hematoxylin and eosin staining (200 primary magnification); (**b**) periodic acid–Schiff and Alcian blue (AB-PAS) staining (50 primary magnification); (**c**) serum LPS; (**d**,**e**) protein expression of FGF15, FXR, MUC2, MUC4, ZO–1, and Claudin–1 in the small intestine ($n = 3$); (**f**) mRNA levels of FGF15, FXR, MUC2, MUC4, ZO–1, and Claudin–1 ($n = 6$). ^{ns} $p > 0.05$, * $p < 0.05$, ** $p < 0.01$, and *** $p < 0.001$ vs. the model group (ANOVA).

3.5. LCJ Improves the Intestinal Barrier and Reduces LPS Entry into the Bloodstream in ALD Mice

The results of HE staining and Alcian blue–periodic acid–Schiff (AB-PAS) staining showed that the crypt structure of the small intestinal tissue in the AG group was severely damaged, with a large number of cell deaths and reduced mucin content. Alcohol had little effect on the mucin content and large intestine anatomy. This may be because alcohol is mainly absorbed and metabolized in the small intestine [27]. The effects of alcohol on intestinal mucus and barrier function were consistent with previous reports [28,29]. The structure and mucin content of the small and large intestines in the HG group, however, were restored to levels comparable to those in the CG group following the administration of LCJ (Figure 3a,b). Intestinal permeability can be reflected in serum LPS [27]. Alcohol intake led to elevated blood levels of LPS in the AG group. A high dose of LCJ reduced the blood LPS content of ALD mice (Figure 3c). We believe that low-dose LCJ cannot significantly alleviate intestinal barrier damage in ALD mice, whereas high-dose LCJ can significantly improve intestinal barrier damage in ALD mice based on the results of HE and AB-PAS staining as well as changes in serum LPS. Therefore, we investigated the mechanism by which LCJ alleviates alcohol-induced intestinal barrier damage in mice, focusing on CG, AG, and HG group mice.

The results showed that high-dose LCJ increased the expression of tight junction proteins (ZO–1, claudin–1) and mucins (MUC2 and MUC4) in the small intestinal tract of ALD mice (Figure 3d–f). The results of intestinal pathological section staining supported these findings. These data imply that LCJ supplementation to improve intestinal barrier dysfunction may be an effective treatment for alcoholic liver injury.

3.6. LCJ Increase the Level of SCFAs in the Cecum of ALD Mice

Based on the biochemical indicators of the liver and serum of mice, as well as the intestinal barrier, intestinal mucin, and bile acid indicators of secondary metabolites of the microbiota, we believe that only high–dose LCJ can improve ALD in mice. Therefore, our study is limited to the CG, AG, and HG groups regarding the effect of LCJ on the intestinal microbiota and its metabolites, short–chain fatty acids, in ALD mice.

SCFAs are acidic byproducts of the fermentation of difficult-to-digest carbohydrates by bacteria in the gut, which strengthen the intestinal barrier [30]. Ethyl alcohol (EtOH) liquid diet caused a significant drop in the amounts of acetic acid, propionic acid, butyric acid, valeric acid, and total SCAFs in the cecum of the AG group (Figure 4a–g). In addition, alcohol consumption has been shown to reduce SCAFs in the cecum of animals [31,32]. Although not statistically significant, ethanol increased isovaleric acid in the AG group. Acetic acid, propionic acid, butyric acid, valeric acid, and total SCAF concentrations in the cecum of mice were significantly higher after administration of high doses of LCJ, while isovaleric acid contents were significantly lower.

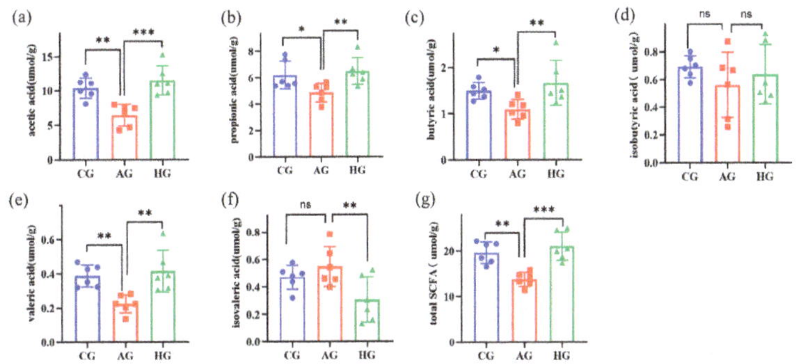

Figure 4. Changes in SCFA levels in the cecum contents of mice. (**a**) Acetic acid; (**b**) propionic acid; (**c**) butyric acid; (**d**) isobutyric acid; (**e**) valeric acid; (**f**) isovaleric acid; (**g**) total SCFAs. $n = 6$. ^{ns} $p > 0.05$, * $p < 0.05$, ** $p < 0.01$, and *** $p < 0.001$ vs. the model group (ANOVA).

3.7. *LCJ Mediates Alcohol-Induced Intestinal Flora Disturbance in the Large and Small Intestine*

As shown in Figure 5a, the dilution curves of each group tended to be smooth, indicating that there was a sufficient amount of sequencing data. Alcohol consumption had minimal effects on the species richness and homogeneity of the small and large intestines in ALD mice. Supplementation with higher doses of LCJ significantly reversed these effects (Figure 5b).

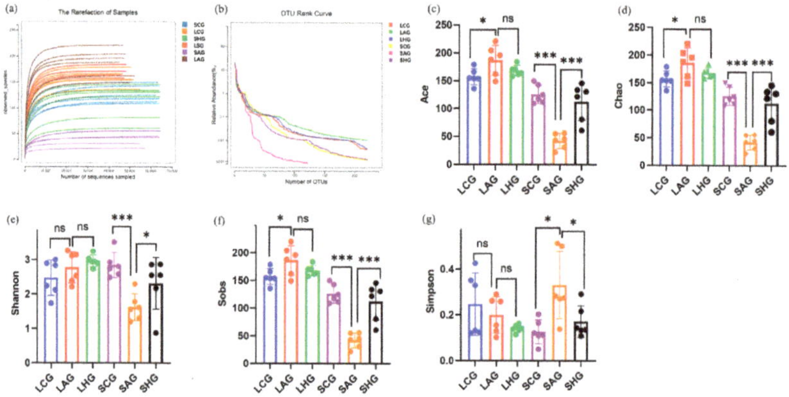

Figure 5. Effects of LCJ on abundance and homogeneity of small intestinal and large intestinal microflora. (**a**) Rarefaction of samples; (**b**) OTU rank curve; (**c**) Ace index; (**d**) Chao index; (**e**) Shannon index; (**f**) Sobs index; (**g**) Simpson index. $n = 6$. ^{ns} $p > 0.05$, * $p < 0.05$ and *** $p < 0.001$ vs. the model group (ANOVA).

The Sobs, Chao, and Ace indexes describe community richness. The Shannon and Simpson indices describe community α diversity (Figure 5c–g). Our data showed that alcohol did not change the diversity of coliforms, but it increased their richness. Besides, alcohol reduced the richness and diversity of microorganisms in the small intestine. LCJ decreased the intestinal microbial richness and significantly increased the intestinal microbial richness and diversity of ALD mice. The findings suggested that LCJ exerted distinct regulatory effects on the α diversity of intestinal flora in ALD mice; however, ultimately, it facilitated the restoration of α diversity of intestinal flora in ALD mice, which was consistent with that in the CG group. Previous studies found that alcohol reduced the Shannon index of the cecum [33]. Xu et al. found that alcohol intake increased the α diversity of mouse feces [34]. This may be because the intestine is highly dynamic; small and large intestines are regulated differently under external influences to maintain overall intestinal homeostasis [35].

β diversity analysis was used to show differences in the microbial community composition among different groups of samples (Figure 6a–e). The large and small intestines of normal mice had different flora structures. The intestinal flora structure in the AG group demonstrated significant alterations in both the small and large intestines compared to that of the CG group. These observations align with the findings of Cao et al.'s study [36]; however, the effect of alcohol on the intestinal flora structure was greater in the small intestine. The structure of the intestinal flora in the small intestine in the HG group recovered to resemble that in the CG group following the administration of high-dose LCJ. Although the intestinal flora of the large intestine in the HG group did not recover to resemble that of the CG group, it was also distinct from that of the AG group.

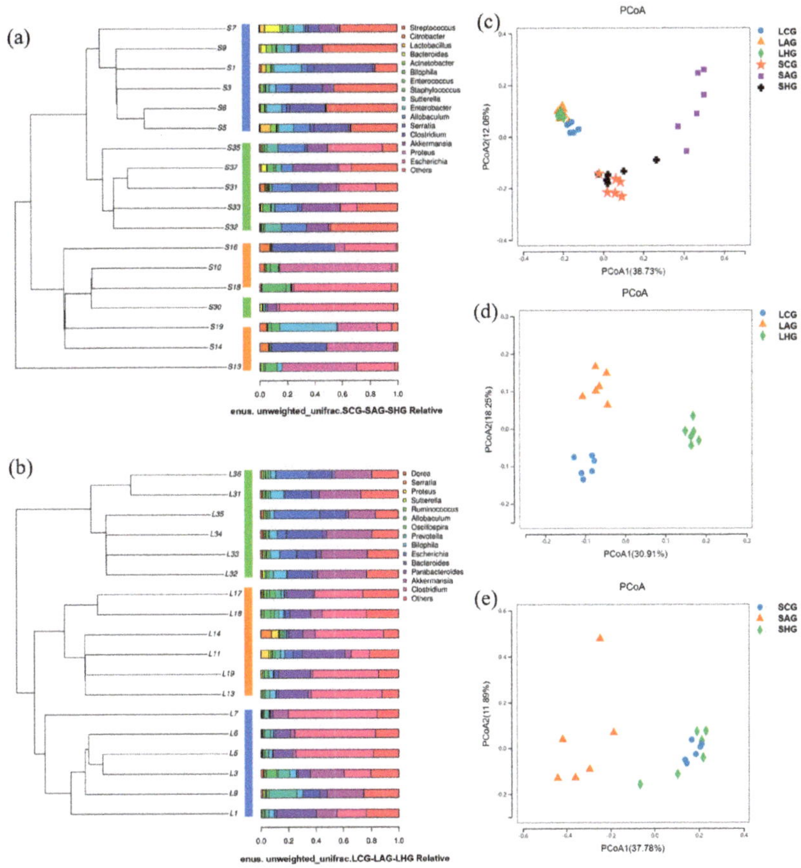

Figure 6. LCJ intervention caused inconsistent changes in gut microbial structure. (**a**) Unweighted pair–group method with arithmetic means cluster tree (small intestine); (**b**) unweighted pair–group method with arithmetic means cluster tree (large intestine); (**c**) principal coordinate analysis (large intestine and small intestine); (**d**) principal coordinate analysis (large intestine); (**e**) principal coordinate analysis (small intestine); $n = 6$.

Bacteroidetes, *Verrucomicrobia*, *Actinobacteria*, *Proteobacteria*, and *Firmicutes* were the dominant phyla in the small and large intestinal contents of normal mice (Figure 7a–d). Alcohol increased the relative abundance of Proteobacteria in the small intestine and decreased the relative abundance of *Firmicutes* and *Bacteroidetes* in the small intestine and *Staphylococcus* in the large intestine. LCJ supplementation restored these levels to those in the CG groups (Figure 7e).

The small intestine had 16 key genera, while the large intestine contained 14 key genera (Figure 7f–i). Alcohol intervention considerably increased the relative abundance of *Staphylococcus* and *Proteus* in the small intestine while significantly decreasing the relative abundance of *Allobaculum* and *Akkermansia*. In the large intestine, alcohol intervention increased the relative abundance of *Parabacteroides*, *Oscillospira*, and *Proteus* and decreased the relative abundance of *Akkermansia*. LCJ administration significantly reversed these intestinal microbiota alterations in ALD mice, restoring them to a level similar to that of the CG group (Figure 7j,k).

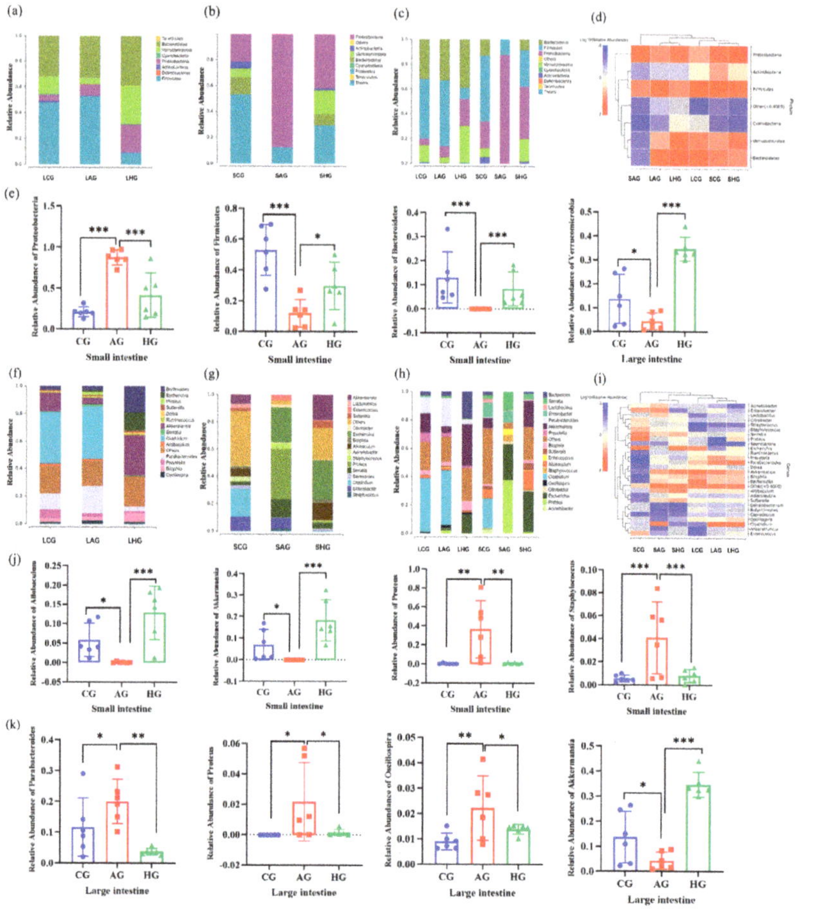

Figure 7. LCJ supplementation altered the composition of microflora in the large and small intestines of ALD mice. (**a**) Large intestine (phylum level); (**b**) Small intestine (phylum level); (**c**) Large intestine and small intestine (phylum level); (**d**) Species composition heat map (phylum level); (**e**) Relative abundance of *Proteobacteria, Firmicutes, Bacteroidetes,* and *Verrucomicrobia* (phylum level); (**f**) Large intestine (genus level); (**g**) Small intestine (genus level); (**h**) Large and small intestine (genus level); (**i**) Species composition heat map (genus level); (**j**) Relative abundance of *Allobaculum, Akkermansia, Proteus,* and *Staphylococcus* (genus level); (**k**) Relative abundance of *Parabacteroides, Proteus, Oscillospira,* and *Akkermansia* (genus level). $n = 6$. * $p < 0.05$, ** $p < 0.01$, and *** $p < 0.001$ vs. the model group (ANOVA).

To explore the effect of LCJ on the microbial structure of the large and small intestines of mice in each group, microbial markers in the large and small intestines of mice in each group were determined using linear discriminant analysis (LDA) effect size (LEfSe) analysis and an LDA analysis score greater than 4. Results are shown in Figure 8a–d. The biomarkers in the small intestine of the AG group included *Enterococcus, Prauseria,* and *Proteus,* whereas those in the large intestine were *Proteus, Ricerysipelorichi, Clostridium, Streptococcus,* and *Parabacteroides*. The biomarkers in the small intestine of HG mice included *Allobaculum, Akkermansia, Bilophila,* and *Sutterella,* whereas those in the large intestine were *Defluviitalea, Bacteroides, Akkermansia, Bilophila,* and *Alistipes*.

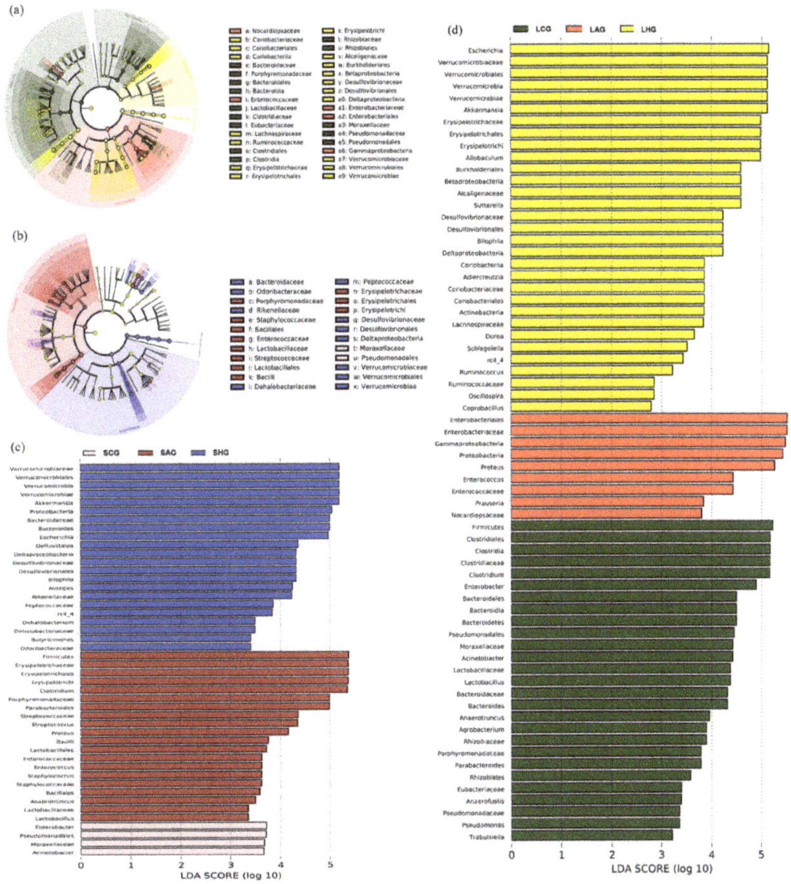

Figure 8. Effect of LCJ on microbial markers (genus level) and metabolic functional pathways. (**a**) Linear discriminant analysis effect size (LEfSe) analysis (small intestine); (**b**) linear discriminant analysis effect size (LEfSe) analysis (large intestine); (**c**) linear discriminant analysis (LDA) in the small intestine; (**d**) linear discriminant analysis (LDA) in the large intestine. $n = 6$.

4. Discussion

ALD is associated with high morbidity and mortality worldwide, which calls for investigations into its pathogenesis to find better treatments. Alcohol consumption can disrupt the composition of the gut microbiota, and resolving the disruption of the gut microbiota caused by alcohol can alleviate ALD. Previous studies have shown that plant natural products and polyphenolic compounds exert beneficial effects on several metabolic disorders associated with ALD via the gut–liver axis [35]. LC is rich in polyphenols such as C3G and can ameliorate non-alcoholic fatty liver disease. We speculate that LC can improve ALD. Meanwhile, the relationship between the modulating effects of LC on the gut microbiota and its protective effects on ALD has not been well studied. Interestingly, LC reduced alcohol-induced liver injury by regulating the FXR–FGF15 signaling pathway mediated by intestinal microbes and inhibiting lipid and bile acid accumulation in our study (Figure 9).

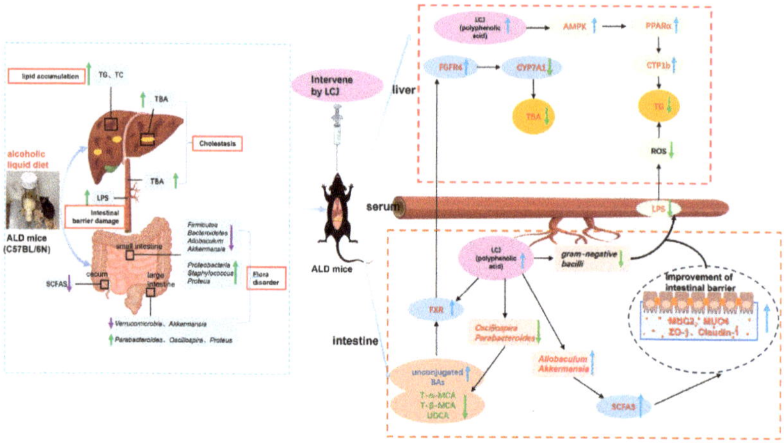

Figure 9. Combination of existing research conclusions and the results of this study. Potential protective mechanism of LCJ against alcohol-induced intestinal injury (Supplementary Table S3). The upward arrow represents an increase, while the downward arrow represents a decrease.

The changes in mouse liver index and serum ALT, AST, and AST/ALT ratio indicated that this alcohol feeding mode affected the normal growth of mice and damaged their liver cells. However, high–dose LCJ can effectively slow down alcohol–induced liver cell damage. The damage to ALD mouse liver cells may be due to the accumulation of liver lipids or inflammatory cell infiltration caused by alcohol metabolism, or bile acid accumulation and peroxidation [5]. This indicates that the protective effect of LCJ on the liver of ALD mice may be mediated through these aspects.

In our study, we observed that LCJ treatment significantly reduced the increase in serum TG and TC as well as the accretion of fat granules in the liver while increasing the protein expression of AMPK, PPARα, and CPT1b in the liver in ALD mice [37]. AMPK is a key upstream factor that regulates lipid synthesis and catabolism. After AMPK is activated, it can upregulate the expression of PPAR and CPT1. Some scholars believe that AMPK regulates PPAR by activating extracellular signal–regulated kinase (ERK) and the p38 pathway [38]. CPT1 is a rate–limiting enzyme during fatty acid β oxidation, including three subtypes: CPT1A, CPT1B, and CPT1C. PPAR can enhance mitochondrial utilization and the oxidation of fatty acids by promoting the transcriptional expression of CPT1 in the liver. Therefore, we believe that high–dose LCJ can promote the β oxidation of fatty acids by increasing the protein expression of AMPK, PPARα, and CPT1b in the liver of ALD mice, thereby reducing the accumulation of TG and TC in the liver of ALD mice. Studies have also shown that LC extract can actively improve nonalcoholic fatty liver disease by increasing CPT–1, which is involved in fatty acid oxidation, and ACC [5], which supports our conclusion. Polyphenol–rich LC has strong antioxidant activity [39], and dietary polyphenols can activate AMPK in the liver [40]. Hence, LCJ may improve lipid peroxidation in the liver by reducing the production of reactive oxygen species.

Long–term ethanol feeding has been found to reduce intestinal FXR activity. This may be because prolonged alcohol exposure increases FXR acetylation and interferes with FXR's ability to bind with the retinoid X receptor alpha (RXR), leading to the inactivation of FXR [41]. FXR is a nuclear receptor for bile acids (BAs) and affects the transport and homeostasis of BA. The natural inhibitors of FXR are T–MCA, T–MCA, and UDCA. The strength of the natural activators of FXR follows the order of CDCA > DCA > CA > LCA. GCDCA, TCA, and TDCA are weak FXR activators. Unconjugated BAs have a greater capacity to activate FXR than conjugated BAs [42]. So, the type and amount of BA produced in the intestine may regulate the expression of FXR. Long–term heavy drinking can lead to an imbalance in the gut microbiota that produces bile salt hydrolase (BSH), which is responsible for dehydroxylation reactions. If it leads to a decrease in bile acids with a strong ability to activate FXR in the intestine, it will further promote liver bile acid synthesis by weakening the expression of intestinal FXR. When intestinal FXR activity decreases, it will cause feedback inhibition of the expression of intestinal FGF15, which then increases the expression of the bile acid synthesis gene CYP7A1 in the liver by binding to FGFR4 in the liver, forming a negative feedback loop to increase bile acid synthesis in the liver [43]. This is consistent with the research results of our AG group. However, high-dose LCJ supplementation can increase the expression of FXR, FGF15, and FGFR4 in the ileum and decrease the expression of CYP7A1 to ameliorate liver damage induced by alcohol–induced bile acid accumulation in the liver. Hesperidin is a flavanone glycoside that can prevent cholestatic liver injury and reduce bile acid toxicity in HepaRG cells by activating FXR [44]. Proanthocyanidins can also activate the transcription activity of FXR and reduce triglyceridemia in vivo in an FXR–dependent manner [45]. Chokeberry polyphenols can reduce the relative contents of CA and DCA by altering the composition of the intestinal flora and can increase the relative content of CDCA [46]. Phenolic compounds can be decomposed and utilized by the intestinal flora, which can regulate the intestinal flora. Therefore, we speculate that LCJ rich in phenols may directly activate the activity of intestinal FXR, or it may also activate FXR activity by regulating the content of unconjugated bile acids by regulating the intestinal flora involved in bile acid metabolism. Thus, it alleviates cholestasis by regulating the FXR-FGF15 axis and reducing the expression of CYP7A1 in the liver.

Due to the fact that more than 90% of alcohol is absorbed in the gut after ingestion, excessive alcohol consumption can cause damage to the intestinal barrier. Intestinal barrier damage increases intestinal permeability, which allows the blood to carry LPS, a byproduct of bacteria found in the intestine, into the liver [27]. After LPS is delivered to the liver through the portal vein, as a central mediator of inflammation, it can activate Kupffer cells and macrophages in the liver, which in turn forces these cells to produce inflammatory cytokines such as TNF–α, IL–6, or ROS [1]. HE staining and mucin AB–PAS staining of intestinal sections revealed that alcohol consumption would have a more detrimental effect on the small intestine's barrier and mucin content than it would on the large intestine. This may be because alcohol is mainly absorbed and metabolized in the small intestine, so the impact on the large intestine is less [7]. Interestingly, high–dose LCJ has the potential to restore alcohol-induced small intestinal barrier damage. Supplementation of LCJ significantly improved alcohol–induced intestinal barrier dysfunction, including the increase of MUC2, MUC4, ZO–1, claudin–1, and the decrease of serum endotoxin levels. These findings supported the findings of the intestinal pathological section staining. These findings imply that supplementing LCJ to improve intestinal barrier function impairments may be an effective way to treat alcohol–induced liver damage.

A high dose of LCJ had different effects on the microbial composition between the large and small intestines. Notably, LCJ increased the relative abundance of *Verrucomicrobia* and *Akkermansia* in the large intestine of ALD mice but decreased the relative abundance of *Parabacteroides, Oscillospira,* and *Proteus.* In the small intestine of ALD mice, LCJ treatment increased the relative abundance of *Firmicutes, Bacteroidetes, Allobaculum,* and *Akkermansia* while reducing the relative abundance of *Proteobacteria, Staphylococcus,* and *Proteus.* A previous study reported that *Proteobacteria* contributes to inflammation [47]. In our study, the abundance of Gram–negative bacteria, *Proteobacteria,* was reduced in the HG group, which was consistent with the reduction in the Gram–negative bacteria product LPS. Furthermore, impairment of the intestinal barrier increases the likelihood of LPS entering the bloodstream, while LCJ has been shown to enhance intestinal barrier function. This suggests that LCJ can improve the intestinal barrier and regulate intestinal flora to decrease the production of LPS, thereby reducing blood LPS to improve alcohol-induced liver injury.

Akkermansia [48] and *Allobaculum* [49] play important roles in the prevention of alcohol–induced liver damage by increasing intestinal mucus and enhancing gut barrier function.

In our study, results showed that the relative richness of *Akkermansia* and *Allobaculum*, the mucin content, and intestinal barrier function in the small intestine of the AG group were significantly reduced, but treatment with LCJ reversed these changes. Moreover, the cecal levels of SCFAs, including acetic acid, propionic acid, butyric acid, and valeric acid, significantly increased in mice with ALD following LCJ intervention. Notably, acetic acid is the most abundant SCFA produced by the gut microbiota and has been shown to improve the function of the intestinal mucosal immune barrier [50]. Butyrate enhances intestinal barrier function by increasing colon mucin and the production of compact junction protein to protect the intestinal mucosal cells from LPS–induced damage [51]. These findings support our results that acetic acid and butyric acid levels in the AG and HG groups were consistent with changes in intestinal compact junction protein levels. *Akkermansia* can produce propionic acid by increasing mucin degradation [52]. *Allobaculum* is physiologically capable of generating butyric acid from carbohydrates [53]. In conclusion, we think that LCJ can increase the abundance of intestinal SCAF–producing bacteria (*Akkermansia* and *Allobaculum*), resulting in high intestinal SCFA production. In this way, it strengthens the intestinal barrier by increasing the expression of intestinal compact junction protein and mucin content to alleviate alcohol–induced liver injury. *Oscillospira* was strongly correlated with the proportion of secondary bile acids in the stool, suggesting that it may contribute to the formation of secondary bile acids in the stool [54]. In the study by Sun et al., it was reported that *Parabacteroides* hydrolyzes a variety of binding cholic acids into secondary cholic acids (stone cholic acid, ursodeoxycholic acid, etc.) through multiple pathways such as bile brine hydrolysase (BSH) [55]. In light of this and the findings of our experiments, we postulate that LCJ, which is abundant in powerful substances like anthocyanins, flavonoids, phenolic acids, and other active polyphenols, can regulate bile acid content as well as FXR and CYP7A1 expression levels in the HG group by improving the abundance of *Oscillospira* and *Parabacteroides* in the intestine. These results demonstrate that LCJ can reduce liver bile acid levels and further ameliorate alcohol-induced liver injury by regulating gut flora via the FXR–FGF15 axis.

Combined with this study, human treatment of ALD needs to pay attention not only to how to reduce the damage to liver cells but also to the changes in intestinal flora and the intestinal barrier. Plant extracts and probiotics with the impact of enhancing the intestinal barrier and improving intestinal flora are substances worthy of further investigation in the treatment of ALD in the future. This is because ALD can also be relieved by improving the intestinal flora. Additionally, LCJ has no side effects, which helps to maintain a healthy intestinal structure and qualifies it as an ingredient in functional foods for relieving alcoholic liver disease. It also raises public awareness of the health advantages and economic value of LC.

5. Conclusions

In conclusion, LCJ can reduce the accumulation of lipids and bile acids in the liver by reducing the levels of serum AST and ALT, thus producing hepatoprotective effects. Moreover, LCJ significantly improved the harm to the intestinal barrier caused by alcohol as well as the imbalance of short–chain fatty acids in the cecum. LCJ was found to reduce alcohol–induced liver injury by regulating gut microbes and its mediated FXR–FGF15 signaling pathway, based on intestinal flora analysis and related gene expression measurements.

Supplementary Materials: The following supporting information can be downloaded at: https://www.mdpi.com/article/10.3390/nu15184025/s1, Table S1: Quantitative RT-PCR primer design [56–66]; Table S2: Phenolic components in LCJ; Table S3: Comparison Table of English Abbreviations; Figure S1: Total chromatograms of 16 types of phenols in LCJ.

Author Contributions: L.N.: methodology, formal analysis, investigation, data curation, writing—original draft, visualization. B.Z.: conceptualization, supervision, resources, writing—review and editing. X.H.: investigation. All authors contributed to the article and approved the submitted version. All authors have read and agreed to the published version of the manuscript.

Funding: This research received no external funding.

Institutional Review Board Statement: All animal experiments were examined and authorized by the Laboratory Animal Ethics Committee of Jiangnan University (Ethical approval number: JN. No20220615c0450803223).

Informed Consent Statement: Not applicable.

Data Availability Statement: The data presented in this study are available in this article.

Acknowledgments: The authors are grateful for the help of Zhuoran Tan and Heilongjiang Fengran Agricultural Group Company for providing Lonicera Caerulea raw materials for the experiment.

Conflicts of Interest: The authors declare no conflict of interest.

References

1. Dunn, W.; Shah, V.H. Pathogenesis of Alcoholic Liver Disease. *Clin. Liver Dis.* **2016**, *20*, 445–456. [CrossRef] [PubMed]
2. Pimpin, L.; Cortez-Pinto, H.; Negro, F.; Corbould, E.; Lazarus, J.V.; Webber, L.; Sheron, N.; The Members of the EASL HEPA-HEALTH Steering Committee. Burden of liver disease in Europe: Epidemiology and analysis of risk factors to identify prevention policies. *J. Hepatol.* **2018**, *69*, 718–735. [CrossRef] [PubMed]
3. Rehm, J.; Samokhvalov, A.V.; Shield, K.D. Global burden of alcoholic liver diseases. *J. Hepatol.* **2013**, *59*, 160–168. [CrossRef] [PubMed]
4. Crabb, D.W.; Matsumoto, M.; Chang, D.; You, M. Overview of the role of alcohol dehydrogenase and aldehyde dehydrogenase and their variants in the genesis of alcohol-related pathology. *Proc. Nutr. Soc.* **2004**, *63*, 49–63. [CrossRef]
5. O'Shea, R.S.; Dasarathy, S.; McCullough, A.J.; Practice Guideline Committee of the American Association for the Study of Liver Diseases and the Practice Parameters Committee of the American College of Gastroenterology. Alcoholic liver disease. *Hepatology* **2009**, *51*, 307–328. [CrossRef]
6. Louvet, A.; Mathurin, P. Alcoholic liver disease: Mechanisms of injury and targeted treatment. *Nat. Rev. Gastroenterol. Hepatol.* **2015**, *12*, 231–242. [CrossRef]
7. Yeluru, A.; Cuthbert, J.A.; Casey, L.; Mitchell, M.C. Alcoholic Hepatitis: Risk Factors, Pathogenesis, and Approach to Treatment. *Alcohol. Clin. Exp. Res.* **2016**, *40*, 246–255. [CrossRef]
8. Choi, Y.R.; Kim, H.S.; Yoon, S.J.; Lee, N.Y.; Gupta, H.; Raja, G.; Gebru, Y.A.; Youn, G.S.; Kim, D.J.; Ham, Y.L.; et al. Nutritional Status and Diet Style Affect Cognitive Function in Alcoholic Liver Disease. *Nutrients* **2021**, *13*, 185. [CrossRef]
9. Chen, S.; Zhu, H.; Luo, Y. The gut-mediated function of polyphenols: Opinions on functional food development for nonalcoholic fatty liver disease. *Curr. Opin. Food Sci.* **2023**, *49*, 100972. [CrossRef]
10. Zhao, L.; Mehmood, A.; Yuan, D.; Usman, M.; Murtaza, M.A.; Yaqoob, S.; Wang, C. Protective Mechanism of Edible Food Plants against Alcoholic Liver Disease with Special Mention to Polyphenolic Compounds. *Nutrients* **2021**, *13*, 1612. [CrossRef]
11. Zhuge, Q.; Zhang, Y.; Liu, B.; Wu, M. Blueberry polyphenols play a preventive effect on alcoholic fatty liver disease C57BL/6 J mice by promoting autophagy to accelerate lipolysis to eliminate excessive TG accumulation in hepatocytes. *Ann. Palliat. Med.* **2020**, *9*, 1045–1054. [CrossRef] [PubMed]
12. Amen, Y.; Sherif, A.E.; Shawky, N.M.; Abdelrahman, R.S.; Wink, M.; Sobeh, M. Grape-Leaf Extract Attenuates Alcohol-Induced Liver Injury via Interference with NF-κB Signaling Pathway. *Biomolecules* **2020**, *10*, 558. [CrossRef] [PubMed]
13. Česonienė, L.; Labokas, J.; Jasutienė, I.; Šarkinas, A.; Kaškonienė, V.; Kaškonas, P.; Kazernavičiūtė, R.; Pažereckaitė, A.; Daubaras, R. Bioactive Compounds, Antioxidant, and Antibacterial Properties of *Lonicera caerulea* Berries: Evaluation of 11 Cultivars. *Plants* **2021**, *10*, 624. [CrossRef] [PubMed]
14. Cheng, Z.; Bao, Y.; Li, Z.; Wang, J.; Wang, M.; Wang, S.; Wang, Y.; Wang, Y.; Li, B. *Lonicera caerulea* (*Haskap* berries): A review of development traceability, functional value, product development status, future opportunities, and challenges. *Crit. Rev. Food Sci. Nutr.* **2022**, *43*, 1–25. [CrossRef]
15. Oszmiański, J.; Wojdyło, A.; Lachowicz, S. Effect of dried powder preparation process on polyphenolic content and antioxidant activity of blue honeysuckle berries (*Lonicera caerulea* L. var. kamtschatica). *Lwt-Food Sci. Technol.* **2016**, *67*, 214–222. [CrossRef]
16. Gavrilova, V.; Kajdžanoska, M.; Gjamovski, V.; Stefova, M. Separation, Characterization and Quantification of Phenolic Compounds in Blueberries and Red and Black Currants by HPLC–DAD–ESI-MS^n. *J. Agric. Food Chem.* **2011**, *59*, 4009–4018. [CrossRef]
17. Wu, S.; Yano, S.; Hisanaga, A.; He, X.; He, J.; Sakao, K.; Hou, D.-X. Polyphenols from *Lonicera caerulea* L. berry attenuate experimental nonalcoholic steatohepatitis by inhibiting proinflammatory cytokines productions and lipid peroxidation. *Mol. Nutr. Food Res.* **2017**, *61*, 1600858. [CrossRef]
18. Wang, Y.; Gao, N.; Nieto-Veloza, A.; Zhou, L.; Sun, X.; Si, X.; Tian, J.; Lin, Y.; Jiao, X.; Li, B. *Lonicera caerulea* polyphenols inhibit fat absorption by regulating Nrf2-ARE pathway mediated epithelial barrier dysfunction and special microbiota. *Food Sci. Hum. Wellness* **2023**, *12*, 1309–1322. [CrossRef]
19. Bursal, E.; Aras, A.; Kılıç, Ö.; Taslimi, P.; Gören, A.C.; Gülçin, I. Phytochemical content, antioxidant activity, and enzyme inhibition effect of *Salvia eriophora* Boiss. & Kotschy against acetylcholinesterase, α-amylase, butyrylcholinesterase, and α-glycosidase enzymes. *J. Food Biochem.* **2019**, *43*, e12776. [CrossRef]

20. Iimuro, Y.U.; Frankenberg, M.V.; Arteel, G.E.; Bradford, B.U.; Wall, C.A.; Thurman, R.G. Female rats exhibit greater susceptibility to early alcohol-induced liver injury than males. *Am. J. Physiol.-Gastrointest. Liver Physiol.* **1997**, *272*, G1186–G1194. [CrossRef]
21. Fulham, M.A.; Mandrekar, P. Sexual Dimorphism in Alcohol Induced Adipose Inflammation Relates to Liver Injury. *PLoS ONE* **2016**, *11*, e0164225. [CrossRef] [PubMed]
22. Osho, S.O.; Wang, T.; Horn, N.L.; Adeola, O. Comparison of goblet cell staining methods in jejunal mucosa of turkey poults. *Poult. Sci.* **2017**, *96*, 556–559. [CrossRef] [PubMed]
23. Guo, C.-F.; Zhang, S.; Yuan, Y.-H.; Li, J.-Y.; Yue, T.-L. Bile Salt Hydrolase and S-Layer Protein are the Key Factors Affecting the Hypocholesterolemic Activity of *Lactobacillus casei*-Fermented Milk in Hamsters. *Mol. Nutr. Food Res.* **2018**, *62*, e1800728. [CrossRef]
24. Reddy, M.V.; Chang, Y.-C. Production of biofuel precursor molecules (monocarboxylic acids, biohydrogen) from apple and pumpkin waste through an anaerobic fermentation process. *Sustain. Energy Fuels* **2021**, *5*, 4133–4140. [CrossRef]
25. Lv, X.-C.; Wu, Q.; Cao, Y.-J.; Lin, Y.-C.; Guo, W.-L.; Rao, P.-F.; Zhang, Y.-Y.; Chen, Y.-T.; Ai, L.-Z.; Ni, L. Ganoderic acid A from *Ganoderma lucidum* protects against alcoholic liver injury through ameliorating the lipid metabolism and modulating the intestinal microbial composition. *Food Funct.* **2022**, *13*, 5820–5837. [CrossRef]
26. Zhou, J.; Zhang, N.; Zhao, L.; Wu, W.; Zhang, L.; Zhou, F.; Li, J. Astragalus Polysaccharides and Saponins Alleviate Liver Injury and Regulate Gut Microbiota in Alcohol Liver Disease Mice. *Foods* **2021**, *10*, 2688. [CrossRef]
27. Bajaj, J.S. Alcohol, liver disease and the gut microbiota. *Nat. Rev. Gastroenterol. Hepatol.* **2019**, *16*, 235–246. [CrossRef]
28. Ge, Y.; Sun, H.; Xu, L.; Zhang, W.; Lv, J.; Chen, Y. The amelioration of alcohol-induced liver and intestinal barrier injury by *Lactobacillus rhamnosus* Gorbach-Goldin (LGG) is dependent on Interleukin 22 (IL-22) expression. *Bioengineered* **2022**, *13*, 12650–12660. [CrossRef]
29. Sangineto, M.; Grander, C.; Grabherr, F.; Mayr, L.; Enrich, B.; Schwärzler, J.; Dallio, M.; Bukke, V.N.; Moola, A.; Moschetta, A.; et al. Recovery of *Bacteroides thetaiotaomicron* ameliorates hepatic steatosis in experimental alcohol-related liver disease. *Gut Microbes* **2022**, *14*, 2089006. [CrossRef]
30. Ma, J.; Piao, X.; Mahfuz, S.; Long, S.; Wang, J. The interaction among gut microbes, the intestinal barrier and short chain fatty acids. *Anim. Nutr.* **2022**, *9*, 159–174. [CrossRef]
31. Lin, D.; Jiang, X.; Zhao, Y.; Zhai, X.; Yang, X. *Komagataeibacter hansenii* CGMCC 3917 alleviates alcohol-induced liver injury by regulating fatty acid metabolism and intestinal microbiota diversity in mice. *Food Funct.* **2020**, *11*, 4591–4604. [CrossRef]
32. Du, Y.; Yang, C.; Ren, D.; Shao, H.; Zhao, Y.; Yang, X. Fu brick tea alleviates alcoholic liver injury by modulating the gut microbiota–liver axis and inhibiting the hepatic TLR4/NF-κB signaling pathway. *Food Funct.* **2022**, *13*, 9391–9406. [CrossRef] [PubMed]
33. Sun, S.; Wang, K.; Sun, L.; Cheng, B.; Qiao, S.; Dai, H.; Shi, W.; Ma, J.; Liu, H. Therapeutic manipulation of gut microbiota by polysaccharides of *Wolfiporia cocos* reveals the contribution of the gut fungi-induced PGE_2 to alcoholic hepatic steatosis. *Gut Microbes* **2020**, *12*, 1830693. [CrossRef] [PubMed]
34. Xu, L.; Li, W.; Chen, S.-Y.; Deng, X.-W.; Deng, W.-F.; Liu, G.; Chen, Y.-J.; Cao, Y. Oenothein B ameliorates hepatic injury in alcoholic liver disease mice by improving oxidative stress and inflammation and modulating the gut microbiota. *Front. Nutr.* **2022**, *9*, 1053718. [CrossRef] [PubMed]
35. Albillos, A.; de Gottardi, A.; Rescigno, M. The gut-liver axis in liver disease: Pathophysiological basis for therapy. *J. Hepatol.* **2020**, *72*, 558–577. [CrossRef]
36. Cao, Y.-J.; Huang, Z.-R.; You, S.-Z.; Guo, W.-L.; Zhang, F.; Liu, B.; Lv, X.-C.; Lin, Z.-X.; Liu, P.-H. The Protective Effects of Ganoderic Acids from *Ganoderma lucidum* Fruiting Body on Alcoholic Liver Injury and Intestinal Microflora Disturbance in Mice with Excessive Alcohol Intake. *Foods* **2022**, *11*, 949. [CrossRef]
37. Wu, L.; Liu, C.; Chang, D.-Y.; Zhan, R.; Zhao, M.; Lam, S.M.; Shui, G.; Zhao, M.-H.; Zheng, L.; Chen, M. The Attenuation of Diabetic Nephropathy by Annexin A1 via Regulation of Lipid Metabolism through AMPK/PPARα/CPT1b Pathway. *Diabetes* **2021**, *70*, 2192–2203. [CrossRef]
38. Lee, W.H.; Kim, S.G. AMPK-Dependent Metabolic Regulation by PPAR Agonists. *PPAR Res.* **2010**, *2010*, 549101. [CrossRef]
39. Wang, Y.; Li, B.; Lin, Y.; Ma, Y.; Zhang, Q.; Meng, X. Effects of *Lonicera caerulea* berry extract on lipopolysaccharide-induced toxicity in rat liver cells: Antioxidant, anti-inflammatory, and anti-apoptotic activities. *J. Funct. Foods* **2017**, *33*, 217–226. [CrossRef]
40. Hu, Z.; Li, M.; Cao, Y.; Akan, O.D.; Guo, T.; Luo, F. Targeting AMPK Signaling by Dietary Polyphenols in Cancer Prevention. *Mol. Nutr. Food Res.* **2022**, *66*, 2100732. [CrossRef]
41. Wu, W.; Zhu, B.; Peng, X.; Zhou, M.; Jia, D.; Gu, J. Activation of farnesoid X receptor attenuates hepatic injury in a murine model of alcoholic liver disease. *Biochem. Biophys. Res. Commun.* **2014**, *443*, 68–73. [CrossRef]
42. Xiang, J.; Zhang, Z.; Xie, H.; Zhang, C.; Bai, Y.; Cao, H.; Che, Q.; Guo, J.; Su, Z. Effect of different bile acids on the intestine through enterohepatic circulation based on FXR. *Gut Microbes* **2021**, *13*, 1949095. [CrossRef]
43. Chiang, J.Y.L.; Ferrell, J.M. Bile acid receptors FXR and TGR5 signaling in fatty liver diseases and therapy. *Am. J. Physiol.-Gastrointest. Liver Physiol.* **2020**, *318*, G554–G573. [CrossRef]
44. Zhang, G.; Sun, X.; Wen, Y.; Shi, A.; Zhang, J.; Wei, Y.; Wu, X. Hesperidin alleviates cholestasis via activation of the farnesoid X receptor in vitro and in vivo. *Eur. J. Pharmacol.* **2020**, *885*, 173498. [CrossRef]
45. She, J.; Gu, T.; Pang, X.; Liu, Y.; Tang, L.; Zhou, X. Natural Products Targeting Liver X Receptors or Farnesoid X Receptor. *Front. Pharmacol.* **2022**, *12*, 772435. [CrossRef] [PubMed]

46. Liu, L.; Liu, Z.; Li, H.; Cao, Z.; Li, W.; Song, Z.; Li, X.; Lu, A.; Lu, C.; Liu, Y. Naturally Occurring TPE-CA Maintains Gut Microbiota and Bile Acids Homeostasis via FXR Signaling Modulation of the Liver–Gut Axis. *Front. Pharmacol.* **2020**, *11*, 12. [CrossRef] [PubMed]
47. Rolig, A.S.; Mittge, E.K.; Ganz, J.; Troll, J.V.; Melancon, E.; Wiles, T.J.; Alligood, K.; Stephens, W.Z.; Eisen, J.S.; Guillemin, K. The enteric nervous system promotes intestinal health by constraining microbiota composition. *PLoS Biol.* **2017**, *15*, e2000689. [CrossRef] [PubMed]
48. Wu, W.; Lv, L.; Shi, D.; Ye, J.; Fang, D.; Guo, F.; Li, Y.; He, X.; Li, L. Protective Effect of *Akkermansia muciniphila* against Immune-Mediated Liver Injury in a Mouse Model. *Front. Microbiol.* **2017**, *8*, 1804. [CrossRef]
49. van Muijlwijk, G.H.; van Mierlo, G.; Jansen, P.W.; Vermeulen, M.; Bleumink-Pluym, N.M.; Palm, N.W.; van Putten, J.P.; de Zoete, M.R. Identification of *Allobaculum mucolyticum* as a novel human intestinal mucin degrader. *Gut Microbes* **2021**, *13*, 1966278. [CrossRef]
50. Lavelle, A.; Sokol, H. Gut microbiota-derived metabolites as key actors in inflammatory bowel disease. *Nat. Rev. Gastroenterol. Hepatol.* **2020**, *17*, 223–237. [CrossRef]
51. Yan, H.; Ajuwon, K.M. Butyrate modifies intestinal barrier function in IPEC-J2 cells through a selective upregulation of tight junction proteins and activation of the Akt signaling pathway. *PLoS ONE* **2017**, *12*, e0179586. [CrossRef] [PubMed]
52. Kim, S.; Shin, Y.-C.; Kim, T.-Y.; Kim, Y.; Lee, Y.-S.; Lee, S.-H.; Kim, M.-N.; Eunju, O.; Kim, K.S.; Kweon, M.-N. Mucin degrader *Akkermansia muciniphila* accelerates intestinal stem cell-mediated epithelial development. *Gut Microbes* **2021**, *13*, 1892441. [CrossRef] [PubMed]
53. Zhou, W.; Yang, T.; Xu, W.; Huang, Y.; Ran, L.; Yan, Y.; Mi, J.; Lu, L.; Sun, Y.; Zeng, X.; et al. The polysaccharides from the fruits of *Lycium barbarum* L. confer anti-diabetic effect by regulating gut microbiota and intestinal barrier. *Carbohydr. Polym.* **2022**, *291*, 119626. [CrossRef] [PubMed]
54. Konikoff, T.; Gophna, U. Oscillospira: A Central, Enigmatic Component of the Human Gut Microbiota. *Trends Microbiol.* **2016**, *24*, 523–524. [CrossRef] [PubMed]
55. Sun, H.; Guo, Y.; Wang, H.; Yin, A.; Hu, J.; Yuan, T.; Zhou, S.; Xu, W.; Wei, P.; Yin, S.; et al. Gut commensal *Parabacteroides distasonis* alleviates inflammatory arthritis. *Gut* **2023**, *72*, 1664–1677. [CrossRef]
56. Li, C.; Yu, S.; Li, X.; Cao, Y.; Li, M.; Ji, G.; Zhang, L. Medicinal Formula Huazhi-Rougan Attenuates Non-Alcoholic Steatohepatitis through Enhancing Fecal Bile Acid Excretion in Mice. *Front. Pharmacol.* **2022**, *13*, 833414. [CrossRef]
57. Li, J.; Zhang, L.; Li, Y.; Wu, Y.; Wu, T.; Feng, H.; Xu, Z.; Liu, Y.; Ruan, Z.; Zhou, S. Puerarin improves intestinal barrier function through enhancing goblet cells and mucus barrier. *J. Funct. Foods* **2020**, *75*, 104246. [CrossRef]
58. Lu, F.; Li, Y.; Zhou, B.; Guo, Q.; Zhang, Y. Early-life supplementation of grape polyphenol extract promotes polyphenol absorption and modulates the intestinal microbiota in association with the increase in mRNA expression of the key intestinal barrier genes. *Food Funct.* **2021**, *12*, 602–613. [CrossRef]
59. Wang, H.H.; Afdhal, N.H.; Gendler, S.J.; Wang, D.Q.-H. Targeted disruption of the murine mucin gene 1 decreases susceptibility to cholesterol gallstone formation. *J. Lipid Res.* **2004**, *45*, 438–447. [CrossRef]
60. Jiang, S.; Xue, D.; Zhang, M.; Li, Q.; Liu, H.; Zhao, D.; Zhou, G.; Li, C. Myoglobin diet affected the colonic mucus layer and barrier by increasing the abundance of several beneficial gut bacteria. *Food Funct.* **2022**, *13*, 9060–9077. [CrossRef]
61. Danjo, T.; Eiraku, M.; Muguruma, K.; Watanabe, K.; Kawada, M.; Yanagawa, Y.; Rubenstein, J.L.R.; Sasai, Y. Subregional specification of embryonic stem cell-derived ventral telencephalic tissues by timed and combinatory treatment with extrinsic signals. *J. Neurosci.* **2011**, *31*, 1919–1933. [CrossRef] [PubMed]
62. Sun, C.-C.; Zhang, C.-Y.; Duan, J.-X.; Guan, X.-X.; Yang, H.-H.; Jiang, H.-L.; Hammock, B.D.; Hwang, S.H.; Zhou, Y.; Guan, C.-X.; et al. PTUPB ameliorates high-fat diet-induced non-alcoholic fatty liver disease via inhibiting NLRP3 inflammasome activation in mice. *Biochem. Biophys. Res. Commun.* **2020**, *523*, 1020–1026. [CrossRef] [PubMed]
63. Zhao, Y.; Meng, C.; Wang, Y.; Huang, H.; Liu, W.; Zhang, J.-F.; Zhao, H.; Feng, B.; Leung, P.S.; Xia, Y. IL-1β inhibits β-Klotho expression and FGF19 signaling in hepatocytes. *Am. J. Physiol.-Endocrinol. Metab.* **2016**, *310*, E289–E300. [CrossRef] [PubMed]
64. Zhang, L.; Duan, Y.; Guo, Q.; Wang, W.; Li, F. A selectively suppressing amino acid transporter: Sodium-coupled neutral amino acid transporter 2 inhibits cell growth and mammalian target of rapamycin complex 1 pathway in skeletal muscle cells. *Anim. Nutr.* **2020**, *6*, 513–520. [CrossRef]
65. Lee, J.; Narayan, V.P.; Hong, E.Y.; Whang, W.K.; Park, T. *Artemisia Iwayomogi* Extract Attenuates High-Fat Diet-Induced Hypertriglyceridemia in Mice: Potential Involvement of the Adiponectin-AMPK Pathway and Very Low Density Lipoprotein Assembly in the Liver. *Int. J. Mol. Sci.* **2017**, *18*, 1762. [CrossRef]
66. Bak, E.-J.; Kim, J.; Choi, Y.H.; Kim, J.-H.; Lee, D.-E.; Woo, G.-H.; Cha, J.-H.; Yoo, Y.-J. Wogonin ameliorates hyperglycemia and dyslipidemia via PPARα activation in db/db mice. *Clin. Nutr.* **2014**, *33*, 156–163. [CrossRef]

nutrients

Review

Overview on the Polyphenol Avenanthramide in Oats (*Avena sativa* Linn.) as Regulators of PI3K Signaling in the Management of Neurodegenerative Diseases

Nitu L. Wankhede [1], Mayur B. Kale [1], Ashwini K. Bawankule [1], Manish M. Aglawe [1], Brijesh G. Taksande [1], Rashmi V. Trivedi [1], Milind J. Umekar [1], Ankush Jamadagni [2], Prathamesh Walse [2], Sushruta Koppula [3] and Spandana Rajendra Kopalli [4,*]

[1] Department of Pharmacology, Smt. Kishoritai Bhoyar College of Pharmacy, Nagpur 441002, Maharashtra, India
[2] Fortem Bioscience Private Limited, Bangalore 560064, Karnataka, India
[3] College of Biomedical and Health Sciences, Konkuk University, Chungju-si 27478, Republic of Korea
[4] Department of Bioscience and Biotechnology, Sejong University, Gwangjin-gu, Seoul 05006, Republic of Korea
* Correspondence: spandanak@sejong.ac.kr; Tel.: +82-2-6935-2619

Abstract: Avenanthramides (Avns) and their derivatives, a group of polyphenolic compounds found abundantly in oats (*Avena sativa* Linn.), have emerged as promising candidates for neuroprotection due to their immense antioxidant, anti-inflammatory, and anti-apoptotic properties. Neurodegenerative diseases (NDDs), characterized by the progressive degeneration of neurons, present a significant global health burden with limited therapeutic options. The phosphoinositide 3-kinase (PI3K) signaling pathway plays a crucial role in cell survival, growth, and metabolism, making it an attractive target for therapeutic intervention. The dysregulation of PI3K signaling has been implicated in the pathogenesis of various NDDs including Alzheimer's and Parkinson's disease. Avns have been shown to modulate PI3K/AKT signaling, leading to increased neuronal survival, reduced oxidative stress, and improved cognitive function. This review explores the potential of Avn polyphenols as modulators of the PI3K signaling pathway, focusing on their beneficial effects against NDDs. Further, we outline the need for clinical exploration to elucidate the specific mechanisms of Avn action on the PI3K/AKT pathway and its potential interactions with other signaling cascades involved in neurodegeneration. Based on the available literature, using relevant keywords from Google Scholar, PubMed, Scopus, Science Direct, and Web of Science, our review emphasizes the potential of using Avns as a therapeutic strategy for NDDs and warrants further investigation and clinical exploration.

Keywords: avenanthramide; AKT-protein kinase B; Alzheimer's disease; Parkinson's disease; oxidative stress; neuroprotection

Citation: Wankhede, N.L.; Kale, M.B.; Bawankule, A.K.; Aglawe, M.M.; Taksande, B.G.; Trivedi, R.V.; Umekar, M.J.; Jamadagni, A.; Walse, P.; Koppula, S.; et al. Overview on the Polyphenol Avenanthramide in Oats (*Avena sativa* Linn.) as Regulators of PI3K Signaling in the Management of Neurodegenerative Diseases. *Nutrients* **2023**, *15*, 3751. https://doi.org/10.3390/nu15173751

Academic Editor: Abdelouahed Khalil

Received: 30 July 2023
Revised: 23 August 2023
Accepted: 25 August 2023
Published: 27 August 2023

1. Introduction

Since ancient times, cultivated oats have been an important cereal crop for human consumption worldwide. Recently, interest in oats has been increased due to their nutritional properties and associated health benefits [1]. In the world grain ranking, oats were placed sixth for their nutritional values and regarded as most suitable for cultivation under variable climatic and soil conditions [2,3]. Cereal oats contain numerous bioactive compounds with high nutritive values including cellulose, arabinoxylan, β-glucans, proteins, unsaturated fatty acids, vitamins, minerals, phenols, and dietary fiber [4,5]. Studies suggest that the regular consumption of oat grains is associated with a lower risk of various chronic diseases [6,7]. Meta-analysis studies also revealed that regular oat intake lowers blood cholesterol and improves insulin sensitivity and post-prandial glycemic control [6,7]. The water-soluble β-glucan contained in oat seeds mainly lowers cholesterol levels, leading to a reduced risk of heart disease in humans [8,9]. Further, pharmacological studies indicate that the polyphenolic active constituents in oats exhibit strong anti-inflammatory, anti-bacterial, anti-carcinogenic, cytotoxic, and anti-proliferative properties [4,10]. For this reason, the human demand for oat consumption has been drastically increased.

In the central nervous system (CNS), different signaling pathways regulate various cellular activities and physiological functions. In particular, the protective roles of the phosphoinositide 3-kinase (PI3K) signaling pathway and protein kinase B (AKT) have been widely reported in neurodegenerative conditions. The PI3K/AKT signaling pathway is functional in many CNS processes such as synaptic plasticity, neurogenesis, proliferation and differentiation, aging, and neuronal autophagy [11]. In addition, PI3K also plays a predominant role in molecular trafficking [12]. The PI3K/AKT pathway constitutes a major signaling cascade, which consists of PI3K, a multifaceted protein, and its downstream molecules including glycogen synthase kinase-3 beta (GSK-3β), the mammalian target of rapamycin (mTOR), and the nuclear factor erythroid 2–related factor (Nrf) [13].

In a growing number of studies, neuroprotective agents from natural products are involved in targeting PI3K/AKT signaling, thus contributing to the prevention and treatment of neurodegenerative diseases (NDDs) including Alzheimer's disease (AD) and Parkinson's disease (PD) [14,15]. Although great efforts have been made to understand the pathogenesis of NDDs and the design of an effective treatment to delay progression, there is still no potential therapy available. Thus, the interest in derived natural compounds with potential benefits in CNS disorders has been substantially growing. The mechanisms of action of natural compounds are variable, suggesting that they could be highly effective and specific in improving neuroprotective capacity.

Avenanthramides (Avns, Figure 1) are the major water-soluble phenolamides obtained exclusively from oats. Several chronic conditions, including cardiovascular disease, cancer, diabetes, and neurological and skin disorders, are among those for which the therapeutic potential of Avns has been widely reported [1,7]. Recently, Avns and their derivatives were reported to possess neuroprotective effects in experimental models and also improve AD and PD pathologies including memory and behavioral impairments [16–19].

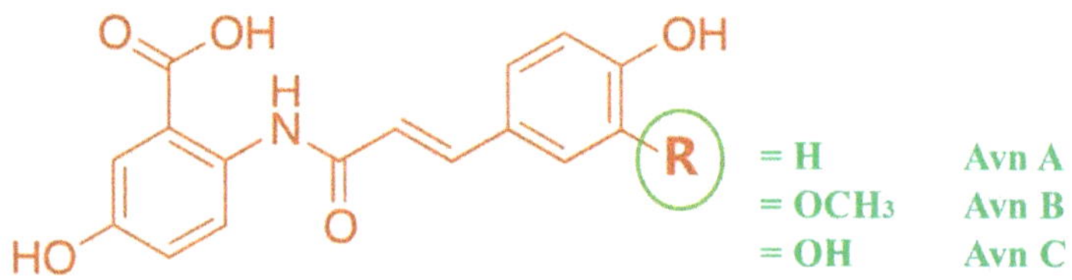

Figure 1. Structure of avenanthramides.

Among the 25 polyphenolic Avns identified, avenanthramide c (Avn-c), an amide conjugate of anthranilic acid and hydroxycinnamic acid, was found to be protective in experimental models of cerebral ischemia [10,20,21]. However, the roles of Avns and their derivatives against NDDs and their mechanistic basis have not been fully understood. In the present review, we highlight the potential of Avns as novel therapeutics for the treatment of NDDs and the influence of Avns on the regulation of the PI3K signaling pathway in the pathophysiology of NDDs including AD and PD. The relevant literature, using the keywords "avenanthramide", "PI3K/AKT", "neurodegenerative diseases", "AD", and "PD" from Google Scholar, PubMed, Scopus, Science Direct, and Web of Science were selected. The most appropriate and recent articles were prioritized based on the purpose of this review.

2. Modulation of PI3K Signaling Pathway in NDDs

It is well documented that the dysregulation of PI3K signal transduction significantly contributes in the pathogenesis of various NDDs including AD, PD, and Huntington's disease [22]. Further, signal transduction involving AKT along with PI3K is known to mediate neuronal survival. Therefore, understanding the role and regulatory aspects of the PI3K and AKT pathways might help in the development of suitable therapeutic agents against NDDs targeting PI3K/AKT signaling. Although the roles of PI3K/Akt and the mechanisms involved in NDDs were thoroughly reviewed (Rai et al. 2019) [23], in the following sections, the key aspects of PI3K/AKT activation and other related signaling molecular pathways in relation to NDDs are discussed.

2.1. PI3K

According to the structure, regulation, and substrate specificity, PI3Ks have been divided into three subcategories: classes I, II, and III [24]. Among these, the class I isoform has been studied extensively, which is activated by surface receptors composed of regulatory and catalytic subunits. These are further subcategorized into class IA and class IB based on the mode of regulation. Class II PI3K enzymes require an additional signal for activation, whereas class III PI3K enzymes are vacuolar protein sorting 34 (Vps34) required mainly for membrane trafficking [25].

2.2. AKT-Protein Kinase B (PKB)

AKT has been considered as a core effector of the PI3K downstream signaling pathway. AKT is a serine/threonine kinase, categorized into three homologous isoforms, AKT1/PKBα, AKT2/PKBβ, and AKT3 [14,23]. It consists of three different domains, including the pleckstrin homology (PH) domain at the N-terminal, which is responsible for membrane translocation after activation by PI3K; the central fragment; and the regulatory domain at C-terminal, which contains the phosphorylation site required for the activation of AKT. Among the isoforms, AKT3/PKBγ is crucial for the development of the brain and microglial survival.

2.3. Activation of PI3K/AKT Pathway

The signaling pathway begins with the interaction between the ligand and transmembrane receptor, mainly tyrosine kinase (RTK), which results in receptor dimerization and the autophosphorylation of the intracellular tyrosine domain, leading to the recruitment of PI3. The P85, a regulatory domain in class I PI3K, binds to the phosphorylated tyrosine residue following the recruitment of p110, a catalytic domain on PI3K that is responsible for the complete activation of the PI3K enzyme. The p110 also recruits the inactive AKT and phosphatidylinositol-dependent protein kinase 1 (PDK1) from the cytoplasm on the cell membrane that brings conformational changes in the AKT, exposing the phosphorylation site on serine (473) and threonine (308). The activated PI3K also phosphorylates phosphatidylinositol 4,5 biphosphate (PtdIns,4,5, P2) to phosphatidylinositol 3,4,5 triphosphate (PtdIns,3,4,5, P3), which becomes dephosphorylated with phosphatase and tensin homologue on chromosome 10 (PTEN), which regulates the inactivation of AKT [26].

The phosphorylated PI3K/AKT-Ser473 forms are complex with a wide range of downstream signaling molecules including GSK-3β, mTOR, ERK, NF-κB, Hsp, etc., to execute diverse cellular activities [14,27]. The downstream target mTOR regulates the metabolism. The activation of PI3K/AKT also inhibits the process of apoptosis by interacting with different signaling molecules including the Bcl2 antagonist of cell death (Bad), the bcl-2-like protein 11 (BIM), and caspase-9. NF-κB, one of the downstream modulators of PI3K/AKT, regulates the inflammatory response by modulating the expression of inflammatory markers IL-1β, IL-6, TNF-α, iNOS, and COX-2. At the resting state, the activity of NF-κB is repressed by inhibitory kappa B (IκB). The PI3K/AKT interacts with nuclear factor erythrocyte two related factors (Nrf2) to regulate oxidative stress [14]. In NDDs including AD and PD, the PI3K/AKT signaling is altered, resulting in the disruption of cellular function including autophagy and synaptic plasticity, thus indicating that targeting the PI3K/AKT pathway or its downstream regulator could be a novel strategy to treat NDDs.

3. PI3K/AKT Pathway in Neurodegeneration

NDDs are the category of conditions with the selective and progressive loss of structure or function of neurons. These disorders are characterized by the gradual degeneration of neurons, resulting in behavioral, learning, emotional, and cognitive abnormalities [28]. The prevalence report suggests a steady increase in the number of people suffering from neurodegenerative conditions. The treatment is targeted to alleviate symptoms and to prevent progression to improve the quality of life. Many studies have been carried out to understand the pathophysiological cause of the progressive neurodegeneration associated with AD and PD, which highlighted several possible targets for the development of neuroprotective strategies for effective treatment [29]. In this section, we discuss the evidence for the involvement of PI3K/AKT downstream targets in AD and PD. In addition, we highlight the potential targets of the PI3K/AKT pathway for the treatment of NDDs.

Reactive oxygen species (ROS) are produced in normal physiological settings as a bioproduct of cellular metabolism. However, ROS levels are firmly regulated via redox homeostasis. The production of ROS is increased in certain pathophysiological conditions due to several risk factors, including environmental stress, mutation, or genetic factors. Failing to regulate the excessive production of ROS causes structural and functional problems and the loss of cellular function. Neurons are particularly vulnerable to oxidative stress conditions due to a high consumption of oxygen, low antioxidant levels, high polyunsaturated membrane fatty acids, as well as a post-mitotic high accumulation of oxidized molecules [30]. In NDDs such as AD and PD in particular, oxidative stress has been considered to be a crucial risk factor. Thus, increased oxidative stress markers as well as deficient antioxidant defense systems are considered to be the common hallmarks in these conditions [30–33].

One of the main neuroprotective factors that regulate antioxidant and anti-inflammatory genes is Nrf2. Nrf2 modulates oxidative stress by regulating the synthesis of antioxidant enzymes through binding with the antioxidant response element (ARE). Under a physiological setting, the activity of Nrf2 is repressed by a protein, Kelch-like ECH-associated protein 1 (Keap1), which is an adaptor for the E3 ligase in the ubiquitin-proteasome pathway [34,35]. However, under stressful conditions, Keap1 is inactivated due to post-translational modification, leading to migration and the accumulation of active Nrf2 inside the nucleus, leading to heterodimerization with small protein designated as sMaf, which stimulates the expression of antioxidants through binding at ARE on the target gene [33,36]. Besides Keap1, the activation of the PI3K/AKT pathway can also indirectly regulate Nrf2-ARE signaling through a serine/threonine protein kinase designated as glycogen synthase kinase 3β (GSK-3β) via phosphorylating Nrf2 [37,38]. Mounting evidence suggests that the negative regulation of GSK-3β via PI3K/AKT signaling can regulate the activity of Nrf2 through stabilizing and regulating the gene expression of Nrf2 [39–41]. Under physiological conditions, the activation of PI3K/AKT inhibits GSK-3β through phosphorylation at Ser21-GSK-3α or Ser9-GSK-3β [42].

Several studies investigated that the PI3K/AKT/Nrf2/GSK-3β pathway has been impaired in AD patients and pre-clinical mouse models [43–46]. Oxidative stress in AD causes the downregulation of phosphorylated PI3K, which results in the inactivation of the PI3K/AKT pathway, leading to GSK-3β activation, which translocates Nrf2 from the nucleus into cytosol, resulting in low levels of antioxidant enzymes [47–50]. In addition, amyloid-β oligomers were found to activate GSK-3β, blocking PI3K/AKT/mTOR signaling, which increases the phosphorylation of Tau, inducing neurofibrillary tangle formation, which is a pathophysiological hallmark in AD. These findings suggest the involvement of different downstream molecules of PI3K/AKT in the pathophysiology of AD.

Oxidative stress-induced neuronal death also contributes to the pathology of PD. PI3K/AKT signaling influences oxidative stress by modulating downstream molecules such as GSK-3β and mTOR. The abnormal expression of GSK-3β has been reported in PD. Several findings indicate that Nrf2 signaling is compromised with age; thus, reduced Nrf2 expression in different brain regions has been associated with increased age [51,52], though the expression of Nrf2 has been suggested to be restricted in astrocytes in substantia nigra (SN), which is still sufficient to protect neurons from 1-Methyl-4-phenyl1,2,3,6-tetrahydropyridine (MPTP)-induced neurotoxicity [53]. In addition, data from post-mortem brains indicated Nrf2 dysfunction in PD [52,54]. Also, the levels of downstream targets in PI3K/AKT/mTOR signaling were found to be significantly reduced in PD patients. In PD, the imbalance of oxidative stress has been reported due to disrupted FoxO3a, a downstream target of mTOR [55,56]. Thus, the molecules that activate the PI3K/AKT pathway through activation of p-GSK-3β (Ser9) or mTOR can prevent oxidative-stress-induced neuronal death, which can be beneficial in preserving neuronal structure and function.

4. Neuroprotective Role of Avns

Avns derived from oats have been shown to exert neuroprotective effects by modulating signaling pathways involved in cell survival, neuronal growth, synaptic plasticity, apoptosis, and neuroinflammation. These mechanisms can enhance neuronal resilience and support the maintenance of cognitive function. Oat polyphenols have been well reported for their pharmacological benefits including antioxidant activities. The polyphenolic combination with different groups such as phenolamide tends to potentiate its action [57–59]. Avns have been found to exhibit antioxidant and anti-inflammatory effects in the CNS, which can be beneficial for reducing stress and neuroinflammation associated with various neurological conditions.

In a preclinical study, Avns was found to reduce the area of tissue damage and improved the neurological outcomes following stroke in the middle cerebral artery occlusion (MCAO) in mice models [17]. These effects were attributed to the ability of Avns to enhance blood flow, inhibit inflammatory responses, and protect against oxidative damage. Furthermore, Avns have been found to promote neurogenesis and enhance synaptic plasticity, which are crucial for brain repair and functional recovery after stroke [16]. While the research on Avns in neurological conditions is still emerging, the available evidence suggests their potential as neuroprotective agents. In the following sections, we discuss the beneficial roles of Avns against NDDs including AD and PD pathology.

4.1. Alzheimer's Disease

Globally, most geriatric people are suffering from this common condition, which is associated with dementia, memory loss, and cognition impairment [60]. According to the annual report of the Alzheimer's Association, about 6 million Americans belonging to the age group of 65–75 years most commonly have AD, and the incidence will increase day by day. Nerve cell destruction, extracellular amyloid plaques, and neurofibrillary tangles (NFTs) within the cell seem to be the markers of this complex NDD. These plaques are composed of amyloid beta (Aβ), a cleavage by-product of the amyloid precursor protein (APP) [61]. Moreover, the gradual formation of oligomers, fibrils, and insoluble amyloid plaques from Aβ monomers results in a reduction in the plasticity of neurons in the synaptic domain [62]. In addition, tau protein has been hyperphosphorylated, which forms NFTs. In healthy conditions, tau encourages microtubule stabilization. But as paired helical filaments connect, hyperphosphorylated tau builds up and eventually produces NFTs. The accumulation of Aβ leads to the dysregulation of synaptic and neuronal activities, which further generates intracellular conditions for NFT production and ultimately results in neuronal death and the disruption of neurotransmitter functions [63].

The antioxidant potential of Avns has been confirmed through a variety of in vivo and in vitro studies. The hydroxy and amide group in its structure are mainly responsible for scavenging free radical species formed during a variety of physiological as well as pathological conditions. This phenolamide is also able to induce the synthesis of a variety of cytoprotective enzymes. In addition, Avn contains an unsaturated amide group, which mainly interacts with the cysteine residue in Keap1, thus effectively preventing the phosphorylation and accumulation of Nrf2 in the cytosol and promoting the upregulation of the transcriptive action of Nrf2 [64,65].

Synaptic dysfunction has a major contribution to memory deficits in AD pathophysiology. Thus, molecules that strengthen the synapse could be beneficial for the treatment of AD. Alteration in synaptic plasticity due to the accumulation of amyloid-β, leading to memory loss and dementia, is a key feature of AD. The methanolic extract of Avns was found to restore and alter long-term potentiation in CA3 to the CA1 region of the hippocampus in the Tg2576 AD mouse brain [19]. In AD, amyloid-β aggregates were found to act through activating GSK3β of the PI3K/AKT pathway [66,67]. The study revealed that Avn-C, in particular, modulates the S9-GSKβ, leading to an improvement in brain function that was altered due to amyloid-β accumulation. The inhibition of S9-GSKβ effectively upregulates Nrf2-ARE activity, which may be responsible for the neuroprotective action of Avn-C in Tg2576 mice [19].

In addition, it was also reported that Avn-C acts as an amyloid inhibitor, thereby preventing Aβ protein aggregation [68]. The study showed that Avn-C improved the memory deficits associated with the Aβ and was reported to strengthen the synapse indicated via an improvement in LTP. The study also revealed that Avn-C effectively reduced the caspase-3 concentration and thereby inhibited the activation of pro-inflammatory markers, and in contrast, increased the levels of anti-inflammatory markers. Further, Avn-C activated the S9GSK3β and improved the antioxidant defense through the activation of Nrf2 in a different model of AD [69]. A possible role of Avn in the modulation of the PI3K/AKT/Nrf2/GSK-3β signaling pathways is shown in Figure 2.

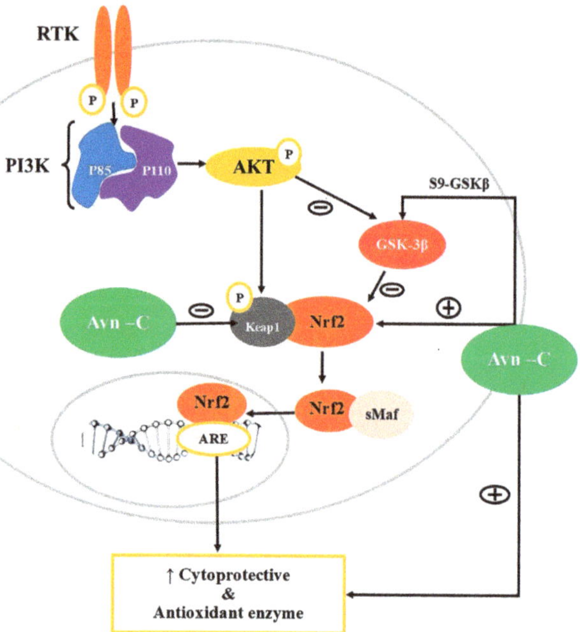

Figure 2. Modulation of PI3K/AKT/Nrf2/GSK-3β signaling pathways by Avn. PI3K: Phosphatidylinositol-3-kinases (PI3Ks); AKT: Ak strain transforming-Phosphokinase B; GSK3β: Glycogen synthase kinase 3β; Nrf2: Nuclear factor erythroid 2-related factor 2; ARE: Antioxidant response element; Keap1: Kelch-like ECH-associated protein 1; RTK: Receptor tyrosine kinase. Oxidative stress limits the action of PI3K/AKT, and thereby activates GSK3β, thus inhibiting Nrf2 translocation into the nucleus, leading to downregulation of expression of cytoprotective enzymes. Avn-C acts as an electrophile that interacts with Keap, resulting in loss of inhibitory control and translocation of Nrf2 into the nucleus and activating ARE for transcription of the antioxidant enzyme. Avn-C can also activate Nrf2 indirectly by inhibiting GSK3β.

4.2. Parkinson's Disease

PD is the second most common neurodegenerative condition, affecting 1% of the old age population. Clinical presentation in PD is associated with motor disturbances including bradykinesia, rigidity, and tremors [70]. Majorly, PD is characterized by a progressive loss of dopaminergic neurons in SN, but the neuropathology indicates a widespread association between different regions of the brain, where the involvement of SN occurs in the middle stage of the disease. The pathological hallmark of the disease involves the deposition of Lewy bodies, which are round eosinophilic protein aggregates composed of α-synuclein (α-syn) and synphililin-1 [71,72]. A variety of factors are responsible for the alteration in transcription in the sporadic form of PD such as environmental factors, oxidative stress, exposures, and aging [73,74], whereas the familial type of PD is associated with a mutation in genes encoding α-syn [73–77]. Studies have suggested that among the list of factors, oxidative stress and ROS are mainly responsible for neuronal loss in the SN region in the PD pathophysiology [78,79]. Further, previous reports also indicated that several flavonoids and polyphenolic compounds that form natural products exhibited neuroprotective effects by suppressing neuroinflammation, oxidative stress, and improved cognition by regulating the PI3K/AKT signaling pathways in PD models [80–82].

In a recent study, Bisavenanthramide-B (Bis-B), a synthetic analog of Avn-C, appears to protect from oxidative stress [83]. The molecule was discovered as a product formed during the reaction of Avn with ROS. Bis-B contains an electrophilic group designated as a Michael acceptor, which is capable of covalently conjugating with cysteine residues of Keap1, thereby inducing conformational changes and leading to the translocation of Nrf2 in the nucleus, which results in cytoprotective gene expression through Nrf2-ARE interaction [83].

The oxidative stress induced via dysfunctional mitochondria has been a well-known pathology in NDDs. These synthetic Avns have also shown neuroprotective effects against rotenone/oligomycin-induced oxidative damage, which was considerably similar to the PD pathology. Thus, Bis-Avns could be the effective therapeutic approach in preventing selective SN dopaminergic degeneration in PD. Further, the same study revealed that Avn has excellent neuroprotective action through directly and indirectly scavenging ROS via Nrf2 stimulation against okadaic-acid-induced Tau hyperphosphorylation and oxidative stress in SH-SY5Y cells, which has been linked critically with the neurodegenerative tauopathies in AD and PD [84,85]. In addition, Avn-C can directly protect from oxidative stress through the activation of Nrf2-ARE [18]. The Avn-2c isoform efficiently translocates Nrf2 into the nucleus in PC12 cells and thereby stimulates the expression of cytoprotective enzymes, and thus could be beneficial in the treatment of NDDs including PD. The available literature on the therapeutic potential of Avn and its derivatives are shown in Table 1.

Table 1. Therapeutic potential of avenanthramides in neurodegenerative disease.

No.	Study Design	Mechanism	Methodology	Reference
1	Electrophysiological study on hippocampal LTP in Tg2576 male mice	Alteration in p-GSK-3β-S9 levels and reduction in Caspase 3	Avn-C extracted from germinated oats using column chromatography	[19]
2	Evaluation of neuropathologies and behavioral impairments associated with AD	Modulates p-GSK-3β-S9 Reduced caspase 3 and neuroinflammation Binds to α1A adrenergic receptors to stimulate phospho-AMPK levels	For electrophysiological studies on the hippocampal slices of WT, Tg2576 and 5XFAD mice were treated with Aβ42 oligomers in the presence of Avn-C (10, 25, and 50 μM)	[16]
			Tg2576 and 5XFAD mice were administered with Avn-C (2, 4, and 6 mg/kg. p.o.) for evaluation of long-term potentiation	

Table 1. *Cont.*

No.	Study Design	Mechanism	Methodology	Reference
3	Effect on protein aggregation using spectroscopy techniques	Inhibition of BSA oligomerization showing anti-amyloid effect	Protein aggregation in bovine serum albumin was initiated by incubating the protein monomer at an elevated temperature and the aggregation kinetics was monitored by incubating with Avn-C (100, 250, and 500 μM) and was analyzed using ThT-fluorescence assay	[68]
4	Evaluation of bisavenanthramide analogues of Nrf2 inductors and neuroprotectors in in vitro models	Nrf2-ARE-dependent protein expression	The antioxidant activity was measured using DPPH scavenging assay and FRAP method, and AChE inhibition assay was performed. Neuroprotection potential against tau hyperphosphorylation was evaluated using SH-SY5Y cell line	[83]
5	Evaluation of cytoprotective activity against oxidative-stress-induced PC12 cell injuries	Activating Nrf2-ARE pathway	Rat PC12 were used to study antioxidant effect of Avns. The antioxidant and cytoprotective activities of Avn-2c, Avn-2f, and Avn-2p were measured in vitro using the ABTS•+ and DPPH scavenging assay, MTT, and LDH release assay	[18]
6	Cognitive dysfunction induced by repeated propofol anesthesia in aging rats	Activating Nrf2/ARE pathway	Aging rat model was established by continuous 200 mg/kg propofol anesthesia	[69]
7	Protective effect on titanium dioxide nanoparticles induced neurotoxicity in SD rats	Decreases oxidative stress and TNF-α Increases the total antioxidant and GSH levels	TiO2 NPs (150 mg/kg b.w.) was administered orally for six weeks and Avn was administered daily at a dose of 20 mg/kg via gastric tube	[86]

Abbreviations: Avns, Avenanthramides; LTP, Long-term potentiation; AMPK, AMP-activated protein kinase; GSK-3β, Glycogen synthase kinase 3 beta; TNF-α, Tumor necrosis factor alpha; WT, Wild type; SD, Sprague Dawley; GSH, Glutathione; Nrf2/ARE, NF-E2-related factor 2/antioxidant response element; TiO2 NPs, Titanium dioxide nanoparticles; BSA, Bovine serum albumin; PC12, Pheochromocytoma cells; DPPH, 2,2-diphenylpicrylhydrazyl; MTT, 3-(4,5-Dimethylthiazol-2-yl)-2,5-Diphenyltetrazolium Bromide; LDH, Lactate dehydrogenase; SH-SY5Y, Neuroblastoma cell line; ABTS, 2,2′-azino-bis (3-ethylbenzothiazoline-6-sulfonic acid; AChE, Acetylcholinesterase; FRAP, Ferric reducing antioxidant power; ThT, Thioflavin T fluorescence.

5. Modulation of Other Targets by Avns

Phytochemicals are mainly associated with neuroprotective benefits against numerous risk factors including chemical toxicity and oxidative stress in neurodegenerative conditions. Oat extract enriched with Avns possesses a neuroprotective role against toxicity and oxidative stress induced by titanium dioxide nanoparticles (Tio2NPs) in rats [86]. Biochemical and histopathological studies revealed that the combination of Avns with thymoquinone exerts antioxidant and anti-inflammatory action, thus preventing neuronal degeneration induced with Tio2NPs [86]. The Avns improved the deleterious effect by altering the oxidative stress and improving antioxidant concentration.

Numerous studies suggested that the levels of pro-inflammatory markers including TNF-α, IL-1β, and IL-6 were significantly increased in the cerebrospinal fluid of AD patients and were correlated with impairment in LTP in animal models of AD [87–89]. Thus, the inhibition of the inflammatory pathway would be an effective approach to improve memory impairment. Ramasamy et al. reported that treatment with Avn-C significantly improved memory impairment in Tg2576 and 5XFAD mice brains [16]. In addition, the study also reported that Avn-C reduced the level of inflammatory markers, which could be a potential mechanism of action. Several studies also suggested that the levels of TNF-α were significantly reduced with Avn-c treatment [90–92].

In addition, Avns were found to reduce inflammation via altering the TNF-α concentration. The Avns were found to limit the TNF-α-induced neurogenic inflammation via modulating the NF-κB signaling and IL-8 levels [90,93]. Thus, the anti-inflammatory mechanism of Avn-c could be beneficial in the treatment of neurodegenerative conditions (Figure 3). The levels of the neurotransmitter acetylcholine are reduced due to hydrolysis due to the overstimulation of acetylcholinesterase in AD patients. Thus, targeting acetylcholinesterase could be beneficial in AD treatment [94]. Avn was also found to inhibit acetylcholinesterase and can thus improve memory in neurodegenerative conditions like AD [83,95].

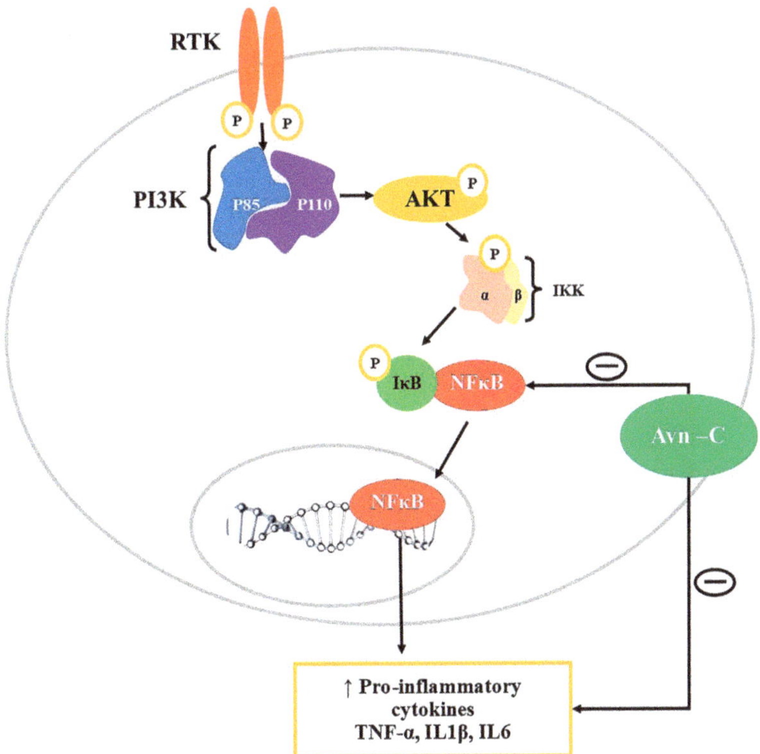

Figure 3. Modulation of PI3K/AKT/NF-κB signaling pathway by Avn-c. PI3K: Phosphatidylinositol-3-kinases (PI3Ks); AKT: Ak strain transforming-Phosphokinase B; RTK: Receptor tyrosine kinase; NF-κB: Nuclear factor kappa B; IKK: Inhibitor of NF-κB kinase; IκB: Inhibitor of nuclear factor-κB; TNFα: Tumor necrosis factor; IL-1β: Interleukin 1β; IL6: Interleukin 6. Avn-C downregulates the expression of inflammatory cytokines by inhibiting NF-κB and prevents its translocation into the nucleus thereby. Avn-C can directly inhibit the action of inflammatory mediators including TNFα, IL6, and IL1β.

6. Future Perspectives

Oats are a useful grain with distinct constituents and a unique source of Avns possessing abundant nutritional benefits. Recently, focus has been given to the biological activities of naturally produced Avns as well as their derivatives in the treatment of NDDs. Avenanthramides' anti-inflammatory effects make them attractive candidates for mitigating neuroinflammation and its detrimental impact on the progression of these diseases.

Future studies could focus on investigating their specific anti-inflammatory mechanisms and assessing their effectiveness in reducing neuroinflammation. Further, the antioxidant properties of Avns make them potential candidates for managing oxidative stress and preventing neuronal damage. Research could explore their potential in slowing or halting the progression of NDDs including AD and PD.

While Avns show promise, it is important to note that further research is necessary to fully understand their mechanisms of action, optimal dosages, and long-term effects in the context of NDDs. Further, the beneficial effects of Avns using synthetic Avn analogs and Avns obtained via recombinant techniques should also be explored in the treatments of NDDs. For this purpose, collaboration between researchers and clinicians will be crucial in translating the potential of Avns into effective therapeutic interventions for NDDs.

7. Conclusions

The PI3K/AKT signaling pathway in NDDs has been proven to be a successful hypothesis for the development of an effective treatment approach and more detailed research is being conducted to understand the involvement of downstream targets for the treatment of NDDs. Avn holds potential as a treatment for NDDs caused by oxidative stress. The protective effects of Avns are mediated mainly through the activation of the Nrf2, a downstream target of PI3K/AKT, which upregulates transcription genes encoding antioxidant enzymes. Avn-C and its synthetic analog were also found to modulate other targets PI3K/AKT including GSKβ and NF-κB in CNS. The molecule also possesses anti-inflammatory and anti-apoptotic action in wide pathological states including cancer and skin inflammation, indicating that its mechanism could be beneficial for the treatment of NDDs. In conclusion, the need for in depth pre-clinical studies in various models of neurodegeneration and clinical exploration to elucidate the specific mechanisms of Avn action and its potential interactions with other signaling cascades involved in neurodegeneration is quite essential.

Author Contributions: N.L.W., M.B.K., A.K.B., M.M.A., S.R.K. and S.K. conceptualized, designed, and wrote the manuscript; N.L.W., M.B.K., S.K., B.G.T., R.V.T., M.J.U., A.K.B., P.W., A.J. and S.R.K., revised the manuscript; N.L.W., M.B.K., S.K., A.J. and S.R.K. participated in drafting the article and revising it critically; and S.K. and S.R.K. were involved in funding acquisition. All authors have read and agreed to the published version of the manuscript.

Funding: This research received no external funding.

Institutional Review Board Statement: Not applicable.

Informed Consent Statement: Not applicable.

Data Availability Statement: Not applicable.

Acknowledgments: This work was supported by Sejong university, Republic of Korea.

Conflicts of Interest: The authors declare no conflict of interest.

References

1. Perrelli, A.; Goitre, L.; Salzano, A.M.; Moglia, A.; Scaloni, A.; Retta, S.F. Biological Activities, Health Benefits, and Therapeutic Properties of Avenanthramides: From Skin Protection to Prevention and Treatment of Cerebrovascular Diseases. *Oxid. Med. Cell. Longev.* **2018**, *2018*, 6015351. [CrossRef]
2. Hoffmann, L.A. World production and use of oats. In *The Oat Crop Production and Utilization*; Springer Science & Business Media: Berlin/Heidelberg, Germany, 1995; pp. 34–61.
3. Strychar, R.; Webster, F.H.; Wood, P.J. World oat production, trade, and usage. In *Oats: Chemistry and Technology*; Academic Press: Cambridge, MA, USA, 2011; pp. 77–94.
4. Singh, R.; De, S.; Belkheir, A. Avena sativa (Oat), A Potential Neutraceutical and Therapeutic Agent: An Overview. *Crit. Rev. Food Sci. Nutr.* **2013**, *53*, 126–144. [CrossRef] [PubMed]
5. Menon, R.; Gonzalez, T.; Ferruzzi, M.; Jackson, E.; Winderl, D.; Watson, J. Oats—From farm to fork. In *Advances in Food and Nutrition Research*; Academic Press: Cambridge, MA, USA, 2016; Volume 77, pp. 1–55.

6. Ho, H.V.T.; Sievenpiper, J.L.; Zurbau, A.; Blanco Mejia, S.; Jovanovski, E.; Au-Yeung, F.; Jenkins, A.L.; Vuksan, V. The effect of oat β-glucan on LDL-cholesterol, non-HDL-cholesterol and apoB for CVD risk reduction: A systematic review and meta-analysis of randomised-controlled trials. *Br. J. Nutr.* **2016**, *116*, 1369–1382. [CrossRef] [PubMed]
7. Reynolds, A.; Mann, J.; Cummings, J.; Winter, N.; Mete, E.; Te Morenga, L. Carbohydrate quality and human health: A series of systematic reviews and meta-analyses. *Lancet* **2019**, *393*, 434–445. [CrossRef]
8. Ripsin, C.M.; Keenan, J.M.; Jacobs, D.R.; Elmer, P.J.; Welch, R.R.; Van Horn, L.; Liu, K.; Turnbull, W.H.; Thye, F.W.; Kestin, M.; et al. Oat Products and Lipid Lowering: A Meta-analysis. *JAMA* **1992**, *267*, 3317–3325. [CrossRef] [PubMed]
9. Maki, K.C.; Galant, R.; Samuel, P.; Tesser, J.; Witchger, M.S.; Ribaya-Mercado, J.D.; Blumberg, J.B.; Geohas, J. Effects of consuming foods containing oat β-glucan on blood pressure, carbohydrate metabolism and biomarkers of oxidative stress in men and women with elevated blood pressure. *Eur. J. Clin. Nutr.* **2007**, *61*, 786–795. [CrossRef] [PubMed]
10. Sang, S.; Chu, Y. Whole grain oats, more than just a fiber: Role of unique phytochemicals. *Mol. Nutr. Food Res.* **2017**, *61*, 1600715. [CrossRef]
11. Matsuda, S.; Ikeda, Y.; Murakami, M.; Nakagawa, Y.; Tsuji, A.; Kitagishi, Y. Roles of PI3K/AKT/GSK3 Pathway Involved in Psychiatric Illnesses. *Diseases* **2019**, *7*, 22. [CrossRef] [PubMed]
12. Johansen, K.H.; Golec, D.P.; Thomsen, J.H.; Schwartzberg, P.L.; Okkenhaug, K. PI3K in T Cell Adhesion and Trafficking. *Front. Immunol.* **2021**, *12*, 708908. [CrossRef]
13. Bilanges, B.; Posor, Y.; Vanhaesebroeck, B. PI3K isoforms in cell signalling and vesicle trafficking. *Nat. Rev. Mol. Cell Biol.* **2019**, *20*, 515–534. [CrossRef]
14. Long, H.-Z.; Cheng, Y.; Zhou, Z.-W.; Luo, H.-Y.; Wen, D.-D.; Gao, L.-C. PI3K/AKT Signal Pathway: A Target of Natural Products in the Prevention and Treatment of Alzheimer's Disease and Parkinson's Disease. *Front. Pharmacol.* **2021**, *12*, 648636. [CrossRef] [PubMed]
15. Chen, Y.; Hsu, C.; Chen, X.; Zhang, H.; Peng, W. Editorial: Regulation of PI3K/Akt signaling pathway: A feasible approach for natural neuroprotective agents to treat various neuron injury-related diseases. *Front. Pharmacol.* **2023**, *14*, 1134989. [CrossRef] [PubMed]
16. Ramasamy, V.S.; Samidurai, M.; Park, H.J.; Wang, M.; Park, R.Y.; Yu, S.Y.; Kang, H.K.; Hong, S.; Choi, W.S.; Lee, Y.Y.; et al. Avenanthramide-C Restores Impaired Plasticity and Cognition in Alzheimer's Disease Model Mice. *Mol. Neurobiol.* **2020**, *57*, 315–330. [CrossRef]
17. Jin, B.; Kim, H.; Choi, J.-I.; Bae, H.-B.; Jeong, S. Avenanthramide C Prevents Neuronal Apoptosis via PI3K/Akt/GSK3β Signaling Pathway Following Middle Cerebral Artery Occlusion. *Brain Sci.* **2020**, *10*, 878. [CrossRef]
18. Hou, Y.; Peng, S.; Song, Z.; Bai, F.; Li, X.; Fang, J. Oat polyphenol avenanthramide-2c confers protection from oxidative stress by regulating the Nrf2-ARE signaling pathway in PC12 cells. *Arch. Biochem. Biophys.* **2021**, *706*, 108857. [CrossRef] [PubMed]
19. Lee, Y.-Y.; Wang, M.; Son, Y.; Yang, E.-J.; Kang, M.-S.; Kim, H.-J.; Kim, H.-S.; Jo, J. Oat Extract Avenanthramide-C Reverses Hippocampal Long-Term Potentiation Decline in Tg2576 Mice. *Molecules* **2021**, *26*, 6105. [CrossRef]
20. Peterson, D.M.; Hahn, M.J.; Emmons, C.L. Oat avenanthramides exhibit antioxidant activities in vitro. *Food Chem.* **2002**, *79*, 473–478. [CrossRef]
21. Helnæs, A.; Kyrø, C.; Andersen, I.; Lacoppidan, S.; Overvad, K.; Christensen, J.; Tjønneland, A.; Olsen, A. Intake of whole grains is associated with lower risk of myocardial infarction: The Danish Diet, Cancer and Health Cohort. *Am. J. Clin. Nutr.* **2016**, *103*, 999–1007. [CrossRef]
22. Rai, S.N.; Dilnashin, H.; Birla, H.; Sen Singh, S.; Zahra, W.; Rathore, A.S.; Singh, B.K.; Singh, S.P. The Role of PI3K/Akt and ERK in Neurodegenerative Disorders. *Neurotox. Res.* **2019**, *35*, 775–795. [CrossRef]
23. Hanada, M.; Feng, J.; Hemmings, B.A. Structure, regulation and function of PKB/AKT—A major therapeutic target. *Biochim. Biophys. Acta-Proteins Proteom.* **2004**, *1697*, 3–16. [CrossRef]
24. Jean, S.; Kiger, A.A. Classes of phosphoinositide 3-kinases at a glance. *J. Cell Sci.* **2014**, *127*, 923–928. [CrossRef] [PubMed]
25. Taksande, B.G.; Gawande, D.Y.; Chopde, C.T.; Umekar, M.J.; Kotagale, N.R. Agmatine ameliorates adjuvant induced arthritis and inflammatory cachexia in rats. *Biomed. Pharmacother.* **2017**, *86*, 271–278. [CrossRef]
26. Matsuda, S.; Nakagawa, Y.; Tsuji, A.; Kitagishi, Y.; Nakanishi, A.; Murai, T. Implications of PI3K/AKT/PTEN Signaling on Superoxide Dismutases Expression and in the Pathogenesis of Alzheimer's Disease. *Diseases* **2018**, *6*, 28. [CrossRef] [PubMed]
27. Hakan Kucuksayan, H.; Sakir Akgun, S. PI3K/Akt/NF-κB Signalling Pathway on NSCLC Invasion. *Med. Chem.* **2016**, *6*, 234–238. [CrossRef]
28. Katsnelson, A.; De Strooper, B.; Zoghbi, H.Y. Neurodegeneration: From cellular concepts to clinical applications. *Sci. Transl. Med.* **2016**, *8*, 364ps18. [CrossRef] [PubMed]
29. Peplow, P.V.; Martinez, B.; Gennarelli, T.A. Prevalence, Needs, Strategies, and Risk Factors for Neurodegenerative Diseases. *Neuromethods* **2022**, *173*, 3–8. [CrossRef]
30. Andersen, J.K. Oxidative stress in neurodegeneration: Cause or consequence? *Nat. Med.* **2004**, *10*, S18–S25. [CrossRef]
31. Jenner, P. Oxidative stress in Parkinson's disease. *Ann. Neurol.* **2003**, *53*, S26–S38. [CrossRef]
32. Aslan, M.; Ozben, T. Reactive Oxygen and Nitrogen Species in Alzheimers Disease. *Curr. Alzheimer Res.* **2004**, *1*, 111–119. [CrossRef]
33. Gan, L.; Johnson, J.A. Oxidative damage and the Nrf2-ARE pathway in neurodegenerative diseases. *Biochim. Biophys. Acta-Mol. Basis Dis.* **2014**, *1842*, 1208–1218. [CrossRef]

34. Itoh, K.; Mimura, J.; Yamamoto, M. Discovery of the Negative Regulator of Nrf2, Keap1: A Historical Overview. *Antioxid. Redox Signal.* **2010**, *13*, 1665–1678. [CrossRef]
35. Suzuki, T.; Motohashi, H.; Yamamoto, M. Toward clinical application of the Keap1–Nrf2 pathway. *Trends Pharmacol. Sci.* **2013**, *34*, 340–346. [CrossRef] [PubMed]
36. Yamazaki, H.; Tanji, K.; Wakabayashi, K.; Matsuura, S.; Itoh, K. Role of the Keap1/Nrf2 pathway in neurodegenerative diseases. *Pathol. Int.* **2015**, *65*, 210–219. [CrossRef]
37. Chowdhry, S.; Zhang, Y.; McMahon, M.; Sutherland, C.; Cuadrado, A.; Hayes, J.D. Nrf2 is controlled by two distinct β-TrCP recognition motifs in its Neh6 domain, one of which can be modulated by GSK-3 activity. *Oncogene* **2013**, *32*, 3765–3781. [CrossRef] [PubMed]
38. Hayes, J.D.; Chowdhry, S.; Dinkova-Kostova, A.T.; Sutherland, C. Dual regulation of transcription factor Nrf2 by Keap1 and by the combined actions of β-TrCP and GSK-3. *Biochem. Soc. Trans.* **2015**, *43*, 611–620. [CrossRef] [PubMed]
39. Wang, Z.-F.; Wang, J.; Zhang, H.-Y.; Tang, X.-C. Huperzine A exhibits anti-inflammatory and neuroprotective effects in a rat model of transient focal cerebral ischemia. *J. Neurochem.* **2008**, *106*, 1594–1603. [CrossRef]
40. Armagan, G.; Sevgili, E.; Gürkan, F.T.; Köse, F.A.; Bilgiç, T.; Dagcı, T.; Saso, L. Regulation of the Nrf2 Pathway by Glycogen Synthase Kinase-3β in MPP+-Induced Cell Damage. *Molecules* **2019**, *24*, 1377. [CrossRef]
41. Sotolongo, K.; Ghiso, J.; Rostagno, A. Nrf2 activation through the PI3K/GSK-3 axis protects neuronal cells from Aβ-mediated oxidative and metabolic damage. *Alzheimer's Res. Ther.* **2020**, *12*, 13. [CrossRef]
42. Beurel, E.; Grieco, S.F.; Jope, R.S. Glycogen synthase kinase-3 (GSK3): Regulation, actions, and diseases. *Pharmacol. Ther.* **2015**, *148*, 114–131. [CrossRef]
43. Luo, J. GSK3β in Ethanol Neurotoxicity. *Mol. Neurobiol.* **2009**, *40*, 108–121. [CrossRef]
44. Le Belle, J.E.; Orozco, N.M.; Paucar, A.A.; Saxe, J.P.; Mottahedeh, J.; Pyle, A.D.; Wu, H.; Kornblum, H.I. Proliferative Neural Stem Cells Have High Endogenous ROS Levels that Regulate Self-Renewal and Neurogenesis in a PI3K/Akt-Dependant Manner. *Cell Stem Cell* **2011**, *8*, 59–71. [CrossRef]
45. Ali, T.; Kim, M.O. Melatonin ameliorates amyloid beta-induced memory deficits, tau hyperphosphorylation and neurodegeneration via PI3/Akt/GSk3β pathway in the mouse hippocampus. *J. Pineal Res.* **2015**, *59*, 47–59. [CrossRef]
46. Singh, A.K.; Kashyap, M.P.; Tripathi, V.K.; Singh, S.; Garg, G.; Rizvi, S.I. Neuroprotection Through Rapamycin-Induced Activation of Autophagy and PI3K/Akt1/mTOR/CREB Signaling Against Amyloid-β-Induced Oxidative Stress, Synaptic/Neurotransmission Dysfunction, and Neurodegeneration in Adult Rats. *Mol. Neurobiol.* **2017**, *54*, 5815–5828. [CrossRef]
47. Erdogdu, Ö.; Nathanson, D.; Sjöholm, Å.; Nyström, T.; Zhang, Q. Exendin-4 stimulates proliferation of human coronary artery endothelial cells through eNOS-, PKA- and PI3K/Akt-dependent pathways and requires GLP-1 receptor. *Mol. Cell. Endocrinol.* **2010**, *325*, 26–35. [CrossRef] [PubMed]
48. Kitagishi, Y.; Nakanishi, A.; Ogura, Y.; Matsuda, S. Dietary regulation of PI3K/AKT/GSK-3β pathway in Alzheimer's disease. *Alzheimer's Res. Ther.* **2014**, *6*, 35. [CrossRef]
49. Ali, T.; Yoon, G.H.; Shah, S.A.; Lee, H.Y.; Kim, M.O. Osmotin attenuates amyloid beta-induced memory impairment, tau phosphorylation and neurodegeneration in the mouse hippocampus. *Sci. Rep.* **2015**, *5*, 11708. [CrossRef] [PubMed]
50. Ali, T.; Kim, T.; Rehman, S.U.; Khan, M.S.; Amin, F.U.; Khan, M.; Ikram, M.; Kim, M.O. Natural Dietary Supplementation of Anthocyanins via PI3K/Akt/Nrf2/HO-1 Pathways Mitigate Oxidative Stress, Neurodegeneration, and Memory Impairment in a Mouse Model of Alzheimer's Disease. *Mol. Neurobiol.* **2018**, *55*, 6076–6093. [CrossRef] [PubMed]
51. Silva-Palacios, A.; Ostolga-Chavarría, M.; Zazueta, C.; Königsberg, M. Nrf2: Molecular and epigenetic regulation during aging. *Ageing Res. Rev.* **2018**, *47*, 31–40. [CrossRef]
52. Schmidlin, C.J.; Dodson, M.B.; Madhavan, L.; Zhang, D.D. Redox regulation by NRF2 in aging and disease. *Free Radic. Biol. Med.* **2019**, *134*, 702–707. [CrossRef] [PubMed]
53. Chen, P.-C.; Vargas, M.R.; Pani, A.K.; Smeyne, R.J.; Johnson, D.A.; Kan, Y.W.; Johnson, J.A. Nrf2-mediated neuroprotection in the MPTP mouse model of Parkinson's disease: Critical role for the astrocyte. *Proc. Natl. Acad. Sci. USA* **2009**, *106*, 2933–2938. [CrossRef]
54. Niu, Y.; Zhang, J.; Dong, M. Nrf2 as a potential target for Parkinson's disease therapy. *J. Mol. Med.* **2021**, *99*, 917–931. [CrossRef] [PubMed]
55. Jia, Y.; Mo, S.-J.; Feng, Q.-Q.; Zhan, M.-L.; OuYang, L.-S.; Chen, J.-C.; Ma, Y.-X.; Wu, J.-J.; Lei, W.-L. EPO-Dependent Activation of PI3K/Akt/FoxO3a Signalling Mediates Neuroprotection in In Vitro and In Vivo Models of Parkinson's Disease. *J. Mol. Neurosci.* **2014**, *53*, 117–124. [CrossRef]
56. Gong, J.; Zhang, L.; Zhang, Q.; Li, X.; Xia, X.-J.; Liu, Y.-Y.; Yang, Q.-S. Lentiviral Vector-Mediated SHC3 Silencing Exacerbates Oxidative Stress Injury in Nigral Dopamine Neurons by Regulating the PI3K-AKT-FoxO Signaling Pathway in Rats with Parkinson's Disease. *Cell Physiol. Biochem.* **2018**, *49*, 971–984. [CrossRef]
57. Rice-Evans, C.A.; Miller, N.J.; Paganga, G. Structure-antioxidant activity relationships of flavonoids and phenolic acids. *Free Radic. Biol. Med.* **1996**, *20*, 933–956. [CrossRef]
58. Ha, H.C.; Sirisoma, N.S.; Kuppusamy, P.; Zweier, J.L.; Woster, P.M.; Casero, R.A. The natural polyamine spermine functions directly as a free radical scavenger. *Proc. Natl. Acad. Sci. USA* **1998**, *95*, 11140–11145. [CrossRef]
59. Kabanda, M.M. A theoretical study of the antioxidant properties of phenolic acid amides investigated through the radical-scavenging and metal chelation mechanisms. *Eur. Food Res. Technol.* **2015**, *241*, 553–572. [CrossRef]

60. Palop, J.J.; Mucke, L. Amyloid-β–induced neuronal dysfunction in Alzheimer's disease: From synapses toward neural networks. *Nat. Neurosci.* **2010**, *13*, 812–818. [CrossRef] [PubMed]
61. Uddin, M.S.; Kabir, M.T.; Tewari, D.; Al Mamun, A.; Mathew, B.; Aleya, L.; Barreto, G.E.; Bin-Jumah, M.N.; Abdel-Daim, M.M.; Ashraf, G.M. Revisiting the role of brain and peripheral Aβ in the pathogenesis of Alzheimer's disease. *J. Neurol. Sci.* **2020**, *416*, 116974. [CrossRef] [PubMed]
62. Festa, G.; Mallamace, F.; Sancesario, G.M.; Corsaro, C.; Mallamace, D.; Fazio, E.; Arcidiacono, L.; Garcia Sakai, V.; Senesi, R.; Preziosi, E.; et al. Aggregation States of Aβ1–40, Aβ1–42 and Aβp3–42 Amyloid Beta Peptides: A SANS Study. *Int. J. Mol. Sci.* **2019**, *20*, 4126. [CrossRef]
63. Mamun, A.; Uddin, M.; Mathew, B.; Ashraf, G. Toxic tau: Structural origins of tau aggregation in Alzheimer's disease. *Neural Regen. Res.* **2020**, *15*, 1417. [CrossRef]
64. Fu, J.; Zhu, Y.; Yerke, A.; Wise, M.L.; Johnson, J.; Chu, Y.; Sang, S. Oat avenanthramides induce heme oxygenase-1 expression via Nrf2-mediated signaling in HK-2 cells. *Mol. Nutr. Food Res.* **2015**, *59*, 2471–2479. [CrossRef] [PubMed]
65. Di Maso, M.J.; Nepomuceno, G.M.; St. Peter, M.A.; Gitre, H.H.; Martin, K.S.; Shaw, J.T. Synthesis of (±)-Bisavenanthramide B-6 by an Anionic Anhydride Mannich Reaction. *Org. Lett.* **2016**, *18*, 1740–1743. [CrossRef] [PubMed]
66. Jo, J.; Whitcomb, D.J.; Olsen, K.M.; Kerrigan, T.L.; Lo, S.-C.; Bru-Mercier, G.; Dickinson, B.; Scullion, S.; Sheng, M.; Collingridge, G.; et al. Aβ1–42 inhibition of LTP is mediated by a signaling pathway involving caspase-3, Akt1 and GSK-3β. *Nat. Neurosci.* **2011**, *14*, 545–547. [CrossRef] [PubMed]
67. Llorens-Martín, M.; Jurado, J.; Hernández, F.; Ávila, J. GSK-3β, a pivotal kinase in Alzheimer disease. *Front. Mol. Neurosci.* **2014**, *7*, 46. [CrossRef] [PubMed]
68. Quiroz Vazquez, M.G.; Montiel Condado, D.; Gonzalez Hernandez, B.; Gonzalez-Horta, A. Avenanthramide-C prevents amyloid formation of bovine serum albumin. *Biophys. Chem.* **2020**, *263*, 106391. [CrossRef]
69. Ma, Z.; Ma, Y.; Cao, X.; Zhang, Y.; Song, T. Avenanthramide-C Activates Nrf2/ARE Pathway and Inhibiting Ferroptosis Pathway to Improve Cognitive Dysfunction in Aging Rats. *Neurochem. Res.* **2023**, *48*, 393–403. [CrossRef]
70. Wirdefeldt, K.; Adami, H.-O.; Cole, P.; Trichopoulos, D.; Mandel, J.; Wirdefeldt, K.; Adami, H.-O.; Trichopoulos, Á.D.; Cole, P.; Mandel, J. Epidemiology and etiology of Parkinson's disease: A review of the evidence. *Eur. J. Epidemiol.* **2011**, *26*, 1–58. [CrossRef]
71. Spillantini, M.G.; Crowther, R.A.; Jakes, R.; Hasegawa, M.; Goedert, M. α-Synuclein in filamentous inclusions of Lewy bodies from Parkinson's disease and dementia with Lewy bodies. *Proc. Natl. Acad. Sci. USA* **1998**, *95*, 6469–6473. [CrossRef]
72. Braak, H.; Del Tredici, K.; Rüb, U.; De Vos, R.A.I.; Jansen Steur, E.N.H.; Braak, E. Staging of brain pathology related to sporadic Parkinson's disease. *Neurobiol. Aging* **2003**, *24*, 197–211. [CrossRef]
73. Chu, Y.; Kordower, J.H. Age-associated increases of α-synuclein in monkeys and humans are associated with nigrostriatal dopamine depletion: Is this the target for Parkinson's disease? *Neurobiol. Dis.* **2007**, *25*, 134–149. [CrossRef]
74. Cho, M.-K.; Nodet, G.; Kim, H.-Y.; Jensen, M.R.; Bernado, P.; Fernandez, C.O.; Becker, S.; Blackledge, M.; Zweckstetter, M. Structural characterization of α-synuclein in an aggregation prone state. *Protein Sci.* **2009**, *18*, 1840–1846. [CrossRef]
75. Singleton, A.B.; Farrer, M.; Johnson, J.; Singleton, A.; Hague, S.; Kachergus, J.; Hulihan, M.; Peuralinna, T.; Dutra, A.N.R.; Lincoln, S.; et al. α-synuclein locus triplication causes Parkinson's disease. *Science* **2003**, *302*, 841–842. [CrossRef] [PubMed]
76. Chartier-Harlin, M.C.; Kachergus, J.; Roumier, C.; Mouroux, V.; Douay, X.; Lincoln, S.; Levecque, C.; Larvor, L.; Andrieux, J.; Hulihan, M.; et al. α-synuclein locus duplication as a cause of familial Parkinson's disease. *Lancet* **2004**, *364*, 1167–1169. [CrossRef] [PubMed]
77. Maraganore, D.M.; de Andrade, M.; Elbaz, A.; Farrer, M.J.; Ioannidis, J.P.; Krüger, R.; Rocca, W.A.; Schneider, N.K.; Lesnick, T.G.; Lincoln, S.J.; et al. Collaborative Analysis of α-Synuclein Gene Promoter Variability and Parkinson Disease. *JAMA* **2006**, *296*, 661–670. [CrossRef] [PubMed]
78. Weng, M.; Xie, X.; Liu, C.; Lim, K.-L.; Zhang, C.; Li, L. The Sources of Reactive Oxygen Species and Its Possible Role in the Pathogenesis of Parkinson's Disease. *Parkinson's Dis.* **2018**, *2018*, 9163040. [CrossRef]
79. Chang, K.-H.; Chen, C.-M. The Role of Oxidative Stress in Parkinson's Disease. *Antioxidants* **2020**, *9*, 597. [CrossRef]
80. Hu, M.; Li, F.; Wang, W. Vitexin protects dopaminergic neurons in MPTP-induced Parkinson’s disease through PI3K/Akt signaling pathway. *Drug Des. Dev. Ther.* **2018**, *12*, 565–573. [CrossRef]
81. Vauzour, D.; Vafeiadou, K.; Rodriguez-Mateos, A.; Rendeiro, C.; Spencer, J.P.E. The neuroprotective potential of flavonoids: A multiplicity of effects. *Genes Nutr.* **2008**, *3*, 115–126. [CrossRef]
82. Liu, Y.; Zhang, R.-Y.; Zhao, J.; Dong, Z.; Feng, D.-Y.; Wu, R.; Shi, M.; Zhao, G. Ginsenoside Rd Protects SH-SY5Y Cells against 1-Methyl-4-phenylpyridinium Induced Injury. *Int. J. Mol. Sci.* **2015**, *16*, 14395–14408. [CrossRef]
83. Cores, Á.; Abril, S.; Michalska, P.; Duarte, P.; Olives, A.I.; Martín, M.A.; Villacampa, M.; León, R.; Carlos Menéndez, J. Bisavenathramide analogues as nrf2 inductors and neuroprotectors in in vitro models of oxidative stress and hyperphosphorylation. *Antioxidants* **2021**, *10*, 941. [CrossRef]
84. Xiong, N.; Xiong, J.; Jia, M.; Liu, L.; Zhang, X.; Chen, Z.; Huang, J.; Zhang, Z.; Hou, L.; Luo, Z.; et al. The role of autophagy in Parkinson's disease: Rotenone-based modeling. *Behav. Brain Funct.* **2013**, *9*, 13. [CrossRef]
85. Alavi Naini, S.M.; Soussi-Yanicostas, N. Tau Hyperphosphorylation and Oxidative Stress, a Critical Vicious Circle in Neurodegenerative Tauopathies? *Oxid. Med. Cell. Longev.* **2015**, *2015*, 151979. [CrossRef] [PubMed]
86. Hassanein, K.M.A.; El-Amir, Y.O. Protective effects of thymoquinone and avenanthramides on titanium dioxide nanoparticles induced toxicity in Sprague-Dawley rats. *Pathol.-Res. Pract.* **2017**, *213*, 13–22. [CrossRef] [PubMed]

87. MAGAKI, S.; MUELLER, C.; DICKSON, C.; KIRSCH, W. Increased production of inflammatory cytokines in mild cognitive impairment. *Exp. Gerontol.* **2007**, *42*, 233–240. [CrossRef]
88. Swardfager, W.; Lanctôt, K.; Rothenburg, L.; Wong, A.; Cappell, J.; Herrmann, N. A Meta-Analysis of Cytokines in Alzheimer's Disease. *Biol. Psychiatry* **2010**, *68*, 930–941. [CrossRef] [PubMed]
89. Brosseron, F.; Krauthausen, M.; Kummer, M.; Heneka, M.T. Body Fluid Cytokine Levels in Mild Cognitive Impairment and Alzheimer's Disease: A Comparative Overview. *Mol. Neurobiol.* **2014**, *50*, 534–544. [CrossRef]
90. Sur, R.; Nigam, A.; Grote, D.; Liebel, F.; Southall, M.D. Avenanthramides, polyphenols from oats, exhibit anti-inflammatory and anti-itch activity. *Arch. Dermatol. Res.* **2008**, *300*, 569–574. [CrossRef] [PubMed]
91. Yu, L.; Chen, C.; Wang, L.-F.; Kuang, X.; Liu, K.; Zhang, H.; Du, J.-R. Neuroprotective Effect of Kaempferol Glycosides against Brain Injury and Neuroinflammation by Inhibiting the Activation of NF-κB and STAT3 in Transient Focal Stroke. *PLoS ONE* **2013**, *8*, e55839. [CrossRef] [PubMed]
92. El Amir, Y.O.; Omar, W.; Khabrani, A.Y.; Jahfali, A.E.; Alhakami, S.M.; Dobab, N.M. Protective effect of avenanthramides against cisplatin induced nephrotoxicity in rats. *J. Adv. Vet. Anim. Res.* **2019**, *6*, 521–527. [CrossRef]
93. Kang, C.; Shin, W.S.; Yeo, D.; Lim, W.; Zhang, T.; Ji, L.L. Anti-inflammatory effect of avenanthramides via NF-κB pathways in C2C12 skeletal muscle cells. *Free Radic. Biol. Med.* **2018**, *117*, 30–36. [CrossRef]
94. Taqui, R.; Debnath, M.; Ahmed, S.; Ghosh, A. Advances on plant extracts and phytocompounds with acetylcholinesterase inhibition activity for possible treatment of Alzheimer's disease. *Phytomed. Plus* **2022**, *2*, 100184. [CrossRef]
95. Yang, J.; Ou, B.; Wise, M.L.; Chu, Y. In vitro total antioxidant capacity and anti-inflammatory activity of three common oat-derived avenanthramides. *Food Chem.* **2014**, *160*, 338–345. [CrossRef] [PubMed]

nutrients

Review

Anti-Diabetic Potential of Polyphenol-Rich Fruits from the Maleae Tribe—A Review of In Vitro and In Vivo Animal and Human Trials

Magdalena Rutkowska * and Monika A. Olszewska

Department of Pharmacognosy, Faculty of Pharmacy, Medical University of Lodz, 1 Muszynskiego St., 90-151 Lodz, Poland; monika.olszewska@umed.lodz.pl
* Correspondence: magdalena.rutkowska@umed.lodz.pl

Abstract: The Maleae tribe consists of over one thousand species, including many well-known polyphenol-containing fruit crops with wide-ranging biological properties, e.g., apples (*Malus*), chokeberries (*Aronia*), pears (*Pyrus*), quinces (*Cydonia, Chaenomeles*), saskatoon (*Amelanchier*), loquats (*Eriobotrya*), medlars (*Mespilus*), rowans (*Sorbus*), and hawthorns (*Crataegus*). Considering the current interest in the concept of functional foods and the still-insufficient methods of diabetes management, the anti-diabetic potential of fruits has been studied intensively, including those of the Maleae tribe. This paper is the first comprehensive overview of this selected topic, covering articles published from 2000 to 2023 (131 articles in total). The first part of this review focuses on the potential mechanisms of action of fruits investigated so far (46 species), including their effects on tissue-specific glucose transport and the expression or activity of proteins in the insulin signalling pathway. The second part covers the phytocompounds responsible for particular fruits' activity—primarily polyphenols (e.g., flavonols, dihydrochalcones, proanthocyanidins, anthocyanins, phenolic acids), but also polysaccharides, triterpenes, and their additive and synergistic effects. In summary, fruits from the Maleae tribe seem promising as functional foods and anti-diabetic agents; however, their prospects for more expansive pro-health application require further research, especially more profound in vivo trials.

Keywords: Maleae fruits; diabetes; glucose transport; insulin signalling pathway; carbohydrate digestion; polyphenols; polysaccharides; triterpenes

Citation: Rutkowska, M.; Olszewska, M.A. Anti-Diabetic Potential of Polyphenol-Rich Fruits from the Maleae Tribe—A Review of In Vitro and In Vivo Animal and Human Trials. *Nutrients* **2023**, *15*, 3756. https://doi.org/10.3390/nu15173756

Academic Editor: Peter Pribis

Received: 29 July 2023
Revised: 23 August 2023
Accepted: 26 August 2023
Published: 28 August 2023

1. Introduction

Diabetes is a chronic, progressive disorder characterised by raised blood glucose levels due to insufficient production of the hormone insulin or decreased effectiveness of the insulin that the body produces [1,2]. Long-term hyperglycaemia has become a significant healthcare burden worldwide, leading to life-threatening damage to the blood vessels, heart, nerves, and kidneys [2]. It was estimated that 353 million patients suffered from diabetes in 2021, and it is projected that by 2030 there will be about 643 million people diagnosed with this disorder [1]. Therefore, diabetes has been recognised as one of four non-communicable diseases targeted as a priority by the world's public health leaders [2].

The pharmacotherapy of diabetes includes insulin (when insulin deficiency is seen) or oral hypoglycaemic drugs that exert anti-diabetic effects through different mechanisms. These mechanisms comprise, i.a., stimulation of endogenous insulin secretion from pancreatic β-cells by sulfonylureas, glucagon-like peptide 1 (GLP-1) analogues, or dipeptidyl peptidase-4 (DPP IV) inhibitors; the increase in the insulin sensitivity, the boost of peripheral absorption of glucose, and the reduction in hepatic gluconeogenesis by peroxisome proliferator-activated receptor γ (PPARγ) activators or biguanides; the delay in the absorption of carbohydrates from the intestine by α-glucosidase inhibitors; or the increase in glucose elimination via the kidneys by sodium–glucose cotransporter-2 (SGLT2) inhibitors [3–5]. However, the conventional treatment of diabetes often fails due to drug resistance (reduction in efficiency over time) and, perhaps even more critically, various side effects and, thus, noncompliant behaviour of patients. Therefore, a constant search is underway for better diabetes management, including lifestyle modifications, proper diet, and the use of medicinal plants [2–4].

The anti-diabetic potential of fruits and vegetables has been widely studied, as the concept of functional foods (i.e., dietary products that offer health benefits beyond their nutritional value) has gained importance [6,7]. Research on the anti-hyperglycaemic effects of fruits has revealed their involvement in glucose transport and metabolism through mechanisms of action that seem analogous to synthetic anti-diabetic drugs. These mechanisms include, e.g., the modulation of glucose transporters' activity, the inhibition of carbohydrate digestion, or multiple effects on glycolysis, gluconeogenesis, glycogen synthesis, and glycogenolysis by affecting the activity or expression of proteins in the insulin signalling pathway [7]. Thus, it has been suggested that many dietary fruits might be used not only in the prevention of diabetes, but also as a potential source of new anti-diabetic agents for mono- or combination therapy, enabling the reduction in synthetic drug doses and, thus, the side effects of conventional diabetes treatment [6].

The Maleae tribe comprises over one thousand species, mainly from the Northern Hemisphere [8]. It includes many well-known fruit crops, e.g., apples (*Malus* Mill. sp.), pears (*Pyrus* L. sp.), black chokeberry (*Aronia melanocarpa* (Michx.) Elliott), quince (*Cydonia oblonga* Mill.) and Japanese quince (*Chaenomeles japonica* (Thunb.) Lindl. ex Spach), saskatoon berry (*Amelanchier alnifolia* Nutt.), loquat (*Eriobotrya japonica* (Thunb.) Lindl.), medlar (*Mespilus germanica* L.), rowans (*Sorbus* L. sp.), and hawthorns (*Crataegus* L. sp.). Fruits from Maleae species are widely consumed both unprocessed and as jams, juices, alcoholic beverages (e.g., wines, liqueurs), etc. [9]. According to the United States Department of Agriculture [10], the global annual consumption of only fresh apples and pears was 81.6 and 23.5 Mt in 2021, respectively, making them some of the most willingly consumed fruits in the world. Consequently, the pro-health properties of fruit consumption have been intensively studied in recent decades [6], including their anti-hyperglycaemic activity.

This paper presents the first comprehensive overview of the anti-diabetic potential of the fruits of the Maleae tribe. It covers articles published from 2000 to 2023, including in vitro research, in vivo tests on animal models, and in vivo human trials. The main focus of this review was to indicate the biological effectiveness of the fruits investigated so far, discuss the potential mechanisms of action behind the observed effects, and point out the phytocompounds that may be responsible for the anti-diabetic properties of different Maleae fruits. Apart from summarising the current knowledge and challenges, we discuss the further research directions required for a deeper understanding of the biological properties and potential of Maleae fruits for their wider use as functional products or anti-diabetic phytotherapeutics.

2. Materials and Methods

The literature selection was performed based on the Scopus, Web of Science, and Google Scholar databases, searching for original articles written in English and published (at least electronically) between January 2000 and June 2023. The search was conducted using the following keyword combination pattern: (1) the genus Latin or common nomenclature (i.e., "*Amelanchier*" or "*Amelasorbus*" or "*Aronia*" or "*Chaenomeles*" or "*Chamaemeles*" or "*Cotoneaster*" or "*Crataegus*" or "*Crataemespilus*" or "*Cydonia*" or "*Dichotomanthes*" or "*Docynia*" or "*Eriobotrya*" or "*Eriolobus*" or "*Hesperomeles*" or "*Heteromeles*" or "*Kageneckia*" or "*Lindleya*" or "*Malacomeles*" or "*Malus*" or "*Mespilus*" or "*Osteomeles*" or "*Peraphyllum*" or "*Photinia*" or "*Pseudocydonia*" or "*Pyracantha*" or "*Pyrus*" or "*Rhaphiolepis*" or "*Sorbus*" or "*Sorbaronia*" or "*Sorbocotoneaster*" or "*Stranvaesia*" or "*Vauquelinia*" or "saskatoon" or "chokeberry" or "hawthorn" or "quince" or "apple" or "crabapple" or "medlar" or "pear" or "loquat" or "rowan" or "service tree" or "whitebeam" or "toyon"); (2) description of the plant part/product (i.e., "fruit/-s" or "berry/-ies" or "juice/-s" or "extract/-s"); (3) activity descriptor (i.e., "diabetes" or "diabetic" or "anti-diabetic" or "glucose" or "insulin" or "glycaemia/glycemia"). Only articles covering the topic of fruits' effects on carbohydrate bioavailability/metabolism and direct toxic effects of hyperglycaemia (e.g., AGE formation) were included. Consequently, the studies on the conditions accompanying or resulting from diabetes as an outcome of complex mechanisms (e.g., inflammation, neurodegenerative diseases or cardiovascular complications of diabetes) were not reviewed. Moreover, if the paper included the analysis of both diabetic and diabetes comorbid-disorder-related parameters (e.g., lipid profiles, cytokine levels, etc.), only the first part was included in this review. The further exclusion criteria were as follows: papers covering only ethnobotanical research on plants used in diabetes (without activity studies), the effects of a complex diet or combination of plants/drugs (making it impossible to indicate which component determines the activity), and studies of single, isolated compounds of plant origin, not covering the activity or health impact of whole fruits. The inclusion or exclusion of the articles was validated manually by reading the entire item. The binomial names of the reviewed species were checked and revised according to World Flora Online [11].

3. Results and Discussion

As a result of an in-depth analysis of the literature data covering articles published from 2000 to 2023, 131 studies were included in the present review. The majority of them covered only in vitro tests (67 papers); then, there were in vivo studies on animal models (51 items, including some mixed in vitro/in vivo tests) and human in vivo trials (14 papers, including one in vivo animal and human study) (Figure 1). The selected documents were based on the activity testing of fruits from 46 species belonging to the genera *Amelanchier, Aronia, Chaenomeles, Cotoneaster, Crataegus, Cydonia, Malus, Mespilus, Pyracantha, Pyrus, Sorbus,* and *Vauquelinia.* The most widely studied taxa among this group were *Aronia melanocarpa* (26 studies, including 13 animal and 5 human trials) and *Malus domestica* (26 studies, including 6 animal and 5 human trials) (Figure 1).

3.1. In Vitro Studies

Most in vitro studies of the anti-diabetic effects of Maleae fruits are based on testing their impact on α-glucosidase or α-amylase (Table 1). While these enzymes are involved in the chain reactions of carbohydrates' breakdown to glucose (or fructose), inhibiting one or two of them may reduce the intestinal absorption of saccharides and, thus, lower postprandial hyperglycaemia. The reviewed data suggest that extracts or juices of fruit origin are relatively weak inhibitors of α-amylase; on the other hand, they can strongly inhibit α-glucosidase. For instance, the inhibitory effects of fruit extracts/juices of *Aronia melanocarpa* [12–14], *Aronia prunifolia* [13], *Cotoneaster integerrimus, Cotoneaster zabelii, Cotoneaster bullatus* [15], *Crataegus laevigata* [16], *Crataegus pinnatifida* [17], *Malus domestica* [18], *Pyracantha fortuneana* [19,20], *Pyrus pashia* [21], *Sorbus alnifolia, Sorbus folgneri, Sorbus minima, Sorbus norvegica, Sorbus hybrid, Sorbus aucuparia, Sorbus meinichii, Sorbus torminalis* [22,23], and *Vauquelinia corymbosa* [24] towards α-glucosidase were many times higher than that observed for acarbose—a known anti-diabetic drug and α-glucosidase inhibitor. The synergistic effects of the *Malus domestica* juice and *Sorbus aucuparia* extracts with acarbose were also suggested [25–27].

The second most frequent type of in vitro research into Maleae fruits involves cellular studies on glucose transport and metabolism (Table 1). Stimulating glucose uptake through skeletal muscle or hepatic cells is one of the mechanisms that may lead to the enhancement of glucose metabolism and reduction in blood sugar levels. This effect, associated with the increased expression of insulin-dependent glucose transporter 4 (GLUT-4), has been observed for *Aronia melanocarpa* [28], *Chaenomeles japonica* [29], *Malus pumila* [30], and *Pyrus pyrifolia* [31]. Moreover, the inhibition of intestinal glucose absorption (i.e., transport from the intestinal lumen into enterocytes) via, e.g., the modulation of SGLT1 (sodium–glucose transport protein-1) and GLUT-2 (glucose transporter 2) levels or activity, has been proposed for *Malus domestica* [32–34]. As for the metabolic part, the effects on the expression or activity of the PI3K/Akt pathway proteins (Figure 2a) were observed for *Aronia melanocarpa* [28,35], *Chaenomeles japonica* [29,36], *Crataegus pinnatifida* [17], *Malus domestica* [37,38], *Malus pumila* [30], and *Pyrus pyrifolia* [31,39]. Considering the complexity of the processes involved in insulin and glucose regulation, the results of the studies mentioned above may indicate the ability of the tested extracts to stimulate glycolysis and glycogen synthesis, inhibit gluconeogenesis and glycogenolysis, and lower blood glucose levels. Moreover, fruits from *Amelanchier alnifolia* [40], *Crataegus pinnatifida* [17], *Malus domestica* [41], *Malus sieversii* [42], *Sorbus aucuparia* [43], and *Sorbus domestica* [44] have been suggested to inhibit enzymes involved in the polyol pathway of glucose metabolism, such as aldose reductase (ALR) and sorbitol dehydrogenase (SDH), and to impair the production of advanced glycation end products (AGEs). Thus, they may prevent diabetic complications, primarily oxidative-stress-related damage to the microvascular systems, caused by pathological glucose metabolism (Figure 2b). Furthermore, one of the investigated species, i.e., *Chaenomeles japonica* [36], was proven to have cytoprotective effects on βTC3 pancreatic β-cells (model of induced toxicity), enabling their viability and normal proliferation to be preserved. Detailed information on the accumulated research can be found in Table 1.

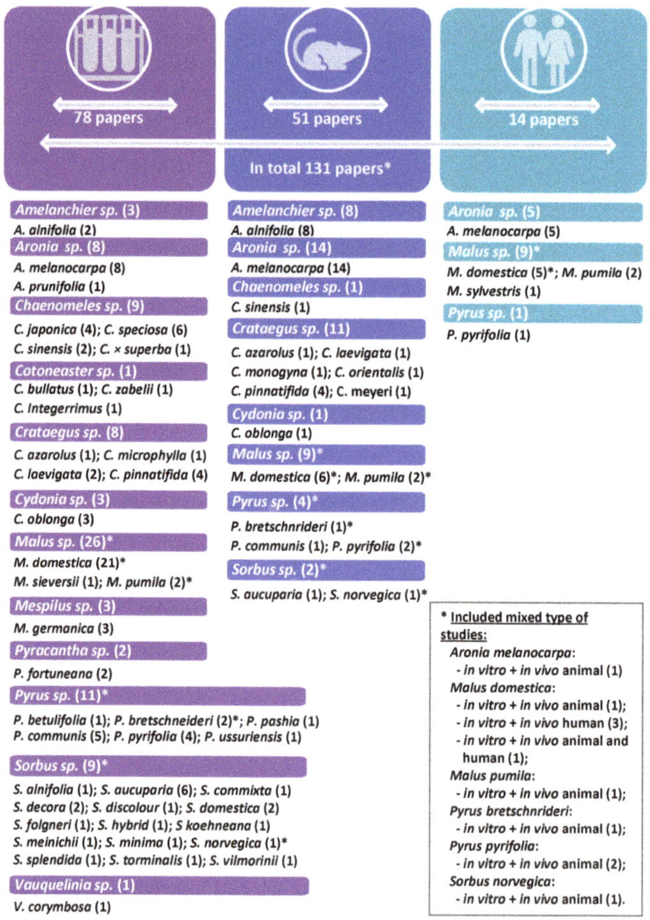

Figure 1. The list of species included in the review, divided into those tested in vitro, in vivo on animal models, and in vivo on human participants. Numbers in parentheses indicate the number of papers on the relevant genus/species (some papers discussed different species, and in some cases there was no indication of the exact species).

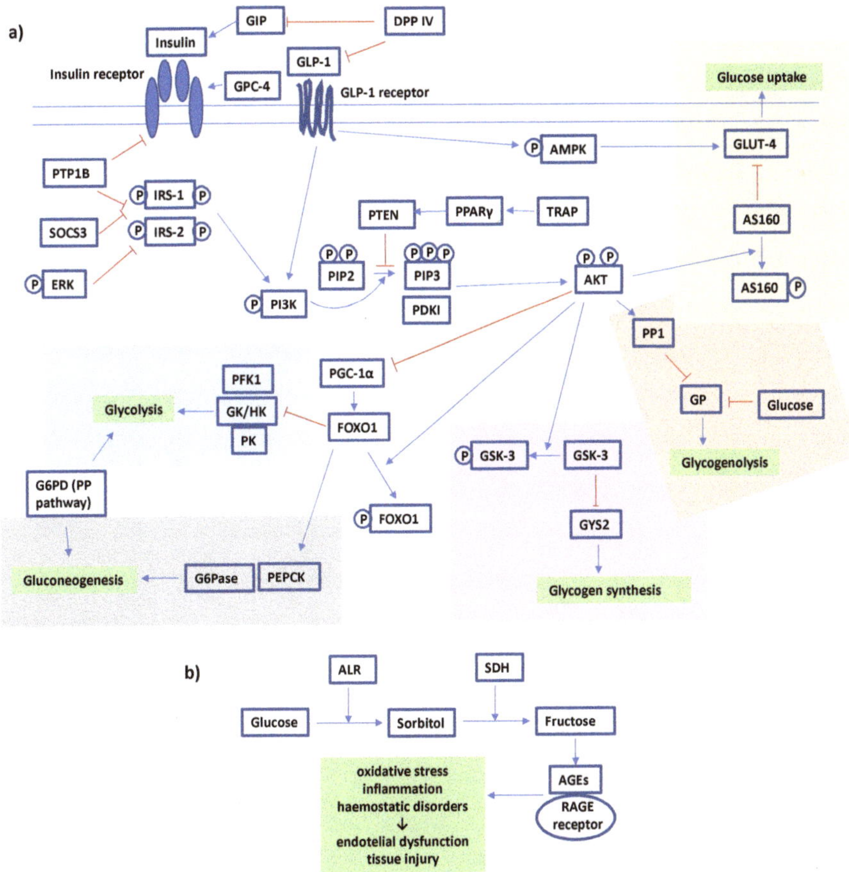

Figure 2. The diagram of glucose metabolism (**a**) associated with the insulin signalling pathway and (**b**) associated with the polyol pathway. The gripping points of the activity tested for different Maleae fruits are described in the manuscript body and tables. The diagram is based on data from the literature [45–54]; blue arrows indicate activation, red Ts indicate inhibition, "P" indicates phosphorylation. Abbreviations: AGEs, advanced glycation end products; AKT, protein kinase B; ALR, aldose reductase; AMPK, AMP-activated protein kinase; AS160, AKT substrate of 160 kDa; DPP IV, dipeptidyl peptidase IV; ERK, extracellular signal-regulated kinases; FOXO1, forkhead box G1; G6Pase, glucose-6-phosphatase; G6PD, glucose-6-phosphate dehydrogenase; GIP, glucose-dependent insulinotropic polypeptide; GK, glucokinase; GLP-1, glucagon-like peptide 1; GLUT-4, glucose transporter-4; GP, glycogen phosphorylase; GPC-4, glypican-4; GSK-3, glycogen synthase kinase 3; GYS2, glycogen synthase 2; HK, hexokinase; IRS-1, insulin receptor substrate 1; IRS-2, insulin receptor substrate 2; PDK1, phosphoinositide-dependent kinase 1; PEPCK, phosphoenolpyruvate carboxykinase; PFK1, phosphofructokinase 1; PGC-1α, peroxisome proliferator-activated receptor gamma, coactivator 1 alpha; PI3K, phosphoinositide 3-kinase; PIP2, phosphatidylinositol 4,5-bisphoshate; PIP3, phosphatidylinositol 3,4,5-triphoshate; PK, pyruvate kinase; PP pathway, pentose phosphate pathway; PP1, protein phosphatase 1; PPARγ, peroxisome proliferator-activated receptor gamma; PTEN, phosphatase and tensin homolog; PTP1B, protein tyrosine phosphatase 1B; RAGE receptor, receptor for advanced glycation end products; SDH, sorbitol dehydrogenase; SOCS3, suppressor of cytokine signalling 3; TRAP, mediator of RNA polymerase II transcription subunit.

Table 1. In vitro studies of the anti-diabetic potential of fruits from the Maleae tribe.

Species	Sample Type, Composition	Model, Study Design	Tested Parameters, Observed Effects *	Ref.
Amelanchier alnifolia Nutt.	crude 80% EtOH extract, water fraction, and EtOAc fraction; detected compounds (HPLC-MS): phenolic acids (chlorogenic, caffeic, hydroxybenzoic acids), anthocyanins (cyanidin monoglycosides, cyanidin 3,5-diglucoside), proanthocyanidins (oligomers, catechin/epicatechin, epicatechin gallate)	cellular studies (L6rat skeletal muscle cells) and in vitro enzyme inhibition; control: 3,3-tetramethylene-glutaric acid (ALR activity), cells without an extract (glycogen accumulation)	ALR inhibition: strong activity (82 ± 0.73% inhibition at 5μg/mL) for EtOAc, other extracts not active; glycogen accumulation in non-insulin-stimulated cells (glucose uptake): 76%, 92%, and 23% changes for 80% EtOH extract, water, and EtOAc fractions, respectively, compared to controls; phenolic acids, anthocyanins, and proanthocyanidins responsible for hypoglycaemic activity (based on literature studies and extracts' composition)	[40]
	70% acetone extract of whole fruits, flesh, and peels from different cultivars; polyphenols 1.11–2.27 g GAE/100 g (fruits), 0.45–0.96 (flesh), 1.11–2.86 (peel) (spectrophotometry), the contents of monophosphate nucleotides and free amino acids were also determined (LC-MS)	in vitro enzyme inhibition; control: -	α-amylase inhibition: IC_{50} = 18.33–31.70 mg/mL (fruits), 22.53–42.15 (flesh), 11.05–18.41 (peels); α-glucosidase inhibition: IC_{50} = 27.83–42.23 mg/mL (fruits), 28.86–43.79 (flesh), 23.60–37.06 (peels); free amino acids responsible for observed effects (PCA analysis)	[55]
Amelanchier Medik. sp.	70% acetone extract, polyphenols 13.8–15.2 mg/g of dry fruits (depending on clone), including hydroxycinnamic acids, anthocyanins, flavonols, and hydroxybenzoic acids (HPLC)	in vitro enzyme inhibition and cellular studies (on βTC3 pancreatic β-cells); control: sodium orthovanadate (for PTP1B)	α-amylase inhibition: IC_{50} = 4.3–5.3 mg/mL depending on clone (better than, i.a., mulberry fruits); α-glucosidase inhibition: IC_{50} = 116.7–134.3 mg/mL depending on clone (very low activity); PTP1B inhibition: IC_{50} = 1.2 mg/mL (better than, i.a., mulberry fruits); cytoprotection of βTC3 pancreatic β-cells: very low effect on cell viability and no effect on cell proliferation	[36]

Table 1. *Cont.*

Species	Sample Type, Composition	Model, Study Design	Tested Parameters, Observed Effects *	Ref.
	acidified EtOH extract; anthocyanins 128.36 μg/mL (HPLC-MS: cyanidin monoglycosides)	cellular studies (HepG2 human hepatoma cell line and C2C12 mouse myoblast cell line with palmitic acid induced insulin resistance (tested) or normal cells (control)	glucose uptake: ↑, optimal extract concentration 40 μg/mL; glycogen level: ↑; protein expression: ↑ GLUT-4, ↑ IRS-1, ↑ *p*-GSK-3β, ↓ SOCS3, ↓ *p*-IRS-1, ↓ GSK-3β compared to insulin-resistant cells (effects on glucose transport and insulin sensitivity); anthocyanins responsible for observed effects (based on literature studies and extracts' composition)	[28]
	standardised *Aronia* berry extract powder (Fort Wayne, IN, USA); polyphenols 40%, anthocyanins 15%	cellular studies (RAW 264.7 and mouse bone-marrow-derived macrophages (BMDMs)); normal or LPS-stimulated cells (control), LPS +50, 100 mg/mL extract (tested)	mRNA expression levels: ↓ GLUT-1, HK1, and G6PD in comparison to LPS-stimulated controls (to the level comparable to normal controls); anthocyanins responsible for observed effects (based on literature studies and extract composition)	[35]
Aronia melanocarpa (Michx.) Elliott	EtOH and 50% EtOH extracts; detected compounds (HPLC-MS): cyanidin monoglycosides; epicatechin; procyanidins B2, B5, and C1	in vitro enzyme inhibition; control: acarbose IC_{50} = 130 ± 20 μg/mL	α-glucosidase inhibition: EtOH inactive; 50% EtOH IC_{50} = 3.5 ± 0.1 μg/mL; cyanidin glycosides IC_{50} = 0.37–5.5 μg/mL (the highest activity for cyanidin arabinoside); procyanidins IC_{50} = 3.8–5.5 μg/mL (the highest activity for procyanidin C1)	[12]
	80% EtOH and acidified MeOH extracts; polyphenols 98–148 mg GAE/g of EtOH extract (1079–1921 mg GAE/100 g of fresh fruits), anthocyanins 249–447 mg/100 g fruits, proanthocyanidins 2.46–3.74 mg PB2E/100 g fruits (HPLC/spectrophotometry)	in vitro enzyme inhibition; control: acarbose IC_{50} = 130 ± 20 μg/mL	α-glucosidase inhibition: 80% EtOH IC_{50} = 0.70–0.88 μg/mL; acidified MeOH IC_{50} = 0.030–0.049 μg/mL (depending on the cultivars); anthocyanins responsible for observed effects (based on previous studies and extract composition)	[13]
	lyophilised wine samples prepared with or without additional sugar and enzyme; detected compounds (HPLC-MS): catechin; epicatechin; protocatechuic, gallic, chlorogenic, caffeic, *p*-coumaric, and ellagic acids	in vitro enzyme inhibition; control: acarbose IC_{50} = 77.8 ± 5.7 μg/mL	α-glucosidase inhibition: IC_{50} = 49–50 μg/mL (without additional sugar); IC_{50} = 28–30 μg/mL (with additional sugar); chlorogenic and caffeic acids' contributions to the IC_{50} values: 16.4–19.5% and 4.78–6.15%, respectively (based on inhibitory curves of pure compounds and statistical analysis)	[14]

Table 1. *Cont.*

Species	Sample Type, Composition	Model, Study Design	Tested Parameters, Observed Effects *	Ref.
	MeOH and water dry extracts	in vitro enzyme inhibition; control: acarbose IC_{50} = 1.31 ± 0.12 μg/mL	α-amylase inhibition: MeOH (2.5 mg of extract/mL) inhibition = 58.6 ± 1.9%; water extract (2.5 mg of extract/mL) inhibition = 49.7 ± 3.8%	[56]
	60% EtOH and water extracts	in vitro enzyme inhibition; control: acarbose IC_{50} = 2.4 ± 0.4 μg/mL	α-amylase inhibition: water extract IC_{50} = 2632 ± 208.5 μg/mL, 60% EtOH IC_{50} = 1130 ± 91.19 μg/mL	[57]
	water, MeOH, and acetic acid extracts; composition (mg/100 mg) of acetic acid extract (HPLC-MS): neochlorogenic acid 4.22 mg, chlorogenic acid 1.26 mg, cyanidin monoglycosides 8.94 mg	in vitro enzyme inhibition; control: -	α-amylase inhibition: MeOH IC_{50} = 10.31 ± 0.04 mg/mL; water 13.55 ± 0.04 mg/mL; acetic acid 14.85 ± 0.06 mg/mL; the activity of pure compounds: the strongest for chlorogenic acid (IC_{50} = 0.57 ± 0.16 mg/mL) and cyanidin-3-glucoside (IC_{50} = 1.74 ± 0.04 mg/mL)	[58]
Aronia prunifolia (Marshall) Rehder	80% EtOH and acidified MeOH extracts; polyphenols 175 mg GAE/g of EtOH extract (2996 mg GAE/100 g of fresh fruits), anthocyanins 737 mg/100 g fruits, proanthocyanidins 4.79 mg PB2E/100 g fruits (HPLC/spectrophotometry)	in vitro enzyme inhibition; control: acarbose IC_{50} = 130 ± 20 μg/mL	α-glucosidase inhibition: 80% EtOH IC_{50} = 0.88 ± 0.08 μg/mL; acidified MeOH IC_{50} = 0.030 ± 0.005 μg/mL; anthocyanins responsible for observed effects (based on previous studies and extract composition)	[13]

Table 1. *Cont.*

Species	Sample Type, Composition	Model, Study Design	Tested Parameters, Observed Effects *	Ref.
	60% EtOH and water extracts before in vitro digestion, water extract after digestion; detected compounds (HPLC-MS): epicatechin, procyanidin B2, procyanidin oligomers, quercetin-*O*-hexoside, epigallocatechin-3-gallate, catechin/epicatechin gallate	in vitro enzyme inhibition; control: acarbose $IC_{50} = 2.4 \pm 0.4$ µg/mL	α-amylase inhibition (before digestion): water extract $IC_{50} = 53.61 \pm 5.074$ µg/mL, 60% EtOH $IC_{50} = 48.69 \pm 4.993$ µg/mL; α-amylase inhibition (after digestion): 55.41–58.48% inhibition at 50 µg/mL after gastric condition and about 50% after intestinal condition; catechin/epicatechin derivatives responsible for observed effects (based on literature studies and extract composition)	[57]
Chaenomeles japonica (Thunb.) Lindl. ex Spach	70% acetone extract; detected compounds (HPLC): epigallocatechin (23.17 mg/g), (+) catechin, procyanidin B1, procyanidin C1, (−) epicatechin, epigallocatechin gallate, epicatechin gallate, hydroxybenzoic acids, hydroxycinnamic acids, and flavonols	in vitro enzyme inhibition and cellular studies (on βTC3 pancreatic β-cells); control: sodium orthovanadate (for PTP1B)	α-amylase inhibition: $IC_{50} = 1.68$–2.41 mg/mL depending on assay (better than, i.a., mulberry fruits); α-glucosidase inhibition: unable to detect; PTP1B inhibition: $IC_{50} = 1.22$ mg/mL (better than, i.a., mulberry fruits); cytoprotection of βTC3 pancreatic β-cells: highly positive effects on cell viability and proliferation; flavan-3-ols responsible for observed effects (based on literature studies and extract composition)	[36]
	70% acetone extract; polyphenols 489.85 mg/g of dry preparation, including (+)-catechin, (−)-epicatechin, epigallocatechingallate, procyanidins B1 and C1, hydroxybenzoic acids, and flavonols (HPLC)	cellular studies on the human hepatoma cell line HepG2; control: -	↑ level of *p*-AMPK compered to cells under normal or hyperglycaemic conditions; genes' expression (5 mg/mL extract vs. cells under normal conditions): ↑ GLUT-4, IRS-2; ↓ PEPCK, PGC-1α, FOXO1, PTP1B; unchanged GLUT-2, G6Pase, GYS2, IRS-1, GSK-3α; genes' expression (hyperglycaemic conditions): ↑ GLUT-2; ↓ PEPCK; unchanged GLUT-4, G6Pase, GYS2, IRS-1, IRS-2, GSK-3α, PTP1B, PGC-1α, FOXO1; glycogen content: ↑ at 5 mg/mL to a higher level than with metformin (at 5 mM); glucose production: ↑ at 5 mg/mL (weaker activity then metformin); glucose uptake: ↑ at 5 mg/mL to a higher level than with metformin (at 5 mM)	[29]

Table 1. *Cont.*

Species	Sample Type, Composition	Model, Study Design	Tested Parameters, Observed Effects *	Ref.
Chaenomeles speciosa (Sweet) Nakai	20, 40, 60, 80, and 100% MeOH; water; 20, 40, 60, 80, and 100% EtOH extracts; and crude polysaccharide fraction; polyphenols 2–318 mg GAE/g (HPLC: catechin; epicatechin; chlorogenic, gallic, caffeic, protocatechuic, *p*-coumaric, syringic, and vanillic acids), triterpenes 23–62 mg OAE/g (oleanolic and ursolic acids), polysaccharides 8–64 mg glucose/g	in vitro enzyme inhibition; control: -	α-glucosidase-inhibitory activity: IC_{50} about 0.2–6.2 mg/mL, the highest activity for 60% EtOH, 20% EtOH, and 60% MeOH extracts; α-glucosidase-inhibitory activity (pure compounds detected in the samples: about 10–100% inhibition at 0.5 mg/mL, the highest activity for the polysaccharide fraction and oleanolic acid); polysaccharides, oleanolic acid, and different polyphenols responsible for observed effects (based on activity studies of pure compounds and PCA analysis)	[59]
	water extract before and after gastric and/or intestinal digestion; polyphenols 7.87–12.7 mg GAE/g (HPLC: phenolic acids, catechin, epicatechin), triterpenes 39–48 mg OAE/g (ursolic and oleanolic acids), anthocyanins 0.6–1.6 mg CyE/g	in vitro enzyme inhibition and influence on starch digestion; control: -	α-glucosidase inhibition: IC_{50} about 0.35–0.7 mg/mL, with the lowest activity after intestinal digestion; effects on the glucose release during in vitro digestion (from corn starch): 98% inhibition, higher then observed for pure native compounds (25–65% inhibition)	[60]
	80% EtOH extracts (different origin of fruits); polyphenols about 19–29 mg GAE/g of plant material, flavonoids about 22–47 mg QE/g, polysaccharides about 20–26 mg glucose/g	in vitro enzyme inhibition; control: quercetin	α-glucosidase inhibition: IC_{50} about 60–210 mg QE/g of plant material (depending on fruit origin); flavonoids and polysaccharides responsible for observed activity (based on correlation studies)	[61]

Table 1. *Cont.*

Species	Sample Type, Composition	Model, Study Design	Tested Parameters, Observed Effects *	Ref.
Chaenomeles speciosa (Sweet) Nakai, *Chaenomeles sinensis* (Thouin) Koehne	60% MeOH extracts from peels, flesh, and endocarps of two species (*C. speciosa*, CSP; *C. sinensis*, CSS); polyphenols 356–405 (CSP) and 345–590 (CSS) mg GAE/g, triterpenes 29–43 (CSP) and 16–40 (CSS) mg OAE/g, detected compounds (HPLC-MS): five phenolic acids, two triterpenes, and three flavonoids	in vitro enzyme inhibition; control: -	α-glucosidase inhibition (*C. speciosa*): IC_{50} = 1.32–2.66 mg/mL, with the highest activity for the extract from flesh; α-glucosidase inhibition (*C. sinensis*): IC_{50} = 0.44–2.28 mg/mL, with the highest activity for the extract from the peel; α-glucosidase inhibition ratio after in vitro digestion: ↓ about 2–20-fold depending on the sample ferulic acid and triterpenes responsible for observed effects (based on correlation studies)	[62]
	60% MeOH extracts from peels and flesh of *C. speciosa* (CSP, 12 varieties) and *C. sinensis* (CSS, 1 variety); polyphenols about 100–270 (CSP, flesh), 170–360 (CSP, peel), 200 (CSS, flesh), and 350 mg GAE/g (CSS, peel); detected compounds (HPLC-MS): catechin; epicatechin; rutin; hyperoside; myricetin; quercetin; kaempferol; chlorogenic, gallic, caffeic, protocatechuic, syringic, oleanolic, and ursolic acids	in vitro enzyme inhibition; control: -	α-glucosidase inhibition (*C. speciosa*): IC_{50} about 0.06–0.35 mg/mL (peel extracts) and about 0.04–0.42 mg/mL (flesh extracts), with the highest activity for the extracts from flesh or peel, depending on the variety; α-glucosidase inhibition (*C. sinensis*): IC_{50} about 0.05 mg/mL (peel extract) and about 0.07 mg/mL (flesh extract); α-glucosidase inhibition of pure compounds detected in the samples: IC_{50} about 0.05–1.8 mg/mL (with the highest activity for hyperoside, quercetin, mirycetin, catechin, epicatechin, protocatechuic acid, chlorogenic acid, and oleanolic acid)	[63]
Chaenomeles japonica (Thunb.) Lindl. ex Spach, *Chaenomeles speciosa* (Sweet) Nakai, *Chaenomeles* × *superba* (Frahm) Rehder.	80% MeOH (1% HCl) extracts of *C. japonica* (5 cultivars), *C. speciosa* (3 cultivars), and *C. superba* (11 cultivars); polyphenols 56–170 mg/g, (HPLC-MS: (+)-catechin; (−)-epicatechin; procyanidins B2, B3, and C1, procyanidin oligomers and polymers; chlorogenic acids); contents of pectins, sugars, and organic acids also determined	in vitro enzyme inhibition; control: -	α-amylase inhibition: 16.11–17.45 mg/mL (*C. japonica*), 16.88–18.48 mg/mL (*C. speciosa*), 13.88–18.25 mg/mL (*C. superba*); α-glucosidase inhibition: 6.09–15.19 mg/mL (*C. japonica*), 5.74–12.48 mg/mL (*C. speciosa*), 2.67–8.54 mg/mL (*C. superba*); sugars, L-ascorbic acid, and flavan-3-ols responsible for α-amylase- and α-glucosidase-inhibitory activity (based on hierarchical clustering analysis)	[64]

Table 1. *Cont.*

Species	Sample Type, Composition	Model, Study Design	Tested Parameters, Observed Effects *	Ref.
Cotoneaster bullatus Bois, *Cotoneaster zabelii* C.K.Schneid., *Cotoneaster integerrimus* Medik.	70% MeOH extract; polyphenols 62.13–81.26 mg GAE/g (HPLC: chlorogenic acids, (−)-epicatechin, procyanidins B2 and C1, quercetin 3-(2″-xylosyl)galactoside, rutin, hyperoside, isoquercitrin, quercitrin)	in vitro enzyme inhibition and effects on AGE formation; Control: acarbose (IC_{50} = 169.52 and 5.78 µg/mL for α-glucosidase and α-amylase, respectively), aminoguanidin (IC_{50} = 71.09 µg/mL for AGEs)	α-glucosidase inhibition: IC_{50} = 48.89, 57.73, 80.12 µg/mL for *C. integerrimus, C. zabelii,* and *C. bullatus* (IC_{50} = 232–416 µg/mL for (−)-epicatechin, hyperoside, quercetin 3-(2″-xylosyl)galactoside, and procyanidin B2); α-amylase inhibition: IC_{50} = 941–1083 µg/mL; inhibition of AGE formation: IC_{50} = 106.36, 118.94, 166.62 µg/mL for *C. integerrimus, C. zabelii,* and *C. bullatus* (IC_{50} = 2.20–15.78 µg/mL for procyanidin B2, (−)-epicatechin, hyperoside, and quercetin 3-(2″-xylosyl)galactoside)	[15]
Crataegus azarolus var. aronia L.	water extract	in vitro enzymatic starch digestion and glucose movement; control: acarbose 0.1 mg/mL (97.6% reduction in starch digestion); guar gum 50 mg/mL (glucose movement assay, iAUC ↓ by 30.8%)	enzymatic starch digestion: dose-dependent ↓ in glucose level, significant effect at 0.5–10 mg/mL (22.0–70.7% reduction, IC_{50} = 3.5 mg/mL); glucose movement (postprandial glucose level in vitro): lack of effects	[65]
Crataegus laevigata (Poir.) DC.	MeOH and water extracts	in vitro enzyme inhibition; control: acarbose IC_{50} = 1.31 ± 0.12 µg/mL	α-amylase inhibition: MeOH (2.5 mg of extract/mL), 35.4 ± 4.7% inhibition; water extract (2.5 mg of extract/mL), 41.1 ± 15.3% inhibition	[56]
	80% EtOH extract and fractions (EtOAc, *n*-butanol); polyphenols 136.54–697.23 mg GAE/g, flavonoids 21.46–154.51 mg QE/g, flavonols 19.21–68.83 mg QE/g	in vitro enzyme inhibition; control: acarbose IC_{50} = 275.43 ± 1.59 µg/mL	α-glucosidase inhibition: IC_{50} = 16.12–30.80 µg/mL, with the highest activity for the *n*-butanol fraction; caffeic acid, epicatechin, naringenin and quercetin responsible for observed effects (based on literature studies and extract composition)	[16]
Crataegus microphylla K. Koch	EtOH, acidified (0.5% HCl) EtOH, 50% EtOH, MeOH, acidified MeOH, 50% MeOH, water, acidified water extracts; polyphenols 5.00–57.28 mg GAE/g	in vitro enzyme inhibition; control: acarbose IC_{50} = 31.92 ± 0.08 µg/mL	α-glucosidase inhibition: IC_{50} = 250.94–731.81 µg/mL, with the highest activity for the acidified MeOH extract	[66]

Table 1. *Cont.*

Species	Sample Type, Composition	Model, Study Design	Tested Parameters, Observed Effects *	Ref.
	MeOH extract and fractions (methylene chloride, EtOAc, *n*-butanol, water fractions); isolated compounds: hyperoside, chlorogenic acid, 3-epicorosolic acid, ursolic acid, oleanolic acid, β-sitosterol, β-sitosterol glucoside	in vitro enzyme inhibition and effects on the formation of AGEs; control: acarbose IC_{50} = 81.65 ± 4.07 μg/mL (α-glucosidase), ursolic acid IC_{50} = 1.00 ± 0.09 μg/mL (PTP1B), quercetin IC_{50} = 0.75 ± 0.07 μg/mL (ALR), aminoguanidine IC_{50} = 127.06 ± 7.10 μg/mL (AGEs)	α-glucosidase inhibition: IC_{50} = 22.70–122.11 μg/mL; PTP1B inhibition: IC_{50} = 1.41–18.75 μg/mL; Rat lens ALR inhibition: IC_{50} = 9.09–160.54 μg/mL; inhibition of AGEs formation: IC_{50} = 65.83–88.90 μg/mL; the highest activity observed for EtOAc (α-glucosidase, PTP1B, ALR inhibition) or MeOH extract (AGE formation); inhibitory potential of 3-epicorosolic acid IC_{50} = 4.08 and 30.18 μg/mL (PTP1B and α-glucosidase tests); different compounds responsible for observed effects (based on extract composition and literature studies)	[17]
Crataegus pinnatifida Bunge	MeOH extract; polyphenols 101.56 mg tannic acid equivalents/g, flavonoids 44.52 mg CE/g	in vitro enzyme inhibition; control: acarbose IC_{50} = 2.15 ± 0.86 μg/mL	α-glucosidase inhibition: IC_{50} = 766.22 ± 8.14 μg/mL	[67]
	water extract before and after gastric and/or intestinal digestion; polyphenols 3.07–11.7 mg GAE/g (HPLC: five phenolic acids, catechin, epicatechin), triterpenes 28.4–47.4 mg OAE/g (ursolic and oleanolic acids), anthocyanins 0.77–1.8 mg CyE/g	in vitro enzyme inhibition and influence on starch digestion; control: -	α-glucosidase inhibition: IC_{50} about 0.3–0.75 mg/mL, with the lowest activity after intestinal digestion; effects on glucose release during in vitro digestion (from corn starch): 83% inhibition, higher then observed for pure native compounds (25–65% inhibition)	[60]
	MeOH, EtOH, EtOAc, acetone, dichloromethane, chloroform, *n*-hexane extracts; polyphenols 12.1–63.5 mg GAE/g (HPLC: chlorogenic acid, hyperoside, epicatechin, procyanidin B2)	in vitro enzyme inhibition and molecular docking studies; control: acarbose IC_{50} = 317.80 ± 16.36 μg/mL	α-glucosidase inhibition: IC_{50} = 42.35–207.46 μg/mL, with the highest activity for acetone, MeOH, and EtOH extracts (IC_{50} = 42.35–58.69 μg/mL); activity of isolated compounds: IC_{50} = 34.98–170.37 μg/mL, with the highest activity for hyperoside (also high affinity for enzyme/molecular docking studies)	[68]
	70% EtOH macerate	in vitro enzyme inhibition; control: acarbose IC_{50} = 275.98 ± 1.57 μg/mL	α-glucosidase inhibition: IC_{50} = 326.48 ± 18.56 μg/mL	[69]
Cydonia oblonga Mill.	70% EtOH extract from pulp fruit callus; polyphenols 10.98 mg/100 g (pulp) and 91.58 mg/100 g (callus), mainly chlorogenic acid, 5-*p*-coumaroylquinic acid, neochlorogenic acid, and (−)-epicatechin (LC-MS, GC-MS)	in vitro enzyme inhibition; control: acarbose about 75% and 50% inhibition at 32 μg (α-amylase and α-glucosidase, respectively)	α-amylase inhibition: about 25% inhibition at 250–1000 μg; α-glucosidase inhibition: about 40% inhibition at 250 μg and about 75% inhibition at 500 μg	[70]

Table 1. *Cont.*

Species	Sample Type, Composition	Model, Study Design	Tested Parameters, Observed Effects *	Ref.
Cydonia oblonga Mill.; *Malus domestica* (Suckow) Borkh.	fruit puree; flavan-3-ols 355.3 and 34.1 mg/kg (*Cydonia* and *Malus*, respectively), flavonols 1269 and 245.1 mg/kg, hydroxycinnamic acid derivatives 306.6 and 140.6 mg/kg, anthocyanins 62.2 and 67.6 mg/kg (HPLC)	in vitro enzyme inhibition; control: acarbose	α-amylase inhibition: IC_{50} = 164 and 161 mg/mL (*Cydonia* and *Malus*, respectively); α-glucosidase inhibition: IC_{50} = 177 and 166 mg/mL (*Cydonia* and *Malus*, respectively)	[71]
Malus domestica (Suckow) Borkh.	80% EtOH extract; 15 phenolics detected (HPLC), with phlorizin and chlorogenic acid as the dominant compounds	cellular studies on the human hepatoma cell line HepG2 with insulin resistance induced by high glucose levels (tested) or normal cells (control)	glucose uptake and glycogen content: ↑ in comparison to insulin-resistant cells (phlorizin and chlorogenic acid activity: comparable); protein levels: ↑ *p*-IRS2/IRS2, ↑ *p*-AKT/AKT, ↑ *p*-GSK3β/GSK3β, ↑ *p*-FOXO1/FOXO1, *p*-IRS1/IRS1 unchanged (insulin-resistant cells); phlorizin activity (10 μg/mL): comparable; chlorogenic acid: not active	[37]
	Phytonutriance® polyphenolic extract Appl'In™; polyphenols min. 80%, phlorizin > 5%	cellular studies on Ishikawa Var 1 endometrial cells; control: genistein; about 30% inhibition at 2.7 μg/mL	glucose uptake: > 60% inhibition at 50 μg/mL (phlorizin equivalent 2.5 μg/mL); phlorizin responsible for observed effects (based on literature studies and extract composition)	[72]
	standardised commercial extract (Appl'In by DIANA Food SAS); polyphenols 67%, including 40% flavonoid monomers and phenolic acids (HPLC: flavan-3-ols, dihydrochalcones, flavonols, hydroxycinnamic acids)	cellular studies on Caco-2 cells and *Xenopus* oocytes injected to express SGLT1; control: cells without extract	total and sodium-independent (GLUT-mediated) glucose uptake (Caco-2 cells): 51% and 46% ↓ at 0.3 mg polyphenols/mL (corresponding to the physiological dose that may be reached after consumption of 600 mg of polyphenols (900 mg of apple extract) in the human study); glucose uptake in oocytes (SGLT1 mediate): ↓ at 0.125–4.0 mg apple polyphenols/mL (dose-independent), for phloretin and phlorizin 59% and 85% ↓ at 0.5 mM, respectively; different compounds responsible for observed effects (based on literature studies and extract composition)	[32]

Table 1. *Cont.*

Species	Sample Type, Composition	Model, Study Design	Tested Parameters, Observed Effects *	Ref.
	standardised commercial extract (Appl'In, Diana Naturals, France); polyphenols min. 80%, phlorizin > 5%	cellular studies on Caco-2 cells; control: dapagliflozin IC_{50} for SGLT1 inhibition ~0.5 μM = 0.2 mg/L	glucose uptake: 90% ↓ of SGLT1 at 0.2 mg/mL; ↓ GLUT-2 IC_{50} = 0.45 mg/mL (in comparison: phlorizin IC_{50} for SGLT1 = 0.13 mg/L, phloretin IC_{50} for GLUT-2 = 22 mg/L)	[34]
	standardised commercial extract (Appl'In, Diana Naturals, France); polyphenols min. 80%, phlorizin > 5%; 14 compounds detected (HPLC), including quercetin glycosides, dihydrochalcones, phenolic acids, and procyanidin oligomers	cellular studies on Caco-2 cells (TC7 subclone); control: cells without extract	glucose uptake: ↓ at 1.12 μg/mL (IC_{50} = 1.19 ± 0.35 mg/mL.); different compounds responsible for observed effects (based on literature studies and extract composition)	[73]
	80% MeOH extract (+0.1% formic acid); polyphenols 3.61 mg/g of fresh fruits, including flavanols, flavones, flavonols, dihydrochalcones, hydroxycinnamic acids, and anthocyanins (18 compounds, HPLC)	cellular studies on Caco-2 cells; control: cells with high glucose (HG) content and normal cells (NC)	formation of AGEs: ↓ compared to HG, the level at 0.8 mmol/L comparable to NC; glycolaldehyde-modified proteins: ↓ compared to HG, the levels at 0.4 and 0.8 mmol/L were comparable to NC; glyoxalase I activity: ↑ in the HG cells; at similar levels in the NC and tested cells (0.4–0.8 mmol/L extract); glyoxalase II activity: unchanged	[74]
	MeOH extract from apple juice; 10 phenolics detected (chlorogenic acid 0.33 mM, procyanidins dimer B type 0.05 mM, (−)-epicatechin 0.06 mM, *p*-coumaroylquinic acid 0.085 mM, phloretin glycosides 0.09 mM, quercetin and kaempferol glycosides 0.09 mM) (HPLC-MS)	cellular studies on Caco-2 cells; control: cells without extract or phenolics	glucose transport: ↓, with greater inhibition under sodium-free conditions (apical GLUT-2, IC_{50} = 136 mg) than under sodium-dependent conditions (SGLT1 and GLUT-2, IC_{50} = 300 mg); contribution of phenolics to observed effect: quercetin-3-*O*-rhamnoside 26% (IC_{50} = 31 μM), phlorizin 52% (IC_{50} = 146 μM), chlorogenic acid 12% (IC_{50} = 2570 μM)	[33]
	commercial extract (BioActive Food GmbH); polyphenols 44% (catechin equivalents), phlorizin 16%, quercetin 12.43%, chlorogenic acid 5.57% (HPLC/spectrophotometry)	cellular studies on *Xenopus* oocytes injected to express SGLT1; control: cells without extract	glucose uptake: ↓, extract IC_{50} = 2.0 ± 0.24 μg/mL; phlorizin IC_{50} = 0.46 ± 0.19 μM, quercetin IC_{50} = 0.62 ± 0.05 mM, chlorogenic acid not active	[75]

Table 1. *Cont.*

Species	Sample Type, Composition	Model, Study Design	Tested Parameters, Observed Effects *	Ref.
	juice: raw, after thermal or ultrasound pasteurisation and before or after in vitro digestion; polyphenols 5.33 mg/L (chlorogenic and *p*-coumaroylquinic acids, phlorizin and phloretin xyloqlucosides, epigallocatechin gallate, quercetin monoglycosides; HPLC-MS)	in vitro enzyme inhibition; control: acarbose IC_{50} = 0.09 µg/mL	α-glucosidase inhibition: IC_{50} = 6.24 mg/mL (raw sample), 0.92 mg/mL (sample after digestion); ↓ after thermal pasteurisation (not digested sample, 18.41 mg/mL) or ↑ (thermal pasteurisation, digested sample, 0.44 mg/mL); effects with acarbose: synergy at combinatory concentration of 2 mg/mL (i.e., up to 40% inhibition), antagonism at 2–9 mg/mL, and additive activity at >9 mg/mL; different compounds responsible for observed effects (based on literature studies and extract composition)	[25, 76]
	DMSO extract; 5% quercetin, 30% phlorizin	in vitro receptor binding; control: rosiglitasone IC_{50} = 0.043 ± 0.004 µg/mL (binding affinity test)	binding affinity to PPARγ receptor: moderate (IC_{50} = 0.49 µg/mL); effects on rosiglitazone-mediated DRIP205/TRAP220 coactivator: strong antagonism (IC_{50} = 0.15 ± 0.03 µg/mL); phlorizin, phloretin, epicatechin and catechin responsible for observed effects (based on literature studies and apple composition)	[38]
	water and 12% EtOH extracts from peel and pulp, with or without different elicitor treatments; polyphenols 0.35–1.44 mg GAE/g of fresh fruits for peel, 0.05–0.32 mg GAE/g for pulp; detected compounds (HPLC): chlorogenic acid, *p*-coumaric acid, gallic acid, quercetin	in vitro enzyme inhibition; control: -	α-glucosidase inhibition: higher activity found for extracts from peel, as well as 12% EtOH extracts; the activity of extracts from peel (without elicitors) ↓ after fruit storage, and the activity of extracts from pulp ↑ and then ↓ depending on the storage time (0–3 months); α-amylase inhibition: higher activity found for extracts from pulp, as well as water extracts; the activity of extracts from peel (without elicitors) ↑ after storage, while the activity of extracts from pulp was similar after storage; chlorogenic acid and quercetin responsible for observed effects (based on literature studies and extract composition)	[77]

Table 1. *Cont.*

Species	Sample Type, Composition	Model, Study Design	Tested Parameters, Observed Effects *	Ref.
	12% EtOH and water extracts from peel and pulp of different cultivars; polyphenols 266–781 μg GAE/g fresh material for peel extracts and 30–143 μg GAE/g for pulp extracts; detected compounds (HPLC): catechin, chlorogenic, and *p*-coumaric acids; quercetin	in vitro enzyme inhibition; control: -	α-amylase inhibition: the highest activity for pulp (about 0–60% at 100 μL) or peel water extracts (about 0–55%); α-glucosidase inhibition: higher activity for pulp water extract (69–83% at 50 μL) than pulp 12% EtOH extract (50–70% at 50 μL); higher or similar activity for peel 12% EtOH extract (54–98% at 50 μL) than peel water extract (52–93% at 50 μL); polyphenols responsible for observed effects (based on correlation studies)	[78]
	acetone–ethanol extract (1:3), purified with 95% MeOH (polyphenols 390.8 μg CE/mg,) and fractionated with acetonitrile (II: polyphenols 61.3 μg CE/mg), EtOAc (III: polyphenols 459.3 μg CE/mg), MeOH (IV: polyphenols 620.6 μg CE/mg), and fraction I residue; 14 compounds detected (HPLC-MS)	in vitro enzyme inhibition; control: acarbose IC_{50} = 840 ± 100 μg/mL, quercetin 661 ± 7 μg/mL	α-glucosidase inhibition: IC_{50} = 19–67 μg/mL (purified 95% MeOH extract and fractions III-IV); chlorogenic acid isomers, flavan-3-ols monomers and oligomers, quercetin and phloretin glycosides responsible for observed effects (based on literature studies and extract composition)	[18]
	70% EtOH extract; polyphenols 534.39 mg GAE/g (HPLC: phlorizin; (−)-epigallocatechin; (−)-epicatechin; (+)-catechin; procyanidins B1 and B2; quercetin glycosides; chlorogenic, *p*-coumaroylquinic, and caffeic acids	in vitro enzyme inhibition and molecular docking studies; control: acarbose IC_{50} = 0.76 μg/mL	α-glucosidase inhibition: IC_{50} = 15 μg/mL; phlorizin, (−)-epicatechin, and tannins responsible for observed effects (based on activity study of pure compounds); inhibition mechanism: competitive or mixed type, potential conformational change of enzyme was suggested	[79]
	apple extract	effects on the formation of AGEs in plasma in vitro; control: -	concentration-dependent (at 100–2000 mmol) inhibition of AGE formation (up to fourfold ↓), depending on the glucose concentration (5.5–50 mmol) and time of study	[80]
	water extract from the juice;	in vitro enzyme inhibition; control: quercetin IC_{50} = 153.85 ± 5.38 μg/mL (ALR), 67.5 ± 0.8 μg/mL (SDH)	ALR inhibition: IC_{50} = 171.63 ± 5.42 μg/mL; SDH inhibition: IC_{50} = 56.52 ± 4.95 μg/mL	[41]

Table 1. *Cont.*

Species	Sample Type, Composition	Model, Study Design	Tested Parameters, Observed Effects *	Ref.
	juice (fermented and unfermented); compounds: 19 polyphenols (HPLC: mainly chlorogenic acid at about 95–145 mg/L, epicatechin 70–105 mg/L, procyanidin B2 70–95 mg/L), sugars, and organic acids	in vitro enzyme inhibition; control: acarbose, with about 60–80% inhibition at 0.025–0.1 mg/mL	α-glucosidase inhibition: about 43% inhibition at 0.1 mL/mL (fermented and unfermented samples); at 0.5 mL/mL, about 67% (unfermented) and 78% (fermented)	[81, 82]
Malus pumila Mill.	80% MeOH extracts (0.5% formic acid) from the peel and flesh of different cultivars; polyphenols 25.12–281.73 mg/100 g, phlorizin 1.10–68.54 mg/100 g (peel > flesh), procyanidins 6.03–78.76 mg/100 g (flesh > peel) (HPLC)	cellular studies on HepG2 human hepatocellular liver carcinoma cells; control: cells without extracts	glucose uptake: ↑, with the highest activity for peel extract from the Red Delicious cultivar, with the highest total polyphenols and phlorizin content; phlorizin responsible for observed effect (based on correlation and literature studies)	[83]
	commercial extract (Exxentia®); polyphenols 57.5%, including phlorizin (9.9%), chlorogenic acid (15.8%), and quercetin (0.4%)	cellular studies on L6.C11 rat skeletal muscle myoblast cells (insulin sensitivity mechanisms) and ex vivo α-glucosidase inhibition on isolated rat intestinal mucosa; control: cells not treated with extract or rosiglitazone (Rosi, 10 μmol/L)	glucose uptake: maximal ↑ by 63.6%, EC_{50} = 4.2 ± 0.7 μg/mL; insulin-stimulated glucose uptake: increasing synergistic effects at 5–25 μg/mL of extract and 50 nM of insulin; GLUT-4 levels: not changed in total membranes, ↑ in a plasma membrane fraction (GLUT-4 translocation, at 25 μg/mL); ↑ *p*-Akt/Akt by about 50% (unchanged for Rosi); PPARγ level and PPARγ-mediated transcription ↑ (comparable to Rosi), ERK1/2 not affected (for extract at 25 μg/mL); α-glucosidase inhibition: IC_{50} 12.54–18.05 μg/mL (three models)	[30]
	fruit water extract	in vitro enzyme inhibition; control: acarbose	α-glucosidase inhibition: 8 mg acarbose equivalents/g	[84]
Malus Mill. sp.	apple juice; polyphenols 0.78 mg GAE/mL	effects on AGE formation in vitro control: -	weak inhibition (about 5%) at 10 μL of juice/mL compared to pomegranate juice (about 95% inhibition)	[85]
	MeOH extract from red or yellow apples; polyphenols about 4 mg GAE/g dry fruits	in vitro enzyme inhibition; control: -	α-amylase inhibition: not active; α-glucosidase inhibition: 49.5% (yellow apple) or 95.4% (red apple) inhibition at 10 mg/mL concentration;	[86]

Table 1. *Cont.*

Species	Sample Type, Composition	Model, Study Design	Tested Parameters, Observed Effects *	Ref.
Malus Mill. sp. (276 *Malus* species/cultivars including *Malus sieversii* and *Malus domestica*)	80% MeOH extract + 0.1% HCl + sonification; polyphenols 0.49–2.61 mg/g fresh fruits (HPLC-MS: phenolic acids, flavan-3-ols, proanthocyanidins, flavonols, dihydrochalcones)	in vitro enzyme inhibition and effects on AGE formation; control: acarbose (IC_{50} = 0.21 mg/mL and 0.58 µg/mL for α-glucosidase and α-amylase), sitagliptin (IC_{50} = 0.044 µg/mL for DPP IV), aminoguanidin (IC_{50} = 24 µg/mL for AGEs)	α-glucosidase inhibition: IC_{50} = 7.1–256 mg/mL (0.01 mg/mL for phlorizin, 0.028–0.073 mg/mL for chlorogenic acid, epicatechin, procyanidin B2, and quercetin 3-*O*-galactoside); α-amylase inhibition: IC_{50} = 5.3–21.5 mg/mL for 10 cultivars with the highest total polyphenols (485 and 749 µg/mL for quercetin 3-*O*-galactoside and epicatechin); DPP IV inhibition: 10.3 mg/mL to inactive for 10 selected cultivars (75 and 90 µg/mL for chlorogenic acid and quercetin 3-*O*-galactoside); inhibition of AGE formation: IC_{50} = 5.2 mg/mL to inactive for 10 selected cultivars (17.1 and 23 µg/mL for quercetin 3-*O*-galactoside and epicatechin)	[42]
Mespilus germanica L.	acidified 80% EtOH extract, along with its water and 80% MeOH fractions; polyphenols 32.05–74.93 mg/g (HPLC: phenolic acids, procyanidin B2, catechin, and epicatechin)	in vitro enzyme inhibition; control: acarbose 100% inhibition at 50 mg/mL	α-amylase inhibition: about 47%, 27%, and 52% inhibition at 5 mg/mL for EtOH, MeOH, and water fractions, respectively; phenolic acids (e.g., gallic, chlorogenic, and ferulic acids) responsible for observed effects (based on literature studies and extract composition)	[87]
	EtOH and water extracts; polyphenols 6.93 mg GAE/g dry fruits	in vitro enzyme inhibition; control: -	α-amylase inhibition: not active; α-glucosidase inhibition: 71.5% at 20 mg/mL for EtOH, 44.9% for water extract (comparable to onion extract)	[88]
	EtOH extract; polyphenols 16.5 mg GAE/g, flavonoids 1.99 mg QE/g	in vitro enzyme inhibition; control: acarbose, with about 85% inhibition at 10 µg/mL (α-amylase), and about 85% inhibition at 25 µg/mL (α-glucosidase)	α-amylase inhibition: about 35% at 10 and 100 µg/mL; α-glucosidase inhibition: about 99% at 25 µg/mL	[89]

Table 1. *Cont.*

Species	Sample Type, Composition	Model, Study Design	Tested Parameters, Observed Effects *	Ref.
Pyracantha fortuneana (Maxim.) H.L.Li	proanthocyanidin fraction from 70% acetone extract; compounds (HPLC): epicatechin, catechin, A-type and B-type procyanidins, procyanidin glucosides	in vitro enzyme inhibition and molecular docking studies; control: acarbose IC_{50} = 307 ± 1 µg/mL	α-glucosidase inhibition: IC_{50} = 0.15 ± 0.01 µg/mL; reversible, non-competitive inhibition; alteration of the catalytic configuration of the enzyme's active site; procyanidins responsible for observed effects (based on extract composition and molecular docking studies)	[20]
	50–90% MeOH, EtOH and acetone extracts; polyphenols 9.67–17.33 mg GAE/g, 25 compounds detected (HPLC-MS: flavonoids and phenolic acids), flavonoids 0.34–1.03 mg QE/g, polysaccharides 72.87–103.65 mg glucose/g	in vitro enzyme inhibition; control: acarbose IC_{50} = 1071 ± 29 µg/mL	α-glucosidase inhibition: IC_{50} = 350–1870 µg/mL, with the highest activity observed for 50% and 70% acetone extracts; polyphenols responsible for observed effects (correlation studies)	[19]
Pyrus bretschneideri Rehder	60% MeOH from peel and pulp; polyphenols about 2.9 and 8.1 mg GAE/g in pulp and peel, respectively (HPLC-MS: catechin; epicatechin; rutin; chlorogenic, *p*-coumaric, vanillic, gallic, and ferulic acids), flavonoids about 1.5 and 6.3 mg RE/g, terpenes about 0.9 and 4.3 mg OAE/g (oleanolic and ursolic acids)	in vitro enzyme inhibition; control: compounds detected in the tested extracts, 2–89% inhibition at 40 µg/mL	α-glucosidase inhibition: IC_{50} 190 µg/mL for peel and 1220 µg/mL for pulp extract; activity of model compounds: the highest activity found for ferulic acid, rutin, and vanillic acid (about 80–89% inhibition at 40 µg/mL)	[90]

Table 1. *Cont.*

Species	Sample Type, Composition	Model, Study Design	Tested Parameters, Observed Effects *	Ref.
Pyrus communis L.	MeOH extracts from peel, flesh, or peel + flesh of different *Pyrus* cultivars	in vitro enzyme inhibition; control: acarbose 79.75 ± 1.86% inhibition at 0.5 mg/mL (α-amylase), 70.16 ± 1.60% inhibition at 0.5 mg/mL (α-glucosidase)	α-amylase inhibition: 1.20–18.49% for peel at 6 mg/mL, flesh or flesh + peel not active; α-glucosidase inhibition: highly dependent on the cultivar, with the highest activity for "Takiša", i.e., 76.50–99.64% inhibition at 0.5 mg/mL; for wild *Pyrus communis*, 13.59% inhibition at 0.5 mg/mL and 63.36% at 1.0 mg/mL (peel)	[91]
	juice before or after fermentation; polyphenols about 0.25–0.6 mg GAE/g fresh fruits (HPLC: catechin; rutin; chlorogenic, *p*-coumaric, protocatechuic, benzoic, and gallic acids)	in vitro enzyme inhibition; control: -	α-amylase inhibition: ↑ after fermentation; α-glucosidase inhibition: ↑, ↑ or unchanged after fermentation (depending on the length, pH, and cultivar)	[92]
	juice from different cultivars before or after fermentation; proteins 3.8–7.8 mg/mL, phenolics about 0.4 mg/mL (HPLC: epicatechin, *p*-coumaric and caffeic acids, quercetin derivatives)	in vitro enzyme inhibition; control: -	α-amylase inhibition: not active; α-glucosidase inhibition: ↑ or ↑ after fermentation (depending on the length, bacteria, and pH), about 5–80% inhibition at 10–50 µL	[93]
	12% EtOH and water extracts from peel and pulp of different cultivars; polyphenols 270–1300 µg GAE/g fresh material for peel and 27–150 µg GAE/g for pulp (HPLC: chlorogenic, caffeic, protocatechuic, p-coumaric, and gallic acids; catechin; quercetin derivatives)	in vitro enzyme inhibition; control: -	α-amylase inhibition: the highest activity for pulp water extract (about 20–50% at 100 µL); α-glucosidase inhibition: depending on the study and cultivar (about 10–60% at 10 µL); polyphenols responsible for observed effects (based on correlation studies or extract composition)	[94,95]
Pyrus pashia Buch.-Ham ex D. Don	EtOAc fraction of 70% EtOH extract before or after NaOH hydrolysis; tannins 780 mg CE/100 g of fresh fruits, sugars 15.93 g/100 g	in vitro enzyme inhibition; control: acarbose IC_{50} = 55 and 440 µg/mL (α-amylase and α-glucosidase)	α-amylase inhibition: IC_{50} = 72–100 µg/mL; α-glucosidase inhibition: IC_{50} = 85–330 µg/mL	[21]

Table 1. *Cont.*

Species	Sample Type, Composition	Model, Study Design	Tested Parameters, Observed Effects *	Ref.
Pyrus pyrifolia (Burm.f.) Nakai	80% EtOH + 70% acetone extract from two cultivars fractionated on Sephadex; polyphenols 20.9–28.5 mg CE/g	in vitro enzyme inhibition; control: -	α-glucosidase inhibition: IC_{50} = 21.3–66.4 µg/mL; oligomeric and polymeric polyphenols responsible for observed effects (based on activity testing of phenolic fractions)	[96]
	water extract	in vitro enzyme activity; control: acarbose about 10–40% inhibition at 0.05–1 mg/mL	GK activity: ↑ at 5 and 10 mg/mL; α-glucosidase inhibition: about 10–30% at 0.25–1 mg/mL	[39]
	50% EtOH extract from pear pomace;	cellular studies on 3T3-L1 mouse cells; control: cells not treated with extract (glucose uptake), rosiglitazone 1 µM (Rosi, for protein expression)	glucose uptake: ↑ at 100 and 250 µg/mL; protein expression: ↑ *p*-IRS-1 (Tyr632) and *p*-Akt at 100 µg/mL (better than Rosi), ↑ GLUT-4 (comparable to Rosi)	[31]
Pyrus pyrifolia (Burm.f.) Nakai, *P. ussuriensis* Maxim. Ex Rupr., *P. betulifolia* Bunge, *P. bretschneideri* Rehder	juices before and after in vitro digestion; polyphenols about 0.18–0.4 mg/mL and 0.18–0.4 mg/mL, polysaccharides about 4.5–8 mg/mL and 4.5–12 mg/mL (before and after digestion, respectively)	in vitro enzyme inhibition; control: acarbose	α-amylase inhibition: about 1.2–1.6 mg acarbose/mL before digestion, 1.2–2.2 mg acarbose/mL after digestion; α-glucosidase inhibition: about 4.5–6.5 mg acarbose/mL before digestion, 5–12.5 mg acarbose/mL after digestion; polyphenols and polysaccharides responsible for observed effects (based on correlation studies or extract composition)	[97]

Table 1. *Cont.*

Species	Sample Type, Composition	Model, Study Design	Tested Parameters, Observed Effects *	Ref.
Sorbus aucuparia L.	80% acetonitrile extract (whole) and fractions (Sephadex); compounds in the whole extract: hydroxycinnamic acids (chlorogenic acids), flavonols, proanthocyanidins, anthocyanins (LC-MS)	in vitro enzyme inhibition; control: acarbose IC_{50} = 0.08 μg/mL	α-amylase inhibition: whole-extract IC_{50} = 4.5 μg GAE/mL; the activity of the proanthocyanidin-rich fraction was comparable to that of the whole extract; synergy with acarbose; proanthocyanidins responsible for observed effects (based on activity testing of phenolic fractions)	[27]
	80% acetonitrile extract (whole) and fractions (Sephadex); compounds in the whole extract: 12 phenolics, including chlorogenic acid (65%), dicaffeoylquinic and coumaroylquinic acids, quercetin-3-*O*-glucoside, and caffeoyl glucose (LC-MS)	in vitro enzyme inhibition; control: acarbose IC_{50} = about 40 μg/mL	α-glucosidase inhibition: whole-extract IC_{50} = about 30 μg GAE/mL, synergy with acarbose, no synergy with other berries (blackcurrant); proanthocyanidin-rich fraction IC_{50} > 100 μg GAE/mL; chlorogenic acids responsible for observed effects (based on extract composition and literature studies)	[26]
	water extract; polyphenols 19.13 mg GAE/g, flavonoids 9.62 mg CE/g	in vitro enzyme inhibition; control: acarbose IC_{50} = 6 ± 0.2 μg/mL (α-amylase), IC_{50} = 86 ± 2.7 μg/mL (α-glucosidase)	α-amylase inhibition: IC_{50} > 800 μg/mL; α-glucosidase inhibition: IC_{50} = 50 μg/mL; polyphenols responsible for observed effects (based on extract composition and literature studies)	[23]
	50% MeOH and 50% acetone extracts and fractions (Et_2O, EtOAc, *n*-butanol, water); polyphenols 1.31–274.79 mg/g (51 compounds (HPLC-MS), including caffeic and ferulic acids pseudodepsides, flavonols, and proanthocyanidins)	effects on the formation of AGEs in vitro; control: aminoguanidine (AG)	inhibition of AGE formation: IC_{50} = 22.37–55.33 μmol AG/mg of extract (2–4-fold higher activity than AG); IC_{50} for chlorogenic acid, quercetin 3-*O*-β-sophoroside, and procyanidin B2: 152, 254, and 486 μmol AG/mg, respectively	[43]
	60% EtOH and water extracts	in vitro enzyme inhibition; control: acarbose IC_{50} = 2.4 ± 0.4 μg/mL	α-amylase inhibition: water extract IC_{50} = 1236 ± 177.0 μg/mL, 60% EtOH IC_{50} = 973.9 ± 61.60 μg/mL	[57]

Table 1. *Cont.*

Species	Sample Type, Composition	Model, Study Design	Tested Parameters, Observed Effects *	Ref.
Sorbus decora (Sarg.) CK Schneid.	80% EtOH extract	effects on the formation of AGEs in vitro; control: quercetin IC_{50} = 6.1 ± 1.8 µM (about 1.84 µg/mL)	inhibition of AGE formation: IC_{50} = 192.7 ± 37.1 µg/mL	[98]
Sorbus domestica L.	water extract	in vitro enzyme inhibition; control: acarbose IC_{50} = 120 ± 23 µg/mL (α-amylase), IC_{50} = 548 ± 21 µg/mL (α-glucosidase)	α-amylase inhibition: IC_{50} = 8768 µg/mL; α-glucosidase inhibition: IC_{50} = 417 µg/mL	[99]
	MeOH extract and fractions (dichloromethane, Et_2O, EtOAc, butanol, water) of fruits at different stages of maturity; polyphenols 2.27–341 mg GAE/g (47 compounds (LC-MS), including flavonoids, benzoic and cinnamic acid derivatives, and biphenyls)	in vitro enzyme inhibition; control: sorbinil, 45% inhibition at 0.25 µM	ALR inhibition: 72–93% for Et_2O and EtOAc fractions at 50 µg/mL, >50% inhibition for dichloromethane fraction, <40% inhibition for butanol and water fractions; flavonoids and hydroxycinnamoyl esters responsible for observed effects (based on extract composition and statistical analysis)	[44]
Sorbus torminalis (L.) Crantz	water extract; polyphenols 24.21 mg GAE/g, flavonoids 15.69 mg CE/g	in vitro enzyme inhibition; control: acarbose IC_{50} = 6 ± 0.2 µg/mL (α-amylase), IC_{50} = 86 ± 2.7 µg/mL (α-glucosidase)	α-amylase inhibition: IC_{50} = 307 µg/mL; α-glucosidase inhibition: IC_{50} = 27 µg/mL; polyphenols responsible for observed effects (based on extract composition and literature studies)	[23]
Sorbus species from subgenus *Aria* ** and *Sorbus* ***	80% acetone extracts and fractions (carbohydrates and phenolics, only from *S. norvegica*); detected compounds (NMR): chlorogenic and neochlorogenic acids, carbohydrates	in vitro enzyme inhibition; control: acarbose IC_{50} = 0.047 ± 0.006 µg/mL (α-amylase), IC_{50} = 742 ± 147 µg/mL (α-glucosidase),	α-amylase inhibition: IC_{50} = 2.5–12.3 µg/mL (*Aria*), the highest activity—*S. norvegica*; IC_{50} = 30.3–2540 µg/mL (*Sorbus*), the highest activity—*S. hybrid*, *S. aucuparia*, *S. meinichii*; α-glucosidase inhibition: IC_{50} = 0.9–3.7 µg/mL (*Aria*), the highest activity—*S. alnifolia*, *S. minima*, *S. norvegica*; IC_{50} = 4.6–300 µg/mL (*Sorbus*), the highest activity—*S. hybrida*, *S. aucuparia*, *S. meinichii*; carbohydrates and polyphenols contributed to the observed effects (based on correlation studies and activity testing of fractions from *S. norvegica*)	[22]

Table 1. *Cont.*

Species	Sample Type, Composition	Model, Study Design	Tested Parameters, Observed Effects *	Ref.
Vauquelinia corymbose Bonpl.	water extract; detected compounds (HPLC): prunasin, (−)-epicatechin, hyperoside, isoquercetin, quercitrin, quercetin-3-O-(6″-benzoyl)-β-galactoside, picein, methylarbutin	in vitro enzyme inhibition and molecular docking studies; control: acarbose IC_{50} = 500 µM (yeast α-glucosidase), IC_{50} = 100 µM (rat small intestinal α-glucosidase)	α-glucosidase inhibition: IC_{50} = 28.6 µg/mL (yeast α-glucosidase); the most active compound: quercetin-3-*O*-(6″-benzoyl)-β-galactoside (IC_{50} = 30 µM for yeast α-glucosidase and 430 µM for rat small-intestinal α-glucosidase; mixed-type inhibitor)	[24]

* Suggestions of what compounds were responsible for the observed effects were taken directly from the cited papers; for a critical discussion, see Section 3.4. ** Tested species from the subgenus Aria: *S. alnifolia* (Siebold & Zucc.) K.Koch., *S. folgneri* Rehder, *S. minima* Hedl., *S. norvegica* Hedl. *** Tested species from the subgenus Sorbus: *S. aucuparia* L., *S. commixta* Hedl., *S. decora* (Serg.) C.K. Schneid., *S. discolor* (Maxim.) Maxim., *S. hybrida* L., *S. meinichii* (Lindeb. ex Hartm.) Sennikov and Kurtto, *S. koehneana* C.K. Schneid., *S. vilmorinii* C.K. Schneid., *S. splendida* Hedl. ↑, increase; ↓, decrease; AGEs, advanced glycation end products; Akt, protein kinase B; ALR, aldose reductase; CE, catechin equivalents; CyE, cyanidin glucoside equivalents; DMSO, dimethyl sulfoxide; DRIP205/TRAP220, mediator of RNA polymerase II transcription subunit 1; ERK 1/2, extracellular signal-regulated kinases 1/2; Et2O, diethyl ether; EtOAc, ethyl acetate; EtOH, ethanol; FOXO1, forkhead box G1; G6Pase, glucose-6-phosphatase; G6PD, glucose-6-phosphate dehydrogenase; GAE, gallic acid equivalents; GK, glucokinase; GLUT-1, glucose transporter-1; GLUT-2, glucose transporter 2; GLUT-4, glucose transporter-4; GSK-3α, glycogen synthase kinase 3 alfa; GSK-3β, glycogen synthase kinase 3 beta; GYS2, glycogen synthase 2; HK1, hexokinase 1; HPLC, high-performance liquid chromatography; HPLC-MS, high-performance liquid chromatography coupled with mass spectrometry; iAUC, incremental area under the curve; IRS-1, insulin receptor substrate 1; IRS-2, insulin receptor substrate 2; LC-MS, liquid chromatography coupled with mass spectrometry; LPS, lipopolysaccharide; MeOH, methanol; OAE, oleanolic acid equivalents; p-Akt, phosphorylated Akt; p-AMPK, phosphorylated AMP-activated protein kinase; PB2E, procyanidin B2 equivalents; PEPCK, phosphoenolpyruvate carboxykinase; p-FOXO1, phosphorylated FOXO-1; PGC-1α, peroxisome proliferator-activated receptor gamma, coactivator 1 alpha; p-GSK-3β, phosphorylated GSK-3β; *p*-IRS-1, phosphorylated IRS-1; *p*-IRS-2, phosphorylated IRS-2; PCA, principal component analysis; PPARγ, peroxisome proliferator-activated receptor gamma; PTP1B, protein tyrosine phosphatase 1B; QE, quercetin equivalents; RE, rutin equivalents; SDH, sorbitol dehydrogenase; SGLT1, sodium–glucose transporter; SOCS3, suppressor of cytokine signalling 3.

3.2. In Vivo Animal Studies

The animal studies included different mouse, rat, or lamb models of diabetes, induced in vivo either chemically (e.g., streptozotocin, alloxan, dexamethasone) or by diet (for detailed information, see Table 2). The results suggested that fruits from *Amelanchier alnifolia, Aronia melanocarpa, Chaenomeles sinensis, Crataegus azarolus, Crataegus laevigata, Crataegus meyeri, Crataegus monogyna, Crataegus orientalis, Crataegus pinnatifida, Malus domestica, Malus pumila, Pyrus bretschneideri, Pyrus communis, Pyrus pyrifolia, Sorbus aucuparia,* and *Sorbus norvegica* can reduce blood/plasma glucose levels to varying degrees, depending on the animal model, plant species, and fruit preparation (e.g., berry powder, pomace, juices, extracts based on water, ethanol–water, or acetone–water, or their organic fractions). In the case of *Aronia melanocarpa* acidified 80% ethanol extract at 300 mg/kg/day [100], *Crataegus laevigata* 70% ethanol extract at 1200 mg/kg/day [101], *Crataegus pinnatifida* acidified 70% ethanol extract at 300 mg/kg/day [102], *Pyrus communis* 80% ethanol and ethyl acetate extracts at 200 mg/kg/day [103], and *Sorbus norvegica* 80% acetone extract at 900–1250 mg/kg/day [22], the blood glucose levels after extract supplementation were comparable with the effects of the reference anti-diabetic drugs, i.e., metformin (150–200 mg/kg/day), glipizide (10 mg/kg/day), glibenclamide (5 mg/kg/day), and acarbose (25 mg/kg/day). Moreover, the HOMA-IR index (homeostatic model assessment for insulin resistance, calculated based on glucose and insulin measurements) and QUICKI (quantitative insulin sensitivity check index) observed for *Amelanchier alnifolia* [104–107], *Cydonia oblonga* [108], *Malus domestica* [82,109], and *Pyrus pyrifolia* [31] suggested that the hypoglycaemic effects of fruits are driven by the improvement of insulin sensitivity rather than stimulation of insulin secretion. This was consistent with the blood/plasma insulin levels, which were reduced for all species and studies mentioned above, as well as for *Aronia melanocarpa* [100] and *Crataegus pinnatifida* [102,110]. Moreover, the ability of *Aronia melanocarpa* [100,111], *Crataegus monogyna* [112], *Crataegus pinnatifida* [102,110], and *Malus pumila* [113] fruits to ameliorate histological changes in pancreatic or liver cells, like hypertrophy or degradation, was also evidenced.

There were several mechanisms behind the activity observed in vivo that were tested for Maleae fruits (Table 2), including the following:

(1) The effects on intestinal absorption of glucose;
(2) The effects on skeletal, hepatic, or adipose transport of glucose;
(3) The changes in the expression of proteins involved in the insulin signalling pathway;
(4) The modulation of the activity of enzymes involved in glucose metabolism;
(5) The inhibition of glucose-derived protein damage.

(1) The impairment of intestinal glucose transporter (SGLT-1), which may decrease the absorption of sugars consumed with foods, was proven for *Malus domestica* [75]. Moreover, modification of the gut microbiota's function was also reported to be involved in reducing intestinal sugar absorption. This effect was observed for fruits from *Amelanchier alnifolia*, which were able to, i.a., alter the α-diversity and β-diversity of gut microbiota and reduce the ratio of *Firmicutes/Bacteroidetes*, which was negatively correlated with carbohydrate digestion [104–106]. Finally, the inhibition of mucosal enzymes' activity, i.e., sucrase and maltase (α-glucosidases that catalyse the hydrolysis of sucrose and maltose *O*-glycosidic bonds), was observed for the *Aronia melanocarpa* juice and extract and proposed as another mechanism of lowering the blood/plasma glucose levels by inhibiting the digestion and absorption of sugars [114,115].

(2) The intensification of hepatic (GLUT-2, GLUT-4), muscle (GLUT-4), and adipose (GLUT-1, GLUT-4) glucose transport was suggested for *Aronia melanocarpa* [35,100,116,117], *Crataegus pinnatifida* [102,118], and *Pyrus pyrifolia* [31] hydroalcoholic fruit extracts. Thus, these may increase tissue glucose uptake for further metabolism and lower the sugar level in the bloodstream.

(3) The homeostasis of glucose metabolism depends on several simultaneous ongoing processes under the control of insulin. The binding of insulin to its receptor leads to a cascade of reactions that promote glucose usage and storage by different tissues (liver, muscle,

or adipose), e.g., the stimulation of glycogen synthesis (i.e., the transformation of glucose to glycogen) and glycolysis (i.e., the conversion of glucose to pyruvate and production of ATP energy). At the same time, de novo synthesis of glucose (gluconeogenesis) and glycogenolysis (metabolism of glycogen to glucose) are suppressed [50]. However, insulin secretion is only the beginning of the chain reactions, and these processes can be controlled at multiple stages (see Figure 2). As mentioned earlier, the results of the hypoglycaemic activity studies of Maleae fruits suggested that the reduction in blood/plasma glucose levels was driven by the improvement of insulin sensitivity rather than the stimulation of insulin secretion. Thus, the effects on the expression of different proteins in the insulin signalling pathway were tested for some species. Depending on the studies, the *Aronia melanocarpa* juices or extracts were able to increase the expression of *p*-PI3K (phosphorylated phosphoinositide 3-kinase), *p*-Akt (phosphorylated protein kinase B), GYS (glycogen synthase), and GLP-1 (glucagon-like peptide 1), as well as the ratios of *p*-IRS-1(2)/IRS-1(2) (phosphorylated insulin receptor substrate 1(2)/ insulin receptor substrate 1(2)) and *p*-GSK-3β/GSK-3β (phosphorylated glycogen synthase kinase 3 beta/ glycogen synthase kinase 3 beta); they were also documented to decrease the levels of PTEN (phosphatase and tensin homolog) and SOCS3 (suppressor of cytokine signalling 3), which may result in the enhancement of glycolysis and glycogen synthesis, as well as the inhibition of gluconeogenesis and glycogenolysis [100,115–117,119]. Similar effects were observed for *Crataegus pinnatifida*—by increasing *p*-Akt, *p*-AMPK (phosphorylated AMP-activated protein kinase), *p*-IRS-1, and *p*-PI3K, and decreasing PEPCK (phosphoenolpyruvate carboxykinase) levels [102,118]; various *Crataegus* spp.—by increasing the GPC-4 (glypican-4) level [120]; *Cydonia oblonga*—by increasing *p*-AMPK and decreasing PPARγ levels [108]; and *Malus pumila*—by increasing the *p*-Akt level [113]. On the other hand, the results of gene expression studies in *Amelanchier alnifolia* berry powder suggested the enhancement of opposing processes of glycolysis and gluconeogenesis, but the final effect on glucose metabolism was still positive, i.e., oral glucose tolerance test parameters were improved in comparison to diabetic controls [121]. In this case, the answer may be not in the levels of particular enzymes, but rather in their activity, which needs further investigation.

(4) Therefore, in addition to protein expression studies, some of the reviewed papers include alternative enzyme activity tests. According to these examinations, the juice or acidified 60% ethanol extracts from *Aronia melanocarpa* [115,116,119] were able to increase the activity of glucokinase (GK) and pyruvate kinase (PK) (which may enhance the glycolysis process), inhibit the activity of enzymes involved in gluconeogenesis (e.g., PEPCK, phosphoenolpyruvate carboxykinase; G6Pase, glucose-6-phosphatase), and regulate glucose homeostasis through enzymatic termination of incretin action (DPP IV inhibition). The inhibition of G6Pase was also observed for 70% methanol macerate from *Malus* sp. [122].

(5) Finally, some mechanisms that may prevent the impairment of the function of various proteins caused by high glucose levels were also revealed. For example, it was observed that the fruit extract from *Chaenomeles sinensis* [123] decreased the levels of intermediate products of glycation, i.e., glyoxal (GO) and methylglyoxal (MG). Moreover, the extract from *Crataegus orientalis* [112] was able to inhibit the activity of the ALR enzyme, which may prevent the development of diabetic complications caused by pathological glucose metabolism (polyol pathway).

The detailed information on the summarised research results can be found in Table 2.

Table 2. In vivo animal studies of the anti-diabetic potential of fruits from the Maleae tribe.

Species	Sample Type, Composition	Model, Study Design	Tested Parameter, Observed Effects *	Ref.
Amelanchier alnifolia (Nutt.) Nutt. ex M. Roem.	berry powder; anthocyanins 5011 mg/kg of dry weight (HPLC-MS: cyanidin-3-*O*-galactoside 74%, cyanidin-3-*O*-glucoside 18%, cyanidin-3-*O*-arabinoside, cyanidin-3-*O*-xyloside)	male diabetic or C57BL/J wild-type mice (n = 5–8/group); duration: 4–5 weeks; tested: diabetic or wild-type mice supplemented with berry powder (0.2, 1, 5, 20%); control: non-supplemented mice (diabetic or wild-type)	blood glucose: ↓ in diabetic mice supplemented with berry powder compared to diabetic non-supplemented mice by 17–41%, depending on the dose (the highest changes for 5% berry powder, i.e., ~8.0 g/kg/day)	[124–126]
	berry powder; anthocyanins 5011 mg/kg of dry weight (HPLC-MS: cyanidin-3-*O*-galactoside 74%, cyanidin-3-*O*-glucoside 18%, cyanidin-3-*O*-arabinoside, cyanidin-3-*O*-xyloside)	C57BL/6J male mice (n = 8/group); duration: 10–15 weeks; tested: mice fed a high-fat, high-sucrose diet supplemented with berry powder (1, 2.5, 5%, HFHS + B1%, HFHS + B2.5%, HFHS + B5%) or cyanidin 3-*O*-glucoside (7.2 mg/kg/day), or non-supplemented mice (HFHS); control: low-fat-diet mice	fasting plasma glucose: ↓ in supplemented mice compared to HFHS, the reduction in HFHS + B2.5% (level comparable to control) was significantly lower than in the HFHS + B1% group; plasma insulin and HOMA-IR: ↓ in supplemented mice compared to HFHS (no variation by dose); effects on gut microbiota: multiple changes, i.a., altered α-diversity and β-diversity of gut microbiota (dose ≥ 2.5%), reduced ratio of *Firmicutes/Bacteroidetes* compared to HFHS; cyanidin 3-*O*-glucoside activity: similar to *Amelanchier* berry powder containing an equal amount of cyanidin 3-*O*-glucoside	[104–107]
	berry powder; flavonoid glycosides 211.79 mg/100 g, quercetin 82.34 mg/100 g, anthocyanins 281 mg/100 g, phenolic acids 108 mg/100 g (HPLC)	male Wistar rats (n = 12/group); duration: 16 weeks; tested: corn starch or high-carbohydrate, high-fat diet + berry powder (26.83 g/kg of food = cyanidin glucoside 8 mg/kg/day) (CS + B or HFHC + B); control: corn starch (CS) or high-carbohydrate, high-fat diet (HFHC)	blood glucose: higher in HFHC and HFHC + B groups compared to CS and CS + B (dependence only on diet; no effect of fruit supplementation); OGTT: ↓ iAUC (0–120 min) of blood glucose in CS + B compared to CS, ↓iAUC in HFHC + B compared to HFHC, higher in both HFHC and HFHC + B compared to CS and CS + B (dependence on both diet and supplementation); gene expression: normalisation of HK1, GP (to a level comparable to controls), and ↑G6Pase	[121]

Table 2. *Cont.*

Species	Sample Type, Composition	Model, Study Design	Tested Parameter, Observed Effects *	Ref.
Aronia melanocarpa (Michx.) Elliott	juice; anthocyanins 5.986 nmol/mL (HPLC: cyanidin 3,5-diglucoside, cyanidin 3-*O*-arabinoside, cyanidin 3-*O*-galactoside, cyanidin 3-*O*-glucoside)	KKAy male mice (n = 5/group); duration: 49 days; tested: *Aronia* juice (A) or cyanidin 3,5-diglucoside (10 µg/mL solution); control: water-drinking mice	blood glucose: ↓ by about 61% (21 days) and by about 42% (49 days); HbA1c: ↓ by about 33% (21 days) and by about 44% (49 days); DPP IV activity in the serum: ↓ by 62% (49 days); serum active GLP-1 level: about 10-fold ↑ (49 days); cyanidin 3,5-diglucoside: similar but weaker effects compared to *Aronia* juice	[119]
	juice; proteins, carbohydrates, fats, minerals, fibres and energy density determined	C57BL/6JmsSlc or KKAy male mice (n = 5/group); duration: 28 days; tested: *Aronia*-drinking mice; control: water-drinking mice	blood glucose: ↓ by about 55% (only KKAy mice); serum insulin: ↓ (only KKAy mice); DPP IV activity in the serum and liver: ↑ (only C57BL/6JmsSlc mice); DPP IV activity in the intestine: ↓ by about 35% in the 'upper small intestine' and about 46% in the 'lower small intestine' (KKAy mice); α-glucosidase activity in the upper region of the small intestine: ↓ by about 58% (KKAy mice); expression level: ↓ GIP in the "upper small intestine" and ↑ GLP-1 expression in the "lower small intestine" (KKAy mice)	[115]
	juice; anthocyanins 9.572 mg/g, procyanidins 5.328 mg/g, flavonols 3.089 mg/g, hydroxycinnamic acids 2.71 mg/g (HPLC)	C57BL/6J male mice (n = 10/group); duration: 12 weeks; tested: low-fat, high-sucrose, and high-fat diet supplemented with *Aronia* juice concentrate (1.44 g anthocyanins/kg of diet); control: non-supplemented mice	plasma glucose and insulin: no differences between groups	[127]
	standardised *Aronia* berry extract powder (Fort Wayne, IN, USA); polyphenols 40%, anthocyanins 15%	C57BL/6J male mice (n = 10/group); duration: 14 weeks; tested: high-fat, high-sucrose diet + 0.2% (*w/w*) *Aronia* extract; control: low-fat (NC) or high-fat, high-sucrose diet (HFHS)	fasting blood glucose: ↓ compared to HFHS group; PPARγ and GLUT-4 mRNA expression levels in adipocyte fraction: ↑ compared to HFHS group (not statistically significant); anthocyanins responsible for observed effects (based on literature studies and extract composition)	[35]

Table 2. *Cont.*

Species	Sample Type, Composition	Model, Study Design	Tested Parameter, Observed Effects *	Ref.
	anthocyanin purified powder extract (80% EtOH + 0.1% HCl + purification); anthocyanins 986.48 mg/g (cyanidin monoglycosides)	C57BL/6J male mice (n = 10/group); STZ-induced diabetes; duration: 5 weeks; tested: high-fat diet, diabetic + 150/300 mg/kg *Aronia* extract (D/A150 and D/A300) or 200 mg/kg metformin (M); control: low-fat (NC) or high-fat diet, diabetic (DC)	blood glucose: dose-dependent ↓ compared to DC; serum insulin and HbA1c: dose-dependent ↓ compared to DC; liver glycogen: ↑ compared to DC (D/A300 and M); hepatic protein expression: ↑ *p*-GSK-3β (all groups), ↑ GLUT-4 (D/A300 and M), ↓ SOCS3, (all groups), unchanged GSK-3β and IRS-1 compared to DC; anthocyanins responsible for observed effects (testing of pure anthocyanin fraction activity)	[100]
	70% EtOH extract; detected compounds (HPLC-MS): gallic acid, chlorogenic acid, quercetin, kaempferol	male ICR mice (n = 8/group); STZ-induced diabetes; duration: 31 days; tested: *Aronia*-extract-treated diabetic mice (10 or 100 mg/kg, D/A10 or D/A100); control: normal (NC) or diabetic (DC) mice	serum glucose: ↓ compared to DC (D/A100 = 208.60 ± 31.05 mg/dL, DC mice = 486.60 ± 81.94 mg/dL, NC mice = 175.67 ± 10.60 mg/dL); serum insulin: ↑ compared to DC (D/A100 = 2.50 ± 0.39 ng/mL, DC mice = 1.34 ± 0.54 ng/mL, NC mice = 2.66 ± 0.36 ng/mL); effect on tissue injury: *Aronia* extract treatment attenuated histological changes induced by STZ in the liver and pancreatic tissues	[111]
	juice; anthocyanins 240 mg/100 mL (HPLC)	male Wistar rats (n = 12/group); duration: 16 weeks; tested: maize starch + *Aronia* juice (9.4 mg anthocyanins/kg/day) or high-carbohydrate, high-fat diet + *Aronia* juice (7.8 mg anthocyanins/kg/day); control: maize starch + water (C), high-carbohydrate, high-fat diet + water (HFHC)	OGTT: iAUC (0–120 min) of blood glucose ↑ in HFHC (786 mmol/L min), ↓ in *Aronia* groups (658 or 648 mmol/L min) compared to C rats (591 mmol/L min); plasma insulin: ↑ in HFHC (4.1 µmol/L), ↓ in *Aronia* groups (2.3 or 1.1 µmol/L) compared to C rats (1.4 µmol/L); anthocyanins responsible for observed effects (based on juice composition and literature studies)	[128]
	60% EtOH commercial extract containing at least 10% anthocyanins	male Wistar rats (n = 6/group); duration: 6 weeks; tested: fructose-rich diet + 100 or 200 mg/kg *Aronia* extract dissolved in water; control: fructose-rich diet + water	blood glucose and insulin: ↓ by about 10% (glucose) and 30% (insulin) regardless of dosage; expression of proteins: ↑ mRNA levels of IRS-1, IRS-2, PI3KR1, GLUT-1, GLUT-4, and GYS (by 1.5–2.3-fold at 200 mg/kg dosage), and ↓ mRNA levels of PTEN and GSK-3β (by 0.61–0.62-fold at 200 mg/kg)	[117]

Table 2. *Cont.*

Species	Sample Type, Composition	Model, Study Design	Tested Parameter, Observed Effects *	Ref.
	juice; ascorbic acid 29 g/L, anthocyanins 1.3 mg/mL, carotenes 97.8 µg/L, polyphenols 31.84 g/L	male Wistar albino rats (n = 10/group); alloxan-induced diabetes; duration: 6 weeks; tested: normal *Aronia*-drinking rats (N/A, 10 mL/kg), diabetic *Aronia*-drinking rats (D/A); control: diabetic water-drinking rats (DC), normal water-drinking rats (NC)	blood glucose: ↓ for D/A rats compared to DC rats (D/A = 136.8 ± 14.6 mg/dL, DC = 220.4 ± 23.5 mg/dL, NC mice = 103.5 ± 11.2 mg/dL)	[129]
	juice; polyphenols 709.3 mg/100 mL; flavonoids 189.4 mg/100 mL; anthocyanins 106.8 mg/100 mL, L-ascorbic acid 3.0 mg/100 ml	male Wistar rats (n = 6/group); STZ-induced diabetes; duration: 6 weeks; tested groups: normal (N/A10 and N/A20) or diabetic (D/A10 and D/A20) rats drinking *Aronia* juice (10 or 20 mL/kg); control: normal (NC) or diabetic (DC) rats drinking water (10 mL/kg)	plasma glucose: DC rats = 17.5 ± 2.9 mmol/l; NC rats = 7.2 ± 0.6 mmol/l; N/A10 and N/A20 rats—no significant changes compared to NC rats; D/A10 and D/A20 rats—↓ to levels comparable to NC rats (by 44% and 42% compared to DC rats, respectively); anthocyanins and flavonoids responsible for observed effects (based on juice composition and literature studies)	[130]
	commercial extract; polyphenols 714.1 mg/g, including anthocyanins 56.6% (cyanidin monoglycosides), flavanols 21.6% (procyanidins, epicatechin), phenolic acids 14.7% (chlorogenic and neochlorogenic acids), flavonols 7.1% (quercetin glycosides) (HPLC)	male Wistar rats (n = 8/group); STZ-induced diabetes; duration: 4 weeks; tested: diabetic rats with diet modified by 8% lard and 65% fructose + *Aronia* extract (0.2%); control: diabetic (DC) or normal rats (NC) fed a standard casein diet enriched with 0.5% cholesterol	activity of microbial enzymes in the caecal digesta: β-glucuronidase activity ↑ in the DC group compared to NC, and ↓ in the *Aronia* group to a level no different vs. NC; the activity of α- and β-glucosidase and α- and β-galactosidase did not differ; mucosal disaccharidase activities: sucrase and maltase activity comparable between the NC and *Aronia* groups and ↑ in DC; in the case of lactase the trend was the opposite; serum glucose: no differences between the *Aronia* and NC groups, remarkably ↑ in the DC group	[114]

Table 2. *Cont.*

Species	Sample Type, Composition	Model, Study Design	Tested Parameter, Observed Effects *	Ref.
	60% EtOH: 0.1% HCl (1:20) extract, purified and concentrated; anthocyanins 254.72 mg/g (cyanidin monoglycosides), flavonoids 11.66 mg/g (quercetin glycosides), phenolics acids 75.97 mg/g (caffeic, chlorogenic, and neochlorogenic acids) (HPLC-MS)	male Wistar rats (n = 6/group); STZ-induced diabetes; duration: 8 weeks; tested: diabetic rats fed a high-fat diet and 100 or 400 mg/kg *Aronia* extract (D/A100 or D/A400); control: normal rats, standard diet (NC) and diabetic rats, high-fat diet (DC)	OGTT: iAUC (0–120 min) of blood glucose ↓ by 24.40% in the D/A400 rats compared to DC rats; blood glucose: ↓ by 2.76 and 4.25 mmol/L compared to DC rats; serum insulin: ↓ compared to DC rats (statistically insignificant); HOMA-IR: 3.26 ± 0.56 (NC), 15.42 ± 4.17 (DC), 11.24 ± 3.10 (D/A100), 8.12 ± 1.94 (D/A400); hepatic glycogen: ↑ compared to DC rats; liver enzyme activity: GK and PK ↑ compared to DC rats, PEPCK and G6Pase ↓ compared to DC rats; hepatic protein expression: ↑ *p*-IRS-2, *p*-PI3K, *p*-Akt, *p*-GSK-3β, and GLUT-2 (by 2.03–4.02-fold for D/A400), and ↓ IRS-2 and GSK-3β (by 1.53–2.76-fold for D/A400), compared to DC	[116]
	water extract; anthocyanins 579.1 mmol/g (cyanidin monoglycosides), (+)-catechin 10.7 mmol/g, chlorogenic acid 32.7 mmol/g, caffeic acid 13.9 mmol/g (HPLC)	male Sprague-Dawley rats (n = 6/group); duration: 4 weeks; tested: high-fat diet + *Aronia* extract (17.4 g extract/kg of diet); control: non-supplemented rats	serum glucose: ↓ (8.20 ± 0.31 mM/dL) compared to controls (9.17 ± 0.38 mM/dL); OGTT: iAUC (0–120 min) of blood glucose ↓ by 20.27%; anthocyanins, proanthocyanidins, and chlorogenic acid responsible for observed effects (based on extract composition and literature studies)	[131]
	pomace	Polish Merino health lambs (n = 8/group); duration: 90 days; tested: 150 g or 300 g (A150 or A300) of *Aronia* pomace/kg of the feed mixture (twice a day, intake monitored and adjusted to the lambs' growing period); control: non-supplemented lambs	serum glucose: A150 = 2.42 ± 0.31 mmol/L, A300 = 1.55 ± 0.66 mmol/L, control = 3.38 ± 0.31 mmol/L; total phenolic contents in the liver or serum: A150 = 7.49 mg GAE/g of liver and 2.41 mg GAE/mL of serum; A300 = 7.46 mg GAE/g of liver and 3.93 mg GAE/mL of serum; control = 4.42 mg GAE/g of liver and 1.77 mg GAE/mL of serum	[132]

Table 2. *Cont.*

Species	Sample Type, Composition	Model, Study Design	Tested Parameter, Observed Effects *	Ref.
Chaenomeles sinensis (Thouin) Koehne	EtOH extract with removed sugars; polyphenols 350 mg GAE/g, including procyanidins (>90%)	KKAy male mice (n = 4–5/group); duration: 4 weeks; tested: diabetic mice fed a high-fat diet + extract (14.3 g/kg diet); control: diabetic mice fed a high-fat diet (DC)	OGTT: ↓ blood glucose after 15 min, iAUC (0–120 min) unchanged; blood glucose: ↓ after 1–4 weeks; HbA1c: unchanged (↓, but not significantly); intermediate products of glycation: 3-DG level not affected, GO and MG levels ↓ about twofold; procyanidins responsible for observed effects (based on extract composition and literature studies)	[123]
Crataegus azarolus var. aronia L.	water decoction of unripe fruits	female Sprague-Dawley rats (n = 10/group); STZ-induced diabetes; duration: 24 days; tested: diabetic (D) or normal (N) rats + *Crataegus* decoction of 50 or 350 mg fruits/mL (D50, D350, N50, N350); controls: normal rats (NC) and diabetic rats (DC)	blood glucose: ↓ in N350 compared to NC rats, ↓ in D50 and D350 compared to DC rats	[133]
Crataegus laevigata (Poir.) DC.	70% EtOH extract	Sprague-Dawley rats (n = 12/group); STZ-induced diabetes; duration: 6 h; tested: diabetic rats + 200, 400, 600, 800, 1000, 1200 mg/kg *Crataegus* extract; controls: normal rats (NC), diabetic rats (DC), and diabetic rats + glipizide (10 mg/kg)	blood glucose: dose-dependent ↓ compared to DC, hypoglycaemic effect of *Crataegus* extract at 1200 mg/mL comparable to glipizide; OGTT: effective ↓ in glucose levels after 30, 60, and 90 min (dosage 800 and 1200 mg/mL)	[101]
Crataegus meyeri Pojark.	water extract from fresh pulp	male Wistar rats (n = 10/group); STZ-induced diabetes; duration: 10 weeks; tested: diabetic rats + 300 mg/kg *Crataegus* extract; controls: normal rats (NC) and diabetic rats (DC)	insulin level and QUICKI: unchanged; glucose: ↓ compared to DC (NC 118 ± 32 mg/dL, DC 538 ± 47 mg/dL, *Crataegus* 348 ± 41 mg/dL); HOMA-IR: ↓ (NC 130 ± 19, DC 177 ± 16, *Crataegus* 119 ± 21)	[134]

Table 2. *Cont.*

Species	Sample Type, Composition	Model, Study Design	Tested Parameter, Observed Effects *	Ref.
Crataegus monogyna Jacq.	70% MeOH extract	male Wistar rats (n = 8/group); STZ-induced diabetes; duration: 3 weeks; tested: diabetic rats + 100, 200 or 400 mg/kg *Crataegus* extract; controls: normal rats (NC) and diabetic rats (DC)	serum glucose: ↓ compared to DC rats (to levels comparable to NC; all dosages); effects on pancreatic tissue: *Crataegus* extract (at 200 and 400 mg/kg) ameliorated the degeneration of pancreatic acinar cells; damage to lobules, acini, and oedema observed in the tissue of DC rats	[135]
Crataegus orientalis subsp. *presliana* K.I.Chr.	90% EtOH extract	Kunming mice (n = 8/group); STZ-induced diabetes; duration: 4 weeks; tested: diabetic mice + extract 1.8 g/kg; controls: normal (NC) or diabetic mice (DC)	blood glucose: ↓ compared to DC (statistically insignificant); ALR activity: ↓ compared to DC (statistically insignificant)	[112]
Crataegus pinnatifida Bunge	proanthocyanidin fraction from 70% EtOH extract; procyanidins 81.85 mg/100 mg (LC-MS: epicatechin 36.12%; procyanidins B2, B5, and C1)	male Wistar rats (n = 12/group); duration: 8 weeks; tested: high-fat diet + 50, 100, or 200 mg/kg *Crataegus* preparation; controls: standard diet (NC) or high-fat diet (HFD)	fasting blood glucose: lack of differences (tested and control groups); OGTT: glucose level ↓, iAUC (0–120 min) ↓ compared to HFD; insulin: dose-dependent ↓ compared to HFD; alleviation of liver histopathological changes compared to HFD; procyanidins responsible for observed effects (based on activity testing of extract fraction)	[110]
	MeOH extract; total phenolics 66.94 mg GAE/g	male Sprague-Dawley rats (n = 9/group); STZ-induced diabetes; duration: 2 weeks; tested: diabetic rats fed a high-fat diet + *Crataegus* extract at 50, 100, or 200 mg/kg; controls: normal rats, standard diet (NC); diabetic rats, high-fat diet (DC); diabetic rats, high-fat diet + orlistat (40 mg/kg)	blood glucose: dose-dependent ↓ compared to DC; 5.62 ± 0.39 mmol/L (NC), 20.25 ± 1.9 mmol/L (DC), 10.5—17.9 mmol/L (*Crataegus*), about 10 mmol/L (orlistat); serum insulin: dose-dependent ↓ compared to DC; 147.74 ± 15.61 pg/mL (NC), 238.59 ± 21.01 pg/mL (DC), 176.82—206.76 (*Crataegus*), about 300 pg/mL (orlistat)	[136]

Table 2. *Cont.*

Species	Sample Type, Composition	Model, Study Design	Tested Parameter, Observed Effects *	Ref.
Crataegus pinnatifida var. major N.E.Br.	70% acidic EtOH (0.1% HCl) extract; polyphenols 594.15 mg/g, including chlorogenic acid 15.2%, procyanidin B2 20.3%, epicatechin, proanthocyanidin B-type oligomers, cyanidin 3-galactoside, and quercetin glycosides (HPLC-MS)	male Wistar rats (n = 8/group); STZ-induced diabetes; duration: 12 weeks; tested: diabetic rats fed a high-fat diet + 300 mg/kg *Crataegus* extract; controls: normal rats, standard diet (NC); diabetic rats, high-fat diet (DC); diabetic rats, high-fat diet + metformin (150 mg/kg)	blood glucose: 44.2% ↓ compared to DC rats (likewise for metformin); OGTT: glucose level ↓ after 30 min (17.32 mmol/L) compared to DC rats (25.80 mmol/L), comparable to metformin (15.73 mmol/L); serum insulin: 34.1% ↓ compared to DC rats (about 50% reduction for metformin); improved histology of skeletal muscle, liver, and aortic vessels; phosphorylation of proteins: ↑ *p*-GLUT-4, *p*-IRS-1, *p*-Akt, and *p*-PI3K in the liver; ↑ *p*-IRS-1 and *p*-Akt but no effects on *p*-PI3K in the skeletal muscle (comparable to metformin)	[102]
	n-butanol fraction of 80% MeOH extract	C57BL/6J male mice (n = 5–9/group); duration: 8 weeks; tested: high-fat diet + *Crataegus* extract at 200, 500, or 1000 mg/kg (CE200–1000); controls: low-fat diet (NC), high-fat diet (HF), and high-fat diet + rosiglitazone (10 mg/kg, Rosi10)	plasma glucose and insulin: ↓ compered to HF, comparable to NC in the CE1000 and Rosi10 groups; insulin resistance test: ↓ compared to HF (significantly in the CE500, CE1000, and Rosi10 groups); GLUT-4 expression (skeletal muscle): the mRNA level ↓ in HF and ↑ in CE200–1000 and Rosi10 compared to NC; protein expression in the liver: ↓ PEPCK (CE500, -1000, and Rosi10) compared to HF, ↑ *p*-AMPK (all groups) compared to NC and HF; OGTT: ↓ in glucose levels after 30 min in low-fat diet mice supplemented with 200 mg/kg extract compared to NC, and after 30, 60, 90, 120, and 180 min in mice supplemented with 1000 and 2000 mg/kg	[118]

Table 2. *Cont.*

Species	Sample Type, Composition	Model, Study Design	Tested Parameter, Observed Effects *	Ref.
Crataegus L. sp.	water extract from dried pulp; polyphenols 20.18 mg GAE/g, flavonoids 8.50 mg QE/g	male Wistar rats (n = 6–8/group); STZ-induced diabetes; duration: 14 days; tested: diabetic rats + extract at 100, 300, or 1000 mg/kg; controls: normal rats (NC) and diabetic rats (DC)	blood glucose: ↓ compared to DC at 100 and 300 mg/kg; unchanged at 1000 mg/kg	[137]
	water extract from dried pulp	male Wistar rats (n = 10/group); STZ-induced diabetes; duration: 10 weeks; tested: diabetic rats + extract at 100 mg/kg ± resistance training; controls: normal rats (NC) and diabetic rats (DC)	blood glucose: unchanged compared to DC; fastening serum insulin: ↑ compared to DC (NC 25.0 µIU/mL; DC 6.93 µIU/mL, extract 8.12 µIU/mL)	[138]
	80% EtOH extract; detected compounds (HPLC): chlorogenic acid, catechin, epigallocatechin gallate, quercetin, kaempferol and apigenin derivatives (including hyperoside and vitexin)	male Wistar rats (n = 8/group); STZ-induced diabetes; duration: 12 weeks; tested: diabetic rats + extract at 100 mg/kg ± resistance training; controls: normal rats (NC) and diabetic rats (DC)	blood glucose: ↓ compared to DC (NC 6.78 mmol/L; DC 22.98 mmol/L, extract 17.96 mmol/L); serum insulin: ↑ compared to DC (NC 10.25 µU/mL; DC 4.98 µU/mL, extract 7.67 µU/mL); GPLD1 serum level: ↓ compared to DC; GPC-4 serum level: ↑ compared to DC (statistically insignificant); HOMA-IS: ↑ compared to DC (statistically insignificant)	[120]
Cydonia oblonga Mill.	30% EtOH extract; chlorogenic acid 0.75 mg/g (HPLC)	C57BL/6N mice (n = 10/group); duration: 8 weeks; tested: high-fat diet + extract at 50, 100, or 200 mg/kg; controls: normal diet (NC) and high-fat diet (HFD),	serum glucose: unchanged; serum insulin: ↓ compared to HFD (NC 2.62 ng/mL, HFD 9.12 ng/mL, *Cydonia* 5.67–6.51 ng/mL); HOMA-IR: ↓ compared to HFD (all dosages); QUICKI: ↑ compared to HFD at 200 mg/kg; *p*-AMPK/AMPK ratio ↑ about twofold compared to HFD (200 mg/kg), PPARγ mRNA expression ↓ about twofold (all dosages) chlorogenic acid responsible for observed effects (based on extract composition and literature studies)	[108]

Table 2. *Cont.*

Species	Sample Type, Composition	Model, Study Design	Tested Parameter, Observed Effects *	Ref.
Malus domestica (Suckow.) Borkh.	commercial extract (Biosearch S.A.); polyphenols 80%, phlorizin min. 5% (HPLC)	male Wistar rats (n = 12/group); duration: 56 days; tested: rats fed a high-fat, high-sucrose diet + extract at 700 mg/kg (HFS + M); controls: standard diet (NC) and high-fat, high-sucrose diet (HFS)	serum glucose: ↓ compared to HFS (HFS + M 5.50 ± 0.22 mmol/L, HFS 6.36 ± 0.28 mmol/L, NC 5.45 ± 0.23 mmol/L); serum insulin: ↓ compared to HFS (HFS + M 14.6 μU/mL, HFS 22.9 μU/mL, NC 9.12 μU/mL); HOMA-IR: ↓ compared to HFS (HFS + M 3.45, HFS 6.65, NC 2.18)	[109]
	juice and peel water extracts; polyphenols 69.68 ± 2.17 mg GAE/100 mL (juice), 673.46 ± 6.90 mg GAE/ 100 g dry extract	male Wistar rats (n = 8/group); STZ-induced diabetes; duration: 21 days; tested: diabetic rats + apple juice (15 mL/kg) or peel extract (1 g/kg); controls: normal rats (NC) and diabetic rats (DC)	blood glucose: ↓ compared to DC after the first, second, and third weeks, to levels comparable to NC (juice: 87.87–134.83 mg/dL, extract: 88.00–105.82 mg/dL, DC: 374.67–432.14 mg/dL, NC: 81.53–89.57 mg/dL)	[139]
	extract dissolved in water	Wistar rats (n = 5/group); fructose-induced diabetes; duration: 28 days; tested: diabetic rats + apple extract (500 mg/kg); controls: normal rats (NC) and diabetic rats (DC)	blood glucose: ↓ compared to DC, to levels comparable to NC (apple: 81.00 ± 3.61 mg/dL, DC: 116.70 ± 9.13 mg/dL, NC: 76.33 ± 3.84 mg/dL)	[140]
	commercial extract (BioActive Food GmbH); polyphenols 44% (catechin equivalents), phlorizin 16%, quercetin 12.43%, chlorogenic acid 5.57% (HPLC/spectrophotometry)	male C57BL/6N mice (n = 10/group); duration: acute consumption; tested: mice fed a high-fat diet + 12.24 mg of extract or 1.96 mg of phlorizin; controls: normal diet (NC) and high-fat diet (HFC)	OGTT: iAUC (0–15, 0–30, 0–60 min) ↓ compared to HFC (to levels comparable to NC), with the highest reduction after 15 min (likewise for phlorizin); glucose uptake in everted jejunal rings: inhibited (EC_{50} = 8.9 ± 2.2 μg/mL for extract and 4.2 ± 0.6 μM for phlorizin); glucose uptake (everted jejunal sacs): SGLT1 inhibited (reversible manner)	[75]

Table 2. *Cont.*

Species	Sample Type, Composition	Model, Study Design	Tested Parameter, Observed Effects *	Ref.
	peel extract, purified, sugar-free; polyphenols 606 mg GAE/g (HPLC: chlorogenic acid, epicatechin, quercetin glycosides), anthocyanins 7.17 mg/g	C57BL/6J male mice (n = 8/group); duration: 10 weeks; tested: high-fat diet + 0.2% (*w/w*) extract or quercetin; controls: low-fat diet (NC) and high-fat diet (HFC)	fasting blood glucose: ↓ compared to HFC (comparable to NC); IPGTT: ↓ iAUC (0–2 h) of blood glucose compared to HFC; serum insulin: ↓ compered to HFC; activity of quercetin: comparable to apple extract	[141]
	juice (fermented or not); compounds: 19 polyphenols (HPLC: mainly chlorogenic acid about 95–145 mg/L, epicatechin 70–105 mg/L, procyanidin B2 70–95 mg/L), sugars, and organic acids	C57BL/6J mice; STZ-induced diabetes; duration: 4 weeks; tested: high-fat diet + juice (fermented or unfermented); controls: normal mice, normal diet (NC); diabetic mice, high-fat diet (DC); diabetic mice, acarbose (A)	fasting blood glucose: ↓ compared to DC (both fermented and unfermented juice), for fermented juice and acarbose to levels comparable to NC; insulin and HOMA-IR: similar trend as for glucose levels; QUICKI: ↑ compared to DC (fermented and unfermented juice), for fermented juice and acarbose to levels comparable to NC	[82]
Malus Mill. sp.	70% MeOH macerate	albino mice; duration: 22 days; tested: plant extract 1 g/kg, 2 g/kg, or 3 g/kg; control: non-supplemented mice	G6Pase activity: dose-dependent inhibition (activity at 1 g/kg = 20.7–22.0, at 2 g/kg = 10.4–13.4, at 3 g/kg = 1.3–2.4, compared to control = 70.35), higher than observed for, e.g., mulberry fruit extract	[122]

Table 2. *Cont.*

Species	Sample Type, Composition	Model, Study Design	Tested Parameter, Observed Effects *	Ref.
Malus pumila Mill.	procyanidin fraction from juice (without chlorogenic acid and phlorizin)	male B6.Cg-Lepob/J mice (C57BL/6J background) (n = 6/group); obese, insulin-resistant, moderate hyperglycaemia; duration: 4 weeks; tested: 0.5% extract dissolved in water, ad libitum, control: non-supplemented obese mice	OGTT (0–120 min): blood glucose ↓ at 15 and 30 min, serum insulin unchanged; insulin TT (0–120 min): blood glucose ↓ at 15, 30, 45, and 60 min; HOMA-IR: ↓ (27.3 ± 7.9) compared to control mice (76.0 ± 13.3); pancreatic islet size: β-cell area ↓ by about 21%, number of islets unchanged (↓ in pancreatic cells' hypertrophy); pyruvate TT: glucose ↓ at 15–30 min (↓ in hepatic gluconeogenesis); protein phosphorylation: ↑ *p*-Akt; suppression of hepatic inflammation resulting in insulin signalling improvement	[113]
	commercial extract (Exxentia®); polyphenols 57.5%, including phlorizin (9.9%), chlorogenic acid (15.8%), quercetin (0.4%)	obese male Zucker fatty rats—insulin-resistant model (n = 10/group); duration: acute consumption (for TT) and 4 weeks; tested: (TT) for first meal maltodextrin + 150 mg extract/kg, for second meal maltodextrin only, (chronic) standard diet + 128 mg extract/kg; control: (TT) maltodextrin, (chronic) standard diet	acute meal TT: iAUC (0–120 min) ↓ for glucose and unchanged for insulin; second acute meal TT: iAUC (0–240 min) for glucose ↓, glucose levels higher compared to tested rats after first meal; chronic effect on insulin sensitivity: iAUC (0–180) ↓ for both glucose and insulin levels, insulin sensitivity ↑ (glucose infusion rate required to establish euglycaemia ↑ by 45%)	[30]
Pyrus bretschneideri Rehder	60% MeOH extract from peel and pulp; polyphenols about 2.9/8.1 mg GAE/g pulp/peel (HPLC-MS: catechin; epicatechin; rutin; chlorogenic, *p*-coumaric, vanillic, gallic, and ferulic acids), flavonoids about 1.5/6.3 mg RE/g, terpenes (oleanolic and ursolic acids) about 0.9/4.3 mg OAE/g	male Kunming mice (n = 10/group); STZ-induced diabetes; duration: 3 weeks; tested: diabetic mice, high-fat diet + 500 mg/kg peel or pulp extract; controls: normal mice, normal diet (NC); diabetic mice, high-fat diet (DC)	blood glucose: ↓ (8.2–8.6 mmol/L) in the peel group after 2–3 weeks compared to DC (14.7–16.0 mmol/L); OGTT: ↓ blood glucose in the peel group compared to DC, ↓ iAUC (0–3 h)	[90]

Table 2. *Cont.*

Species	Sample Type, Composition	Model, Study Design	Tested Parameter, Observed Effects *	Ref.
Pyrus communis L.	EtOAc and 80% EtOH extracts; phytochemical screening: carbohydrates, phenolics, tannins, and flavonoids	Wistar rats (n = 6/group); dexamethasone-induced diabetes; duration: 11 days; tested: diabetic rats + 200 mg/kg EtOAc or 80% EtOH; controls: normal rats (NC), diabetic rats (DC), and diabetic rats + glibenclamide (5 mg/kg)	OGTT: ↓ blood glucose after 60 min compared to DC (comparable to glibenclamide); urine sugar: significant levels in DC rats, trace amounts in tested groups and glibenclamide controls; blood glucose: ↓ from 3rd to 11th day compared to DC (comparable to glibenclamide)	[103]
Pyrus pyrifolia (Burm.f.) Nakai	80% EtOH + 70% acetone extract from different cultivars; polyphenols 20.9–28.5 mg CE/g	male DDY mice (n = 8/group); duration: acute consumption; tested: 250 or 500 mg/kg Pyrus extract; control: non-supplemented mice	oral starch TT: ↓ blood glucose after 30 min at 250 mg/kg and after 30, 60, and 120 min at 500 mg/g; ↓ iAUC (2 h) at 500 mg/kg	[96]
	50% EtOH extract from pomace	C57BL/6J male mice (n = 10/group); duration: 8 weeks; tested: high-fat diet + 200 or 400 mg/kg extract; control: high-fat diet + water	blood glucose: unchanged; insulin: ↓ at 400 mg/kg (not significantly) HOMA-IR: ↓ at 400 mg/kg; protein expression: ↑ *p*-IRS-1 (Tyr632), ↑ GLUT-4, ↓ *p*-IRS-1 (Ser307)	[31]
Sorbus aucuparia L.	EtOH extract	Kunming mice (n = 8/group); STZ-induced diabetes; duration: 4 weeks; tested: high-glucose, high-fat diet mice + extract at 10, 50, or 100 mg/L; controls: normal-diet mice (NC) and high-glucose, high-fat-diet mice (HGFD)	blood glucose: ↓ in the extract-fed mice compared to HGFD (dose-dependent, about twofold ↓ at 100 mg/L); IPGTT: ↓ serum glucose in the extract-fed mice compared to HGFD (dose-dependent, at 50 mg/L significant ↓ after 15 min, at 100 mg/L significant ↓ after 5–120 min)	[142]

Table 2. *Cont.*

Species	Sample Type, Composition	Model, Study Design	Tested Parameter, Observed Effects *	Ref.
Sorbus norvegica Hedl.	80% acetone extract; detected compounds (NMR): chlorogenic and neochlorogenic acids, carbohydrates	C57BL/6J male mice (n = 6–10/group); STZ-induced diabetes; duration: acute consumption (3.5 h); tested: mice fed starch (2 g/kg) and berry extract (600, 900, or 1250 mg/kg) or mice fed glucose (2 g/kg) and berry extract (1250 mg/kg); controls: mice fed starch and acarbose (25 mg/kg, positive control) or mice fed starch/glucose (negative control)	oral starch TT: ↓ maximal blood glucose compared to negative controls; the activity of berry extract at 900–1250 mg/kg was comparable to that of acarbose; for extract at 1250 mg/kg, ↓ iAUC; OGTT: for extract at 1250 mg/kg, ↓ in blood glucose after 30 min compared to the negative control	[22]

* Suggestions of what compounds were responsible for the observed effects were taken directly from the cited papers; for a critical discussion, see Section 3.4. ↑, increase; ↓, decrease; 3-DG, 3-deoxyglucosone; Akt, protein kinase B; ALR, aldose reductase; C3G, cyaniding 3-O-glucoside; CE, catechin equivalents; DPP IV, dipeptidyl peptidase IV; EtOAc, ethyl acetate; EtOH, ethanol; G6Pase, glucose-6-phosphatase; GAE, gallic acid equivalents; GIP, glucose-dependent insulinotropic polypeptide; GK, glucokinase; GLP-1, glucagon-like peptide 1; GLUT-1, glucose transporter 1; GLUT-2, glucose transporter 2; GLUT-4, glucose transporter-4; GO, glyoxal; GP, glycogen phosphorylase; GPC-4, glypican-4; GPLD1, insulin-regulated glycosylphosphatidylinositol-specific phospholipase D; GSK-3β, glycogen synthase kinase 3 beta; GYS, glycogen synthase; GYS2, glycogen synthase 2; HbA1c, glycated haemoglobin; HK1, hexokinase 1; HOMA-IR, the homeostasis model assessment-estimated insulin resistance; HOMA-IS, the homeostasis model assessment of insulin secretion; HPLC, high-performance liquid chromatography; HPLC-MS, high-performance liquid chromatography coupled with mass spectrometry; iAUC, incremental area under the curve; ICR, Institute of Cancer Research; IPGTT, intraperitoneal glucose tolerance test; IRS-1, insulin receptor substrate 1; IRS-2, insulin receptor substrate 2; LC-MS, liquid chromatography coupled with mass spectrometry; MeOH, methanol; MG, methylglyoxal; OAE, oleanolic acid equivalents; OGTT, oral glucose tolerance test; *p*-Akt, phosphorylated Akt; *p*-AMPK, phosphorylated AMP-activated protein kinase; *p*-GLUT-4, phosphorylated GLUT-4; PEPCK, phosphoenolpyruvate carboxykinase; PFK1, phosphofructokinase 1; *p*-GSK-3β, phosphorylated GSK-3β; PI3K, phosphoinositide 3-kinase; PI3KR1, phosphatidylinositol 3 kinase regulatory subunit 1; *p*-IRS-1, phosphorylated IRS-1; *p*-IRS-2, phosphorylated IRS-2; PK, pyruvate kinase; PPARγ, peroxisome proliferator-activated receptor gamma; *p*-PI3K, phosphorylated PI3K; PTEN, phosphatase and tensin homolog; QUICKI, quantitative insulin sensitivity check index; RE, rutin equivalents; SGLT1, sodium–glucose transporter; SOCS3, suppressor of cytokine signalling 3; STZ, streptozotocin; TT, tolerance test.

3.3. Human Studies

The in vivo human studies on the anti-diabetic potential of fruits from the Maleae tribe referred to five species, i.e., *Aronia melanocarpa*, *Malus domestica*, *Malus sylvestris*, *Malus pumila*, and *Pyrus pyrifolia* (Table 3).

The 1–3-month *Aronia melanocarpa* juice or extract supplementation resulted in reduced fasting blood glucose and glycated haemoglobin (HbA1c) levels in diabetic patients [143–145]. Moreover, the study of Simeonov et al. [143] showed that 60 min after ingestion of *Aronia* juice, the blood glucose levels in diabetic patients were reduced; however, this effect was statistically insignificant when the *Aronia* supplementation was combined with the meal. On the other hand, it was suggested that one-time *Aronia* juice supplementation decreased the postprandial blood glucose excursion in healthy people [146].

The anti-diabetic potential of *Malus domestica* fruits was tested only on healthy adults during one-time consumption of the apple powder or commercial apple extracts with the carbohydrate meal (postprandial response tests) or glucose (oral glucose tolerance test (OGTT)) [32,34,73,75,147]. These trials indicated the delay time (T_{max}) and the lower or unchanged maximal glucose and insulin levels (C_{max}), as well as the reduction in the total glucose concentration during the time of the study (iAUC, incremental area under the curve of glycaemic excursion). Moreover, a significant decrease in the glucose-dependent insulinotropic peptide (GIP) and C-peptide (an indicator of pancreatic β-cell function) was observed. The urinary glucose excretion was increased or unchanged.

The positive effects on the OGTT parameters (decreased iAUC and glucose levels after 30 min) [148,149] or postprandial response to carbohydrate meals (decreased iAUC and C_{max}) [148,149] were also observed for *Malus pumila* and *Pyrus pyrifolia*. In the first study, the effect was significant in the patients with high–normal or borderline glucose levels (100–125 mg/dL) who were regularly supplemented with the apple extract for 12 weeks. In the second study, the participants with normal glucose levels received a single apple or pear preload combined with specialised carbohydrate meals. Finally, the fasting blood glucose levels were significantly reduced in diabetes mellitus type II patients who consumed *Malus sylvestris* fruits for 14 days [150].

Table 3. Human studies of the anti-diabetic potential of fruits from the Maleae tribe.

Species	Sample Type, Composition	Model, Study Design	Tested Parameter, Observed Effects *	Ref.
Aronia melanocarpa (Michx.) Elliott	juice; proteins, carbohydrates, fats, minerals, fibres, and energy density determined	open-label, randomised, two-period, one-way crossover study; 37 healthy Japanese patients (women and men, >30 years old); 100 mL of *Aronia* juice (tested) or 100 mL of water (control) + 200 g of rice; duration: acute consumption	postprandial blood glucose: ↓ iAUC (0–150 min) by > 50%	[146]
	juice; low-calorie, sugar-free	(1) 41 diabetic patients (16 insulin-dependent, 25 non-insulin-dependent, women and men, 3–62 years old); glucose at baseline (control) vs. 60 min after ingestion of 200 mL of juice or 200 mL of juice and standard meal (tested); duration: acute consumption; (2) 21 diabetic patients (non-insulin-dependent, women and men, 42–62 years old) drinking 200 mL of juice daily (tested) and 23 diabetic patients (non-insulin-dependent, women and men, 48–67 years) without supplementation (controls); duration: 3 months	fasting blood glucose (1): ↓ from 14.23 ± 1.32 mmol/L at baseline to 11.4 ± 0.89 mmol/L after 60 min; postprandial blood glucose (1): ↓ (statistically insignificant); blood glucose (2): ↓ from 13.28 ± 4.55 mmol/L at baseline to 9.10 ± 3.05 mmol/L after 3 months; HbA1c (2): ↓ from $9.39 \pm 2.16\%$ at baseline to $7.49 \pm 1.33\%$ after 3 months	[143]
	juice; total phenolics 413.0 ± 5.1 mg GAE/100 g of fresh fruits, total anthocyanins 172.7 ± 4.4 mg/100 g	35 diabetic patients (women and men, 35–65 years old, diabetes type 2); 3×50 mL per day of *Aronia* juice (tested), and the same patients not supplemented with *Aronia* juice (control); duration: 3 months for supplementation, next 3 months for self-control	fasting blood glucose: ↓ (statistically insignificant); HbA1c: ↓ from 59.1–59.4 mmol/mol (baseline and control) to 55.1 ± 14.7 mmol/mol	[144]
	juice	11 overweight and 11 normal-weight patients (women and men, 51.9 ± 3.9 years old); 3×50 mL per day of *Aronia* juice (tested) and baseline parameters (control); duration: 3 months	fasting blood glucose and HbA1c: not changed	[151]
	standardised commercial extract (Alixir 400 PROTECT®, polyphenols (431 mg/30 mL), anthocyanins (120 mg/30 mL), potassium sorbate (35.1 mg/30 mL)); detected compounds (HPLC): cyanidin and quercetin glycosides	prospective, open-label, clinical case-series study; 143 patients with metabolic syndrome, with or without diabetes type 2 (women and men, 50–60 years old); 30 mL of *Aronia* extract per day, control: baseline parameters; duration: 28 days	blood glucose: ↓, especially in diabetic groups (6.40–6.82 mmol/L) compared to baseline levels (7.97–8.41 mmol/L)	[145]

Table 3. *Cont.*

Species	Sample Type, Composition	Model, Study Design	Tested Parameter, Observed Effects *	Ref.
Malus domestica (Suckow) Borkh.	standardised commercial extract (Appl'In by DIANA Food SAS); total polyphenols 67%, including 40% flavonoid monomers and phenolic acids (HPLC: flavan-3-ols, dihydrochalcones, flavonols, hydroxycinnamic acids)	randomised, controlled, double-blind crossover study; 20 healthy men and 5 postmenopausal women (20–60 years old); 200 mL drink containing 1200 mg of apple phenolics (tested) or <5 mg of phenolics (control) + a high-carbohydrate meal; duration: acute consumption	postprandial plasma glucose: ↓ iAUC (0–30 min), ↓ C_{max}, ↑ T_{max}; postprandial insulin: ↓ iAUC (0–30 min and 0–120 min), ↓ at 10, 20, 30, and 45 min, ↑ T_{max}; C-peptide: ↓ iAUC (0–30 min), ↓ at 10, 20, 30, and 45 min, ↑ T_{max}; GIP: ↓ iAUC (0–30 min and 0–120 min), ↓at 10, 20, 30, 45, 60, and 75 min, ↓ C_{max} decreased, ↑ T_{max};	[32]
	standardised commercial extract (Appl'In, Diana Naturals, France); min. 80% polyphenols, >5% phlorizin	randomised controlled trial, balanced incomplete block design; 65 healthy adults (women and men, average 20–50 years old), apple extract group: 33 adults; 2 g of plant extract (tested) or not supplemented (control) + rice porridge (~50 g available carbohydrate); duration: acute consumption	postprandial glucose: ↓ iAUC (0–2 h) by about 25.4%, effect comparable to those of mulberry fruit and leaf extracts; postprandial insulin: ↓ iAUC (0–2 h) by about 22.3% urine glucose: no glucosuria observed	[34]
	standardised commercial extract (Appl'In, Diana Naturals, France); polyphenols min. 80%, phlorizin > 5%; 14 compounds detected (HPLC), including quercetin glycosides, dihydrochalcones, phenolic acids, and procyanidin oligomers	randomised, controlled, double-blind crossover study; 30 healthy adults (women and men, 18–68 years old); drink without (control) or with 1.8, 1.35, or 0.9 g of apple extract (tested) + a 75 g starch/sucrose meal; duration: acute consumption	postprandial blood glucose: ↓ iAUC (0–30 min) by about 8.99–15.6 mmol/L per minute (all doses), iAUC (0–120 min and 0–240 min) unchanged, ↑T_{max}, C_{max} unchanged; postprandial insulin and C-peptide: similar trends as for glucose concentration; GIP: ↓ at 30 and 60 min (1.8 and 1.35 g), T_{max} and C_{max} unchanged; urinary glucose: unchanged different compounds responsible for observed effects (based on literature studies, extract composition, and serum metabolite studies)	[73]

Table 3. *Cont.*

Species	Sample Type, Composition	Model, Study Design	Tested Parameter, Observed Effects *	Ref.
	apple powder from unripe apples; total sugars 153.44 g/kg, water-soluble pectins 27.73 g/kg, phlorizin 12.61 g/kg, chlorogenic acid 18.90 g/kg, catechin, epicatechin, quercetin glycosides, phloretin-2-*O*-D-xyloglucoside (HPLC)	open-label, randomised crossover study; six healthy females with increased risk of cardiovascular disease and diabetes; 25 g of apple preparation (tested) or not supplemented (control) + glucose; duration: acute consumption	OGTT: glucose at 15 to 30 min reduced ↓ by about twofold, urinary glucose excretion after 2–4 h ↑ by about fivefold; phlorizin responsible for observed effects (based on extract composition and urine metabolite studies)	[147]
	commercial extract (BioActive Food GmbH); total polyphenols 44% (catechin equivalents), phlorizin 16%, quercetin 12.43%, chlorogenic acid 5.57% (HPLC/spectrophotometry)	randomised crossover study; 10 healthy men (23.5 ± 3.1 years old); 2.8 g of capsuled apple extract (tested) or not supplemented (control) + glucose; duration: acute consumption	OGTT: ↓ iAUC (0–15, 0–30, and 0–45 min), glucose level at 15, 30, and 45 min timepoints not significantly changed insulin: ↓ iAUC (0–30, 0–45, 0–60, and 0–90 min), with the highest reduction after 30 min; urinary glucose: 4.9-fold higher in the first 3 h	[75]
Malus pumila Mill.	polyphenolic extract; 48.9% procyanidins (dimers, trimers, tetramers, pentamers, hexamers, and polymers), 14.1% flavan-3-ols (monomers) and 10.5% phloretin glucosides, including phlorizin (HPLC)	double-blinded, placebo-controlled study; 65 patients (30–60 years old) with normal (<100 mg/dL), high–normal (100–109 mg/dL), and borderline (110–125 mg/dL) glucose; 600 mg of apple polyphenols/day (tested) or placebo (control); duration: 12 weeks	OGTT: glucose ↓ after 30 min in the high–normal and borderline groups (164.0 ± 7.4 mg/dL) compared to controls (194.7 ± 10.4 mg/dL), ↓ iAUC (0–2 h), normal group—lack of changes; fasting plasma insulin, HOMA-I andR, HbA1c: unchanged; procyanidins responsible for observed effects (based on extract composition and literature studies)	[148]
Malus sylvestris (L.) Mill.	fruits	randomised pre-/post-test; 22 diabetes mellitus type II patients (40–55 years old); 300 g of apple/day (tested) or patients not given fruits (control); duration: 14 days	fasting blood glucose: ↓ by 40.27 ± 23.018 mg/dL in the apple group compared to baseline parameters; in the control group ↓, but not statistically	[150]

Table 3. *Cont.*

Species	Sample Type, Composition	Model, Study Design	Tested Parameter, Observed Effects *	Ref.
Malus Mill. sp.	clear or cloudy commercial juice; phlorizin 67/148 µM, chlorogenic acid 378/1304 µM, phloretin xyloglucoside 16/105 µM, D-fructose 6.28/59.3 g/L, D-glucose 22.6/17.9 g/L, sucrose 26.6/27.8 g/L (clear/cloudy juice) (HPLC)	three-way, single-blind, randomised crossover study; nine healthy adults (women and men, 24 ± 3.2 years old); 400 mL of clear or cloudy apple juice (tested) or drink without apple juice (control); duration: acute consumption	plasma glucose: ↓ at 15 and 30 min for clear juice, ↓ at 15 min and ↑ at 45–60 min for cloudy juice, ↓ iAUC (0–30 min), glucose absorption delayed; plasma insulin: ↓ over the first 90 min; GIP: ↓ over the first 90 min, ↓ iAUC (0–30 min); GLP-1: ↑ over the first 90 min (cloudy juice); chlorogenic acid and phlorizin responsible for observed effects (based on extract composition and literature studies)	[152]
Malus pumila Mill.; *Pyrus pyrifolia* (Burm.f.) Nakai	fruits; *Malus:* 58% fructose, 30.7% glucose, 11.3% sucrose; *Pyrus:* 57.3% fructose, 36.7% glucose, 6% sucrose;	randomised, eight-period crossover trial; 14 healthy women (18–25 years old); tested: *Malus/Pyrus* + iso-carbohydrate test meals (M + IC or P + IC) or hyper-carbohydrate test meals (M + HC or P + HC); IC, 50 g of available carbohydrate; HC, 65 g of available carbohydrate; 845–1342 kcal; control: water + iso-carbohydrate or hyper-carbohydrate test meals (W + IC or W + HC); duration: acute consumption	*Malus*: ↓ iAUC (0–180 min) by 45.7% and 19.0% for M + IC and M + HC, respectively; ↓ C_{max} by 51.3% and 27.9% for M + IC and M + HC, respectively; maximum amplitude of glycaemic excursion ↓ by about 30–46%; *Pyrus*: ↓ iAUC (0–180 min) by 30.5% and 18.3% for M + IC and M + HC, respectively; ↓ C_{max} by about 40%; maximum amplitude of glycaemic excursion ↓ by about 35–40%	[149]

* Suggestions of what compounds were responsible for the observed effects were taken directly from the cited papers; for a critical discussion, see Section 3.4. ↑, increase; ↓, decrease; C_{max}, maximal level; GAE, gallic acid equivalents; GIP, glucose-dependent insulinotropic polypeptide; GLP-1, glucagon-like peptide 1; HbA1c, glycated haemoglobin; HOMA-IR, the homeostasis model assessment-estimated insulin resistance; HPLC, high-performance liquid chromatography; iAUC, incremental area under the curve of tested parameter excursion; OGTT, oral glucose tolerance test; T_{max}, time to reach maximal level.

3.4. Polyphenols and Other Chemical Contributors to the Anti-Diabetic Activity of Maleae Fruits

The Maleae fruits contain different bioactive substances, primarily polyphenols (e.g., anthocyanins, flavonoids, proanthocyanidins, phenolic acids), but also terpenoids, proteins, carbohydrates, vitamins, and minerals. In this review, we do not present the detailed chemical composition of all species, since their profiles are highly complex and described in detail in other reviews [153–155], but we discuss the suggestions from the literature on the chemical constituents that might be responsible for the observed anti-diabetic effects of fruits. To this end, we focused not on the hypothetical indication of active markers based only on fruit composition (i.e., the presence of particular components), but on activity confirmation by testing pure compounds or their purified fractions (issued in 26 papers). The details, i.a., the chemical structures in question and the brief chemical profiles of the fruits (if covered in the reviewed articles), are shown in Tables 1–3. In addition, the impacts of various compounds on the anti-diabetic potential of Maleae fruits are discussed below, and summarised in the form of Figures 3 and 4. All in all, as the leading phytochemical constituents of Maleae fruits, polyphenols were most often tested for biological effects and confirmed as being responsible for the anti-diabetic activity of different species through various mechanisms. In addition, some papers pointed out the contribution of polysaccharides or triterpenes to the biological effects of, e.g., *Chaenomeles* or *Sorbus* species, as well as their synergy with polyphenols.

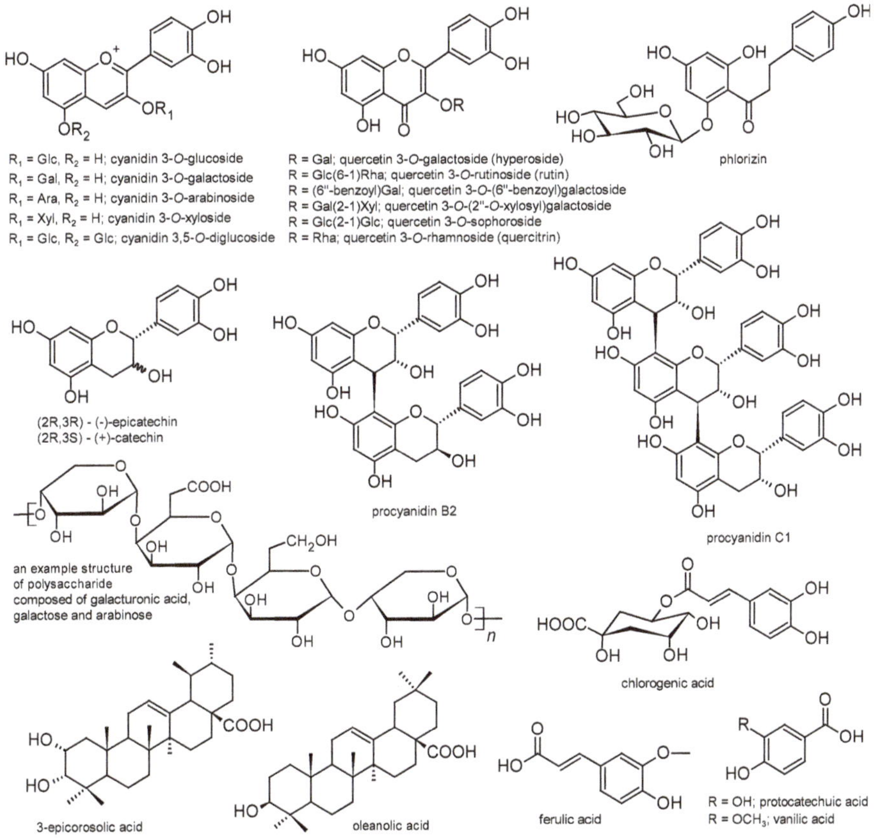

Figure 3. The chemical structures of compounds studied as constituents responsible for the anti-diabetic effects of Malea fruits.

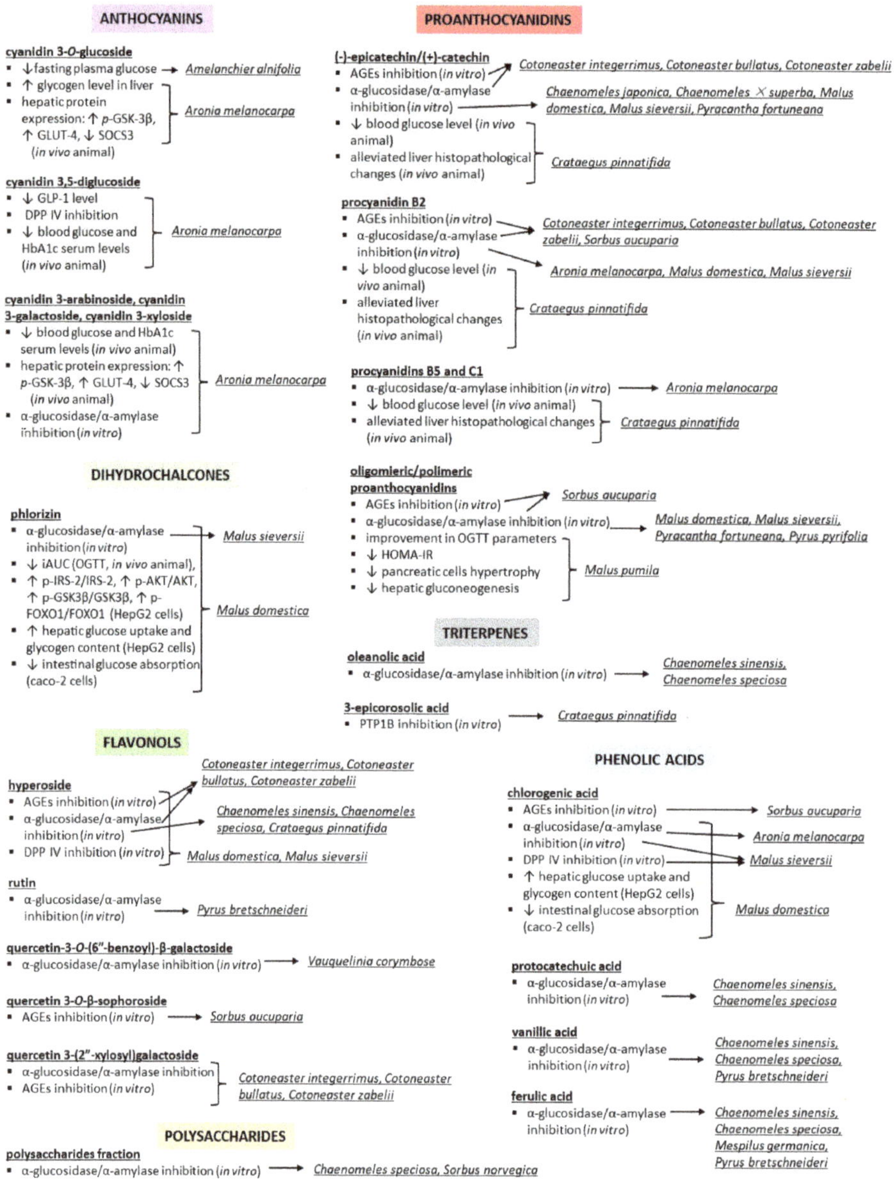

Figure 4. Summarising the contributions of phenolic and non-phenolic compounds to the anti-diabetic effects of Maleae fruits, based on activity studies of pure compounds and their contributions to the fruit samples' composition and activity [12,14,15,17,20,22,24,27,33,37,42,43,58,59,62–64,68,75,79,90,94,100,104,110,113,119]. AGEs, advanced glycation end products; DPP IV, dipeptidyl peptidase-4; GLUT-4, glucose transporter 4; GLP-1, glucagon-like peptide 1; HbA1c, glycated haemoglobin; HOMA-IR, homeostatic model assessment for insulin resistance; OGTT, oral glucose tolerance test; *p*-AKT, phosphorylated protein kinase B; *p*-IRS-2, phosphorylated insulin receptor substrate 2; *p*-FOXO1, phosphorylated forkhead box G1; *p*-GSK-3β, phosphorylated glycogen synthase kinase 3 beta; PTP1B, protein tyrosine phosphatase 1B; SOCS3, suppressor of cytokine signalling 3.

3.4.1. Anthocyanins' Contribution to the *Amelanchier* and *Aronia* Fruits' Activity

Anthocyanins were studied as active constituents of *Amelanchier alnifolia* [104–107] and *Aronia melanocarpa* [12,58,119]. Cyanidin 3-*O*-glucoside (7.2 mg/kg/day) was revealed to reduce the fasting plasma glucose levels in the mice fed a high-fat, high-sucrose diet, reaching levels similar to those observed for *Amelanchier* berry powder containing an equal amount of cyanidin glucoside [104]. Cyanidin monoglycosides isolated from *Aronia melanocarpa* fruits (986.48 mg of cyanidin galactoside, glucoside, arabinoside, and xyloside per g of fraction; 150–300 mg/kg/day) reduced blood glucose and HbA1c serum levels, increased the glycogen levels in the liver, and modulated hepatic protein expression (↑ *p*-GSK-3β, ↑ GLUT-4, ↓ SOCS3) in the diabetic mice [100]. Moreover, the in vitro studies of the α-glucosidase-inhibitory potential of individual anthocyanins from *Aronia* fruits suggested the higher anti-diabetic potential of cyanidin arabinoside and glucoside (IC_{50} = 0.37–0.87 μg/mL) compared to cyanidin galactoside and xyloside (IC_{50} = 1.54–5.5 μg/mL). Still, all cyanidin monoglycosides were considered to be co-responsible for the anti-glucosidase activity of the chokeberry 50% ethanol extract (IC_{50} = 3.5 μg/mL) [12]. The in vivo (animal model) anti-diabetic potential was also tested for cyanidin 3,5-diglucoside (10 μg/mL solution), but its effectiveness was weaker than that of *Aronia* juice [119]. The contribution of anthocyanins to the anti-diabetic potential of fruits may explain the higher activity observed for extracts from peel than from the flesh of *Amelanchier*, as well as that of acidified alcoholic extracts from *Aronia* fruits compared with non-acidified extracts (higher extraction potential and content of anthocyanins) [13,55].

3.4.2. Dihydrochalcone's Contribution to the *Malus* Fruits' Activity

Phlorizin intake (1.96 mg/kg) significantly reduced the iAUC (in vivo animal studies, OGTT) to levels comparable to those achieved by *Malus domestica* commercial extract (12.24 mg/kg) containing the same amount of phlorizin [75]. This observation may be due to the inhibition of intestinal glucose absorption that was reported for phlorizin, e.g., it was calculated that phlorizin contributed to 52% of the glucose transport reduction (Caco-2 cells) noted for *Malus domestica* extract [33]. Phlorizin was also able to inhibit α-glucosidase activity (IC_{50} = 0.01 mg/mL, with significantly stronger effectiveness than extracts from different *Malus sp.* cultivars IC_{50} = 7–256 mg/mL) [42]. Moreover, phlorizin stimulated the glucose uptake by hepatic cells and modified the protein levels in the insulin signalling pathway in vitro, with effects comparable to those achieved with *Malus domestica* 80% ethanolic extract [37].

3.4.3. Flavonols' Contribution to the *Chaenomeles, Cotoneaster, Malus, Pyrus, Sorbus,* and *Vauquelinia* Fruits' Activity

A contribution to the anti-diabetic potential of Maleae fruits was also suggested for flavonols. Hyperoside was found to inhibit α-glucosidase and/or DPP-IV activity, as well as AGE formation, and the activity was higher than or comparable to that observed for corresponding fruit extracts from *Chaenomeles speciosa, Chaenomeles sinensis, Cotoneaster integerrimus, Cotoneaster zabelii, Cotoneaster bullatus, Crataegus pinnatifida, Malus domestica,* and *Malus sieversii* [15,42,63,68]. Moreover, rutin and quercetin-3-*O*-(6′′-benzoyl)-β-galactoside were suggested to be α-glucosidase inhibitors from *Pyrus bretschneideri* [90] and *Vauquelinia corymbose* [24], respectively, while quercetin 3-(2″-xylosyl)galactoside and quercetin 3-*O*-β-sophoroside contributed to the inhibition of AGE formation observed for *Cotoneaster* sp. [15] and *Sorbus aucuparia* extracts [43], respectively. Finally, it was calculated that quercetin-3-*O*-rhamnoside contributed to the reduction in intestinal glucose transport (in vitro cellular studies on Caco-2 cells) by *Malus domestica* methanol extract, to the extent of 26% [33].

3.4.4. Proanthocyanidins' Contribution to the *Aronia, Crataegus, Chaenomeles, Cotoneaster, Malus, Pyracantha, Pyrus,* and *Sorbus* Fruits' Activity

The influence of proanthocyanidins on the anti-diabetic activity of fruits was tested with the use of both purified fractions (composed of monomeric-to-polymeric flavan-3-ols)

and individual compounds (i.e., monomers, dimers, and trimers of flavan-3-ols). The four-week supplementation of the procyanidin fraction from *Malus pumila* juice (purified from other active components, like phlorizin and chlorogenic acid, administered without restriction in drinking water, with 0.5 g of procyanidin fraction per 100 mL of water) resulted in a significant improvement in the OGTT parameters and a reduction in HOMA-IR, pancreatic cell hypertrophy, and hepatic gluconeogenesis in insulin-resistant mice [113]. Moreover, the procyanidin fraction from *Crataegus pinnatifida* (epicatechin and procyanidins B2, B5, and C1; 200 mg/kg) reduced blood glucose levels (OGTT) and alleviated histopathological changes to the liver in high-fat-diet rats [110]. Furthermore, different flavan-3-ols isolated from numerous Maleae fruits were proven to inhibit α-glucosidase/α-amylase activity and, thus, suggested to contribute to the fruits' anti-diabetic activity based on this mechanism; these were (−)-epicatechin/(+)-catechin (*Chaenomeles* sp., *Cotoneaster* sp., *Malus* sp.), procyanidin B2 (*Aronia melanocarpa, Cotoneaster* sp., *Malus* sp., *Sorbus aucuparia*), procyanidin C1 (*Aronia melanocarpa*), oligomeric and polymeric proanthocyanidin fractions (*Pyrus pyrifolia, Sorbus aucuparia*), the sum of (+)-catechin, (−)-epicatechin, A-type and B-type procyanidins, and procyanidin glycosides (*Pyracantha fortuneana*) [12,15,20,27,42,43,63,64,79,96]. In addition, (−)-epicatechin and procyanidin B2 were demonstrated to inhibit AGE formation, and their effectiveness was a dozen to several dozen times higher than that observed for corresponding fruit extracts from *Cotoneaster* sp. and *Sorbus aucuparia* [15,43].

3.4.5. Phenolic Acids' Contribution to the *Aronia, Chaenomeles, Malus, Mespilus, Pyrus,* and *Sorbus* Fruits' Activity

Regarding phenolic acids, their contribution to the anti-diabetic activity of Maleae fruits was best documented for chlorogenic acid. It was reported that chlorogenic acid contributed to the inhibition of intestinal glucose transport (observed for the *Malus domestica extract*, 12% share [33]), the inhibition of α-glucosidase activity (*Aronia melanocarpa* wines, *Chaenomeles sp.* extracts, and *Malus* sp. extracts [14,42,63]), the inhibition of hepatic glucose uptake and DPP-IV activity (*Malus* sp. extracts [37,42]), and the inhibition of AGE formation (*Sorbus aucuparia* extracts [43]). Moreover, the reduction in α-glucosidase activity was also observed for protocatechuic, ferulic, and vanillic acids from *Chaenomeles* sp., *Mespilus germanica*, and *Pyrus bretschneideri* extracts [63,87,90].

3.4.6. Contribution of Non-Phenolic Compounds to the *Crataegus, Chaenomeles,* and *Sorbus* Fruits' Activity, and Their Synergy with Polyphenols

Apart from polyphenols, some triterpenes were also suggested as anti-diabetic agents. For instance, 3-epicorosolic acid isolated from *Crataegus pinnatifida* inhibited protein tyrosine phosphatase 1B (PTP1B) and α-glucosidase to comparable or higher extents than the respective crude methanol extract and its organic fractions [17]. In addition, α-glucosidase-inhibitory activity was observed for oleanolic acid from *Chaenomeles* sp. [59,63].

Last, but not least, the anti-diabetic potential has been confirmed for polysaccharides. The α-glucosidase inhibition observed for the polysaccharide fraction from *Chaenomeles speciosa* fruits was potent (100% inhibition at 0.5 mg/mL) and significantly higher compared to all other compounds isolated from the fruits, including various polyphenols and triterpenes. However, there was no clear relationship between activity parameters and polyphenol/polysaccharide/triterpene concentrations in different *Chaenomeles* extracts/fractions. Consequently, statistical analysis performed using composition data of the tested extracts/fractions and activity of pure compounds suggested that α-glucosidase inhibition of the fruits seems additive or synergic and depends on various chemical constituents and proportions between them [59]. The synergy of polysaccharides with polyphenols in the context of anti-diabetic potential was also confirmed for *Sorbus norvegica*. In this case, the inhibition of α-amylase activity was significantly higher for the whole-fruit extract (IC_{50} = 2.5 μg/mL) than for its two fractions, i.e., polysaccharides (IC_{50} = 48 μg/mL) and polyphenols (IC_{50} = 20 μg/mL) [22]. The structures of both tested polymers were not analysed. Still, another study on the carbohydrates from *Chaenomeles* and *Sorbus* species suggested that galacturonic acid, arabinose, and galactose may be the primary components of these active polysaccharides [156,157].

3.4.7. Impact of Monosaccharides on Fruits' Anti-Diabetic Potential

Considering the possibility of using fruits as potential herbal drug candidates for the prevention or treatment of diabetes, it is worth paying attention to the presence of simple sugars and the related glycaemic index (GI). The GI expresses the capacity of the organism to deal with carbohydrates in foods as a percentage of the response to an equal weight of glucose. Since all fruits from the Maleae tribe are classified as having a low GI (<55), their consumption by diabetic patients is considered to be safe [158–160]. This is especially important for ingesting the whole, fresh fruits or juices, while supplementation of fruit extracts with concentrated contents of only selected compounds is all the more secure.

3.5. The Anti-Diabetic Potential of the Most Promising Maleae Fruits—Concluding Thoughts

Considering all studies on the anti-diabetic effects of Maleae fruits, their number, and the results presented above, the most promising species for more expansive pro-health applications today seem to be *Aronia melanocarpa* and *Malus domestica*, which are among the few that have been tested in human trials. Therefore, in this chapter, we sum up the currently available data on these two species and discuss the potential outcomes for future diabetes management using fruit products. At the same time, we do not forget about other species, research on which is less advanced but still promising and worthy of further attention, e.g., *Amelanchier* sp., *Chaenomeles* sp., *Crataegus* sp., *Pyrus* sp., *Sorbus* sp., and other *Aronia* sp., and *Malus* sp.

3.5.1. The Anti-Diabetic Potential of *Aronia melanocarpa* Fruits

The anti-diabetic potential of *Aronia melanocarpa* seems to be mainly due to the increase in insulin sensitivity and the boost in the peripheral absorption of glucose (likewise, e.g., metformin), as well as the delay in the absorption of carbohydrates from the intestine (likewise, e.g., acarbose).

According to Chen and Meng [100], the five-week supplementation of acidified 80% ethanol *Aronia melanocarpa* fruit extract, at 300 mg/kg/day, resulted in about 1.6-fold lower glucose levels, about 2-fold lower insulin levels, and about 1.8-fold lower HbA1c levels in comparison to diabetic mice, while the effect of metformin (200 mg/kg/day) was about 2-fold lower glucose levels, about 2.2-fold lower insulin levels, and about 2.3-fold lower HbA1c levels. The effectiveness of *Aronia* extract relative to the synthetic anti-diabetic drug was also observed in the enhancement of GLUT-4 and p-GSK-3β expression, resulting in higher glucose uptake by hepatic cells and stimulating glycogen synthesis [100]. The increased tissue-specific glucose uptake and glycogen levels, as well as the modulation of different proteins' expression in the insulin signalling pathway, which led to a decrease in glucose levels, was confirmed by many in vitro and in vivo studies (Tables 1–3). What is essential in this metformin-like mechanism of action is that the sensitisation of tissues to the action of insulin occurs without increasing the insulin level, which reduces the risk of hypoglycaemia as the main disadvantage of sulfonylurea drugs (e.g., glipizide).

The delay in the absorption of carbohydrates from the intestine observed for *Aronia melanocarpa* fruit products relies on glucosidase inhibition, which is therefore another verified mechanism of action of the fruit. According to different studies [114,115], *Aronia* juices and extracts significantly inhibited α-glucosidase, maltase, and sucrase activity in the intestine (in vivo animal models). The effectiveness of *Aronia* extracts towards α-glucosidase compared to acarbose (in vitro studies) was measured as 37–186 times higher, depending on the fruit cultivar, the origin of the sample, and the type of extract (and, thus, chemical composition). It is also advantageous that synthetic glucosidase inhibitors are associated with some slight gastrointestinal side effects, of which fruits seem to be devoid [143,144].

Finally, as mentioned above, the activity of fruit products is closely related to their composition. From the whole chemical pool of *Aronia melanocarpa* fruits (over one hundred phenolic compounds identified), only a few have been tested as contributors to the anti-diabetic potential of fruit. These were mainly anthocyanins (i.e., cyanidin 3-arabinoside, cyanidin 3-glucoside, cyanidin 3-galactoside, cyanidin 3-xyloside, cyanidin 3,5-diglucoside), which accounted for about 10–25% (or 98% in case of the purified fraction) of extracts/juice dry mass (Tables 1–3). Like *Aronia* fruits, these compounds were able to lower blood glucose, HbA1c, and GLP-1 levels, inhibit DPP IV and α-glucosidase activity, increase glycogen levels in the liver, and modulate hepatic protein expression (in vitro and in vivo models) [12,58,100,119]. The contribution to the α-glucosidase-inhibitory activity was also noticed for procyanidin dimers, trimers, and some phenolic acids (i.e., chlorogenic and caffeic acids) [12,14,58]. Moreover, considering the contents of individual compounds and their activity, as well as the activity of relevant plant samples, the synergistic and additive effects of different constituents were suggested [119,131]. Synergy with synthetic anti-diabetic drugs is also possible, but this matter has not been studied so far.

3.5.2. The Anti-Diabetic Potential of *Malus domestica* Fruits

In the case of *Malus domestica* fruits, their effectiveness is suggested to be mainly due to the inhibition of both sodium-dependent and sodium-independent glucose transporters in the intestine (SGLT-1 and GLUT-2), which results in a delay in the absorption of carbohydrates [32,34]. SGLT-1 inhibitors are currently an eagerly studied group of anti-diabetic drugs (synthetic and plant-derived) whose potential is due not only to their glucose-lowering ability but also to their cardioprotective effects. One of the leading natural SGLT-1 inhibitors is phlorizin—a dihydrochalcone isolated from apples [147]. Indeed, the phlorizin-containing apple products were also confirmed to reduce glucose-derived protein damage by inhibiting the activity of ALR and SDH enzymes and the formation of AGEs (polyol pathway); thus, they may prevent the development of diabetic complications [41,74,80].

The second type of *Malus domestica* fruit activity related to the content of phlorizin (5–16%) was suggested to be the increased insulin sensitivity resulting from the enhanced hepatic glucose uptake and expression of proteins involved in glycogen synthesis and glycolysis (e.g., ↑ *p*-GSK3β/GSK3β, ↑ *p*-FOXO1/FOXO1) [147]. Therefore, *Malus domestica* is another species with a metformin-like mechanism of action, i.e., the sensitisation of tissues to the action of insulin without increasing the insulin level. Indeed, the in vivo studies confirmed the *Malus* extracts'/juices' ability to lower both glucose and insulin in diabetic animals to levels comparable to those observed in healthy subjects, as well as to normalise the postprandial and OGTT parameters in healthy people [32,34,73,75,109,139,140,143].

As for anti-diabetic bioactive compounds, aside from phlorizin, there has also been research on the contributions of chlorogenic acid, procyanidins, and quercetin derivatives to hepatic and intestinal glucose transport, as well as AGEs and α-glucosidase inhibition [33,37,42,79]. Interestingly, while the inhibitory potential of apples towards α-glucosidase seems to be less evident (i.e., only extracts with increased contents of polyphenols—about 40–80%—were able to inhibit glucosidase more strongly than acarbose), the synergistic anti-glucosidase activity with acarbose was confirmed for apple juice. Thus, this is another beneficial aspect of using fruit products in combinatory therapy for diabetes.

4. Conclusions

The data reported in this review show that many Maleae fruits indeed have anti-diabetic potential and may be recommended for the prevention and treatment of diabetes. However, from over 1000 species belonging to the Maleae tribe, only 46 have been investigated so far in the context of diabetes. The majority of the conducted studies covered only in vitro tests (67 papers); then, there are in vivo studies on animal models (51 reports, including some mixed in vitro/in vivo tests), as well as in vivo human studies (14 trials on *Aronia melanocarpa, Malus* sp., and *Pyrus pyrifolia*). The species most thoroughly studied in terms of anti-diabetic effects and mechanisms were *Amelanchier alnifolia*, *Aronia melanocarpa*, *Chaenomeles japonica*, *Crataegus pinnatifida*, *Malus domestica*, *Malus pumila*, and *Pyrus pyrifolia*. The reviewed papers indicated the ability of Maleae fruits, e.g., to modulate the expression and activity of the proteins in the insulin-mediated PI3K/Akt pathway, induce incretin-based effects (e.g., GLP-1 and GIP agonism, DPP IV inhibition), regulate the intestinal glucose absorption and tissue-specific glucose uptake by affecting the glucose transporters (GLUTs and SGLT1), and inhibit enzymes involved in carbohydrate digestion (α-glucosidase) or in the polyol pathway of glucose metabolism (ALR, SDH). As for phytochemicals responsible for the anti-diabetic effectiveness of Maleae fruits, some reviewed papers suggested contributions of various compounds to the observed effects—primarily polyphenols (e.g., flavonols, dihydrochalcones, proanthocyanidins, anthocyanins, phenolic acids), but also triterpenes and polysaccharides. Thanks to the additive and synergistic actions of individual phytochemicals, the biological effects of fruits/juices/extracts are significantly higher than those of pure compounds. Therefore, various fruits from the Maleae tribe seem to be advantageous anti-diabetic agents. Still, their potential for functional application depends on various factors, in addition to the obvious plant species/variety, all of affect the composition of fruit products and, thus, may explain some discrepancies in the results of different studies. These include variations in climatic growth conditions, maturity stage, type of sample (i.e., the part of the fruit, like peel or flesh), processing conditions, or ingested form, e.g., if consumed as fresh fruits, juices, or specific extracts with high concentrations of selected active compounds. The latter form seems especially promising, as it reduces excessive intake of ballast constituents, primarily diabetes-promoting free monosaccharides. Finally, the prospects of different fruits and their extracts for more expansive pro-health applications require further research, especially more profound in vivo trials with the establishment of effective doses, and formulation and toxicity studies on the fruit extracts as potential herbal drug candidates.

Author Contributions: Conceptualisation, M.R. and M.A.O.; methodology, M.R.; investigation, M.R.; resources, M.A.O.; data curation, M.R.; writing—original draft preparation, M.R.; writing—review and editing, M.A.O. and M.R.; visualisation, M.R.; funding acquisition, M.A.O. All authors have read and agreed to the published version of the manuscript.

Funding: This research was funded by the Medical University of Lodz, Poland, grant number 503/3-022-01/503-31-001.

Institutional Review Board Statement: Not applicable.

Informed Consent Statement: Not applicable.

Data Availability Statement: Not applicable.

Conflicts of Interest: The authors declare no conflict of interest.

References

1. International Diabetes Federation. *IDF Diabetes Atlas*, 10th ed.; International Diabetes Federation: Brussels, Belgium, 2021; ISBN 978-2-930229-98-0.
2. WHO Library Cataloguing. *Global Report on Diabetes*; WHO: Geneve, Switzerland, 2016; Volume 978, ISBN 9789241565257.
3. Kooti, W.; Farokhipour, M.; Asadzadeh, Z.; Ashtary-Larky, D.; Asadi-Samani, M. The role of medicinal plants in the treatment of diabetes: A systematic review. *Electron. Physician* **2016**, *8*, 1832–1842. [CrossRef]
4. Tan, S.Y.; Mei Wong, J.L.; Sim, Y.J.; Wong, S.S.; Mohamed Elhassan, S.A.; Tan, S.H.; Ling Lim, G.P.; Rong Tay, N.W.; Annan, N.C.; Bhattamisra, S.K.; et al. Type 1 and 2 diabetes mellitus: A review on current treatment approach and gene therapy as potential intervention. *Diabetes Metab. Syndr. Clin. Res. Rev.* **2019**, *13*, 364–372. [CrossRef]
5. Reed, J.; Bain, S.; Kanamarlapudi, V. A review of current trends with type 2 diabetes epidemiology, aetiology, pathogenesis, treatments and future perspectives. *Diabetes Metab. Syndr. Obes. Targets Ther.* **2021**, *14*, 3567–3602. [CrossRef]
6. Fereidoon, S.; Cesarettin, A. (Eds.) *Handbook of Functional Beverages and Human Health*; CRC Press: Boca Raton, FL, USA, 2016; ISBN 9781466596429.
7. Sun, C.; Liu, Y.; Zhan, L.; Rayat, G.R.; Xiao, J.; Jiang, H.; Li, X.; Chen, K. Anti-diabetic effects of natural antioxidants from fruits. *Trends Food Sci. Technol.* **2021**, *117*, 3–14. [CrossRef]

8. Sun, J.; Shi, S.; Li, J.; Yu, J.; Wang, L.; Yang, X.; Guo, L.; Zhou, S. Phylogeny of Maleae (Rosaceae) based on multiple chloroplast regions: Implications to genera circumscription. *Biomed. Res. Int.* **2018**, *2018*, 6–9. [CrossRef]
9. Mandal, D.; Wermund, U.; Phavaphutanon, L.; Cronje, R. (Eds.) *Temperate Fruits*; Series statement: Innovations in horticultural science; Apple Academic Press: Palm Bay, FL, USA, 2020; ISBN 9781003045861.
10. United States Department of Agriculture. Fresh apples, grapes, and pears: World markets and trade. In *Foreign Agricultural Service*; United States Department of Agriculture: Washington, DC, USA, 2019; pp. 1–10.
11. WFO. The World Flora Online. Available online: http://www.worldfloraonline.org/taxon/wfo-0000550491#synonyms (accessed on 20 August 2023).
12. Bräunlich, M.; Slimestad, R.; Wangensteen, H.; Brede, C.; Malterud, K.E.; Barsett, H. Extracts, anthocyanins and procyanidins from *Aronia melanocarpa* as radical scavengers and enzyme inhibitors. *Nutrients* **2013**, *5*, 663–678. [CrossRef]
13. Wangensteen, H.; Bräunlich, M.; Nikolic, V.; Malterud, K.E.; Slimestad, R.; Barsett, H. Anthocyanins, proanthocyanidins and total phenolics in four cultivars of aronia: Antioxidant and enzyme inhibitory effects. *J. Funct. Foods* **2014**, *7*, 746–752. [CrossRef]
14. Čakar, U.; Grozdanić, N.; Pejin, B.; Vasić, V.; Čakar, M.; Petrović, A.; Djordjević, B. Impact of vinification procedure on fruit wine inhibitory activity against α-glucosidase. *Food Biosci.* **2018**, *25*, 1–7. [CrossRef]
15. Kicel, A.; Magiera, A.; Skrzywanek, M.; Malczuk, M.; Olszewska, M.A. The inhibition of α-glucosidase, α-amylase and protein glycation by phenolic extracts of *Cotoneaster bullatus*, *Cotoneaster zabelii*, and *Cotoneaster integerrimus* leaves and fruits: Focus on anti-hyperglycemic activity and kinetic parameters. *Molecules* **2022**, *27*, 7081. [CrossRef]
16. Mecheri, A.; Amrani, A.; Benabderrahmane, W.; Bensouici, C.; Boubekri, N.; Benaissa, O.; Zama, D.; Benayache, F.; Benayache, S. In Vitro Pharmacological Screening of Antioxidant, Photoprotective, Cholinesterase, and α-Glucosidase Inhibitory Activities of Algerian *Crataegus oxyacantha* Fruits and Leaves Extracts. *Pharm. Chem. J.* **2021**, *54*, 1150–1156. [CrossRef]
17. Chowdhury, S.S.; Islam, M.N.; Jung, H.A.; Choi, J.S. In vitro antidiabetic potential of the fruits of *Crataegus pinnatifida*. *Res. Pharm. Sci.* **2014**, *9*, 11–22. [PubMed]
18. de Oliveira Raphaelli, C.; dos Santos Pereira, E.; Camargo, T.M.; Vinholes, J.; Rombaldi, C.V.; Vizzotto, M.; Nora, L. Apple Phenolic Extracts Strongly Inhibit α-Glucosidase Activity. *Plant Foods Hum. Nutr.* **2019**, *74*, 430–435. [CrossRef]
19. Wang, H.; Ye, Y.H.; Wang, H.H.; Liu, J.; Liu, Y.J.; Jiang, B.W. HPLC-QTOF-MS/MS profiling, antioxidant, and α-glucosidase inhibitory activities of *Pyracantha fortuneana* fruit extracts. *J. Food Biochem.* **2019**, *43*, e12821. [CrossRef]
20. Wei, M.; Chai, W.M.; Yang, Q.; Wang, R.; Peng, Y. Novel Insights into the Inhibitory Effect and Mechanism of Proanthocyanidins from *Pyracantha fortuneana* Fruit on α-Glucosidase. *J. Food Sci.* **2017**, *82*, 2260–2268. [CrossRef] [PubMed]
21. Prakash, O.; Bettadiah, B.K.; Kudachikar, V.B. Food Bioscience Chemical composition and in vitro antihyperglycemic potential of Kainth fruit (*Pyrus pashia* Buch.-Ham ex D.Don). *Food Biosci.* **2021**, *42*, 101119. [CrossRef]
22. Broholm, S.L.; Gramsbergen, S.M.; Nyberg, N.T.; Jäger, A.K.; Staerk, D. Potential of *Sorbus* berry extracts for management of type 2 diabetes: Metabolomics investigation of 1H NMR spectra, α-amylase and α-glucosidase inhibitory activities, and in vivo anti-hyperglycaemic activity of S. norvegica. *J. Ethnopharmacol.* **2019**, *242*, 112061. [CrossRef]
23. Hasbal, G.; Yilmaz Ozden, T.; Can, A. In vitro Antidiabetic Activities of Two *Sorbus* Species. *Eur. J. Biol.* **2017**, *76*, 57–60. [CrossRef]
24. Flores-Bocanegra, L.; Pérez-Vásquez, A.; Torres-Piedra, M.; Bye, R.; Linares, E.; Mata, R. α-Glucosidase inhibitors from *Vauquelinia corymbosa*. *Molecules* **2015**, *20*, 15330–15342. [CrossRef] [PubMed]
25. Alongi, M.; Verardo, G.; Gorassini, A.; Anese, M. Effect of pasteurization on in vitro α-glucosidase inhibitory activity of apple juice. *LWT* **2018**, *98*, 366–371. [CrossRef]
26. Boath, A.S.; Stewart, D.; McDougall, G.J. Berry components inhibit α-glucosidase in vitro: Synergies between acarbose and polyphenols from black currant and rowanberry. *Food Chem.* **2012**, *135*, 929–936. [CrossRef]
27. Grussu, D.; Stewart, D.; McDougall, G.J. Berry polyphenols inhibit α-amylase in vitro: Identifying active components in rowanberry and raspberry. *J. Agric. Food Chem.* **2011**, *59*, 2324–2331. [CrossRef]
28. Chen, J.; Zhu, J.; Meng, X. *Aronia melanocarpa* anthocyanin extracts are an effective regulator of suppressor of cytokine signaling 3-dependent insulin resistance in HepG2 and C2C12 cells. *J. Funct. Foods* **2020**, *75*, 104258. [CrossRef]
29. Zakłos-Szyda, M.; Pawlik, N. Japanese quince (*Chaenomeles japonica* L.) fruit polyphenolic extract modulates carbohydrate metabolism in HepG2 cells via AMP-activated protein kinase. *Acta Biochim. Pol.* **2018**, *65*, 67–78. [CrossRef]
30. Manzano, M.; Giron, M.D.; Vilchez, J.D.; Sevillano, N.; El-Azem, N.; Rueda, R.; Salto, R.; Lopez-Pedrosa, J.M. Apple polyphenol extract improves insulin sensitivity in vitro and in vivo in animal models of insulin resistance. *Nutr. Metab.* **2016**, *13*, 32. [CrossRef] [PubMed]
31. You, M.K.; Kim, H.J.; Rhyu, J.; Kim, H.A. Pear pomace ethanol extract improves insulin resistance through enhancement of insulin signaling pathway without lipid accumulation. *Nutr. Res. Pract.* **2017**, *11*, 198–205. [CrossRef] [PubMed]
32. Castro-Acosta, M.L.; Stone, S.G.; Mok, J.E.; Mhajan, R.K.; Fu, C.I.; Lenihan-Geels, G.N.; Corpe, C.P.; Hall, W.L. Apple and blackcurrant polyphenol-rich drinks decrease postprandial glucose, insulin and incretin response to a high-carbohydrate meal in healthy men and women. *J. Nutr. Biochem.* **2017**, *49*, 53–62. [CrossRef]
33. Manzano, S.; Williamson, G. Polyphenols and phenolic acids from strawberry and apple decrease glucose uptake and transport by human intestinal Caco-2 cells. *Mol. Nutr. Food Res.* **2010**, *54*, 1773–1780. [CrossRef]
34. Mela, D.J.; Cao, X.Z.; Dobriyal, R.; Fowler, M.I.; Lin, L.; Joshi, M.; Mulder, T.J.P.; Murray, P.G.; Peters, H.P.F.; Vermeer, M.A.; et al. The effect of 8 plant extracts and combinations on post-prandial blood glucose and insulin responses in healthy adults: A randomized controlled trial. *Nutr. Metab.* **2020**, *17*, 51. [CrossRef] [PubMed]

35. Yu, S.; Kim, M.; Park, Y.; Bae, M.; Kang, H.; Hu, S.; Pham, T.X.; Carpenter, R.; Lee, J.; Lee, O.; et al. Anthocyanin-Rich Aronia Berry Extract Mitigates High-Fat and High-Sucrose Diet-Induced Adipose Tissue Inflammation by Inhibiting Nuclear Factor-j B Activation. *J. Med. Food* **2021**, *24*, 586–594. [CrossRef] [PubMed]
36. Zakłos-Szyda, M.; Majewska, I.; Redzynia, M.; Koziołkiewicz, M. Antidiabetic Effect of Polyphenolic Extracts from Selected Edible Plants as α-Amylase, α-Glucosidase and PTP1B Inhibitors, and β Pancreatic Cells Cytoprotective Agents—A Comparative Study. *Curr. Top. Med. Chem.* **2015**, *15*, 2431–2444. [CrossRef]
37. Li, D.; Yang, Y.; Sun, L.; Fang, Z.; Chen, L.; Zhao, P.; Wang, Z.; Guo, Y. Effect of young apple (*Malus domestica Borkh.* cv. *Red Fuji*) polyphenols on alleviating insulin resistance. *Food Biosci.* **2020**, *36*, 100637. [CrossRef]
38. Mueller, M.; Jungbauer, A. Culinary plants, herbs and spices—A rich source of PPARγ ligands. *Food Chem.* **2009**, *117*, 660–667. [CrossRef]
39. Kim, D.J.; Chung, M.J.; You, J.K.; Seo, D.J.; Kim, J.M.; Choe, M. Effect of medicinal plant water extracts on glucose-regulating enzyme activities in goto-kakizaki rat liver cytosol. *J. Korean Soc. Food Sci. Nutr.* **2009**, *38*, 1331–1335. [CrossRef]
40. Kraft, T.F.B.; Dey, M.; Rogers, R.B.; Ribnicky, D.M.; Gipp, D.M.; Cefalu, W.T.; Raskin, I.; Lila, M.A. Phytochemical composition and metabolic performance-enhancing activity of dietary berries traditionally used by native North Americans. *J. Agric. Food Chem.* **2008**, *56*, 654–660. [CrossRef]
41. Kazeem, M.I.; Adeyemi, A.A.; Adenowo, A.F.; Akinsanya, M.A. *Carica papaya* Linn. fruit extract inhibited the activities of aldose reductase and sorbitol dehydrogenase: Possible mechanism for amelioration of diabetic complications. *Futur. J. Pharm. Sci.* **2020**, *6*, 96. [CrossRef]
42. Yu, C.H.J.; Migicovsky, Z.; Song, J.; Rupasinghe, H.P.V. (Poly)phenols of apples contribute to in vitro antidiabetic properties: Assessment of Canada's Apple Biodiversity Collection. *Plants People Planet.* **2022**, *5*, 225–240. [CrossRef]
43. Rutkowska, M.; Kolodziejczyk-Czepas, J.; Owczarek, A.; Zakrzewska, A.; Magiera, A.; Olszewska, M.A. Novel insight into biological activity and phytochemical composition of *Sorbus aucuparia* L. fruits: Fractionated extracts as inhibitors of protein glycation and oxidative/nitrative damage of human plasma components. *Food Res. Int.* **2021**, *147*, 110526. [CrossRef]
44. Termentzi, A.; Alexiou, P.; Demopoulos, V.J.; Kokkalou, E. The aldose reductase inhibitory capacity of *Sorbus domestica* fruit extracts depends on their phenolic content and may be useful for the control of diabetic complications. *Pharmazie* **2008**, *63*, 693–696. [CrossRef]
45. Cohen, G.N. Biosynthesis of the Amino Acids of the Glutamic Acid Family and Its Regulation. In *Microbial Biochemistry*; Springer: Dordrecht, The Netherland, 2014; ISBN 9789048194360.
46. Gupta, P.; Taiyab, A.; Hassan, M.I. Emerging role of protein kinases in diabetes mellitus: From mechanism to therapy. *Adv. Protein Chem. Struct. Biol.* **2021**, *124*, 47–85. [CrossRef]
47. Han, H.S.; Kang, G.; Kim, J.S.; Choi, B.H.; Koo, S.H. Regulation of glucose metabolism from a liver-centric perspective. *Exp. Mol. Med.* **2016**, *48*, e218. [CrossRef]
48. Jang, H.R.; Lee, H.-Y. Mechanisms linking gut microbial metabolites to insulin resistance. *World J. Diabetes* **2021**, *12*, 730–744. [CrossRef]
49. Mathebula, S.D. Polyol pathway: A possible mechanism of diabetes complications in the eye. *Afr. Vis. Eye Health* **2015**, *74*, 13. [CrossRef]
50. Mirra, P.; Nigro, C.; Prevenzano, I.; Leone, A.; Raciti, G.A.; Formisano, P.; Beguinot, F.; Miele, C. The destiny of glucose from a MicroRNA perspective. *Front. Endocrinol.* **2018**, *9*, 46. [CrossRef] [PubMed]
51. Schultze, S.M.; Hemmings, B.A.; Niessen, M.; Tschopp, O. PI3K/AKT, MAPK and AMPK signalling: Protein kinases in glucose homeostasis. *Expert. Rev. Mol. Med.* **2012**, *14*, e1. [CrossRef]
52. Soto-Avellaneda, A.; Morrison, B.E. Signaling and other functions of lipids in autophagy: A review. *Lipids Health Dis.* **2020**, *19*, 214. [CrossRef]
53. Titchenell, P.M.; Lazar, M.A.; Birnbaum, M.J. Unraveling the Regulation of Hepatic Metabolism by Insulin. *Trends Endocrinol. Metab.* **2017**, *28*, 497–505. [CrossRef] [PubMed]
54. Wang, X.C.; Gusdon, A.M.; Liu, H.; Qu, S. Effects of glucagon-like peptide-1 receptor agonists on non-alcoholic fatty liver disease and inflammation. *World J. Gastroenterol.* **2014**, *20*, 14821–14830. [CrossRef]
55. Lachowicz, S.; Wiśniewski, R.; Ochmian, I.; Drzymała, K.; Pluta, S. Anti-microbiological, anti-hyperglycemic and anti-obesity potency of natural antioxidants in fruit fractions of saskatoon berry. *Antioxidants* **2019**, *8*, 397. [CrossRef] [PubMed]
56. Buchholz, T.; Melzig, M.F. Medicinal Plants Traditionally Used for Treatment of Obesity and Diabetes Mellitus—Screening for Pancreatic Lipase and α-Amylase Inhibition. *Phyther. Res.* **2016**, *30*, 260–266. [CrossRef]
57. Siegień, J.; Buchholz, T.; Popowski, D.; Granica, S.; Osińska, E.; Melzig, M.F.; Czerwińska, M.E. Pancreatic lipase and α-amylase inhibitory activity of extracts from selected plant materials after gastrointestinal digestion in vitro. *Food Chem.* **2021**, *355*, 129414. [CrossRef]
58. Worsztynowicz, P.; Napierała, M.; Białas, W.; Grajek, W.; Olkowicz, M. Pancreatic α-amylase and lipase inhibitory activity of polyphenolic compounds present in the extract of black chokeberry (*Aronia melanocarpa* L.). *Process Biochem.* **2014**, *49*, 1457–1463. [CrossRef]
59. Miao, J.; Li, X.; Zhao, C.; Gao, X.; Wang, Y.; Cheng, K.; Gao, W. Solvents effect on active chemicals and activities of antioxidant, anti-α-glucosidase and inhibit effect on smooth muscle contraction of isolated rat jejunum of Chaenomeles speciosa. *J. Funct. Foods* **2018**, *40*, 146–155. [CrossRef]

60. Zhao, C.; Miao, J.; Li, X.; Chen, X.; Mao, X.; Wang, Y.; Hua, X.; Gao, W. Impact of in vitro simulated digestion on the chemical composition and potential health benefits of *Chaenomeles speciosa* and *Crataegus pinnatifida*. *Food Biosci.* **2020**, *35*, 100511. [CrossRef]
61. Zheng, X.; Wang, H.; Zhang, P.; Gao, L.; Yan, N.; Li, P.; Liu, X.; Du, Y.; Shen, G. Chemical composition, antioxidant activity and α-Glucosidase inhibitory activity of *chaenomeles speciosa* from four production areas in China. *Molecules* **2018**, *23*, 2518. [CrossRef] [PubMed]
62. Miao, J.; Zhao, C.; Li, X.; Chen, X.; Mao, X.; Huang, H.; Wang, T.; Gao, W. Chemical Composition and Bioactivities of Two Common Chaenomeles Fruits in China: *Chaenomeles speciosa* and *Chaenomeles sinensis*. *J. Food Sci.* **2016**, *81*, H2049–H2058. [CrossRef]
63. Miao, J.; Li, X.; Zhao, C.; Gao, X.; Wang, Y.; Gao, W. Active compounds, antioxidant activity and α-glucosidase inhibitory activity of different varieties of *Chaenomeles* fruits. *Food Chem.* **2018**, *248*, 330–339. [CrossRef]
64. Turkiewicz, I.P.; Wojdyło, A.; Tkacz, K.; Nowicka, P.; Golis, T.; Bąbelewski, P. ABTS on-line antioxidant, α-amylase, α-glucosidase, pancreatic lipase, acetyl-and butyrylcholinesterase inhibition activity of chaenomeles fruits determined by polyphenols and other chemical compounds. *Antioxidants* **2020**, *9*, 60. [CrossRef]
65. Al-Hallaq, E.K.; Kasabri, V.; Abdalla, S.S.; Bustanji, Y.K.; Afifi, F.U. Anti-Obesity and Antihyperglycemic Effects of *Crataegus aronia* Extracts: In Vitro and in Vivo Evaluations. *Food Nutr. Sci.* **2013**, *04*, 972–983. [CrossRef]
66. Renda, G.; Özel, A.; Barut, B.; Korkmaz, B.; Yayli, N. In vitro protection by crataegus microphylla extracts against oxidative damage and enzyme inhibition effects. *Turk. J. Pharm. Sci.* **2018**, *15*, 77–84. [CrossRef] [PubMed]
67. Li, C.; Son, H.J.; Huang, C.; Lee, S.K.; Lohakare, J.; Wang, M.H. Comparison of *Crataegus pinnatifida* Bunge var. *typica* Schneider and C. pinnatifida Bunge fruits for antioxidant, anti-α-glucosidase, and anti-inflammatory activities. *Food Sci. Biotechnol.* **2010**, *19*, 769–775. [CrossRef]
68. Lin, Y.T.; Lin, H.R.; Yang, C.S.; Liaw, C.C.; Sung, P.J.; Kuo, Y.H.; Cheng, M.J.; Chen, J.J. Antioxidant and Anti-α-Glucosidase Activities of Various Solvent Extracts and Major Bioactive Components from the Fruits of *Crataegus pinnatifida*. *Antioxidants* **2022**, *11*, 320. [CrossRef]
69. Sakhri, F.Z.; Zerizer, S.; Bensouici, C. Evaluation of the Antioxidant, Antidiabetic and Immunomodulatory Activity of *Cydonia oblonga* Fruit Extract. *Chiang Mai Univ. J. Nat. Sci.* **2021**, *20*, e2021052. [CrossRef]
70. De Bellis, R.; Chiarantini, L.; Potenza, L.; Gorassini, A.; Verardo, G.; De Marco, R.; Benayada, L.; Stocchi, V.; Cristina Albertini, M.; Fraternale, D. High production of secondary metabolites and biological activities of *Cydonia oblonga* Mill. pulp fruit callus. *J. Funct. Foods* **2022**, *94*, 105133. [CrossRef]
71. Cano-Lamadrid, M.; Tkacz, K.; Turkiewicz, I.P.; Nowicka, P.; Hernández, F.; Lech, K.; Carbonell-Barrachina, Á.A.; Wojdyło, A. Inhibition of enzymes associated with metabolic and neurological disorder by dried pomegranate sheets as a function of pomegranate cultivar and fruit puree. *J. Sci. Food Agric.* **2021**, *101*, 2294–2303. [CrossRef] [PubMed]
72. Besnard, M.; Megard, D.; Rousseau, I.; Zaragoza, M.C.; Martinez, N.; Mit Javila, M.T.; Inisan, C. Polyphenolic apple extract: Characterisation, safety and potential effect on human glucose metabolism. *Agro Food Ind. Hi Tech.* **2008**, *19*, 16–19.
73. Prpa, E.J.; Corpe, C.P.; Atkinson, B.; Blackstone, B.; Leftley, E.S.; Parekh, P.; Philo, M.; Kroon, P.A.; Hall, W.L. Apple polyphenol-rich drinks dose-dependently decrease early-phase postprandial glucose concentrations following a high-carbohydrate meal: A randomized controlled trial in healthy adults and in vitro studies. *J. Nutr. Biochem.* **2020**, *85*, 108466. [CrossRef]
74. Cianfruglia, L.; Morresi, C.; Bacchetti, T.; Armeni, T.; Ferretti, G. Protection of polyphenols against glyco-oxidative stress: Involvement of glyoxalase pathway. *Antioxidants* **2020**, *9*, 1006. [CrossRef]
75. Schulze, C.; Bangert, A.; Kottra, G.; Geillinger, K.E.; Schwanck, B.; Vollert, H.; Blaschek, W.; Daniel, H. Inhibition of the intestinal sodium-coupled glucose transporter 1 (SGLT1) by extracts and polyphenols from apple reduces postprandial blood glucose levels in mice and humans. *Mol. Nutr. Food Res.* **2014**, *58*, 1795–1808. [CrossRef] [PubMed]
76. Alongi, M.; Verardo, G.; Gorassini, A.; Lemos, M.A.; Hungerford, G.; Cortella, G.; Anese, M. Phenolic content and potential bioactivity of apple juice as affected by thermal and ultrasound pasteurization. *Food Funct.* **2019**, *10*, 7366–7377. [CrossRef]
77. Ankolekar, C.; Sarkar, D.; Greene, D.; Shetty, K. Using Biological Elicitation to Improve Type 2 Diabetes Targeted Food Quality of Stored Apple. *Front. Sustain. Food Syst.* **2021**, *5*, 709384. [CrossRef]
78. Barbosa, A.C.L.; Pinto, M.D.S.; Sarkar, D.; Ankolekar, C.; Greene, D.; Shetty, K. Influence of varietal and ph variation on antihyperglycemia and antihypertension properties of long-term stored apples using in vitro assay models. *J. Food Biochem.* **2012**, *36*, 479–493. [CrossRef]
79. Gong, T.; Yang, X.; Bai, F.; Li, D.; Zhao, T.; Zhang, J.; Sun, L.; Guo, Y. Young apple polyphenols as natural α-glucosidase inhibitors: In vitro and in silico studies. *Bioorg. Chem.* **2020**, *96*, 103625. [CrossRef] [PubMed]
80. Ahmad, R.; Aslam, N.; Sheikh, A. *Malus domestica* as an Inhibitor of Glycation. *Sch. Acad. J. Biosci. Sch. Acad. J. Biosci.* **2014**, *2*, 2321–68831.
81. Wang, X.; Han, M.; Zhang, M.; Wang, Y.; Ren, Y.; Yue, T.; Gao, Z. In vitro evaluation of the hypoglycemic properties of lactic acid bacteria and its fermentation adaptability in apple juice. *LWT* **2021**, *136*, 110363. [CrossRef]
82. Wang, X.; Wang, Y.; Han, M.; Liang, J.; Zhang, M.; Bai, X.; Yue, T.; Gao, Z. Evaluating the changes in phytochemical composition, hypoglycemic effect, and influence on mice intestinal microbiota of fermented apple juice. *Food Res. Int.* **2022**, *155*, 110998. [CrossRef] [PubMed]
83. Tenore, G.C.; Campiglia, P.; Stiuso, P.; Ritieni, A.; Novellino, E. Nutraceutical potential of polyphenolic fractions from Annurca apple (*M. pumila* Miller cv Annurca). *Food Chem.* **2013**, *140*, 614–622. [CrossRef] [PubMed]

84. Nasu, R.; Miura, M.; Gomyo, T. Effects of fruit, spices and herbs on α glucosidase activity and glycemic index. *Food Sci. Technol. Res.* **2005**, *11*, 77–81. [CrossRef]
85. Dorsey, P.G.; Greenspan, P. Inhibition of nonenzymatic protein glycation by pomegranate and other fruit juices. *J. Med. Food* **2014**, *17*, 447–454. [CrossRef]
86. Alu'datt, M.H.; Rababah, T.; Alhamad, M.N.; Johargy, A.; Gammoh, S.; Ereifej, K.; Almajoul, A.; Al-Karaki, G.; Kubow, S.; Ghozlan, K.A. Phenolic contents, in vitro antioxidant activities and biological properties, and HPLC profiles of free and conjugated phenolics extracted from onion, pomegranate, grape, and apple. *Int. J. Food Prop.* **2017**, *20*, 1823–1837. [CrossRef]
87. Żołnierczyk, A.K.; Ciałek, S.; Styczyńska, M.; Oziembłowski, M. Functional properties of fruits of common medlar (*Mespilus germanica* L.) extract. *Appl. Sci.* **2021**, *11*, 7528. [CrossRef]
88. Wu, H.; Xu, B. Inhibitory effects of onion against α-glucosidase activity and its correlation with phenolic antioxidants. *Int. J. Food Prop.* **2014**, *17*, 599–609. [CrossRef]
89. Isbilir, S.S.; Kabala, S.I.; Yagar, H. Assessment of in vitro antioxidant and antidiabetic capacities of medlar (*Mespilus germanica*). *Not. Bot. Horti Agrobot. Cluj-Napoca* **2019**, *47*, 384–389. [CrossRef]
90. Wang, T.; Li, X.; Zhou, B.; Li, H.; Zeng, J.; Gao, W. Anti-diabetic activity in type 2 diabetic mice and α-glucosidase inhibitory, antioxidant and anti-inflammatory potential of chemically profiled pear peel and pulp extracts (*Pyrus* spp.). *J. Funct. Foods* **2015**, *13*, 276–288. [CrossRef]
91. Pavlović, M.O.; Aradski, A.A.; Savić, A.; Janković, S.; Milutinović, M.; Marin, P.D.; Duletić-Lausević, S. Traditional varieties and wild pear from Serbia: A link among antioxidant, antidiabetic and cytotoxic activities of fruit peel and flesh. *Bot. Serbica* **2021**, *45*, 203–213. [CrossRef]
92. Pucel, N.; Sarkar, D.; Labbe, R.G.; Khanongnuch, C.; Shetty, K. Improving Health Targeted Food Quality of Blackberry: Pear Fruit Synergy Using Lactic Acid Bacterial Fermentation. *Front. Sustain. Food Syst.* **2021**, *5*, 703672. [CrossRef]
93. Ankolekar, C.; Pinto, M.; Greene, D.; Shetty, K. In vitro bioassay based screening of antihyperglycemia and antihypertensive activities of Lactobacillus acidophilus fermented pear juice. *Innov. Food Sci. Emerg. Technol.* **2012**, *13*, 221–230. [CrossRef]
94. Barbosa, A.C.L.; Sarkar, D.; Pinto, M.D.S.; Ankolekar, C.; Greene, D.; Shetty, K. Type 2 diabetes relevant bioactive potential of freshly harvested and long-term stored pears using in vitro assay models. *J. Food Biochem.* **2013**, *37*, 677–686. [CrossRef]
95. Sarkar, D.; Ankolekar, C.; Pinto, M.; Shetty, K. Dietary functional benefits of Bartlett and Starkrimson pears for potential management of hyperglycemia, hypertension and ulcer bacteria *Helicobacter pylori* while supporting beneficial probiotic bacterial response. *Food Res. Int.* **2015**, *69*, 80–90. [CrossRef]
96. Ci, Z.; Jiang, C.; Cui, Y.; Kojima, M. Suppressive Effect of Polyphenols from Immature Pear Fruits on Blood Glucose Levels. *J. Food Nutr. Res.* **2018**, *6*, 445–449. [CrossRef]
97. He, M.; Zeng, J.; Zhai, L.; Liu, Y.; Wu, H.; Zhang, R.; Li, Z.; Xia, E. *Effect of In Vitro Simulated Gastrointestinal Digestion on Polyphenol and Polysaccharide Content and Their Biological Activities among 22 Fruit Juices*; Elsevier Ltd.: Amsterdam, The Netherlands, 2017; Volume 102, ISBN 8676922896.
98. Harris, C.S.; Cuerrier, A.; Lamont, E.; Haddad, P.S.; Arnason, J.T.; Bennett, S.A.L.; Johns, T. Investigating Wild Berries as a Dietary Approach to Reducing the Formation of Advanced Glycation Endproducts: Chemical Correlates of In Vitro Antiglycation Activity. *Plant Foods Hum. Nutr.* **2014**, *69*, 71–77. [CrossRef] [PubMed]
99. Hasbal, G.; Yilmaz-Ozden, T.; Sen, M.; Yanardag, R.; Can, A. In vitro investigation of *Sorbus* domestica as an enzyme inhibitor. *Istanb. J. Pharm.* **2020**, *50*, 28–32.
100. Chen, J.; Meng, X. *Aronia melanocarpa* Anthocyanin Extracts Improve Hepatic Structure and Function in High-Fat Diet-/Streptozotocin-Induced T2DM Mice. *J. Agric. Food Chem.* **2022**, *70*, 11531–11543. [CrossRef] [PubMed]
101. Alaghawani, W.; Naser, I. Study the hypoglycemic effect of *Crataegus laevigata* in diabetic rats. *Int. J. Pharm. Clin. Res.* **2013**, *5*, 145–149.
102. Liu, S.; Yu, J.; Fu, M.; Wang, X.; Chang, X. Regulatory effects of hawthorn polyphenols on hyperglycemic, inflammatory, insulin resistance responses, and alleviation of aortic injury in type 2 diabetic rats. *Food Res. Int.* **2021**, *142*, 110239. [CrossRef] [PubMed]
103. Velmurugan, C.; Bhargava, A. Anti-diabetic and hypolipidemic activity of fruits of *Pyrus communis* L. in hyperglycemic rats. *Asian J. Pharm. Clin. Res.* **2013**, *6*, 108–111.
104. Huang, F.; Zhao, R.; Xia, M.; Shen, G.X. Impact of cyanidin-3-glucoside on gut microbiota and relationship with metabolism and inflammation in high fat-high sucrose diet-induced insulin resistant mice. *Microorganisms* **2020**, *8*, 1238. [CrossRef]
105. Zhao, R.; Khafipour, E.; Sepehri, S.; Huang, F.; Beta, T.; Shen, G.X. Impact of Saskatoon berry powder on insulin resistance and relationship with intestinal microbiota in high fat–high sucrose diet-induced obese mice. *J. Nutr. Biochem.* **2019**, *69*, 130–138. [CrossRef]
106. Zhao, R.; Huang, F.; Shen, G.X. Dose-responses relationship in glucose lowering and gut dysbiosis to saskatoon berry powder supplementation in high fat-high sucrose diet-induced insulin resistant mice. *Microorganisms* **2021**, *9*, 1553. [CrossRef]
107. Zhao, R.; Xiang, B.; Dolinsky, V.W.; Xia, M.; Shen, G.X. Saskatoon berry powder reduces hepatic steatosis and insulin resistance in high fat-high sucrose diet-induced obese mice. *J. Nutr. Biochem.* **2021**, *95*, 108778. [CrossRef]
108. Lee, H.S.; Jeon, Y.E.; Jung, J.I.; Kim, S.M.; Hong, S.H.; Lee, J.; Hwang, J.S.; Hwang, M.O.; Kwon, K.; Kim, E.J. Anti-obesity effect of *Cydonia oblonga* Miller extract in high-fat diet-induced obese C57BL/6 mice. *J. Funct. Foods* **2022**, *89*, 104945. [CrossRef]

109. Boqué, N.; Campión, J.; de la Iglesia, R.; de la Garza, A.L.; Milagro, F.I.; San Román, B.; Bañuelos, Ó.; Martínez, J.A. Screening of polyphenolic plant extracts for anti-obesity properties in Wistar rats. *J. Sci. Food Agric.* **2013**, *93*, 1226–1232. [CrossRef]
110. Han, X.; Zhao, W.; Zhou, Q.; Chen, H.; Yuan, J.; Zhang, X.F.; Zhang, Z. Procyanidins from Hawthorn (*Crataegus pinnatifida*) Alleviates Lipid Metabolism Disorder via Inhibiting Insulin Resistance and Oxidative Stress, Normalizing Gut Microbiota Structure and Intestinal Barrier, Further Suppressing Hepatic Inflammation and Li. *Food Funct.* **2022**, *13*, 7901–7917. [CrossRef] [PubMed]
111. Jeon, Y.D.; Kang, S.H.; Moon, K.H.; Lee, J.H.; Kim, D.G.; Kim, W.; Kim, J.S.; Ahn, B.Y.; Jin, J.S. The Effect of Aronia Berry on Type 1 Diabetes In Vivo and In Vitro. *J. Med. Food* **2018**, *21*, 244–253. [CrossRef]
112. He, K.; Li, X.; Chen, X.; Ye, X.; Huang, J.; Jin, Y.; Li, P.; Deng, Y.; Jin, Q.; Shi, Q.; et al. Evaluation of antidiabetic potential of selected traditional Chinese medicines in STZ-induced diabetic mice. *J. Ethnopharmacol.* **2011**, *137*, 1135–1142. [CrossRef]
113. Ogura, K.; Ogura, M.; Shoji, T.; Sato, Y.; Tahara, Y.; Yamano, G.; Sato, H.; Sugizaki, K.; Fujita, N.; Tatsuoka, H.; et al. Oral Administration of Apple Procyanidins Ameliorates Insulin Resistance via Suppression of Pro-Inflammatory Cytokine Expression in Liver of Diabetic ob/ob Mice. *J. Agric. Food Chem.* **2016**, *64*, 8857–8865. [CrossRef]
114. Jurgoński, A.; Juśkiewicz, J.; Zduńczyk, Z. Ingestion of black chokeberry fruit extract leads to intestinal and systemic changes in a rat model of prediabetes and hyperlipidemia. *Plant Foods Hum. Nutr.* **2008**, *63*, 176–182. [CrossRef]
115. Yamane, T.; Kozuka, M.; Konda, D.; Nakano, Y.; Nakagaki, T.; Ohkubo, I.; Ariga, H. Improvement of blood glucose levels and obesity in mice given aronia juice by inhibition of dipeptidyl peptidase IV and α-glucosidase. *J. Nutr. Biochem.* **2016**, *31*, 106–112. [CrossRef] [PubMed]
116. Mu, J.; Xin, G.; Zhang, B.; Wang, Y.; Ning, C.; Meng, X. Beneficial effects of *Aronia melanocarpa* berry extract on hepatic insulin resistance in type 2 diabetes mellitus rats. *J. Food Sci.* **2020**, *85*, 1307–1318. [CrossRef] [PubMed]
117. Qin, B.; Anderson, R.A. An extract of chokeberry attenuates weight gain and modulates insulin, adipogenic and inflammatory signalling pathways in epididymal adipose tissue of rats fed a fructose-rich diet. *Br. J. Nutr.* **2012**, *108*, 581–587. [CrossRef]
118. Shih, C.C.; Lin, C.H.; Lin, Y.J.; Wu, J.B. Validation of the antidiabetic and hypolipidemic effects of hawthorn by assessment of gluconeogenesis and lipogenesis related genes and AMP-activated protein kinase phosphorylation. *Evid. Based Complement. Altern. Med.* **2013**, *2013*, 597067. [CrossRef]
119. Yamane, T.; Imai, M.; Handa, S.; Yamada, K.; Sakamoto, T.; Ishida, T.; Inui, H.; Yamamoto, Y.; Nakagaki, T.; Nakano, Y. Reduction of blood glucose and HbA1c levels by cyanidin 3,5-diglucoside in KKAy mice. *J. Funct. Foods* **2019**, *58*, 21–26. [CrossRef]
120. Heidarianpour, A.; Keshvari, M.; Shahidi, S.; Zarei, M. Modulation of GPC-4 and GPLD1 serum levels by improving glycemic indices in type 2 diabetes: Resistance training and hawthorn extract intervention. *Heliyon* **2023**, *9*, e15537. [CrossRef] [PubMed]
121. du Preez, R.; Wanyonyi, S.; Mouatt, P.; Panchal, S.K.; Brown, L. Saskatoon berry *Amelanchier alnifolia* regulates glucose metabolism and improves cardiovascular and liver signs of diet-induced metabolic syndrome in rats. *Nutrients* **2020**, *12*, 931. [CrossRef]
122. Zahoor, M.; Jan, M.R.; Naz, S. Investigation of repressive and enhancive effects of fruit extracts on the activity of glucose-6-phophatase. *Pak. J. Pharm. Sci.* **2016**, *29*, 1985–1991.
123. Nagahora, N.; Ito, Y.; Nagasawa, T. Dietary Chinese quince polyphenols suppress generation of α-dicarbonyl compounds in diabetic KK-Ay mice. *J. Agric. Food Chem.* **2013**, *61*, 6629–6635. [CrossRef]
124. Zhao, R.; Xie, X.; Le, K.; Li, W.; Moghadasian, M.H.; Beta, T.; Shen, G.X. Endoplasmic reticulum stress in diabetic mouse or glycated LDL-treated endothelial cells: Protective effect of Saskatoon berry powder and cyanidin glycans. *J. Nutr. Biochem.* **2015**, *26*, 1248–1253. [CrossRef]
125. Zhao, R.; Le, K.; Li, W.; Ren, S.; Moghadasian, M.H.; Beta, T.; Shen, G.X. Effects of Saskatoon berry powder on monocyte adhesion to vascular wall of leptin receptor-deficient diabetic mice. *J. Nutr. Biochem.* **2014**, *25*, 851–857. [CrossRef]
126. Moghadasian, M.H.; Masisi, K.; Le, K.; Beta, T.; Shen, G.; Fischer, G. The Potential Anti-Diabetic Effects of Saskatoon Berry in Experimental Mouse Models. *Austin J. Nutr. Food Sci.* **2019**, *7*, 1111. [CrossRef]
127. Baum, J.I.; Howard, L.R.; Prior, R.L.; Lee, S.O. Effect of *Aronia melanocarpa* (Black Chokeberry) supplementation on the development of obesity in mice fed a high-fat diet. *J. Berry Res.* **2016**, *6*, 203–212. [CrossRef]
128. Bhaswant, M.; Shafie, S.R.; Mathai, M.L.; Mouatt, P.; Brown, L. Anthocyanins in chokeberry and purple maize attenuate diet-induced metabolic syndrome in rats. *Nutrition* **2017**, *41*, 24–31. [CrossRef]
129. Oprea, E.; Manolescu, B.N.; Fărcăşanu, I.C.; Mladin, P.; Mihele, D. Studies concerning antioxidant and hypoglycaemic activity of *Aronia melanocarpa* fruits. *Farmacia* **2014**, *62*, 254–263.
130. Valcheva-Kuzmanova, S.; Kuzmanov, K.; Tancheva, S.; Belcheva, A. Hypoglycemic and hypolipidemic effects of *Aronia melanocarpa* fruit juice in streptozotocin-induced diabetic rats. *Methods Find. Exp. Clin. Pharmacol.* **2007**, *29*, 101–105. [CrossRef] [PubMed]
131. Takahashi, A.; Shimizu, H.; Okazaki, Y.; Sakaguchi, H.; Taira, T.; Suzuki, T.; Chiji, H. Anthocyanin-rich phytochemicals from aronia fruits inhibit visceral fat accumulation and hyperglycemia in high-fat diet-induced dietary obese rats. *J. Oleo Sci.* **2015**, *64*, 1243–1250. [CrossRef]
132. Lipińska, P.; Jóźwik, A. Hepatoprotective, Hypoglycemic, and Hypolipidemic Effect of Chokeberry Pomace on Polish Merino Lambs. *Anim. Biotechnol.* **2018**, *29*, 136–141. [CrossRef] [PubMed]
133. Ljubuncic, P.; Azaizeh, H.; Cogan, U.; Bomzon, A. The effects of a decoction prepared from the leaves and unripe fruits of crataegus aronia in streptozotocin-induced diabetic rats. *J. Complement. Integr. Med.* **2006**, *3*, 6. [CrossRef]

134. Asgari, M.; Salehi, I.; Ranjbar, K.; Khosravi, M. Biomedicine & Pharmacotherapy Interval training and *Crataegus persica* ameliorate diabetic nephropathy via miR-126/Nrf-2 mediated inhibition of stress oxidative in rats with diabetes after myocardial ischemia-reperfusion injury. *Biomed. Pharmacother.* **2022**, *153*, 113411. [CrossRef] [PubMed]
135. Valipour Chahardahcharic, S.; Setorki, M. The effect of hydroalcoholic extract of crataegus monogyna on hyperglycemia, oxidative stress and pancreatic tissue damage in streptozotocin-induced diabetic rats. *J. Herbmed Pharmacol.* **2018**, *7*, 294–299. [CrossRef]
136. Aierken, A.; Buchholz, T.; Chen, C.; Zhang, X.; Melzig, M.F. Hypoglycemic effect of hawthorn in type II diabetes mellitus rat model. *J. Sci. Food Agric.* **2017**, *97*, 4557–4561. [CrossRef]
137. Pirmoghani, A.; Salehi, I.; Moradkhani, S.; Karimi, S.A.; Salehi, S. Effect of Crataegus extract supplementation on diabetes induced memory deficits and serum biochemical parameters in male rats. *IBRO Rep.* **2019**, *7*, 90–96. [CrossRef]
138. Zarrinkalam, E.; Ranjbar, K.; Salehi, I.; Kheiripour, N.; Komaki, A. Resistance training and hawthorn extract ameliorate cognitive deficits in streptozotocin-induced diabetic rats. *Biomed. Pharmacother.* **2018**, *97*, 503–510. [CrossRef]
139. Fathy, S.M.; Drees, E.A. Protective effects of Egyptian cloudy apple juice and apple peel extract on lipid peroxidation, antioxidant enzymes and inflammatory status in diabetic rat pancreas. *BMC Complement. Altern. Med.* **2016**, *16*, 8. [CrossRef]
140. Wusu, D.; Kazeem, M.; Lawal, O.; Opoku, A. Antidiabetic Effects of Some Tropical Fruit Extracts in Fructose Induced Insulin Resistant Wistar Rats. *Br. J. Pharm. Res.* **2015**, *7*, 230–235. [CrossRef]
141. Snyder, S.M.; Zhao, B.; Luo, T.; Kaiser, C.; Cavender, G.; Hamilton-Reeves, J.; Sullivan, D.K.; Shay, N.F. Consumption of quercetin and quercetin-containing apple and cherry extracts affects blood glucose concentration, hepatic metabolism, and gene expression patterns in obese C57BL/6J high fat-fed mice. *J. Nutr.* **2016**, *146*, 1001–1007. [CrossRef]
142. Wei, J.; Zhang, G.; Zhang, X.; Gao, J.; Zhou, Z.; Fan, J. Polyphenols from *Sorbus aucuparia* ameliorate insulin resistance and metabolic disorders in diabetic mice. *Curr. Top. Nutraceutical Res.* **2016**, *14*, 227–233.
143. Simeonov, S.B.; Botushanov, N.P.; Karahanian, E.B.; Pavlova, M.B.; Husianitis, H.K.; Troev, D.M. Effects of *Aronia melanocarpa* juice as part of the dietary regimen in patients with diabetes mellitus. *Folia Med.* **2002**, *44*, 20–23.
144. Milutinović, M.; Radovanović, R.V.; Šavikin, K.; Radenković, S.; Arvandi, M.; Pešić, M.; Kostić, M.; Miladinović, B.; Branković, S.; Kitić, D. Chokeberry juice supplementation in type 2 diabetic patients—Impact on health status. *J. Appl. Biomed.* **2019**, *17*, 218–224. [CrossRef]
145. Tasic, N.; Jakovljevic, V.L.J.; Mitrovic, M.; Djindjic, B.; Tasic, D.; Dragisic, D.; Citakovic, Z.; Kovacevic, Z.; Radoman, K.; Zivkovic, V.; et al. Black chokeberry *Aronia melanocarpa* extract reduces blood pressure, glycemia and lipid profile in patients with metabolic syndrome: A prospective controlled trial. *Mol. Cell. Biochem.* **2021**, *476*, 2663–2673. [CrossRef]
146. Yamane, T.; Kozuka, M.; Wada-Yoneta, M.; Sakamoto, T.; Nakagaki, T.; Nakano, Y.; Ohkubo, I. Aronia juice suppresses the elevation of postprandial blood glucose levels in adult healthy Japanese. *Clin. Nutr. Exp.* **2017**, *12*, 20–26. [CrossRef]
147. Makarova, E.; Górnaś, P.; Konrade, I.; Tirzite, D.; Cirule, H.; Gulbe, A.; Pugajeva, I.; Seglina, D.; Dambrova, M. Acute anti-hyperglycaemic effects of an unripe apple preparation containing phlorizin in healthy volunteers: A preliminary study. *J. Sci. Food Agric.* **2014**, *95*, 560–568. [CrossRef]
148. Shoji, T.; Yamada, M.; Miura, T.; Nagashima, K.; Ogura, K.; Inagaki, N.; Maeda-Yamamoto, M. Chronic administration of apple polyphenols ameliorates hyperglycaemia in high-normal and borderline subjects: A randomised, placebo-controlled trial. *Diabetes Res. Clin. Pract.* **2017**, *129*, 43–51. [CrossRef]
149. Lu, X.; Lu, J.; Fan, Z.; Liu, A.; Zhao, W.; Wu, Y.; Zhu, R. Both isocarbohydrate and hypercarbohydrate fruit preloads curbed postprandial glycemic excursion in healthy subjects. *Nutrients* **2021**, *13*, 2470. [CrossRef] [PubMed]
150. Yuliwati, N.; Nugroho, R.F. The Potential of Strawberry, Rome Beauty Apple, and New Combination on Fasting Blood as Supporting Diet Therapy in Patients with Type II Diabetes Mellitus. *Glob. Med. Health Commun.* **2021**, *9*, 69–75. [CrossRef]
151. Gancheva, S.; Ivanova, I.; Atanassova, A.; Gancheva-tomova, D.; Eftimov, M.; Moneva, K.; Zhelyazkova-savova, M. Effects of *Aronia melanocarpa* fruit juice on oxidative stress, energy homeostasis, and liver function in overweight and healthy—Weight individuals. *Scr. Sci. Med.* **2021**, *53*, 39–46. [CrossRef]
152. Johnston, K.L.; Clifford, M.N.; Morgan, L.M. Possible role for apple juice phenolic compounds in the acute modification of glucose tolerance and gastrointestinal hormone secretion in humans. *J. Sci. Food Agric.* **2002**, *82*, 1800–1805. [CrossRef]
153. Sołtys, A.; Galanty, A.; Podolak, I. *Ethnopharmacologically Important But Underestimated Genus Sorbus: A Comprehensive Review*; Springer: Amsterdam, The Netherlands, 2020; Volume 19, ISBN 0123456789.
154. Sidor, A.; Gramza-Michałowska, A. Gramza-Michałowska Black Chokeberry *Aronia melanocarpa* L.—A Qualitative Composition, Phenolic Profile and Antioxidant Potential. *Molecules* **2019**, *24*, 3710. [CrossRef] [PubMed]
155. Patocka, J.; Bhardwaj, K.; Klimova, B.; Nepovimova, E.; Wu, Q.; Landi, M.; Kuca, K.; Valis, M.; Wu, W. Malus domestica: A Review on Nutritional Features, Chemical Composition, Traditional and Medicinal Value. *Plants* **2020**, *9*, 1408. [CrossRef] [PubMed]
156. Zhou, L.; Li, Y.; Gong, X.; Li, Z.; Wang, H.; Ma, L.; Tuerhong, M.; Abudukeremu, M.; Ohizumi, Y.; Xu, J.; et al. Preparation, characterization, and antitumor activity of *Chaenomeles speciosa* polysaccharide-based selenium nanoparticles. *Arab. J. Chem.* **2022**, *15*, 103943. [CrossRef]
157. Zlobin, A.A.; Martinson, E.A.; Litvinets, S.G.; Ovechkina, I.A.; Durnev, E.A.; Ovodova, R.G. Pectin polysaccharides of rowan *Sorbus aucuparia* L. *Russ. J. Bioorganic Chem.* **2012**, *38*, 702–706. [CrossRef]
158. Gyurova, D.K.; Enikova, R.K. Dried Fruits—Brief Characteristics of Their Nutritional Values. Author's Own Data for Dietary Fibers Content. *J. Food Nutr. Sci.* **2014**, *2*, 105–109. [CrossRef]

159. Jenkins, D.J.; Srichaikul, K.; Kendall, C.W.; Sievenpiper, J.L.; Abdulnour, S.; Mirrahimi, A.; Meneses, C.; Nishi, S.; He, X.; Lee, S.; et al. The Relation of Low Glycaemic Index Fruit Consumption to Glycaemic Control and Risk Factors for Coronary Heart Disease in Type 2 Diabetes. *Diabetologia* **2011**, *54*, 271–279. [CrossRef]
160. Monro, J.A. Glycaemic Glucose Equivalent: Combining Carbohydrate Content, Quantity and Glycaemic Index of Foods for Precision in Glycaemia Management. *Asia Pac. J. Clin. Nutr.* **2002**, *11*, 217–225. [CrossRef] [PubMed]

nutrients

Article

Effects of Fermented *Artemisia annua* L. and *Salicornia herbacea* L. on Inhibition of Obesity In Vitro and In Mice

Jeong-Yeon On [1,†], Su-Hyun Kim [2,†], Jeong-Mee Kim [3], Sungkwon Park [4], Ki-Hyun Kim [5], Choong-Hwan Lee [2,6,*] and Soo-Ki Kim [1,3,*]

1 Department of Animal Science and Technology, Konkuk University, 120 Neungdong-ro, Gwangjin-gu, Seoul 05029, Republic of Korea; on7701@naver.com
2 Department of Bioscience and Biotechnology, Konkuk University, Seoul 05029, Republic of Korea
3 Institute of Animal Resource Center, Konkuk University, Seoul 05029, Republic of Korea
4 Department of Food Science and Biotechnology, Sejong University, Seoul 05006, Republic of Korea
5 Animal Welfare Research Team, National Institute of Animal Science, RDA, Wanju 55365, Republic of Korea
6 Research Institute for Bioactive-Metabolome Network, Konkuk University, Seoul 05029, Republic of Korea
* Correspondence: chlee123@konkuk.ac.kr (C.-H.L.); sookikim@konkuk.ac.kr (S.-K.K.); Tel.: +82-2-2049-6177 (C.-H.L.); +82-2-450-3728 (S.-K.K.); Fax: +82-2-458-3728 (S.K.K.)
† These authors contributed equally to this work.

Abstract: Plant extracts including secondary metabolites have anti-inflammatory and anti-obesity activities. This study was conducted to investigate the anti-obesity properties of fermented *Artemisia annua* (AW) and *Salicornia herbacea* (GW) in vitro and in mice. The metabolite profiling of AW and GW extracts was performed using UHPLC−LTQ−Orbitrap–MS/MS, and gene expression was analyzed using real-time PCR for adipocyte difference factors. The anti-obesity effects in mice were measured using serum AST, ALT, glucose, TG, and cholesterol levels. Metabolites of the plant extracts after fermentation showed distinct differences with increasing anti-obesity active substances. The efficacy of inhibitory differentiation adipogenesis of 3T3-L1 adipocytes was better for GW than AW in a concentration-dependent manner. RT-PCR showed that the GW extract significantly reduced the expression of genes involved in adipocyte differentiation and fat accumulation (C/EBPα, PPARγ, and Fas). In C57BL/6 mice fed the HFD, the group supplemented with AW and GW showed reduced liver weight, NAS value, and fatty liver by suppressing liver fat accumulation. The GW group significantly reduced ALT, blood glucose, TG, total cholesterol, and LDL-cholesterol. This study displayed significant metabolite changes through biotransformation in vitro and the increasing anti-obesity effects of GW and AW in mice. GW may be applicable as functional additives for the prevention and treatment of obesity.

Keywords: annual wormwood; glasswort; metabolites; fermentation; anti-obesity; mouse

Citation: On, J.-Y.; Kim, S.-H.; Kim, J.-M.; Park, S.; Kim, K.-H.; Lee, C.-H.; Kim, S.-K. Effects of Fermented *Artemisia annua* L. and *Salicornia herbacea* L. on Inhibition of Obesity In Vitro and In Mice. *Nutrients* **2023**, *15*, 2022. https://doi.org/10.3390/nu15092022

Academic Editors: Sonia de Pascual-Teresa, Lindsay Brown and Luis Goya

Received: 22 March 2023
Revised: 19 April 2023
Accepted: 19 April 2023
Published: 22 April 2023

1. Introduction

Obesity, which is considered a major public health problem, ranks as the fifth leading cause of death worldwide and has tripled in the last 40 years [1]. The World Obesity Atlas published in 2022 predicts that one billion people worldwide will live with obesity by 2030 [2]. The main cause of obesity is the increase in body fat storage efficiency because of a high-fat diet and the accumulation of fat due to adipogenesis by the differentiation of adipocytes [3]. Increased fat accumulation accompanies various diseases such as obesity, insulin resistance, type 2 diabetes, hyperglycemia, dyslipidemia, and metabolic syndrome due to changes in body weight, increased fasting blood sugar, and cholesterol [4,5]. In addition, mRNA expression of genes, such as peroxisome proliferator-activated receptor-γ (PPARγ), CCAAT/enhancer-binding protein-α (C/EBPα), and leptin, is known increase in obesity [6]. Adipose tissue plays an important role in the immune response, contributing to a chronic low-grade inflammation process linked to tumor development through adipokine secretion molecules, hormone (leptin), growth factors, and proinflammatory cytokines, proliferation, angiogenesis, and expression process [7]. Leptin participates in endocrine metabolism as well as the regulation of appetite and energy expenditure. Recent data have emphasized the brain's ability to control the complex mechanisms of food intake and storage [8]. The highlight is the value of adipocyte-secreted hormones in obesity and cancer prevention through risk reduction and specific adapted therapies [9]. Recently, in order to prevent and treat obesity, methods such as diet and exercise, as well as next-generation anti-obesity drugs, are being developed [10]. Traditionally, various medicinal plants have been used as treatments for diseases, and various medicinal plants were known to have an effect on obesity [11]. Many studies have scientifically proven that plant extracts can be potential preventive and therapeutic agents for obesity because physiologically active compounds in plant extracts possess anti-obesity effects [12]. Medicinal plants with antioxidant and anti-obesity effects include *Moringa oleifera*, *Solenostemma argel*, and *Salvia* species [13]. The use of natural therapies for weight loss has increased based on their reliability, safety, and cost, compared to synthetic drugs or surgical procedures that have side effects [14]. Among many medicinal plants, annual wormwood (AW: *Artemisia annua* L.) is an annual plant belonging to the *Asteraceae* family that is known for its anti-inflammatory, anti-cancer, antibacterial, and anti-obesity effects and can be found in some parts of Asia, including Korea and China [15]. Compounds, such as ketone, camphor, and 1,8-cineole, exist in AW essential oil [16], and extracts of AW are reported to have excellent antioxidant capacity [17]. In addition, AW has been reported to contribute to the prevention of obesity, including compounds with various anti-obesity activities, such as artemisinic acid, rhamnetin, myricetin, and sabinene [18–20]. AW is used as livestock feed and raw material for medicine and cosmetics [21,22].

Glasswort (GW: *Salicornia europaea*) is a halophyte belonging to the pemphigus family *Chenopodiaceae*, which grows on the coasts of temperate and subtropical regions [23]. GW is known to have various functionalities such as antioxidant, anti-cancer, and anti-obesity [24,25]. Chemical constituents such as sterols, quinic acid derivatives, flavonoid derivatives, triterpenoid saponins, and pentadecyl ferulate are present in GW, and some of them have antioxidant activity [26]. In addition, there are various anti-obesity compounds such as trans-ferulic acid (TFA) and isorhamnetin 3-O-β-D-glucopyranoside in GW, which are known to regulate adipogenesis and differentiation inhibition [27,28]. GW is applied as a sodium substitute to various foods such as in baking and sausage making [29,30]. Some of the medicinal herbs used in traditional medicine were biologically activated through the biotransformation of bacteria through fermentation [31]. It is known that fermentation induces the structural destruction of plant cell walls in plant foods and induces the release or synthesis of various antioxidant compounds, especially increasing the amount of phenols and flavonoids due to microbial hydrolysis [32]. As such, AW and GW have been studied for their various physiological activities, but most of them focused on antioxidant and anti-cancer effects. In particular, there are fewer studies on the anti-obesity effects of fermented natural products, and studies comparing anti-obesity effects, physiological activities, and metabolomes before and after fermentation have not been reported.

This study was conducted to elucidate the prospective substances for anti-obesity and to investigate the effects of these medicinal plants as functional additives, as well as natural medicaments. Therefore, the metabolome changes in the medicinal plants AW and GW before and after fermentation, and the differentiation inhibitory ability using 3T3-L1 adipocytes in vitro, were confirmed. In addition, to evaluate the possibility of obesity prevention, male C57BL/6 mice were fed the HFD supplemented with AW and GW.

2. Materials and Methods

2.1. Preparation of Medicinal Plants

Annual wormwood (AW) and glasswort (GW) were purchased in powder form from a Korean food company and stored in a refrigerator at 4 °C before use (AW: Ingreen Co., Ltd.,

Gangwha-do, Republic of Korea; GW: Suncheon Bay Hamcho Agricultural Cooperative Corporation, Suncheon-si, Jeollanam-do, Republic of Korea).

2.2. Fermentation of Medicinal Plants

The medicinal plant powder, 5% (*w/v*) of AW and GW was added separately to one-fifth diluted Man Rogosa Sharpe (MRS) broth and Bacillus minimal medium (BMM) for the fermentation. Then, pH was adjusted to 7.0 ± 0.5 and sterilized at 121 °C for 15 min. For fermentation, 1% (*v/v*) of *Lactobacillus plantarum* SK3494 and *Enterococcus faecium* SK4369, which was isolated from natural extract of each plant [33,34], was inoculated into the respective medium. The AW and GW were fermented for 16 h at 37 °C using a shaking incubator (BF-60SIRL, Biofree, Seoul, Republic of Korea) under shaking of 100 rpm.

2.3. Extraction of Fermented Plants

Before and after fermentation, solution was collected at each time point (0 and 16 h) and stored at −20 °C. After fermentation, ultra-pure water was added to each solution at a rate of 70% and then boiled at 100 °C for 15 min using a water bath (Wise Bath, Seoul, Republic of Korea). Extracted solutions were cooled to room temperature and centrifuged at 14,500 rpm for 15 min (Mega 17R; Hanil, Seoul, Republic of Korea). The collected supernatants were freeze-dried using a freeze dryer (UniFreeze FD-8, Daihan Scientific Ltd., Seoul, Republic of Korea) and stored in a deep freezer at −80 °C for future experiments.

2.4. Chemicals and Reagents

The HPLC-grade water and methanol were purchased from Fisher Scientific (Pittsburgh, PA, USA). Formic acid was purchased from Sigma-Aldrich (St. Louis, MO, USA).

2.5. Preparation of Metabolic Extracts

The freeze-dried supernatants of non-fermented and fermented samples from AW and GW were extracted using 75% methanol (50 mg/mL). The mixtures were sonicated for 30 min and incubated for 24 h under deep-freezing (−20 °C) conditions. Each of the samples was centrifuged at 11,000 rpm for 10 min, and the collected supernatant was filtered through a 0.22 μm polytetrafluoroethylene filter and dried using a speed vacuum concentrator (Biotron, Seoul, Republic of Korea). The dried samples were reconstituted with 75% methanol to a final concentration of 20 mg/mL, to be used for instrument analysis.

2.6. UHPLC–LTQ–Orbitrap–MS Profiling

Metabolite profiling of non-fermented and fermented extracts of AW and GW was preformed using UHPLC–LTQ–Orbitrap–MS/MS. The 2-chloro-phenylalanine (1.5 μg/mL) was used as an internal standard (IS). A sample of 5 μL was injected into a UHPLC system equipped with a Vanquish binary pump H system (Thermo Fisher Scientific, Waltham, MA, USA) coupled with auto-sampler and column compartment at the flow rate of 3 mL/min. Chromatographic separation was performed on a Phenomenex KINETEX® C18 column (100 mm × 2.1 mm, 1.7 μm particle size: Torrance, CA, USA). The mobile phase, 0.1% formic acid in water (A) and 0.1% formic acid in acetonitrile (B), was used in ESI negative mode. The gradient parameters were set as follows: 5% solvent B was maintained initially for 1 min, followed by a linear increase to 100% solvent B over 9 min and then sustained at 100% solvent B for 1 min, with a gradual decrease to 5% solvent B over 3 min. The total run time was 14 min. The column temperature was set to 40 °C, the flow rate was 0.3 mL/min, and the injection volume was 5 μL. The MS data were collected in the range of 100–1000 *m/z* (under negative- and positive-ion modes) using an Orbitrap Velos ProTM system, which was combined with an ion trap mass spectrometer (Thermo Fisher Scientific, Waltham, MA, USA) coupled with an HESI-II probe. The probe heater and capillary temperatures were set to 300 °C and 350 °C, respectively. The capillary voltage was set to 2.5 kV in negative mode (positive mode: 3.7 Kv).

2.7. Cell Viability Assay

Cell viability of the plant extracts in 3T3-L1 preadipocytes was measured using cell count kit-8 (WST-8/CCK8, Abcam, ab228554), according to the manufacturer's instructions. The 3T3-L1 preadipocytes were seeded at 1×10^4 cells in a 96-well plate (SPL, SPL30096) and cultured overnight in a CO_2 incubator at 37 °C. After incubation, the cells were treated with various concentrations (0.2~50 μg/mL) of plant extracts dissolved in 10 mg/mL DMSO solution and cultured for 72 h. Then, 10 μL of cell count kit-8 (WST-8/CCK8, Abcam, ab228554) reagent was added, and after 30 min, absorbance was measured at 460 nm using a microplate reader. The cell viability was calculated using the following equation:

$$\text{Cell viability (\%)} = (\text{OD}_{\text{Sample}} - \text{OD}_{\text{Media}})/(\text{OD}_{\text{DMSO solution}} - \text{OD}_{\text{Media}}) \times 100$$

2.8. Oil Red O Staining

Oil Red O [35] staining was performed to confirm intracellular lipid droplet production. Differentiated 3T3-L1 cells were washed with PBS and fixed with 10% formalin for 30 min. After washing, 60% isopropanol and 0.5% ORO solution (Sigma-Aldrich, St. Louis, MO, USA) were added to the fixed cells and stained for 20 min at room temperature. Then, the cell was washed with distilled water, dried at 37 °C, and added to 200 μL of isopropanol to dissolve intracellular ORO solution; then, 100 μL of each was transferred to a 96-well plate, and absorbance was measured at 540 nm using a microplate reader (Synergy 2, BioTek Instruments Inc., Winooski, VT, USA). The ability to inhibit adipocyte differentiation of fermented AW and GW was confirmed for each concentration of 3.1~50 μg/mL.

2.9. Reverse Transcription-Polymerase Chain Reaction (RT-PCR)

Reverse transcription-polymerase chain reaction (RT-PCR) was performed to confirm the expression of regulators in fat metabolism using test substances in differentiated 3T3-L1 cells. Changes in mRNA levels of CCAAT/enhancer-binding protein-alpha (C/EBPα), peroxisome proliferator-activated receptor-γ (PPARγ), leptin, and Fas, regulators involved in fat metabolism, were analyzed using RT-PCR and shown in Table 1. Total RNA was isolated using TRI reagent (Sigma, St. Louis, MO, USA), cDNA was synthesized using Power cDNA synthesis kit (iNtRON, Seoul, Republic of Korea), according to the manufacturer's instructions, and real-time PCR was performed using SYBR green and each primer. PCR was performed using a real-time PCR machine (Corbett, Mortlake, Australia), and gene expression was quantified and analyzed using Rotor-Gene Q Series Software 2.3.1. (Qiagen, Hilden, Germany).

Table 1. Primer sequences of target genes used in the PCR.

Gene	Name	Primer Sequence	Tm (°C)
C/EBPα	mC/EBPα_F	CAA GAA GTC GGT GGA CAA G	55.2
	mC/EBPα_R	GCT TTA TCT CGG CTC TTG C	55.2
PPARγ	mPPARγ_F	GAC ATC CAA GAC AAC CTG CT	55.4
	mPPARγ_R	TGT CAT CTT CTG GAG CAC CT	55.4
Leptin	mLeptin_F	TGA CAC CAA AAC CCT CAT CA	53.4
	mLeptin_R	AGC CCA GGA ATG AAG TCC A	55.2
Fas	mFas_F	AGA GAT CCC GAG ACG CTT CT	57.4
	mFas_R	GCT TGG TCC TTT GAA GTC GAA GA	58.2

2.10. Animals and Diet

The animal experimentation was approved by the NDIC Co., Ltd. Institutional Animal Care and Use Committee (IACUC), Gyeonggi-do, Republic of Korea, in accordance with the guidelines of IACUC (N2021001). Six-week-old male C57BL/6 mice (20~21 g) were obtained from ORIENTBIO Inc., Republic of Korea. The animals were housed in a polycarbonate breeding box under a controlled environment (temperature: 21 ± 2 °C;

humidity: 50 ± 20%; lighting time: 12 h/day) and were freely fed food and water. Mice were placed into 6 groups of 8 mice each: normal diet (ND, TEKLAD Certified Global 18% Protein Rodent DIET 2918C, Harlan TEKLAD, Madison, WI, USA); high-fat diet (HFD, D12492; 5.24 kcal/g with 60% of fat-, 20% of protein-, and 20% of carbohydrate-derived calories, Research DIETS Inc., New Brunswick, NJ, USA); HFD supplemented with fermented 100 mg/kg AW extract (AW 100); HFD supplemented with fermented 300 mg/kg AW extract (AW 300); HFD supplemented with fermented 100 mg/kg GW extract (GW 100); and HFD supplemented with fermented 300 mg/kg GW extract (GW 300). Oral administration was performed once a day, and the body weight of the experimental animals was measured once a week for 12 weeks. Feed intake was measured once a week after the test substance was administered. The food efficiency ratio was calculated as FER = total body weight gain (g)/total food intake (g).

2.11. Sample Preparation and Treatment

For autopsy, after the inhalational anesthesia with isoflurane, blood was collected through the abdominal vena cava and the liver was removed. The extracted liver was washed with physiological saline, dried with a filter paper, and then the absolute weight was measured using an electronic balance. After calculating the absolute weights, a portion of liver was fixed in 10% neutral buffered formalin for histological examination. The remaining liver was rapidly frozen for hepatic triglyceride (Hepatic TG) analysis and placed in deep freezer (Model 706, Thermo Fisher Scientific Inc., Clevelang, OH, USA) below −70 °C. Blood collected from the abdominal vena cava was placed in a 0.6 mL SST tube (Microtainer, BD, USA), completely solidified, centrifuged at 4 °C at 5000 rpm for 15 min, put into a 1.5 mL tube, and stored in deep freezer (Model 706, Thermo Fisher Scientific Inc., USA) below −70 °C.

2.12. Histological Analysis

For histological analysis, liver was fixed in 10% neutral buffered formalin. The fixed tissue was prepared as a paraffin block through a general tissue-processing process and sectioned. The remaining tissues were stored in 10% neutral buffered formalin solution. Hematoxylin and Eosin (H&E) staining was performed on thin slices of liver tissue. The accumulation of lipid droplets in the liver tissue and the degree of adipose inflammation were analyzed using the evaluation method used in the non-alcoholic steatosis model. Non-alcoholic fatty liver disease activity score [13] evaluation, which is a synthesis of five broad categories (steatosis, inflammation, hepatocellular injury, fibrosis, and miscellaneous features), was evaluated on H&E-stained liver tissue slides using the method described in a previous study [36].

2.13. Serum and Hepatic Triglyceride Analysis

At the time of autopsy, serum was collected from the abdominal vena cava. Aspartate aminotransferase (AST), alanine aminotransferase (ALT), glucose [24], total cholesterol (T-Chol), triglycerides (TGs), HDL-cholesterol (HDL-C), and LDL-cholesterol (LDL-C) were analyzed using a blood chemistry analyzer (AU480, Beckman Coulter, Germany) using serum stored in deep freezer. Fasting blood insulin levels were measured using the Rat/Mouse Insulin ELISA Kit (EZRMI-13K, Millipore, MA, USA), according to the manufacturer's instructions, with the separated serum collected from the abdominal vena cava during autopsy and placed in a 0.6 mL SST tube (Microtainer, BD, USA). The remaining serum was stored in deep freezer. At necropsy, 200 mg of liver tissue collected was weighed and washed with normal physiological saline. After adding 1 mL of PBS containing 1% Triton X-100 per 200 mg of liver tissue, homogenization was performed. The liver homogenate was centrifuged at 10,000 rpm at 4 °C for 10 min to separate the supernatant, and then the TG content in the liver tissue was measured with a microplate reader (570 nm) using a TG quantification kit (Cell Biolabs, STA-396, San Diego, CA, USA).

2.14. Data Processing and Statistical Analysis

UHPLC–LTQ–Orbitrap–MS raw data files were converted to NetCDF (*.cdf) using Thermo X caliber software (version 2.1, Thermo Fisher Scientific). Retention time correction, peak detection, and alignment were processed using the Metalign software package (http://www.metalign.nl, accessed on 15 November 2021). Metabolites were tentatively identified based on various data comparing mass fragment patterns, retention time, and MS analysis data with standard compounds under identical conditions and in commercial databases, such as the National Institutes of Standards and Technology Library (version 2.0, 2011, FairCom, Gaithersburg, MD, USA), pubchem (https://pubchem.ncbi.nlm.nih.gov/accessed on 1 December 2021), chemspider (http://www.chemspider.com/ on 1 December 2021), and the Human Metabolome Database (HMDB; https://hmdb.ca/ accessed on 1 December 2021). Statistical analysis was performed using SIMCA-P+ software (version 12.0, Umetrics, Umea, Sweden) based on partial least squares discriminant analysis (PLS-DA) modeling to determine metabolite differences between different incubation times. The peaks were selected based on variable importance in the projection (VIP) values and significant differences were determined using analysis of variance (ANOVA). In cell, experiment results were expressed as the mean values ± standard deviations, and significance was tested using the paired-comparison t-test of IBM SPSS Statistics 25 for Windows (IBM, New York, NY, USA) ($p < 0.05$). For multiple comparison, Duncan's multiple range test was used after comparison test with one-way ANOVA, and significance was verified at $p < 0.05$ level. All results obtained in the mouse experiment were expressed as mean values ± SD, and body weight and weight gain were tested using one-way ANOVA using IBM SPSS Statistics 25 for Windows (IBM, New York, NY, USA), and Fisher's Least Significant Difference (LSD) test was performed ($p \leq 0.05$; $p \leq 0.01$).

3. Results

3.1. Multivariate Analysis in Annual Wormwood and Glasswort with LAB-Mediated Fermentation

Non-fermented and fermented extracts of AW and GW were analyzed with UHPLC–MS/MS combined with multivariate analysis on the ingredients to investigate the metabolic change using LAB-mediated fermentation. As a result of principal component analysis (PCA), PC1 (43.4%) and PC2 (17.4%) showed a clear separation of four kinds of extracts, reflecting a metabolic difference between plant species and fermentation (Figure 1A). Similar cluster patterns were also observed in the partial least squares discriminant analysis (PLS-DA) model with a high predictive ability (Q2 = 0.976) as well as a considerable significance metric ($p = 1.87 \times 10^{-5}$) (Figure 1B).

Based on the PLS-DA model for profiling the dataset, we selected the significantly discriminant metabolites using variable importance in projection (VIP) values > 1.0 and $p > 0.05$. A total of forty-two significantly discriminant metabolites were selected, and thirty-nine of them were tentatively identified including three peptides, two organic acids, six fatty acids, eight phenolic acids and derivatives, seven quinic acids and derivatives, seven benzoic acids and derivatives, and four flavonoids, among two others (Table 2). Plant extracts are rich sources of volatile terpenoids and phenolic compounds with complex mixtures of organic framework-added functional groups including alcohol, aldehydes, esters, ethers, ketones, and phenols. It is used as an analgesic, antibacterial, antidepressant, antimicrobial, and antioxidant agent [22,26]. Polyphenolic compounds are one of the most critical ingredients related to free radical scavenging activity in medicinal plants. They exhibit various biological properties, such as antioxidant, cardioprotective, anti-mutagenic, antibacterial, and anti-inflammatory activities [24].

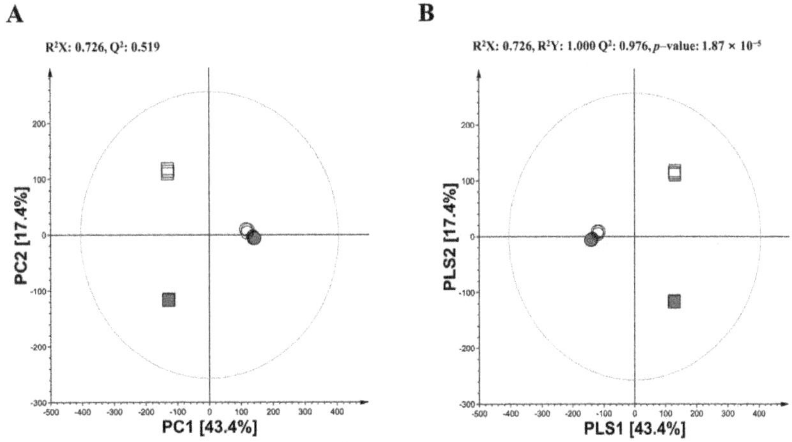

Figure 1. (**A**) PCA score plot and (**B**) PLS-DA score plot derived from UHPLC–LTQ–Orbitrap–MS/MS datasets for non-fermented and fermented extracts of annual wormwood (AW) and glasswort (GW). (■: AN, non-fermentation of AW; □: AF, fermentation of AW with *L. plantarum;* •: GN, non-fermentation of GW; and ○: GF, fermentation of GW with *Ec. Faecium*).

Table 2. Tentatively identified metabolites in non-fermented and fermented extracts of AW and GW based on the UHPLC–LTQ–Orbitrap–MS/MS analyses.

No.	Tentative Metabolite	VIP [a] 1	VIP 2	RT (min) [b]	MW	Measured Mass Negative Mode [c] (m/z)	MS/MS Fragments	Molecular Formula	Delta ppm
Peptides									
1	Pyroglutamyl-valine	1.47	1.12	2.18	228	227.1046	227 > 183 > 155, 127, 82	$C_{10}H_{15}O_4N_2$	3.698
2	Pyroglutamyl-leucine	1.43	1.15	4.17	242	241.1202	241 > 197 > 169, 141	$C_{11}H_{17}O_4N_2$	3.441
3	Lactoyl-tryptophan	1.45	1.13	4.53	276	275.1049	275 > 231, 127 > 109	$C_{14}H_{15}O_4N_2$	4.325
Organic acid									
4	Hydroxyglutaric acid	0.97	1.46	1.06	148	147.0307	147 > 129 > 101, 85	$C_5H_7O_5$	5.124
5	Succinic Acid	1.17	1.35	1.06	118	117.0200	117 > 99, 73	$C_4H_5O_4$	6.307
Fatty acid									
6	Hydroxyisocaproic acid	0.91	1.50	4.02	132	131.0722	131 > 113, 85	$C_6H_{11}O_3$	6.275
7	9,12,13-TriHOME	1.09	0.79	6.59	330	329.2343	329 > 229 > 211, 125	$C_{18}H_{33}O_5$	2.742
8	9-DiHODE	0.96	1.42	7.29	312	311.2241	311 > 293 > 275, 185	$C_{18}H_{31}O_4$	4.297
9	9,10-DHOME	1.36	0.97	7.99	314	313.2393	313 > 295 > 277, 195	$C_{18}H_{33}O_4$	2.737
10	Hydroxymyristic acid	0.92	1.13	8.99	244	243.1974	243 > 225 > 207, 181	$C_{14}H_{27}O_3$	3.257
11	Hydroxystearic acid	0.18	1.03	9.64	300	299.2598	299 > 281, 253 > 249, 225	$C_{18}H_{35}O_3$	2.211
Phenolic acids and deriatives									
12	Hydroxyphenyllactic acid	0.89	1.51	2.37	182	181.0522	181 > 163 > 119	$C_9H_9O_4$	4.462
13	Dihydrocaffeic acid	0.90	1.50	3.54	182	181.0515	181 > 137 > 119, 109	$C_9H_9O_4$	4.849
14	Caffeic acid	0.60	1.61	4.02	180	179.0360	179 > 135 > 107, 91	$C_9H_7O_4$	5.574
15	Phenyllactic acid	0.91	1.50	4.56	166	165.0571	165 > 147 > 121,97	$C_9H_9O_3$	5.226
16	Coumaric acid	0.60	1.60	4.69	164	163.0409	163 > 119 > 91	$C_9H_7O_3$	5.168
17	Scopoletin	1.42	1.16	4.89	192	191.0358	191 > 177 > 104	$C_{10}H_7O_4$	4.177
18	Ferulic acid	1.42	1.01	4.96	194	193.0516	193 > 178, 149 > 134	$C_{10}H_9O_4$	5.117
19	Phloretic acid	1.08	1.41	4.97	166	165.0564	165 > 147, 121 > 106, 93	$C_9H_9O_3$	4.68
Quinic acid and deriatives									
20	Quinic acid	1.29	1.26	0.85	192	191.0569	191 > 173, 127, 111, 85	$C_7H_{11}O_6$	4.128
21	3-caffeoylquinic acid	1.50	1.08	2.29	354	353.0894	353 > 191 > 173, 127, 85	$C_{16}H_{17}O_9$	2.222
22	5-caffeoylquinic acid	1.51	1.07	3.85	354	353.0894	353 > 191 > 173, 127, 85	$C_{16}H_{17}O_9$	4.573
23	3,4-di-*O*-caffeoylquinic acid	1.50	1.08	4.39	516	515.1215	515 > 353 > 191, 179	$C_{25}H_{23}O_{12}$	3.923
24	3-feruloyl-4-caffeoylquinic acid	1.50	1.09	4.78	530	529.1362	529 > 367 > 193	$C_{26}H_{25}O_{12}$	4.688

Table 2. *Cont.*

No.	Tentative Metabolite	VIP [a] 1	VIP 2	RT (min) [b]	MW	Measured Mass Negative Mode [c] (*m/z*)	MS/MS Fragments	Molecular Formula	Delta ppm
25	3,5-di-*O*-caffeoylquinic acid	1.18	0.83	5.03	516	515.1213	515 > 353 > 191, 179, 135	$C_{25}H_{23}O_{12}$	2.389
26	3-feruloyl-5-caffeoylquinic acid	1.44	1.15	5.48	530	529.1367	529 > 367 > 191, 173	$C_{26}H_{25}O_{12}$	2.855
Benzoic acid and derivatives									
27	Protocatechuic acid-*O*-glucoside	0.77	1.56	1.51	316	315.0735	315 > 153 > 123, 109	$C_{13}H_{15}O_9$	4.332
28	Vanillic acid	0.11	1.66	1.61	168	167.0359	167 > 152, 149 > 121	$C_8H_7O_4$	5.675
29	Protocatechuic acid	1.13	1.37	1.96	154	153.0201	153 > 138, 109 > 81	$C_7H_5O_4$	5.281
30	Hydroxybenzoic acid	1.42	1.15	2.83	138	137.0251	137 > 93	$C_7H_5O_3$	5.201
31	4-vinylcatechol	1.14	1.36	5.29	136	135.0452	135 > 107, 91	$C_8H_7O_2$	5.163
32	4-ethylcatechol	0.88	1.50	5.56	138	137.0615	137 > 93	$C_8H_9O_2$	4.721
33	Syringic aldehyde	1.05	1.43	6.09	182	181.0513	181 > 166 > 138	$C_9H_9O_4$	3.689
Flavonoids									
34	Rutin	0.84	1.09	4.81	610	609.1485	609 > 301 > 271, 179	$C_{27}H_{29}O_{16}$	3.73
35	Quercetin-3-glycoside	1.47	1.09	4.95	464	463.0905	463 > 301 > 271, 179, 151	$C_{21}H_{19}O_{12}$	5.012
36	Isorhamnetin 3-*O*-β-d-glucoside	0.57	1.61	5.23	478	477.1061	477 > 314 > 300, 285	$C_{22}H_{21}O_{12}$	4.613
37	Casticin	1.44	1.15	7.42	374	373.0937	375 > 358 > 343 > 328	$C_{19}H_{17}O_8$	2.142
Etc									
38	Shikimic acid	1.17	1.34	1.31	174	173.0463	173 > 155, 129	$C_7H_9O_5$	4.238
39	Pantothenic acid	1.37	1.21	1.46	219	218.1044	218 > 187 > 143, 130	$C_9H_{16}O_5N$	4.604
N.I									
40	N.I 1	1.51	1.08	6.63	174	173.1190	173 > 127 > 123, 97	-	-
41	N.I 2	1.50	1.06	6.86	926	925.4467	939 > 808 > 645	-	-
42	N.I 3	1.51	1.07	6.90	216	215.1298	215 > 173 > 127	-	-

[a] VIP: variable in projection; [b] RT: retention time; [c] mass to charge ratio for $[M-H]^-$.

3.2. Relative Metabolite Abundance in Annual Wormwood and Glasswort with LAB-Mediated Fermentation

The relative abundance of the significantly discriminant metabolites are shown in the box-and-whisker plots (Figure 2). We observed distinct metabolic change from LAB-mediated fermentation in AW and GW. Several metabolites showed increased or decreased patterns from fermentation in both AF and GF.

The relative contents of several metabolites in both AF and SF groups were higher than each non-fermented group, including succinic acid (5), hydroxyisocaproic acid (6), 9,12,13-TriHOME (7), hydroxystearic acid (11), hydroxyphenyllactic acid (12), phenyllactic acid (15), vanillic acid (27), protocatechuic acid (28), rutin (34), and pantothenic acid (39) (Figure 2A). On the other hand, some metabolites showed lower concentrations from fermentation, that is hydroxymyristic acid (10), caffeic acid (14), coumaric acid (16), ferulic acid (18), quinic acid (20), 5-caffeoylquinic acid (22), 3-5-dicaffeoylquinic acid (25), 3,5-feruloyl-5-caffeoylquinic acid (26), protocatechuic acid-O-glucoside (27), quercetin-3-glycoside (35), and isorhamnetin-3-glucoside (36) (Figure 2B). Some metabolites showed different patterns for each fermented group. Unlike the GF group, the AF group showed a decrease in casticin (37) and an increase in pyroglutamyl-valine (1), pyroglutamyl-leucine (2), lactotyl-tryptophan (3), hydroxyglutaric acid (5), dihydrocaffeic acid (13), scopoletin (17), phloretic acid (19), 3-caffeoylquinic acid (21), 3,4-dicaffeoylquinic acid (23), 3-feruloyl-4-caffeoylquinic acid (24), hydroxybenzoic acid (30), syringic aldehyde (33), 4-ethylcatechol (32), 4-vinylcatechol (31), shikimic acid (38), and two non-identified metabolites (40, 42) (Figure 2C). Relatedly, 9-DiHODE (8), 9,10-DiHOME (9), and N.I 2 (41) increased only in SF (Figure 2D).

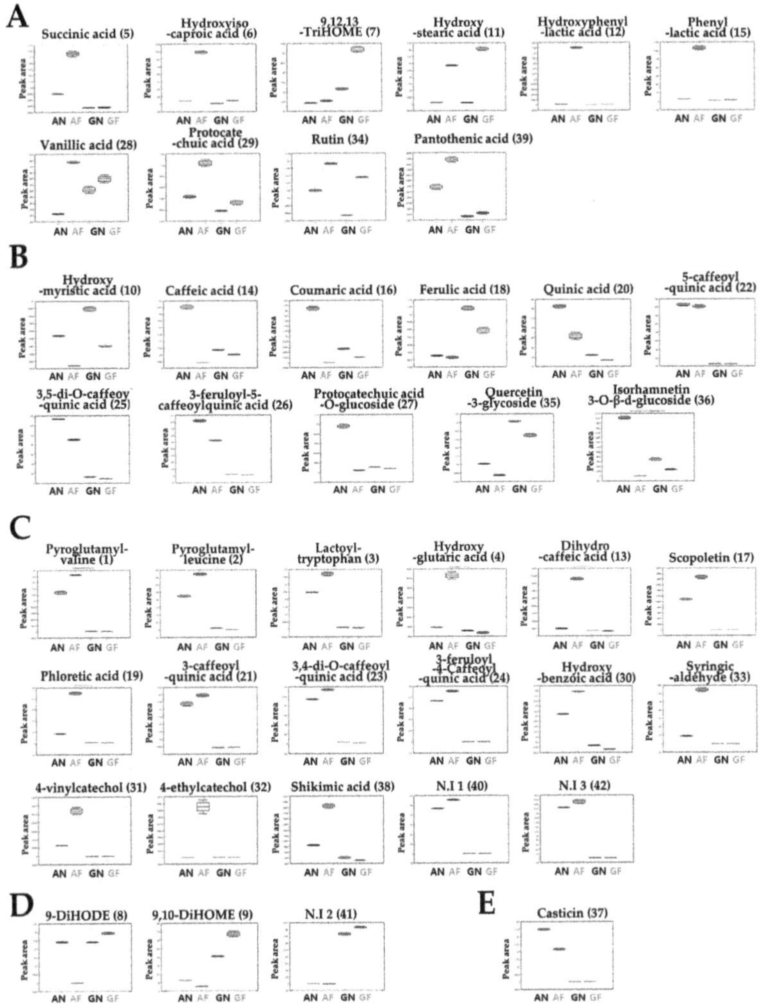

Figure 2. Box-and-whisker plots illustrating the relative abundance of metabolite levels from the fermentation of AW and GW. (A) Increased and (B) decreased metabolites in both AF and GF. (C) Increased and (E) decreased metabolites only in the AF group. (D) Increased metabolites only in the GF group. (Line: mean; box: standard error; whisker: standard deviation.)

3.3. Phenolic Acid Degradation Pathways by Biotransformation of Plant Substrates Using LAB

To investigate the biotransformation of plant substrates, we focused on the phenolic acid degradation pathway in annual wormwood and glasswort. Among the significantly discriminant metabolites, thirteen compounds related to phenolic acid degradation from LAB were selected to construct a metabolic pathway. Chlorogenic-acid-derived caffeic acid, quinic acid, *p*-coumaric acid, and ferulic acid showed a common decrease in both fermentation groups. The contents of metabolites produced by phenolic acid degradation in the AF group were all increased, including shikimic acid, dihydrocaffeic acid, vinyl catechol, ethyl catechol, phloretic acid, hydroxybenzoic acid, protocatechuic acid, and vanillic acid. The SF group also showed an increase in vanillic acid and protocatechuic acid in the phenolic acid degradation pathway (Figure 3).

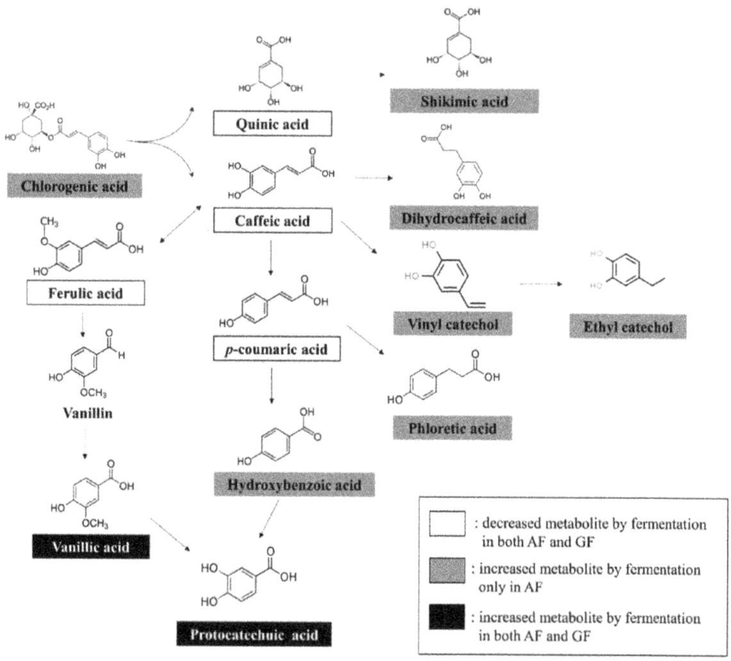

Figure 3. Schematic representation of the degradation pathways of phenolic acids from fermentation of AW and GW.

3.4. Inhibition of 3T3-L1 Adipocyte Differentiation

To evaluate the anti-obesity effects of these plants, the adipogenesis and fat accumulation were investigated in vitro. Figure 4A shows the adipocyte differentiation inhibitory activity of fermented AW and GW at various concentrations (3.1~50 μg/mL). If the fat accumulation of the control group was 100%, the fermented AW extract showed concentration-dependent adipocyte differentiation inhibitory activity, and at 50 μg/mL, fat accumulation was suppressed as much as the undifferentiated adipocyte control. The GW extract had lower fat accumulation than the control at all concentrations except 3.1 μg/mL and showed great anti-differentiation activity even at low concentrations. Figure 4B shows the results of observing differentiated 3T3-L1 adipocytes treated with fermented AW and GW at various concentrations of 3.1~50 μg/mL under a 100× optical microscope. Both extracts inhibited the differentiation of adipocytes in a concentration-dependent manner, and in the morphology it was observed that the size and number of lipid droplets decreased. Oil Red O staining was performed in vitro, after which the physiological data and effects on adipogenesis were observed based on the histology. In the micrographs, it can be seen that the area stained with Oil Red O was significantly reduced when 3T3-L1 adipocytes were treated with the highest concentration of 50 μg/mL of the AW extract. On the other hand, the GW extract was observed to inhibit the formation of lipid droplets even at low concentrations.

Cells were treated at concentrations ranging from 0.2 to 50 μg/mL to the maximal concentration, showing high survival rates in all treatment groups. AW and GW before and after fermentation at concentrations of 50 μg/mL or less in 3T3-L1 cells did not show cytotoxicity.

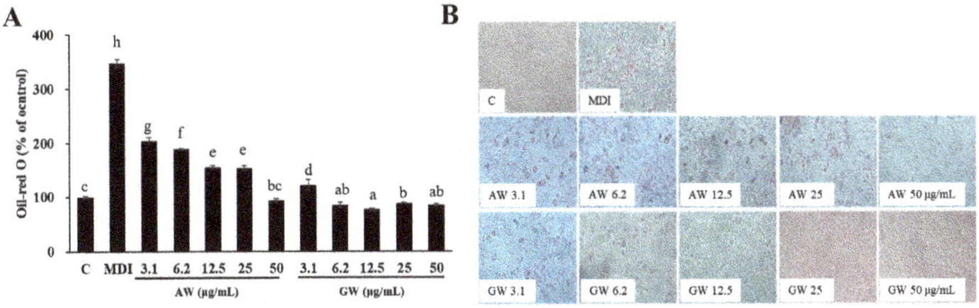

Figure 4. Inhibitory ability of 3T3-L1 adipocyte differentiation from fermented AW and GW extracts. (**A**) Test with oil red O staining method. (**B**) Using an optical microscope. All data are expressed as mean ± SD. Significant differences were determined by Duncan's test after one-way ANOVA and expressed in letters (a–h) ($p < 0.05$). C: Control; MDI: differentiated control; AW: annual wormwood; GW: glasswort.

3.5. Gene Expression Analyzed Using Real-Time PCR (RT-PCR)

To test the anti-obesity effect of plant extracts on lipogenesis in adipose tissue, the real-time PCR was analyzed. The effect of the AW and GW extracts on the expression of adipocyte differentiation-related factors (C/EBPα, PPARγ, leptin, and Fas) in 3T3-L1 cells was experimented (Figure 5). PPARγ is a nuclear hormone receptor that plays a major role in regulating the expression of proteins necessary for the development of functional mature adipocytes. C/EBPα plays a synergistic role in terminal adipocyte differentiation. C/EBPα and PPARγ are integrated in the control of lipid and carbohydrate metabolism. In the real-time PCR results, the gene expression of C/EBPα and PPARγ for adipogenesis decreased compared to the MDI group. After adipocyte differentiation (MDI), the expression of all genes, except leptin, increased significantly compared to before differentiation (control). There was no significant difference in gene expression of adipocytes treated with AW extract compared to MDI. On the other hand, when the GW extract was treated at a concentration of 6.2 μg/mL or more, the expression of differentiation-related genes, except for leptin, was significantly reduced compared to MDI. In addition, C/EBPα and PPARγ, which are closely related to each other, showed a similar decrease pattern, and Fas involved in adipogenesis was reduced. In the case of leptin, GW tended to be less produced than AW, but there was no significant difference. As a result, since the GW extract of 6.2 μg/mL or more reduces the expression of adipocyte genes to the same level as the control, it can be assumed to have an inhibitory activity on the differentiation of 3T3-L1 preadipocytes. The effects of the GW extract on the inhibition of adipogenesis were observed more apparently.

3.6. Changes in Body and Organ Weight in C57BL/6 Mice

Figure 6A shows the body weight and organ weight changes for 12 weeks of C57BL/6 mice fed a high-fat diet containing fermented AW and GW extracts at 100 and 300 mg/kg. The body weight of the GW 300 significantly decreased at 10 and 12 weeks compared to the HFD ($p < 0.05$). The final body weight of the ND was 29.73 ± 3.59 g, an increase of 8.86 ± 2.25 g from the initial body weight, whereas the final body weight of the HFD was 45.50 ± 2.15 g, an increase of 24.43 ± 2.26 g from the initial weight. The HFD significantly increased body weight compared to the ND from the second week of the experiment, and a significant weight increase was observed continuously until the end of the test ($p < 0.01$), indicating that obesity was induced in mice due to the high-fat diet. The final body weight of AW 100 among the AW extract treatment groups was 42.26 ± 3.80 g, and there was no significant difference compared to the HFD, but the weight gain was significantly reduced to 21.13 ± 2.26 g ($p < 0.05$). Relatedly, among the GW extract treatment groups, GW 300 had a final body weight of 41.71 ± 4.27 g and a body weight gain of 20.71 ± 3.68 g, which was significantly reduced compared to the HFD ($p < 0.05$). In addition, the body weight of the GW 300 significantly decreased at 10 and 12 weeks compared to the HFD ($p < 0.05$), and the body weight gain significantly decreased at 7, 10, 11, and 12 weeks ($p < 0.05$). The weights of the initial mice were similarly set in the range of 20~21 g. The FER was lowest in the ND and highest in the HFD. Compared to the HFD, the FER of the AW 300 and GW 300 groups was significantly lower ($p < 0.05$). In particular, AW 300 had a small FER even though its weight gain was the second highest after the HFD, suggesting that weight gain was small even though it consumed a lot of food. Therefore, it was considered that the AW and GW extracts affected the weight loss and FERs of C57BL/6 mice. Figure 6B shows the liver weights of C57BL/6 mice fed a high-fat diet containing fermented AW and GW extracts of 100 and 300 mg/kg. Liver weights in the ND were significantly decreased compared to the HFD ($p < 0.01$). The AW 100 and AW 300 had no significant differences. Compared to the HFD, the liver weights of the GW 100 and GW 300 decreased significantly ($p < 0.05$). In conclusion, GW inhibited the accumulation of triglycerides in the liver.

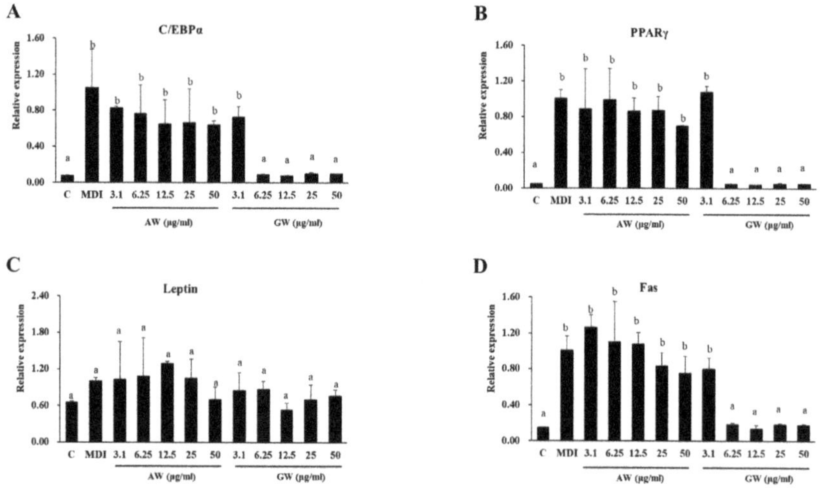

Figure 5. Effect of fermented AW and GW extracts on the mRNA expressions of (**A**) C/EBPα, (**B**) PPARγ, (**C**) leptin, and (**D**) Fas, regulatory factors involved in 3T3-L1 preadipocytes. All data are expressed as mean ± SD. Significant differences were determined by Duncan's test after one-way ANOVA and expressed in letters ($p < 0.05$). C: Control; MDI: differentiated control; AW: annual wormwood; GW: glasswort.

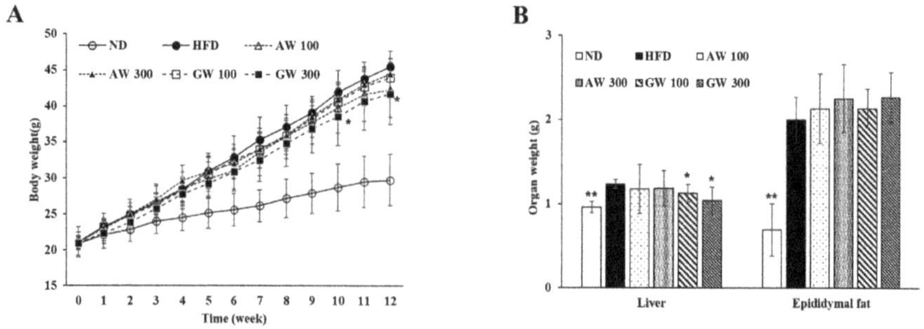

Figure 6. Changes in (**A**) body weight and (**B**) organ weight of C57BL/6 mice during the experiment period. Data are presented as mean ± SD (n = 8). Significant difference from HFD using one-way ANOVA followed by the LSD test: * $p < 0.05$, ** $p < 0.01$. ND: Normal diet; HFD: 60% high-fat diet; AW 100: HFD + fermented annual wormwood (AW) 100 mg/kg; AW 300: HFD + fermented AW 300 mg/kg; GW 100: HFD + fermented glasswort (GW) 100 mg/kg; GW 300: HFD + fermented GW 300 mg/kg.

3.7. Histological Analysis

In order to analyze non-alcoholic steatosis, using image J program photographed the lipid droplet and inflammatory symptoms in liver tissue. Figure 7A showed the histological feature of the liver based on hematoxylin and cosin staining in rats fed control and HFD. Figure 7A displayed liver tissue images of C57BL/6 mice fed a high-fat diet containing fermented AW and GW extracts (100 and 300 mg/kg). Compared to the ND, it was confirmed that the number of lipid droplets produced in the liver tissue of obese HFD increased. Lipid droplets were less in the liver tissue of C57BL/6 mice fed with the GW extract than with the AW extract. Therefore, it was considered that GW inhibits lipid accumulation in the liver more than AW. The most common cause of chronic disease by excessive fat accumulation in the liver is non-alcoholic fatty liver disease (NAFLD). NAFLD caused that high carbohydrate diet induced serum free fatty acid, thereafter increased the absorption of FFA biosynthesis of TG in liver. NAFLD includes a wide spectrum of liver damage ranging from simple steatosis to non-alcoholic steatohepatitis (NASH), advanced fibrosis, and hepatocellular carcinoma [37]. Figure 7B shows the NAFLD activity score measuring the histological characteristics of non-alcoholic fatty liver disease (NAFLD) in C57BL/6 mice fed a high-fat diet containing fermented AW and GW extracts (100 and 300 mg/kg). The NAS of the C57BL/6 mice fed normal diet (ND) in which obesity was not induced was 0, and NAFLD did not appear, and it was significantly lower than that of the HFD (3.00 ± 0.19) ($p < 0.01$). AW 100 (2.5 ± 0.38) and AW 300 (2.25 ± 0.49) decreased concentration dependently compared to the HFD, but no significant difference occurred. GW 100 (1.75 ± 0.41) and GW 300 (1.13 ± 0.13) also decreased in a concentration-dependent manner with significant differences compared to the HFD ($p < 0.05$, $p < 0.01$).

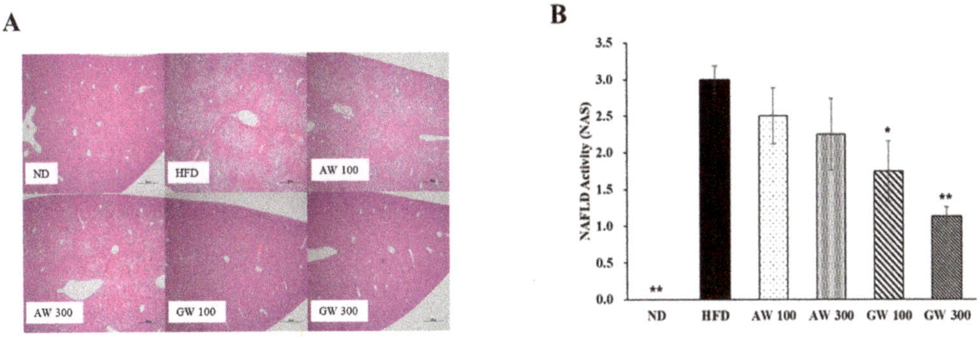

Figure 7. Effect of fermented AW and GW extracts on (**A**) histological features of the liver and (**B**) non-alcoholic fatty liver disease activity score in liver of C57BL/6 mice. Original magnifications: ×40; scale bar: 500 μm. Data are presented as mean ± SD (n = 8). Significant difference from HFD using *t*-test: * $p < 0.05$, ** $p < 0.01$. ND: Normal diet; HFD: 60% high-fat diet; AW 100: HFD + fermented annual wormwood (AW) 100 mg/kg; AW 300: HFD + fermented AW 300 mg/kg; GW 100: HFD + fermented glasswort (GW) 100 mg/kg; GW 300: HFD + fermented GW 300 mg/kg.

3.8. Serum and Hepatic Triglycerie Analysis

Serum analysis was performed in C57BL/6 mice fed a high-fat diet containing fermented AW and GW extracts at 100 and 300 mg/kg (Table 3). Serum AST and ALT levels are indicators of the liver dysfunction condition, caused by the HFD. AST (163.38 ± 20.60) and ALT (122.50 ± 31.01) values of AW 100 were significantly increased, compared to the HFD ($p < 0.01$, $p < 0.05$). The level of AST in the normal group was 94 U/L, and the HFD G2 group significantly increased; in contrast, in the supplemented AW 300 groups was significantly decreased. The ALT values of the supplemented GW groups were significantly decreased; in particular, the G6HFD 300 (46.63 ± 3.30) groups showed the lowest value of all the other HFD groups (p < 0.05). In the case of blood Glu, those of the ND (118.00 ± 10.49), GW 100 (212.63 ± 12.40), and GW 300 (210.50 ± 17.02) were decreased when compared to the HFD at 226.00 ± 4.76 mg/dL ($p < 0.01$). Blood T-Chol significantly decreased in all groups compared to the HFD at 210.75 ± 8.39 mg/dL. Blood TGs significantly decreased at the GW 300 at 80.00 ± 3.33 mg/dL, compared to the HFD at 97.75 ± 5.31 mg/dL ($p < 0.01$). HDL-C was significantly reduced, compared to the HFD at 97.13 ± 2.73 mg/dL, in the AW 300 and GW 300 at 89.25 ± 2.94 and 91.50 ± 2.17 mg/dL, respectively, treated with high-concentration extract ($p < 0.05$). The GW extract treatment groups, GW 100 14.75 ± 0.70 and GW 300 13.63 ± 0.91 mg/dL, showed reduced LDL-C, compared to the HFD at 18.88 ± 1.33 mg/dL ($p < 0.05$). As a result of the blood test, when obesity was induced, the values of all blood analysis indicators increased. On the other hand, the AW extract reduced T-Chol and HDL-C in C57BL/6 mice, and the fermented GW extract reduced blood ALT, Glu, T-Chol, TGs, HDL-C, and LDL-C. As a result, since AW and GW reduce the levels of ALT, Glu, T-Chol, TGs, HDL-C, and LDL-C in the blood, it is thought that they have the possibility of preventing obesity. Insulin levels in the HFD, in which obesity was induced by eating a high-fat diet, were increased, compared to the ND. Among the plant extract treatment groups, only GW 300 at 2.70 ± 0.66 ng/mL significantly reduced blood insulin levels ($p < 0.01$), compared to the HFD at 5.55 ± 0.96 ng/mL. When obesity was induced, hepatic TGs increased, and the AW 300 at 103.31 ± 89.76 µg/mg tissue decreased, compared to the HFD at 145.66 ± 56.7 µg/mg tissue, but there was no significant difference. Hepatic TG levels tended to decrease in a dose-dependent manner in the GW extract treatment group, and GW 300 decreased significantly, compared to the HFD ($p < 0.05$). Therefore, it can be assumed that GW 300 suppressed the accumulation of triglycerides in the liver.

Table 3. Blood chemistry in control and HFD-fed C57BL/6 rats treated with AW and GW.

Group			Blood Chemistry						
			AST (U/L)	ALT (U/L)	GLU (mg/dL)	TG (mg/dL)	T-Chol (mg/dL)	HDL-C (mg/dL)	LDL-C (mg/dL)
ND	N	Mean	94.00	29.38 *	118.00 **	74.00 **	113.13 **	81.88 **	9.63 **
	8	SE	8.80	1.05	10.49	3.56	5.11	1.14	0.26
HFD Control	N	Mean	122.75	76.75	226.00	97.75	210.75	97.13	18.88
	8	SE	5.57	8.07	4.76	5.31	8.39	2.73	1.33
AW 100	N	Mean	163.38 *	122.50 *	212.63	88.13	195.75	96.50	20.00
	8	SE	20.60	31.01	12.40	2.40	7.57	2.28	1.46
AW 300	N	Mean	96.63	67.00	210.50	101.13	177.50 **	89.25 *	16.13
	8	SE	7.25	8.61	17.02	7.17	9.60	2.94	1.42
GW 100	N	Mean	117.63	58.88	192.25 *	92.13	182.13 **	95.88	14.75 *
	8	SE	12.06	7.49	5.58	1.54	5.40	2.84	0.70
GW 300	N	Mean	137.88	46.63	203.13	80.00 **	163.25 **	91.50	13.63 **
	8	SE	8.01	3.30	7.90	3.33	6.30	2.17	0.91

ND: Normal diet; HFD: 60% high-fat diet; AW 100: HFD + fermented annual wormwood (AW) extracts 100 mg/Kg; AW 300: HFD + fermented AW extracts 300 mg/Kg; GW 100: HFD + fermented glasswort (GW) extracts 100 mg/Kg; GW 300: HFD + fermented glasswort extracts 300 mg/Kg. Data are presented as mean ± SD. Significant difference from HFD using *t*-test: * $p < 0.05$, ** $p < 0.01$.

4. Discussion

Obesity is a major public health problem with high prevalence worldwide and is accompanied by various diseases such as diabetes, hyperglycemia, dyslipidemia, and metabolic syndrome [5]. Treatments are being developed to prevent and treat obesity, but side effects such as nausea and diarrhea exist [9]. In order to develop plant-derived functional materials with fewer side effects and superior anti-obesity effects than drugs, this study aimed to reveal the anti-obesity activity of fermented annual wormwood (AW) and glasswort (GW) at both the in vitro and in vivo levels. In addition, metabolome changes were verified for the difference before and after fermentation. AW (*Artemisia annua* L.)

is a common type of wormwood native to temperate Asia but naturalized worldwide and belonging to the family *Asteraceae* [15]. Its bioactive derivatives control ROS, DNA damage, DNA repair, cell cycle arrest, apoptosis, inflammatory response, angiogenesis, and multiple signaling pathways, showing an anti-cancer effect [38]. In addition, it has an anti-hyperglycemic effect and reduces insulin resistance by increasing adiponectin secretion in adipocytes [35].

GW (*Salicornia herbacea* L.) belongs to *Chenopodiaceae* and has been used as a food ingredient or as a folk remedy for improving intestinal function, indigestion, gastrointestinal disease, kidney disease, and diabetes [39]. GW is seaweed that contains physiologically active substances, such as dietary fiber, and minerals such as potassium, magnesium, and calcium that stimulate intestinal movement. GW is known to have beneficial effects on obesity by reducing body weight and fat levels and improving blood lipids [40].

The bacteria used for AW fermentation are *Lactobacillus plantarum*, a lactic acid bacterium [41]. It is known that pH decreases by producing lactic acid as a fermentation metabolite [42]. At this time, as the fermentation progresses, the pH decreases and the acidity increases, which is considered to reduce the activity of *Lactobacillus plantarum*. GW showed a small change in pH during fermentation, an increase in viable cell count, and then a tendency to die after a certain period of time, which was similar to previous studies [33]. Since *Enterococcus faecium*, the strain used for GW fermentation, mainly uses carbohydrates as an energy source, it is thought that the death occurred due to a lack of carbohydrates as fermentation proceeded [43].

AW and GW contain phenolic compounds, and there is no difference in physiological activity before and after fermentation. It is true that fermentation is known as one of the ways to improve the physiological activity of natural products, but it is difficult to see that some phenols are unconditionally improved due to small changes [44].

Untargeted metabolomics were employed to investigate metabolic change using LAB-mediated fermentation in AW and GW. Among the discriminant metabolites, we observed the relative abundance of several metabolites related to bacterial biotransformation of plant-derived compounds such as flavonoids and phenolic acids. Flavonoids in plants are major secondary metabolites, mainly in the form of flavonoid glycosides [45]. Flavonoid glycosides can be decomposed into aglycone and phenolic acid by microorganisms. Some LAB have enzymes that convert flavonoid glycosides such as α-l-rhamnosidase and β-d-glucosidase, but these abilities vary among specific species and strains. In this study, the reduction of quecetin-3-glycoside and isorhamnetin-3-glycoside using fermentation was observed, but the types of aglycone and flavonol produced from bioconversion could not be identified. On the other hand, the content of rutin tended to increase through fermentation. It has been reported that there is a difference in the conversion ability of rutin for each strain. In the study of buckwheat fermented by 14 LAB strains, most of them decreased the content of the rutin, but some strains increased [46]. Rutin is known to have various pharmacological activities such as antimicrobial, anti-inflammatory, and anti-cancer effects [47,48]. In addition, rutin administration in high-fat diet mice was affected, affecting the intestinal bacteria and anti-obesity effect [49].

Bacterial degradation of phenolic compounds has also been reported along with metabolic changes. The reaction of phenolic acid decarboxylase (PAD) and reductase in LAB decreased phenolic acid as a substrate and increased some benzoic acid as a product [50–52]. We also observed the fermentation patterns of metabolites in the degradation pathways of quinic acid, caffeic acid, *p*-coumaric acid, and ferulic acid. As a clear reduction of phenolic acid, we identified the products of eight benzoic acid related to four kinds of phenolic acid in the AF group and only two compounds (vanillic acid and protocatechuic acid) related to ferulic acid in the GF group. The converted metabolites derived from phenolic acid showed high bioavailability in human and health benefits [53,54]. Dihydrocaffeic acid and the catechol group converted from caffeic acid were reported as potent antioxidant properties caused by the scavenging of intracellular reactive oxygen species [41,55,56]. Relatedly, phloretic acid converted from p-coumaric acid was one of the substances that recently attracted attention as physiological activities, such as protecting the host from influenza infection by regulating the immune balance, that have been reported [57,58]. Hydroxybenzoic acid and its derivatives are known to exhibit various physiological activities such as anti-obesity, antibacterial, anti-mutagenic, anti-algae, anti-inflammatory, and antioxidant activities [59–61]. In addition, protocatechuic acid was also reported to have various effects such as anti-obesity, anti-inflammatory, and antioxidant effects through in vivo and in vitro tests [62,63]. Vanillic acid is a substance used pharmacologically with various physiological activities such as anti-obesity, antioxidant, anti-cancer, antifungal, and liver protection [64,65].

Additionally, we observed the production of several metabolites related to amino acid and fatty acid metabolism of LAB. These metabolites were presumed to be produced using nutrients in the medium rather than compounds derived from plants. Phenyllactic acid, hydroxyphenyllactic acid, and hydroxyisocaproic acid were produced by amino acid catabolism, which was phenylalanine, tyrosine, and leucine, respectively [66,67]. In fatty acid metabolism, oleic acid and linoleic acid were metabolized to hydroxystearic acid, azelaic acid, and hydroxyoctadecanoic acid [68,69]. Among these increased metabolites, hydroxy fatty acids and phenylpropanoids produced by LAB have been reported to have antifungal activity that can be useful to biopreservation [70–72]. On the other hand, some dipeptides (pyroglutamyl-valine, pyroglutamyl-leucine, and lactoyl-tryptophan) increased only in the AF group. Those compounds included the taste active peptide derivatives in food fermentation of LAB [73]. An anti-inflammatory effect and attenuating dysbiosis in vivo has been reported for pyroglutamyl-leucine [74,75]. Similarly, some oxylipins (9-DiHODE, 9,10-DiHOME, and 9,12,13-TriHOME) dramatically increased in the GF group, which are well-known as bioactive lipids for their anti-inflammatory and antioxidant effects [76,77].

Relatedly, the anti-obesity effect of fermented AW and GW was confirmed in vitro using 3T3-L1 cells, which can differentiate between fibroblasts and adipocytes and are widely used in adipogenesis and biochemical studies of adipocytes [78]. There was no cytotoxicity because both AW and GW showed a high survival rate up to 50 μg/mL. Lipid droplet generation can be confirmed by staining triglyceride and cholesterol ester accumulated due to lipolysis using Oil Red O staining reagent, and the degree of fat differentiation can be quantified by dissolving it and measuring absorbance [79,80]. Both the fermented AW and GW extracts inhibited the differentiation of adipocytes in a concentration-dependent manner, and it was morphologically observed that the size and number of lipid droplets produced when adipocytes differentiated and accumulated fat decreased. Excessive lipid accumulation and formation of lipid droplets cause obesity [81], and it was confirmed that GW suppressed the formation of lipid droplets even at low concentrations when compared to AW. Therefore, it can be said that fermented GW has the potential for preventing obesity.

During adipogenesis, several types of genes and transcription factors were regulated. In this experiment, the effects of fermented AW and GW extracts on the expression of adipocyte differentiation-related factors C/EBPα, PPARγ, leptin, and Fas in 3T3-L1 cells were investigated. C/EBPα and PPARγ are specific transcription factors expressed in adipocytes, and the activities of the two regulators were enhanced in the early stages of adipogenesis and promote adipogenesis [82]. When adipocytes were compared before and after differentiation, C/EBPα and PPARγ significantly increased after differentiation, and transcription factors were suppressed in the GW group, similarly to before differentiation. Leptin, a protein produced and secreted by adipocytes, increased as fat accumulation increased and is a hormone that induces appetite suppression to control obesity. Leptin levels increased after differentiation, but there was no significant difference. The level of leptin in AW was similar to that after differentiation, but in the case of GW, it slightly decreased. Since leptin secretion increased with adipocyte size and fat accumulation, it was thought that the level of leptin was low in GW where fat accumulation was suppressed [83]. When the expression of Fas, which is involved in lipid synthesis, transport, and storage, was inhibited, adipogenesis and triglyceride accumulation were reduced [84]. The expression of Fas increased after differentiation, and GW suppressed the expression to the same level as the control group. Demineralized glasswort and annual wormwood leaves, which have been shown to have an inhibitory effect on the differentiation of adipocytes, and fermented lemons showed similar gene expression inhibitory activity [28,85,86]. Therefore, 3T3-L1 cells increase the expression of all genes as fat accumulates after differentiation, but GW extract reduces the expression of adipocyte-expressed genes C/EBPα, PPARγ, and Fas, suggesting that the differentiation of adipocytes is inhibited.

The anti-obesity activity was investigated through weight change, histological change, and serum analysis. C57BL/6 mice were used as experimental animals, and they are often used as obesity-prevention models because they lack the leptin gene and cannot control appetite, causing obesity [87]. In this experiment, it was also confirmed that the weight of the high-fat diet (HFD) group continued to increase, and obesity was caused by the high-fat diet. After 12 weeks, the final weight and weight gain (g) of GW 300 decreased compared to the HFD, and the weight gain (g) of AW 100 decreased ($p < 0.05$). The food efficiency ratios (FERs) of the AW 300 group and GW 300 group significantly decreased compared to the HFD ($p < 0.05$). If the FER was small, it can be predicted that obesity was prevented in the high-concentration treatment group of natural products [88].

Obesity refers to an increase in body fat caused by the accumulation of triglycerides in adipose tissue, and it has been reported that the risk of metabolic diseases increases as the abdominal fat content increases [89]. In this experiment, the weight of the liver was measured to confirm the formation of abdominal fat. Similar to the previous study [86], in which C57BL/6 mice fed the HFD gained weight as cholesterol and triglycerides accumulated in the liver, it was confirmed that the liver weight increased in this experiment. The liver weight of mice fed the GW extract significantly decreased compared to the HFD ($p < 0.05$). The GW extract inhibited liver fat accumulation. In addition, when lipids accumulated in the liver as obesity progressed, lipid droplets also accumulated [90]. An obesity-induced HFD increased the number of lipid droplets in liver tissue compared to the ND, and it was confirmed that the number of lipid droplets in GW was smaller than that in AW. Non-alcoholic fatty liver diseases (NAFLDs) are known to be liver damage diseases caused mainly by insulin resistance, accompanied by obesity and dyslipidemia due to the accumulation of triglycerides in hepatocytes [91]. The non-alcoholic fatty liver disease activity score [13] is a method of expressing NAFLD as a score for histological characteristics. Non-alcoholic steatohepatitis (NASH) can be diagnosed if the NAS is five or higher, and "not NASH" if the NAS is less than three [36]. The HFD group has a high risk of NASH with a score of three, and all groups except the HFD group can be diagnosed as not NASH, so it can be said that there is no liver damage. In addition, the content of useful substances having various activities increased by this decomposition. Serum aspartate aminotransferase (AST) and alanine aminotransferase (ALT), biomarkers that can estimate liver health, increased in obesity. Similarly, AST and ALT of an obesity-induced HFD increased significantly compared to the ND. The ALT of the GW 300 group significantly decreased compared to the HFD ($p < 0.01$). On the other hand, AW 100 increased both the AST ($p < 0.01$) and ALT ($p < 0.05$), compared to the HFD, which was contrary to previous studies in which the AST and ALT of the AW extract were lower than those fed a high-fat diet [15,25]. When AST and ALT increased, it can be said that liver damage had occurred. In the present study, two of the 100 AW groups had higher AST and ALT than other animals, suggesting that liver damage may have occurred. Blood glucose [24] was the highest in an obesity-induced HFD and significantly decreased in the GW-treated group, compared to the HFD group ($p < 0.01$). The GLU-lowering effect of GW was similar to a previous study that revealed a hyperglycemia-preventive effect [92]. The total cholesterol (T-Chol) and high-density lipoprotein-cholesterol (HDL-C) of mice fed fermented AW extract significantly decreased when compared to the HFD ($p < 0.05$). The fermented GW extract significantly reduced GUL, T-Chol, triglycerides (TGs) ($p < 0.01$), HDL-C, and LDL-C ($p < 0.05$). GW lowered the lipid concentration in the blood than AW. This result was similar to a previous experiment in which T-Chol, low-density lipoprotein-cholesterol (LDL-C), and TG levels were reduced in mice fed high-fat diets and seaweed [93]. The occurrence of cardiovascular disease can be confirmed by the TC/HDL-C ratio, and it is known that the occurrence of the disease is low when the ratio is low. The TC/HDL-C ratios were ND (1.38), HFD (2.17), AW 100 (2.03), AW 300 (1.99), GW 100 (1.90), and GW 300 (1.78), and ND and GW 300 were low. Therefore, the fermented GW extract can reduce the risk of obesity and cardiovascular disease by reducing blood ALT, GUL, T-Chol, TGs, LDL-C, blood fat, and cholesterol.

Obesity increases insulin resistance, which does not normalize blood glucose levels, and increases the risk of developing type 2 diabetes [94]. In the case of the HFD, insulin resistance was high, but blood sugar increased, so it can be judged that insulin resistance occurred. Among the experimental groups, GW 300 showed a decrease in insulin level ($p < 0.01$), and even though they ate a high-fat diet, the rate of insulin secretion was regulated, and it is thought that the GW extract lowered insulin resistance. When obesity was induced, hepatic triglycerides (hepatic TGs) increased, and among them, the GW 300 group significantly decreased compared to the HFD ($p < 0.05$), and it was found that obesity was suppressed. Glasswort supplementation showed similar results to a previous study [95] that significantly reduced hepatic triglycerides compared to mice fed a high-fat diet.

5. Conclusions

In this study, in vitro experiments using 3T3-L1 adipocyte cells and in vivo experiments using C57BL/6 mice were conducted to investigate the anti-obesity effects of fermented annual wormwood (AW) and glasswort (GW). In addition, an analysis of compound changes was investigated comparing before and after fermentation. In particular, phenolic compounds were abundant in AW, and antioxidant activity was also high. In both AW and GW plants, significant metabolite changes from fermentation were observed. The metabolites increased from fermentation were metabolites produced by decomposing the plant substrate and substances produced by strain biosynthesis. Specifically, several substances with reported physiological activity were identified from the degradation of phenolic acid derived from plants using lactic acid bacteria. It is suggested that the compounds involved in the anti-obesity activity of GW can be activated by the biotransformation of fermentation, but detailed research is needed for the exact mechanism.

The AW and GW extracts were not toxic to 3T3-L1 adipocytes up to the maximum concentration of 50 μg/mL, and the plants after fermentation had better ability to inhibit adipocyte differentiation. The fermented AW and GW extracts inhibited the differentiation of adipocytes in a concentration-dependent manner and also reduced the size and number of lipid droplets. In particular, fermented GW inhibited lipid droplet accumulation even at low concentrations and significantly reduced the expression of C/EBPα, PPARγ, and Fas, transcription factors and genes involved in adipogenesis that were expressed during adipocyte differentiation. When C57BL/6 mice were supplemented with fermented AW extract, the body weight, food efficiency ratio (FER), and total blood cholesterol (T-Chol) reduced compared to obese mice. The supplementation of GW extract to C57BL/6 mice prevented liver lipid accumulation and fatty liver as well as weight loss. GW extract also prevented liver disease, reduced blood lipids, and regulated insulin.

Through this study, the anti-obesity activity of glasswort was revealed in vitro and in vivo experiments such as weight loss and lipid accumulation inhibition. Compounds with anti-obesity activity were activated through biotransformation using fermentation. As seen in the results, it is suggested that glasswort and annual wormwood be used for companion animals as well as human health as supplementary food additives for the control of obesity and metabolic disorders. The metabolic extracts from the fermented plants have shown promising anti-obesity effects which need to be substantiated further with large-scale animal studies and cross-sectional population trials with physiological studies.

Author Contributions: S.-K.K. designed this research and acquired funds and supervision; C.-H.L. and S.-H.K. performed the experiments, data interpretation, and data analyzer; J.-Y.O. investigated; J.-M.K. reviewed and edited; S.P. supervised and acquired funds; K.-H.K. found resources. All authors have read and agreed to the published version of the manuscript.

Funding: This paper was supported by the Cooperative Research Program for Agriculture Science & Technology Development (Project No. PJ 01689101), Rural Development Administration, Republic of Korea.

Institutional Review Board Statement: The animal experiment protocol approval No.: N 2021001 was obtained from the NDIC animal care and use committee.

Informed Consent Statement: Not applicable.

Data Availability Statement: The data presented in this research are available in this article.

Acknowledgments: The authors are thankful to the NDIC Corporation for their continuous support.

Conflicts of Interest: The authors declare no conflict of interest.

References

1. Loos, R.J.; Yeo, G.S. The genetics of obesity: From discovery to biology. *Nat. Rev. Genet.* **2022**, *23*, 120–133. [CrossRef] [PubMed]
2. Choi, S.-Y.; Lee, S.Y.; hye Jang, D.; Lee, S.J.; Cho, J.-Y.; Kim, S.-H. Inhibitory effects of *Porphyra dentata* extract on 3T3-L1 adipocyte differentiation. *J. Anim. Sci. Technol.* **2020**, *62*, 854–863. [CrossRef] [PubMed]
3. Hariri, N.; Thibault, L. High-fat diet-induced obesity in animal models. *Nutr. Res. Rev.* **2010**, *23*, 270–299. [CrossRef]
4. Kajikawa, M.; Higashi, Y. Obesity and Endothelial Function. *Biomedicines* **2022**, *10*, 1745. [CrossRef] [PubMed]
5. Klein, S.; Gastaldelli, A.; Yki-Järvinen, H.; Scherer, P.E. Why does obesity cause diabetes? *Cell Metab.* **2022**, *34*, 11–20. [CrossRef] [PubMed]
6. Krempler, F.; Breban, D.; Oberkofler, H.; Esterbauer, H.; Hell, E.; Paulweber, B.; Patsch, W. Leptin, peroxisome proliferator-activated receptor-γ, and CCAAT/enhancer binding protein-α mRNA expression in adipose tissue of humans and their relation to cardiovascular risk factors. *Arterioscler. Thromb. Vasc. Biol.* **2000**, *20*, 443–449. [CrossRef]
7. Riondino, S.; Roselli, M.; Palmirotta, R.; Ferroni, P.; Gaudagni, F. Obesity and colorectal cancer role of adipokines in tumor initiation and progression. *World J. Gastroenterol.* **2014**, *20*, 5177–5190. [CrossRef]
8. Duquenne, M.; Folgueira, C.; Bourouh, C.; Millet, M.; Dam, J.; Prevot, V. Leptin brain entry via a tanycyclic LepR : EGFR shuttle controls lipid metabolism and pancreas function. *Nat. Metab.* **2021**, *3*, 1071–1090. [CrossRef]
9. Ryan, D.H. Next generation antiobesity medications: Setmelanotide, semaglutide, tirzepatide and bimagrumab: What do they mean for clinical practice? *J. Obes. Metab. Syndr.* **2021**, *30*, 196–208. [CrossRef]
10. Paul, A.K.; Jahan, R.; Paul, A.; Mahboob, T.; Bondhon, T.A.; Jannat, K.; Hasan, A.; Nissapatorn, V.; Wilairatana, P.; de Lourdes Pereira, M. The role of medicinal and aromatic plants against obesity and arthritis: A review. *Nutrients* **2022**, *14*, 985. [CrossRef]
11. Wang, H.-N.; Xiang, J.-Z.; Qi, Z.; Du, M. Plant extracts in prevention of obesity. *Crit. Rev. Food Sci. Nutr.* **2022**, *62*, 2221–2234. [CrossRef] [PubMed]
12. El-Shiekh, R.; Al-Mahdy, D.; Hifnawy, M.; Abdel-Sattar, E. In-vitro screening of selected traditional medicinal plants for their anti-obesity and anti-oxidant activities. *S. Afr. J. Bot.* **2019**, *123*, 43–50. [CrossRef]
13. Redha, A.A.; Perna, S.; Riva, A.; Petrangolini, G.; Peroni, G.; Nichetti, M.; Iannello, G.; Naso, M.; Faliva, M.A.; Rondanelli, M. Novel insights on anti-obesity potential of the miracle tree, Moringa oleifera: A systematic review. *J. Funct. Foods* **2021**, *84*, 104600. [CrossRef]
14. Kazemipoor, M.; Radzi, C.W.J.W.M.; Cordell, G.A.; Yaze, I. Potential of traditional medicinal plants for treating obesity: A review. *arXiv* **2012**, arXiv:1208.1923.
15. Choi, E.-Y.; Park, C.Y.; Ho, S.H.; Park, S.-J.; Kim, D.; Han, B.; Kim, S.-H. Anti-Obesity Effects of *Artemisia annua* Extract in Zucker Fatty Rats and High-Fat Diet Sprague Dawley Rats through Upregulation of Uncoupling Protein 1. *J. Obes. Metab. Syndr.* **2021**, *30*, 32–43. [CrossRef]
16. Şenkal, B.C.; Kiralan, M.; Yaman, C. The effect of different harvest stages on chemical composition and antioxidant capacity of essential oil from *Artemisia annua* L. *J. Agric. Sci.* **2015**, *21*, 71–77. [CrossRef]
17. Gouveia, S.C.; Castilho, P.C. *Artemisia annua* L.: Essential oil and acetone extract composition and antioxidant capacity. *Ind. Crops Prod.* **2013**, *45*, 170–181. [CrossRef]
18. Baek, H.K.; Shim, H.; Lim, H.; Shim, M.; Kim, C.-K.; Park, S.-K.; Lee, Y.S.; Song, K.-D.; Kim, S.-J.; Yi, S.S. Anti-adipogenic effect of Artemisia annua in diet-induced-obesity mice model. *J. Vet. Sci.* **2015**, *16*, 389–396. [CrossRef]
19. Darusman, L.; Batubara, I.; Utami, M. Fractionation of active components from Piper cf. fragile essential oil as aromatherapy for anti-obesity. In Proceedings of the International Symposium on Medicinal and Aromatic Plants 1023, Chiang Mai, Thailand, 15 December 2011; pp. 23–28.
20. Kim, M.J.; Jeon, D.; Kwak, C.; Ryoo, S.; Kim, Y. Rhamnetin Exhibits Anti-Tuberculosis Activity and Protects against Lung Inflammation. *Bull. Korean Chem. Soc.* **2016**, *37*, 1703–1709. [CrossRef]
21. Beigh, Y.A.; Ganai, A.M. Potential of wormwood (*Artemisia absinthium* Linn.) herb for use as additive in livestock feeding: A review. *Pharma Innov.* **2017**, *6*, 176–187.
22. Ekiert, H.; Klimek-Szczykutowicz, M.; Rzepiela, A.; Klin, P.; Szopa, A. Artemisia species with high biological values as a potential source of medicinal and cosmetic raw materials. *Molecules* **2022**, *27*, 6427. [CrossRef]

23. Ozawa, T.; Wu, J.; Fujii, S. Effect of inoculation with a strain of *Pseudomonas pseudoalcaligenes* isolated from the endorhizosphere of *Salicornia europea* on salt tolerance of the glasswort. *Soil Sci. Plant Nutr.* **2007**, *53*, 12–16. [CrossRef]
24. Altay, A.; Celep, G.S.; Yaprak, A.E.; Baskose, I.; Bozoglu, F. Glassworts as possible anticancer agents against human colorectal adenocarcinoma cells with their nutritive, antioxidant and phytochemical profiles. *Chem. Biodivers.* **2017**, *14*, e1600290. [CrossRef]
25. Park, Y.-H.; Lee, J.-J.; Son, H.-K.; Kim, B.-H.; Byun, J.; Ha, J.-H. Antiobesity effects of extract from Spergularia marina Griseb in adipocytes and high-fat diet-induced obese rats. *Nutrients* **2020**, *12*, 336. [CrossRef]
26. Cho, J.-Y.; Kim, J.Y.; Lee, Y.G.; Lee, H.J.; Shim, H.J.; Lee, J.H.; Kim, S.-J.; Ham, K.-S.; Moon, J.-H. Four new dicaffeoylquinic acid derivatives from glasswort (*Salicornia herbacea* L.) and their antioxidative activity. *Molecules* **2016**, *21*, 1097. [CrossRef] [PubMed]
27. Kong, C.-S.; Seo, Y. Antiadipogenic activity of isohamnetin 3-O-β-D-glucopyranoside from *Salicornia herbacea*. *Immunopharm. Immunotoxicol.* **2012**, *34*, 907–911. [CrossRef]
28. Rahman, M.M.; Kim, M.-J.; Kim, J.-H.; Kim, S.-H.; Go, H.-K.; Kweon, M.-H.; Kim, D.-H. Desalted *Salicornia europaea* powder and its active constituent, trans-ferulic acid, exert anti-obesity effects by suppressing adipogenic-related factors. *Pharm. Biol.* **2018**, *56*, 183–191. [CrossRef]
29. Kim, H.-W.; Hwang, K.-E.; Song, D.-H.; Kim, Y.-J.; Lim, Y.-B.; Ham, Y.-K.; Yeo, E.-J.; Chang, S.-J.; Choi, Y.-S.; Kim, C.-J. Effect of glasswort (*Salicornia herbacea* L.) on the texture of frankfurters. *Meat Sci.* **2014**, *97*, 513–517. [CrossRef] [PubMed]
30. Lopes, M.; Cavaleiro, C.; Ramos, F. Sodium reduction in bread: A role for glasswort (*Salicornia ramosissima* J. Woods). *Compr. Rev. Food Sci. Food Saf.* **2017**, *16*, 1056–1071. [CrossRef]
31. Lee, J.-H.; Lee, J.-H.; Jin, J.-S. Fermentation of traditional medicine: Present and future. *Orient. Pharm. Exp. Med.* **2012**, *12*, 163–165. [CrossRef]
32. Hur, S.J.; Lee, S.Y.; Kim, Y.-C.; Choi, I.; Kim, G.-B. Effect of fermentation on the antioxidant activity in plant-based foods. *Food Chem.* **2014**, *160*, 346–356. [CrossRef] [PubMed]
33. Park, D.H.; Kothari, D.; Niu, K.-M.; Han, S.G.; Yoon, J.E.; Lee, H.-G.; Kim, S.-K. Effect of fermented medicinal plants as dietary additives on food preference and fecal microbial quality in dogs. *Animals* **2019**, *9*, 690. [CrossRef] [PubMed]
34. Lee, A.-R.; Niu, K.-M.; Kang, S.-K.; Han, S.-G.; Lee, B.-J.; Kim, S.-K. Antioxidant and antibacterial activities of *Lactobacillus*-fermented *Artemisia annua* L. as a potential fish feed additive. *J. Life Sci.* **2017**, *27*, 652–660.
35. Ghanbari, M.; Lamuki, M.S.; Habibi, E.; Sadeghimahalli, F. Artemisia annua L. Extracts Improved Insulin Resistance via Changing Adiponectin, Leptin and Resistin Production in HFD/STZ Diabetic Mice. *J. Pharmacopunct.* **2022**, *25*, 130–138. [CrossRef]
36. Kleiner, D.E.; Brunt, E.M.; Van Natta, M.; Behling, C.; Contos, M.J.; Cummings, O.W.; Ferrell, L.D.; Liu, Y.C.; Torbenson, M.S.; Unalp-Arida, A. Design and validation of a histological scoring system for nonalcoholic fatty liver disease. *Hepatology* **2005**, *41*, 1313–1321. [CrossRef]
37. Younossi, Z.N.; Konenig, A.B.; Abdeltif, D.; Wymer, M. Global epidemiology of NAFLD meta analytic assessment of prevalence, incidence, and outcome. *Hepatology* **2016**, *64*, 73–84. [CrossRef]
38. Jung, E.J.; Paramanantham, A.; Kim, H.J.; Shin, S.C.; Kim, G.S.; Jung, J.-M.; Ryu, C.H.; Hong, S.C.; Chung, K.H.; Kim, C.W. Artemisia annua L. Polyphenol-Induced Cell Death Is ROS-Independently Enhanced by Inhibition of JNK in HCT116 Colorectal Cancer Cells. *Int. J. Mol. Sci.* **2021**, *22*, 1366. [CrossRef]
39. Kang, S.; Choi, Y.; Hong, J. Modulation of arachidonic acid metabolism and inflammatory process in macrophages by different solvent fractions of Glasswort (*Salicornia herbacea* L.) extract. *Korean J. Food Sci. Technol.* **2018**, *50*, 671–679.
40. Na, E.-J.; Kim, D.-J.; Kim, J.-H.; Kim, G.-R. Recent trends in anti-obesity and anti-inflammatory studies in modern health care. *Technol. Health Care* **2019**, *27*, 519–530. [CrossRef]
41. De La Cruz, J.; Ruiz-Moreno, M.; Guerrero, A.; López-Villodres, J.; Reyes, J.; Espartero, J.; Labajos, M.; González-Correa, J. Role of the catechol group in the antioxidant and neuroprotective effects of virgin olive oil components in rat brain. *J. Nutr. Biochem.* **2015**, *26*, 549–555. [CrossRef]
42. da Silva Sabo, S.; Vitolo, M.; González, J.M.D.; de Souza Oliveira, R.P. Overview of *Lactobacillus plantarum* as a promising bacteriocin producer among lactic acid bacteria. *Food Res. Int.* **2014**, *64*, 527–536. [CrossRef] [PubMed]
43. Audisio, M.C.; Oliver, G.; Apella, M.C. Effect of different complex carbon sources on growth and bacteriocin synthesis of Enterococcus faecium. *Int. J. Food Microbiol.* **2001**, *63*, 235–241. [CrossRef] [PubMed]
44. Leonard, W.; Zhang, P.; Ying, D.; Adhikari, B.; Fang, Z. Fermentation transforms the phenolic profiles and bioactivities of plant-based foods. *Biotechnol. Adv.* **2021**, *49*, 107763. [CrossRef]
45. Park, C.-M.; Kim, G.-M.; Cha, G.-S. Biotransformation of flavonoids by newly isolated and characterized *Lactobacillus pentosus* NGI01 strain from kimchi. *Microorganisms* **2021**, *9*, 1075. [CrossRef]
46. Zieliński, H.; Wiczkowski, W.; Honke, J.; Piskuła, M.K. In vitro expanded bioaccessibility of quercetin-3-rutinoside and quercetin aglycone from buckwheat biscuits formulated from flours fermented by lactic acid bacteria. *Antioxidants* **2021**, *10*, 571. [CrossRef] [PubMed]
47. Ganeshpurkar, A.; Saluja, A.K. The pharmacological potential of rutin. *Saudi Pharm. J.* **2017**, *25*, 149–164. [CrossRef]
48. Gullon, B.; Lú-Chau, T.A.; Moreira, M.T.; Lema, J.M.; Eibes, G. Rutin: A review on extraction, identification and purification methods, biological activities and approaches to enhance its bioavailability. *Trends Food Sci. Technol.* **2017**, *67*, 220–235. [CrossRef]
49. Yan, X.; Zhai, Y.; Zhou, W.; Qiao, Y.; Guan, L.; Liu, H.; Jiang, J.; Peng, L. Intestinal Flora Mediates Antiobesity Effect of Rutin in High-Fat-Diet Mice. *Mol. Nutr. Food Res.* **2022**, *66*, 2100948. [CrossRef]
50. Filannino, P.; Bai, Y.; Di Cagno, R.; Gobbetti, M.; Gänzle, M.G. Metabolism of phenolic compounds by *Lactobacillus* spp. during fermentation of cherry juice and broccoli puree. *Food Microbiol.* **2015**, *46*, 272–279. [CrossRef]

51. Rogozinska, M.; Korsak, D.; Mroczek, J.; Biesaga, M. Catabolism of hydroxycinnamic acids in contact with probiotic Lactobacillus. *J. Appl. Microbiol.* **2021**, *131*, 1464–1473. [CrossRef]
52. Ryu, J.Y.; Kang, H.R.; Cho, S.K. Changes over the fermentation period in phenolic compounds and antioxidant and anticancer activities of blueberries fermented by *Lactobacillus plantarum*. *J. Food Sci.* **2019**, *84*, 2347–2356. [CrossRef] [PubMed]
53. Hussain, M.B.; Hassan, S.; Waheed, M.; Javed, A.; Farooq, M.A.; Tahir, A. Bioavailability and metabolic pathway of phenolic compounds. In *Plant Physiological Aspects of Phenolic Compounds*; IntechOpen: London, UK, 2019. [CrossRef]
54. Hsiao, Y.-H.; Ho, C.-T.; Pan, M.-H. Bioavailability and health benefits of major isoflavone aglycones and their metabolites. *J. Funct. Foods* **2020**, *74*, 104164. [CrossRef]
55. Huang, J.; de Paulis, T.; May, J.M. Antioxidant effects of dihydrocaffeic acid in human EA. hy926 endothelial cells. *J. Nutr. Biochem.* **2004**, *15*, 722–729. [CrossRef]
56. Moon, J.-H.; Terao, J. Antioxidant activity of caffeic acid and dihydrocaffeic acid in lard and human low-density lipoprotein. *J. Agric. Food Chem.* **1998**, *46*, 5062–5065. [CrossRef]
57. Li, F.; Wang, L.; Cai, Y.; Luo, Y.; Shi, X. Safety assessment of desaminotyrosine: Acute, subchronic oral toxicity, and its effects on intestinal microbiota in rats. *Toxicol. Appl. Pharmacol.* **2021**, *417*, 115464. [CrossRef] [PubMed]
58. Steed, A.L.; Christophi, G.P.; Kaiko, G.E.; Sun, L.; Goodwin, V.M.; Jain, U.; Esaulova, E.; Artyomov, M.N.; Morales, D.J.; Holtzman, M.J. The microbial metabolite desaminotyrosine protects from influenza through type I interferon. *Science* **2017**, *357*, 498–502. [CrossRef] [PubMed]
59. Velika, B.; Kron, I. Antioxidant properties of benzoic acid derivatives against superoxide radical. *Free Radic. Antioxid.* **2012**, *2*, 62–67. [CrossRef]
60. Khadem, S.; Marles, R.J. Monocyclic phenolic acids; hydroxy-and polyhydroxybenzoic acids: Occurrence and recent bioactivity studies. *Molecules* **2010**, *15*, 7985–8005. [CrossRef]
61. Jung, U.J.; Choi, M.S. Obesity and its metabolic complications: The role of adipokines and the relationship between obesity, inflammation, insulin resistance, dyslipidemia and nonalcoholic fatty liver disease. *Int. J. Mol. Sci.* **2014**, *15*, 6184–6223. [CrossRef]
62. Son, J.H.; Kim, S.-Y.; Jang, H.H.; Lee, S.N.; Ahn, K.J. Protective effect of protocatechuic acid against inflammatory stress induced in human dermal fibroblasts. *Biomed. Dermatol.* **2018**, *2*, 1–5. [CrossRef]
63. Nour, O.A.; Ghoniem, H.A.; Nader, M.A.; Suddek, G.M. Impact of protocatechuic acid on high fat diet-induced metabolic syndrome sequelae in rats. *Eur. J. Pharmacol.* **2021**, *907*, 174257. [CrossRef] [PubMed]
64. Sharma, N.; Tiwari, N.; Vyas, M.; Khurana, N.; Muthuraman, A.; Utreja, P. An overview of therapeutic effects of vanillic acid. *Plant Arch.* **2020**, *20*, 3053–3059.
65. Kaur, J.; Gulati, M.; Singh, S.K.; Kuppusamy, G.; Kapoor, B.; Mishra, V.; Gupta, S.; Arshad, M.F.; Porwal, O.; Jha, N.K. Discovering multifaceted role of vanillic acid beyond flavours: Nutraceutical and therapeutic potential. *Trends Food Sci. Technol.* **2022**, *122*, 187–200. [CrossRef]
66. Fernandez, M.; Zuniga, M. Amino acid catabolic pathways of lactic acid bacteria. *Crit. Rev. Microbiol.* **2006**, *32*, 155–183. [CrossRef]
67. Averesch, N.J.; Krömer, J.O. Metabolic engineering of the shikimate pathway for production of aromatics and derived compounds—Present and future strain construction strategies. *Front. Bioeng. Biotechnol.* **2018**, *6*, 32–51. [CrossRef]
68. Serra, S.; De Simeis, D.; Castagna, A.; Valentino, M. The fatty-acid hydratase activity of the most common probiotic microorganisms. *Catalysts* **2020**, *10*, 154. [CrossRef]
69. Hirata, A.; Kishino, S.; Park, S.-B.; Takeuchi, M.; Kitamura, N.; Ogawa, J. A novel unsaturated fatty acid hydratase toward C16 to C22 fatty acids from *Lactobacillus acidophilus*. *J. Lipid Res.* **2015**, *56*, 1340–1350. [CrossRef]
70. Sjögren, J.; Magnusson, J.; Broberg, A.; Schnürer, J.; Kenne, L. Antifungal 3-hydroxy fatty acids from *Lactobacillus plantarum* MiLAB 14. *Appl. Environ. Microbiol.* **2003**, *69*, 7554–7557. [CrossRef]
71. Schnürer, J.; Magnusson, J. Antifungal lactic acid bacteria as biopreservatives. *Trends Food Sci. Technol.* **2005**, *16*, 70–78. [CrossRef]
72. Yoo, J.A.; Lee, C.-J.; Kim, Y.-G.; Lee, B.-E.; Yoon, M.-H. Antifungal Effect of Phenyllactic Acid Produced by *Lactobacillus casei* Isolated from Button Mushroom. *J. Mushroom* **2016**, *14*, 162–167. [CrossRef]
73. Goh, Y.J.; Klaenhammer, T.R. Genomic features of *Lactobacillus* species. *Front. Biosci. -Landmark* **2009**, *14*, 1362–1386. [CrossRef]
74. Hirai, S.; Horii, S.; Matsuzaki, Y.; Ono, S.; Shimmura, Y.; Sato, K.; Egashira, Y. Anti-inflammatory effect of pyroglutamyl-leucine on lipopolysaccharide-stimulated RAW 264.7 macrophages. *Life Sci.* **2014**, *117*, 1–6. [CrossRef] [PubMed]
75. Shirako, S.; Kojima, Y.; Tomari, N.; Nakamura, Y.; Matsumura, Y.; Ikeda, K.; Inagaki, N.; Sato, K. Pyroglutamyl leucine, a peptide in fermented foods, attenuates dysbiosis by increasing host antimicrobial peptide. *npj Sci. Food* **2019**, *3*, 18–27. [CrossRef]
76. Chistyakov, D.V.; Gavrish, G.E.; Goriainov, S.V.; Chistyakov, V.V.; Astakhova, A.A.; Azbukina, N.V.; Sergeeva, M.G. Oxylipin profiles as functional characteristics of acute inflammatory responses in astrocytes pre-treated with IL-4, IL-10, or LPS. *Int. J. Mol. Sci.* **2020**, *21950*, 1780. [CrossRef] [PubMed]
77. Gosset, V.; Goëbel, C.; Laine, G.; Delaplace, P.; Du Jardin, P.; Feussner, I.; Fauconnier, M.-L. The role of oxylipins and antioxidants on off-flavor precursor formation during potato flake processing. *J. Agric. Food Chem.* **2008**, *56*, 11285–11292. [CrossRef]
78. Zebisch, K.; Voigt, V.; Wabitsch, M.; Brandsch, M. Protocol for effective differentiation of 3T3-L1 cells to adipocytes. *Anal. Biochem.* **2012**, *425*, 88–90. [CrossRef] [PubMed]
79. Kim, M.J.; Jun, H.Y.; Kim, J.H. Antiadipogenic effect of Korean glasswort (*Salicornia herbacea* L.) water extract on 3T3-L1 adipocytes. *J. Korean Soc. Food Sci. Nutr.* **2014**, *43*, 814–821. [CrossRef]

80. Rizzatti, V.; Boschi, F.; Pedrotti, M.; Zoico, E.; Sbarbati, A.; Zamboni, M. Lipid droplets characterization in adipocyte differentiated 3T3-L1 cells: Size and optical density distribution. *Eur. J. Histochem. EJH* **2013**, *57*, e24–e29. [CrossRef] [PubMed]
81. Walther, T.C.; Farese, R.V., Jr. Lipid droplets and cellular lipid metabolism. *Annu. Rev. Biochem.* **2012**, *81*, 687–714. [CrossRef] [PubMed]
82. Prusty, D.; Park, B.-H.; Davis, K.E.; Farmer, S.R. Activation of MEK/ERK signaling promotes adipogenesis by enhancing peroxisome proliferator-activated receptor γ (PPARγ) and C/EBPα gene expression during the differentiation of 3T3-L1 preadipocytes. *J. Biol. Chem.* **2002**, *277*, 46226–46232. [CrossRef]
83. Poeggeler, B.; Schulz, C.; Pappolla, M.A.; Bodó, E.; Tiede, S.; Lehnert, H.; Paus, R. Leptin and the skin: A new frontier. *Exp. Dermatol.* **2010**, *19*, 12–18. [CrossRef]
84. Jeong, H.J.; Park, J.H.; Kim, M.-J. Ethanol extract of *Hippophae rhamnoides* L. leaves inhibits adipogenesis through AMP-activated protein kinase (AMPK) activation in 3T3-L1 preadipocytes. *Korean J. Plant Resour.* **2015**, *28*, 582–590. [CrossRef]
85. Song, Y.; Lee, S.-J.; Jang, S.-H.; Kim, T.H.; Kim, H.-D.; Kim, S.-W.; Won, C.-K.; Cho, J.-H. Annual wormwood leaf inhibits the adipogenesis of 3T3-L1 and obesity in high-fat diet-induced obese rats. *Nutrients* **2017**, *9*, 554. [CrossRef]
86. Wu, T.; Yu, Z.; Tang, Q.; Song, H.; Gao, Z.; Chen, W.; Zheng, X. Honeysuckle anthocyanin supplementation prevents diet-induced obesity in C57BL/6 mice. *Food Funct.* **2013**, *4*, 1654–1661. [CrossRef] [PubMed]
87. Canter, R.J.; Le, C.T.; Beerthuijzen, J.M.; Murphy, W.J. Obesity as an immune-modifying factor in cancer immunotherapy. *J. Leukoc. Biol.* **2018**, *104*, 487–497. [CrossRef]
88. Lee, J.S.; Kim, K.-J.; Kim, Y.-H.; Kim, D.-B.; Shin, G.-H.; Cho, J.-H.; Kim, B.K.; Lee, B.-Y.; Lee, O.-H. Codonopsis lanceolata extract prevents diet-induced obesity in C57BL/6 mice. *Nutrients* **2014**, *6*, 4663–4677. [CrossRef] [PubMed]
89. Lee, H.-S.; Choi, J.-H.; Kim, Y.-G.; Lee, C.-H. Effect of dietary intake of Salicornia herbacea L. hot water extract on anti-obesity in diet-induced obese rats. *J. Korean Soc. Food Sci. Nutr.* **2012**, *41*, 950–956. [CrossRef]
90. Jia, Y.; Kim, S.; Kim, J.; Kim, B.; Wu, C.; Lee, J.H.; Jun, H.j.; Kim, N.; Lee, D.; Lee, S.J. Ursolic acid improves lipid and glucose metabolism in high-fat-fed C57BL/6J mice by activating peroxisome proliferator-activated receptor alpha and hepatic autophagy. *Mol. Nutr. Food Res.* **2015**, *59*, 344–354. [CrossRef] [PubMed]
91. Angulo, P. Obesity and nonalcoholic fatty liver disease. *Nutr. Rev.* **2007**, *65*, S57–S63. [CrossRef]
92. Seo, H.-B.; Kwak, Y.-Y.; Nam, J.-O.; Song, Y.-J.; Kim, B.-O.; Ryu, S.-P. Glasswort powder diet activates lipid metabolism in rat. *J. Life Sci.* **2012**, *22*, 478–485. [CrossRef]
93. Park, S.H.; Ko, S.K.; Choi, J.G.; Chung, S.H. Salicornia herbacea prevents high fat diet-induced hyperglycemia and hyperlipidemia in ICR mice. *Arch. Pharmacal Res.* **2006**, *29*, 256–264. [CrossRef] [PubMed]
94. Al-Goblan, A.S.; Al-Alfi, M.A.; Khan, M.Z. Mechanism linking diabetes mellitus and obesity. *Diabetes Metab. Syndr. Obes. Targets Ther.* **2014**, *7*, 587–591. [CrossRef] [PubMed]
95. Pichiah, P.T.; Cha, Y.S. Salicornia herbacea prevents weight gain and hepatic lipid accumulation in obese ICR mice fed a high-fat diet. *J. Sci. Food Agric.* **2015**, *95*, 3150–3159. [CrossRef] [PubMed]

nutrients

MDPI

Article

Dietary Sources of Anthocyanins and Their Association with Metabolome Biomarkers and Cardiometabolic Risk Factors in an Observational Study

Hamza Mostafa [1,2,†], Tomás Meroño [1,2,†], Antonio Miñarro [1,2,3], Alex Sánchez-Pla [1,2,3], Fabián Lanuza [1,2], Raul Zamora-Ros [1,4], Agnetha Linn Rostgaard-Hansen [5], Núria Estanyol-Torres [1], Marta Cubedo-Culleré [1,3], Anne Tjønneland [5], Rikard Landberg [6], Jytte Halkjær [5,*] and Cristina Andres-Lacueva [1,2,*]

1 Biomarkers and Nutrimetabolomics Laboratory, Department of Nutrition, Food Sciences and Gastronomy, Xarxa d'Innovació Alimentària (XIA), Nutrition and Food Safety Research Institute (INSA), Facultat de Farmàcia i Ciències de l'Alimentació, Universitat de Barcelona (UB), 08028 Barcelona, Spain

2 Centro de Investigación Biomédica en Red de Fragilidad y Envejecimiento Saludable (CIBERFES), Instituto de Salud Carlos III, 28029 Madrid, Spain

3 Departament de Genètica, Microbiologia i Estadística, Facultat de Biologia, Universitat de Barcelona (UB), 08028 Barcelona, Spain

4 Unit of Nutrition and Cancer, Cancer Epidemiology Research Program, Catalan Institute of Oncology (ICO), Bellvitge Biomedical Research Institute (IDIBELL), 08908 L'Hospitalet de Llobregat, Spain

5 Danish Cancer Society Research Center, Strandboulevarden 49, DK 2100 Copenhagen, Denmark

6 Department of Biology and Biological Engineering, Division of Food and Nutrition Science, Chalmers University of Technology, 412 96 Gothenburg, Sweden

* Correspondence: jytteh@cancer.dk (J.H.); candres@ub.edu (C.A.-L.); Tel.: +45-35257659 (J.H.); +34-934034840 (C.A.-L.)

† These authors contributed equally to this work.

Citation: Mostafa, H.; Meroño, T.; Miñarro, A.; Sánchez-Pla, A.; Lanuza, F.; Zamora-Ros, R.; Rostgaard-Hansen, A.L.; Estanyol-Torres, N.; Cubedo-Culleré, M.; Tjønneland, A.; et al. Dietary Sources of Anthocyanins and Their Association with Metabolome Biomarkers and Cardiometabolic Risk Factors in an Observational Study. *Nutrients* **2023**, *15*, 1208. https://doi.org/10.3390/nu15051208

Academic Editors: Sonia de Pascual-Teresa and Luis Goya

Received: 9 February 2023
Revised: 24 February 2023
Accepted: 25 February 2023
Published: 28 February 2023

Abstract: Anthocyanins (ACNs) are (poly)phenols associated with reduced cardiometabolic risk. Associations between dietary intake, microbial metabolism, and cardiometabolic health benefits of ACNs have not been fully characterized. Our aims were to study the association between ACN intake, considering its dietary sources, and plasma metabolites, and to relate them with cardiometabolic risk factors in an observational study. A total of 1351 samples from 624 participants (55% female, mean age: 45 ± 12 years old) enrolled in the DCH-NG MAX study were studied using a targeted metabolomic analysis. Twenty-four-hour dietary recalls were used to collect dietary data at baseline, six, and twelve months. ACN content of foods was calculated using Phenol Explorer and foods were categorized into food groups. The median intake of total ACNs was 1.6mg/day. Using mixed graphical models, ACNs from different foods showed specific associations with plasma metabolome biomarkers. Combining these results with censored regression analysis, metabolites associated with ACNs intake were: salsolinol sulfate, 4-methylcatechol sulfate, linoleoyl carnitine, 3,4-dihydroxyphenylacetic acid, and one valerolactone. Salsolinol sulfate and 4-methylcatechol sulfate, both related to the intake of ACNs mainly from berries, were inversely associated with visceral adipose tissue. In conclusion, plasma metabolome biomarkers of dietary ACNs depended on the dietary source and some of them, such as salsolinol sulfate and 4-methylcatechol sulfate may link berry intake with cardiometabolic health benefits.

Keywords: metabolomics; anthocyanins; food matrix; diet; gut microbiota; berries; cardiometabolic health

1. Introduction

Anthocyanins (ACNs) are phytochemical compounds of the subclass of flavonoids in the broader (poly)phenol class, highly present in plant foods, such as berries, grapes, eggplants, and many other colored fruits and vegetables [1,2]. Most of the dietary ACNs reach the large intestine unaffected where they may affect both gut microbial composition and microbial metabolism of ACNs [2,3]. Dietary ACNs and their microbial metabolites are suggested to play roles in the prevention and treatment of cardiometabolic diseases [4,5]. Microbial metabolites of ACNs have been shown to reach higher concentrations in the systemic circulation and may be more bioactive than the consumed ACNs per se [6]. Nonetheless, studies evaluating the association between ACN dietary intake and plasma concentrations of ACN-derived microbial metabolites in observational studies are lacking.

Sources of variability in ACN metabolism by the host and gut microbiota could be related to differences in consumed types and quantity of ACNs, as well as to a food matrix effect. Ultimately, these differences might be translated into different health effects of ACNs in response to their intake. Up to the moment, besides parent ACNs, such as cyanidin, delphinidin, malvidin, and petunidin, various metabolites derived from microbial and host metabolism (i.e., protocatechuic acid, syringic acid, and 4-hydroxybenzoic acid) have been associated with the consumption of berries in dietary intervention studies [7–9]. However, to the best of our knowledge, associations between the intake of ACNs, annotated according to their dietary sources, microbial metabolites, and cardiometabolic risk factors have not been studied in an observational study. Our hypothesis is that specific sets of microbial metabolites would be associated with dietary ACNs from different food sources and that these will display differential associations with cardiometabolic risk factors. The aims of this study were to evaluate the association between the intake of ACNs, considering their different dietary sources, and plasma metabolites. The aim was further to explore the associations between ACN-related metabolites and cardiometabolic risk factors in a subsample of the Danish Diet, Cancer, and Health-Next Generations (DCH-NG) MAX study [10].

2. Materials and Methods

2.1. Study Design and Subjects

We studied a validation subsample within the Diet, Cancer, and Health-Next Generations (DCH-NG) cohort: the DCH-NG MAX study. The DCH-NG was an extension of the previous cohort the Diet, Cancer, and Health (DCH) [10]. A sample of 39,554 participants was included in the DCH-NG involving biological children, their spouses, and grandchildren of the DCH cohort [11]. The DCH-NG MAX study recruited 720 volunteers with residency in Copenhagen, aged 18 years old or more, between August 2017 and January 2019. The major aims of the MAX study were to validate a semi-quantitative food frequency questionnaire against the twenty-four-hour dietary recalls (24-HDR) and to examine the plasma and urine metabolome reproducibility as well as gut microbial stability on a long-term scale. Biological samples, health examinations such as anthropometric and blood pressure measurements, and two questionnaires about lifestyle and dietary habits were collected at baseline, 6, and 12 months.

The DCH-NG cohort study was approved by the Danish Data Protection Agency (journal number 2013-41-2043/2014-231-0094) and by the Committee on Health Research Ethics for the Capital Region of Denmark (journal number H-15001257). The volunteers provided their written informed consent to participate in the study. All the details about clinical measurements, dietary and metabolomics data were previously detailed [11].

2.2. Anthropometric Measurements

Participants were asked to wear underwear and be barefoot for measuring height and weight using a wireless stadiometer and a body composition analyzer, respectively (SECA mBCA515, Hamburg, Germany). Height and weight were measured to the nearest 0.1 cm and 0.01 kg, respectively, and body mass index (BMI) was calculated. The waist circumference was measured twice at the midpoint between the lower rib margin and the iliac crest. A third measurement for the waist circumference was measured if the difference between the first two was more than 1 cm. Blood pressure and pulse rate were measured 3 times using the left arm, considering the measurement with the lower systolic blood pressure and its corresponding diastolic blood pressure value as valid. DEXA-validated bioimpedance instrument (SECA mBCA515, Germany) was used to estimate visceral adipose tissue volume.

2.3. Dietary Data

The 24-h dietary recalls (24-HDR) were recorded at baseline, 6 and 12 months using a Danish version of the web-based tool myfood24 (www.myfood24.org/) (7 February 2023) from Leeds University [12], containing almost 1600 Danish food items. All foods consumed the day before the examinations were reported by the participants in either grams or in standard portion size. The percentage of calories using the energy equivalents for carbohydrates, proteins, and fat was used to indicate the intake of macronutrients. Complex food products were appointed as recipes or dishes. The McCance and Widdowson's Food Composition Table [13], or recipes from the food frequency questionnaires in the DCH were used to have the standardized recipes [14].

Dietary Intake of Anthocyanins

Estimation of the intake of polyphenols from 24-HDRs was completed by a protocol using "in-house" software developed by the University of Barcelona, the Bellvitge Biomedical Research Institute (IDIBELL), and the Centro de Investigación Biomédica en Red (CIBER) ©. A link between all 24-HDR food items or ingredients and the foods from the Phenol-Explorer database was created [15]. The intake of individual (poly)phenols in mg/day was obtained and ACN consumption from separate foods were estimated as the sum of 71 individual ACNs included in Phenol-Explorer database. The estimated intake of dietary (poly)phenols of the DCH-NG MAX study has previously been described [16]. A total of 147 food items that contain ACNs were used to estimate the total dietary ACN intake as shown in the Supplementary Table S1. Intake of berries was estimated as the sum of foods with at least 50% of its composition or recipe made by berries. These include raw and frozen berries, berries marmalades or jams, and stewed berries. Dietary ACN intake related with the other foods were classified and added up according to the following food groups: dairy products with berries (including ice cream and yogurt), other fruits (i.e., plums, cherries, apples, etc.), non-alcoholic drinks (including fruit smoothies and juices), wines, vegetables, mixed dishes (meat or fish dishes with vegetables containing ACNs), and bakery (including pastry, biscuits, desserts, and waffles with berries or other ACN-containing preparations). (Supplementary Table S2). Intakes of foods not containing ACN were disregarded.

2.4. Blood Sampling, Analysis of Cardiometabolic Risk Factors and Metabolomics

Participants were instructed to maintain a fasting time of 1–9 h (mean fasting time: 5 h) during all the examination days. Blood samples were taken into Vacutainer tubes containing lithium heparin at baseline time 0 (n = 624), 6 months (n = 380), and 12 months (n = 349). Within 2 h of blood draw, plasma was obtained by centrifugation, and samples were stored at −80 °C. After that, plasma samples were delivered to the Danish National Biobank (DNB), where plasma was divided into aliquots and sent to University of Barcelona and kept at −80 °C until metabolomic analysis. Other blood measurements such as hemoglobin, A1c (HbA1c), serum lipids, and high sensitivity C reactive protein (hsCRP) were measured as described before [17].

2.4.1. Metabolomics Analysis of Plasma Samples

Repeated measures of the plasma metabolome at all three time points were used for metabolomics analysis. All the samples were prepared and analyzed using the targeted UPLC-MS/MS method described previously, with slight modifications [18,19]. Briefly, 100 µL of plasma was added into protein precipitation plates together with 500 µL cold acetonitrile containing 1.5 M formic acid and 10 mM of ammonium formate and were kept at −20 °C for 10 min to enhance protein precipitation. Then, positive pressure was applied to recover the extracts, which were taken to dryness and reconstituted with 100 µL of an 80:20 *v/v* water:aceto nitrile solution containing 0.1% *v/v* formic acid and 100 ppb of a mixture of 13 internal standards. Samples were then transferred to 96-well plates and analyzed by a targeted metabolomic analysis using an Agilent 1290 Infinity UPLC system coupled to a Sciex QTRAP 6500 mass spectrometer, using the operating conditions described elsewhere [18]. The Sciex OS 2.1.6 software (Sciex, Framingham, MA, USA) was used for data processing.

2.4.2. Metabolomics Data Pre-Processing

The POMA R/Bioconductor package (https://github.com/nutrimetabolomics/POMA) (7 February 2023) was used for the pre-processing of metabolomics data [20]. Metabolites with more than 40% missing values, and those with a coefficient of variation (CV) > 30% in internal quality control were removed. K-nearest neighbor (KNN) algorithm and correction of batch effects using the ComBat function ("sva" R package) were used to impute the remaining missing values [21], while auto-scaling and Euclidean distances (±1.5× Interquartile range) were used to normalize the data and remove the outliers, respectively. The final metabolomics dataset included the concentration of 408 plasma metabolites.

2.5. Statistical Analyses

For descriptive statistics, intake of total ACNs (irrespective of their dietary source) was categorized by tertiles using 0.3–8.9 mg as thresholds. Continuous variables following a normal distribution are shown as mean ± SD, and those following a skewed distribution are shown as median (p25–p75). Sociodemographic and clinical characteristics were compared across tertiles of ACNs intake using linear mixed models in a random intercepts model adjusted for age and sex. Associations between intake of ACN dietary sources and their association with cardiometabolic risk factors were tested using linear mixed models in random intercepts models adjusted for age, sex, and BMI.

First, associations between intake of total ACNs and metabolome biomarkers were analyzed using a censored regression for panel data with "censReg" and "plm" R-packages [22]. Censored regression models were applied due to the right-skewed distribution of total ACN intake and the considerable proportion of zero values (24% of non-consumers of ACNs). Covariates included in the models were age, sex, and BMI. *p*-values were adjusted for multiple comparisons using the Benjamini–Hochberg method, and adjusted *p*-values <0.05 were considered statistically significant. Second, associations between ACNs from different dietary sources and metabolites were assessed using Mixed Graphical Models (MGM) with the "mgm" R-package [23]. MGMs are undirected probabilistic graphical models able to represent associations between nodes adjusted for all the other variables in the model. MGM specifications were set to allow the maximum number of interactions in the network. Variables in the model were dietary ACN intakes by food categories (8 food groups), and the whole metabolomics set of variables. The agreement between repeated measurements for total dietary intake of ACNs and for ACN intake from different dietary sources was poor across the study evaluations (intra-class correlation coefficient < 0.15). Therefore, all observations of the study were considered independent and were included in MGM analysis (k = 1351). For visual clarity, only the first-order neighborhood of ACNs food sources was plotted.

To evaluate the associations between metabolites and cardiometabolic risk factors linear mixed models were used in random intercepts models adjusted for age, sex, and BMI. Metabolites were selected based on the combination of both analyses, censored regression, and MGM. Standardized coefficients were plotted in a heatmap built using the "pheatmap" R-package (Kolde R (2019). pheatmap: Pretty Heatmaps).

All statistical analyses were performed using R, version 4.1.3. (R foundation, Austria).

3. Results

3.1. Sociodemographic, Clinical, and Dietary Characteristics

At baseline, out of the 720 volunteers who agreed to participate in the study, 624 had completed clinical, dietary, and plasma metabolomics data. Of the 624 participants included, 55% were female, aged (mean ± SD) 45 ± 12 years old, and had a BMI of 25 ± 4 kg/m^2. At 6 months, 380 participants had completed clinical, dietary, and metabolomics data and at

12 months completed data were available for 349 participants. Only, 287 participants had completed clinical dietary and plasma metabolomics data available at all three time points.

The distribution of total ACN intake was right-skewed with a median value of 1.6 (p25–p75: 0.0–26.9) mg/day and a mean value of 26.4 (SD: 60.4) mg/day. Berries were the highest contributors to total ACN intake with a mean contribution of 34%, followed by wines with 33%, and non-alcoholic drinks (which included fruit smoothies and juices) with 20% of the total reported intake. Other fruits (i.e., cherries, apples, and plums) and vegetables were minor contributors with 4% and 2%, respectively. Bakery (pastry, biscuits, and desserts), dairy products (yogurts and strawberry ice creams or ice creams with berries), and other mixed dishes (dishes including vegetables with ACNs) contributed within a similar range between 2 and 3% (Supplementary Table S1).

Participants were divided into tertiles based on the consumed reported intakes of total ACNs as shown in Table 1. There were no significant differences in clinical characteristics across tertiles of ACN intake. Consistently, there were no statistically significant associations between total ACN intake and cardiometabolic risk factors (data not shown). Dietary characteristics are illustrated in Supplementary Table S2. Participants in the highest compared to the ones at the lowest tertile of ACN intake showed statistically significant higher consumption of total protein, saturated fatty acids (SFA), monounsaturated fatty acids (MUFA), polyunsaturated fatty acids (PUFA), alcohol, fruits, and berries.

Table 1. Sociodemographic and clinical characteristics according to tertiles of total ACN intake.

	All ***n* = 624** **k = 1351**	**Tertile 1** **<0.3 mg ACN/Day** **k = 453**	**Tertile 2** **0.3–8.9 mg ACN/Day** **k = 448**	**Tertile 3** **>8.9 mg ACN/Day** **k = 450**
Age (years)	44.7 ± 12.3	43.7 ± 12.6	44.2 ±12.5	46.1 ± 11.8
Gender, female (*n*, %)	745 (55)	236 (52)	250 (55)	256 (57)
BMI (kg/m²)	25 ± 4	25 ± 4	24 ± 3	25 ± 4
WC (cm)	87.5 ± 12.1	88.7 ± 12.4	86.0 ± 11.6	87.7 ± 12
VAT (L)	1.3 (0.7–2.5)	1.5 (0.8–2.6)	1.2 (0.6–2.1)	1.4 (0.8–2.5)
		Physical activity (*n*, %)		
Not regular	114 (17)	41 (19)	25 (12)	37 (18)
Once/month last 6 months	52 (89	15 (79	22 (11)	12 (6)
Once/month last 12 months	510 (75)	162 (74)	159 (77)	152 (76)
		Smoking status (*n*, %)		
Never	353 (52.2)	114 (52.3)	117 (56.89	97 (48.3)
Former	186 (27.5)	53 (24.3)	53 (25.79)	69 (34.39)
Current	137 (20.3)	51 (23.4)	36 (17.59)	35 (17.4)
SBP (mmHg)	117 ± 16	117 ± 15	116 ± 15	116 ± 16
DBP (mmHg)	81 ± 11	80 ± 10	79 ± 10	80 ± 11
HbA1c (mmol/mol)	34.5 ± 6	34.6 ± 7	33.8 ± 5	34.6 ± 6

Table 1. *Cont.*

	All *n* = 624 k = 1351	Tertile 1 <0.3 mg ACN/Day k = 453	Tertile 2 0.3–8.9 mg ACN/Day k = 448	Tertile 3 >8.9 mg ACN/Day k = 450
TG (mmol/L)	1.1 (0.8–1.6)	1.1 (0.8–1.7)	1.0 (0.8–1.4)	1.1 (0.8–1.6)
TC (mmol/L)	4.9 ± 1.0	4.9 ± 0.9	4.9 ± 0.9	5.1 ± 1.0
HDL (mmol/L)	1.6 ± 0.4	1.5 ± 0.4	1.6 ± 0.4	1.6 ± 0.5
LDL (mmol/L)	3.0 ± 0.9	3.0 ± 0.9	2.9 ± 0.9	3.1 ± 0.9
hsCRP (mg/L)	0.7 (0.3–1.6)	0.8 (0.3–1.6)	0.7 (0.3–1.5)	0.7 (0.3–1.6)

BMI, body mass index; WC, waist circumference; VAT, visceral adipose tissue; SBP, systolic blood pressure, DBP, diastolic blood pressure; HbA1c, hemoglobin A1c; TG, triglycerides; TC, total cholesterol; HDL, high-density lipoproteins; LDL, low-density lipoproteins; hsCRP, high-sensitivity C-reactive protein. Variables following a normal distribution are shown as mean ± SD, and those with a skewed distribution are shown as median (p25–p75).

3.2. Association between Intake of ACN Dietary Sources and Cardiometabolic Risk Factors

Several inverse and direct associations between self-reported intake of ACN-containing food groups and cardiometabolic risk factors were observed (Figure 1). For example, intake of berries, dairy products with berries, and ACN-containing vegetables had inverse associations with visceral adipose tissue volume, while wine had direct associations with total cholesterol, HDL-C, and systolic blood pressure. Other direct associations were found between the intake of berries and hemoglobin A1c, and between ACN-containing drinks with hsCRP.

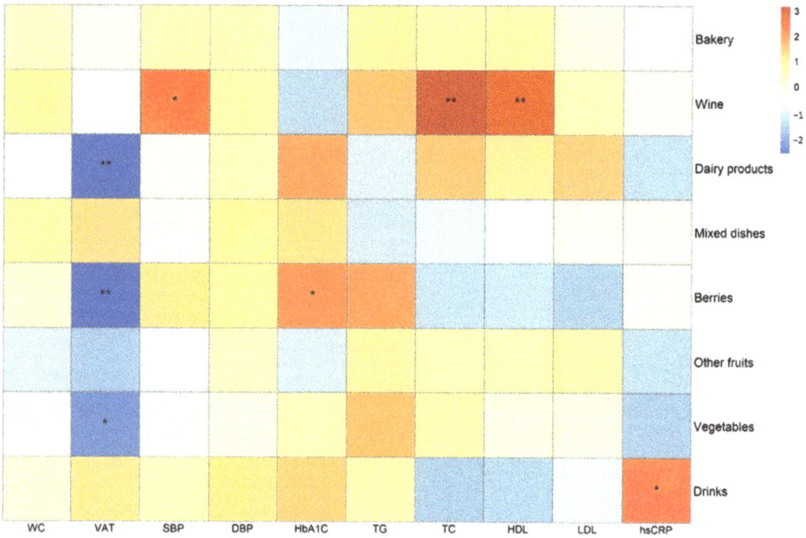

Figure 1. Association between different self-reported food groups and cardiometabolic risk factors in the DCH-NG MAX study (n = 624, k = 1351). Standardized coefficients according to linear mixed models with random intercepts adjusting for age, sex, and BMI. * $p < 0.05$, ** $p < 0.01$. n = number of subjects, k = total number of observations. TG, triglycerides; SBP, systolic blood pressure; DBP, diastolic blood pressure; WC, waist circumference; HbA1c, hemoglobin A1c; hsCRP, high-sensitivity C-reactive protein; VAT, visceral adipose tissue; TC, total cholesterol; HDL, high-density lipoproteins; TC, total cholesterol; LDL, low-density lipoproteins.

3.3. Metabolome Biomarkers Associated with Total ACN Intake

In censored regression analysis, 10 metabolites were positively associated with total ACN intake (Figure 2). Among them, three were exogenous metabolites, hypaphorine, salsolinol sulfate, and ethyl glucuronide, two were endogenous metabolites, including linoleoyl carnitine and glycerol, and five were gut microbial metabolites, 4-methylcatechol sulfate, 4′-hydroxy-3′-methoxyphenyl-γ-valerolactone-sulfate (MHPV-S), 5-(4-hydroxy(3,4-dihydroxyphenyl)-valeric acid sulfate (3,4-DHPHVA-S), 3,4-dihydroxyphenylacetic acid sulfate (3,4-DHPA-3S) and indolepropionic acid. On the other hand, only oleoyl carnitine, another endogenous metabolite, was inversely associated with total ACN dietary intake.

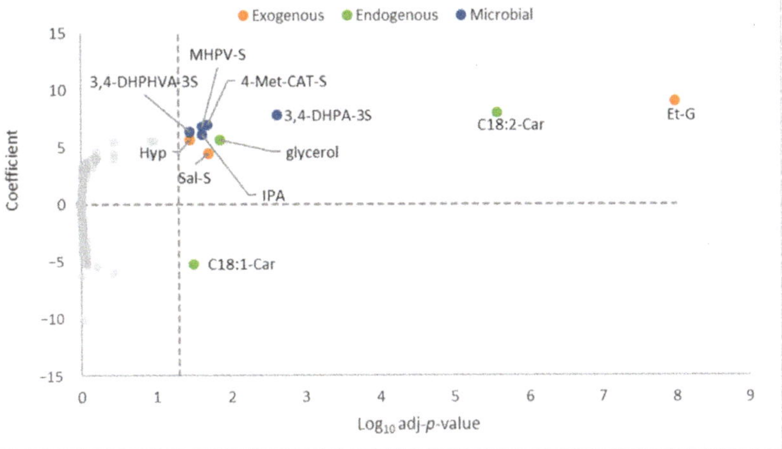

Figure 2. Association between metabolome biomarkers and total intake of ACNs. Censored regression for panel data adjusting for age, sex, and BMI (n = 624, k = 1351). p-values were calculated and adjusted by Benjamini–Hochberg procedure. Adjusted p-values < 0.05 were considered statistically significant. Hyp, hypaphorine; Sal-S, salsolinol sulfate; Et-G, ethyl glucuronide; 4-Met-Cat-S, 4-methylcatechol sulfate; MHPV-S, 4′-hydroxy-3′-methoxyphenyl-γ-valerolactone sulfate; 3,4-DHPHVA-3S, 5-(4-hydroxy(3,4-dihydroxyphenyl)-valeric acid sulfate; 3,4-DHPA-3S, 3,4-dihydroxyphenylacetic acid sulfate; IPA, indolepropionic acid; C18,2-Car, linoleoyl carnitine; C18,1-Car, oleoyl carnitine.

3.4. Metabolome Biomarkers Associated with Intake of ACNs Related to Different ACN Dietary Sources

MGM analysis showed associations between self-reported ACN intake from different dietary sources and 16 metabolites (Figure 3). ACNs derived from dairy products were associated with plasma asparagine, epicatechin sulfate, urolithin C-glucuronide, and acesulfame K. ACNs from the intake of berries were associated with linoleoyl carnitine, salsolinol sulfate, glycochenodeoxycholic-3-sulfate (GCDCA-3S) and 4-methylcatechol sulfate. ACNs from wine consumption were linked with methylpyrogallol sulfate (Met-Pyr-S) and ethyl glucuronide. ACNs from vegetable intake were associated with 2-hydroxybenzoic acid and bergaptol glucuronide. ACNs from other fruits were associated with 3,4-DHPHVA-3S, 5-(3′-hydroxyphenyl)-γ-valerolactone 3′-sulfate (3-HPV-S) and 3,4-dihydroxyphenylacetic acid sulfate (3,4-DHPA-3S). Lastly, the consumption of ACNs from mixed dishes was associated with 1-methylhistidine and 2-hydroxybenzoic acid. Overall, not all the metabolites selected in the MGM analysis were related to ACNs or its microbial metabolites, but to other food components such as acesulfame K, ethyl glucuronide, etc. Therefore, metabolome biomarkers were selected considering both statistical analyses, censored regression, and MGM, to be used for the study of its association with cardiometabolic risk factors.

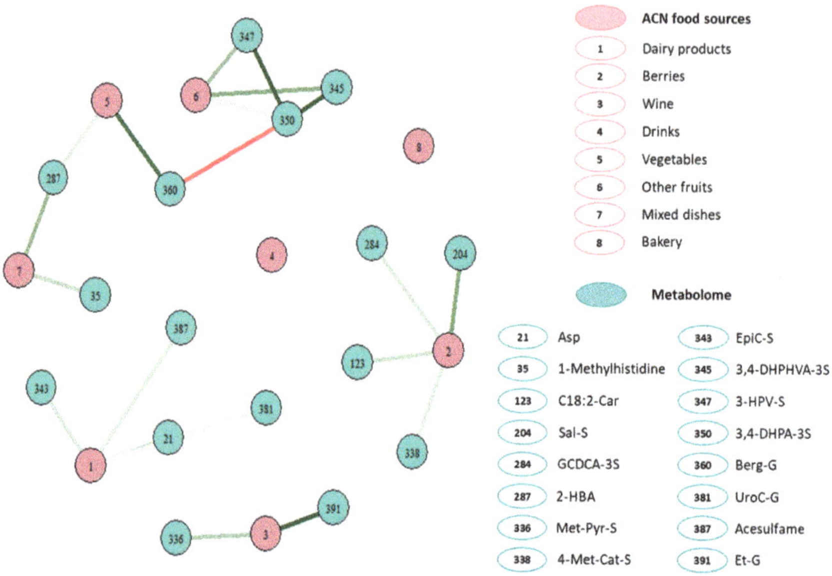

Figure 3. First-order neighborhood of ACNs intake related to different self-reported food groups with plasma metabolome biomarkers according to Mixed Graphical Models in the DCH-NG MAX study (n = 624, k = 1351). Edge intensity reflects the strength of the association from strong direct (dark green) to strong inverse association (dark red). Variables included in the mixed graphical model were ACN intake related to self-reported intake of dairy, berries, wines, non-alcoholic drinks (smoothies and fruit juices), vegetables, other fruits, mixed dishes, and bakery, and all the 408 plasma metabolites quantified with our targeted metabolomics method. n = number of subjects, k = total number of observations. For a detailed list of foods within each category go to Supplementary Table S1. Asp, asparagine; EpiC-S, epicatechin sulfate; UroC-G, urolithin C-glucuronide; GCDCA-3S, glycochenodeoxycholic 3-sulfate; Sal-S, salsolinol sulfate; 4-Met-Cat-S, 4-methylcatechol sulfate; C18:2-Car, linoleoyl carnitine; 2-HBA, 2-hydroxybenzoic acid; Berg-G, bergaptol glucuronide; Met-Pyr-S, methylpyrogallol sulfate; 3-HPV-S, 5-(3′-hydroxyphenyl)-γ-valerolactone 3′-sulfate; 3,4-DHPHVA-S, 5-(4-hydroxy(3,4-dihydroxyphenyl)-valeric acid sulfate; 3,4-DHPA-3S, 3,4-dihydroxyphenylacetic acid sulfate; Et-G, ethyl glucuronide.

3.5. Associations between Selected ACN-Related Metabolome Biomarkers and Cardiometabolic Risk Factors

Metabolites associated with ACN intake in both of the previous analyses were: salsolinol sulfate, 4-methylcatechol sulfate, linoleoyl carnitine, 3,4-DHPHVA-3S, and 3,4-DHPA-S. Figure 4 shows the associations between these metabolites and cardiometabolic risk factors. Out of the metabolites associated with berries' ACNs, salsolinol sulfate and 4-methylcatechol sulfate were inversely associated with visceral adipose tissue volume. In addition, inverse associations were also found between salsolinol sulfate and LDL-C and diastolic blood pressure. Conversely, there was a direct association between salsolinol sulfate and triglyceride levels (Figure 4). Linoleoyl carnitine, 3,4-DHPHVA-3S, 3,4-DHPA-S did not show any statistically significant association with cardiometabolic risk factors.

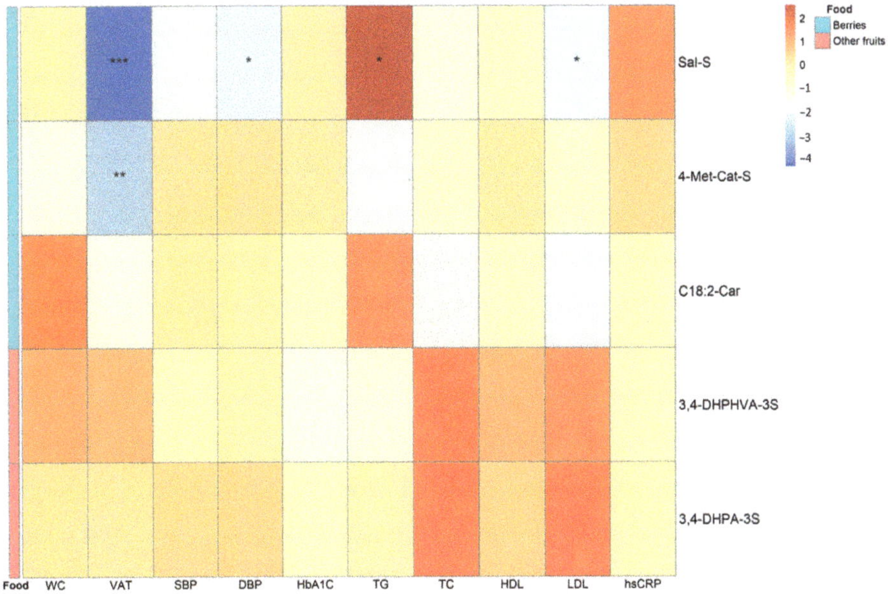

Figure 4. Association between ACN-related selected metabolites and cardiometabolic risk factors in the DCH-NG MAX study (n = 624, k = 1351). Standardized coefficients according to linear mixed models with random intercepts adjusting for age, sex, and BMI. Foods associated with the metabolites according to MGM analysis are displayed by colors in the food column. * $p < 0.05$, ** $p < 0.01$, *** $p < 0.001$. n = number of subjects, k = total number of observations. Sal-S, salsolinol sulfate; 4-Met-Cat-S, 4-methylcatechol sulfate; C18:2-Car, linoleoyl carnitine; 3,4-DHPHVA-3S, 5-(4-hydroxy(3,4-dihydroxyphenyl)-valeric acid sulfate; 3,4-DHPA-3S, 3,4-dihydroxyphenylacetic acid sulfate; TG, triglycerides; SBP, systolic blood pressure; DBP, diastolic blood pressure; WC, waist circumference; HbA1c, hemoglobin A1c; hsCRP, high-sensitivity C-reactive protein; VAT, visceral adipose tissue; TC, total cholesterol; HDL, high-density lipoproteins; TC, total cholesterol; LDL, low-density lipoproteins.

4. Discussion

The present study shows for the first time the specific associations between ACNs related to different dietary sources, and plasma metabolome biomarkers and their association with cardiometabolic risk factors in a free-living population. These results may take into account not only the quantitative and qualitative heterogeneity of ACNs presence in foods but also the internal dose of specific microbial metabolites generated from ACNs which could have been affected by the food matrix. Indeed, food matrices have been shown to influence the microbial metabolism of (poly)phenols [24]. Ultimately, we observed different associations between ACN-related metabolites and cardiometabolic risk factors in relationship with specific foods suggesting a stronger cardiometabolic benefit associated with the consumption of berries.

Up to 80% of the total intake of dietary ACNs came from the consumption of berries, wines, and non-alcoholic drinks in this observational study. Minor contributors were dairy foods, other fruits, and vegetables. While the MGM analysis revealed different metabolomic fingerprints associated with different dietary sources of ACNs, the resultant metabolites were not specific to ACNs. Therefore, we selected metabolites that were also significantly associated with the censored regression analysis. This was a strict criterion but in the context of such low levels of ACN intake in the overall population (median 1.6 mg/day), it is justified. After applying this selection criterion, only metabolites related to ACNs from berries and other fruits (according to MGM) were tested for their association with cardiometabolic risk factors. Metabolites specifically related to ACNs from other major food sources, such as wines, were excluded. Nonetheless, other studies showed that for example 4-methylcatechol sulfate was increased after a 15-day moderate red wine intervention trial [25]. Therefore, we cannot be fully certain that in our study the same metabolites could be related to other ACN dietary sources. Future randomized controlled trials using single foods are warranted to validate the present results.

Regarding the association between metabolome biomarkers and cardiometabolic risk factors, 4-methylcatechol sulfate showed an inverse association with visceral adipose tissue volume. According to our MGM analysis, 4-methylcatechol sulfate was associated with the intake of ACNs from berries. Similarly, another metabolite associated with ACNs from berries was salsolinol sulfate. Salsolinol sulfate is an alkaloid that has been suggested as a biomarker of banana intake [26]. However, salsolinol can be produced endogenously through dopamine oxidative metabolism [27,28] and may have a role in modulating dopamine neurons activity in the striatum region of the brain [21]. In fact, patients with obesity showed impaired dopamine brain activity, underscoring a potential role for low dopamine activity in obesity (lower reward associated with food intake) [29]. Hence, we speculate that the inverse association between salsolinol sulfate and visceral adipose tissue could be mediated by brain dopamine activity. An animal study showed that a blackberry extract intervention reversed the effects of a high-fat diet increasing dopamine turnover in the brain striatum region [30]. The role of berries on brain dopamine metabolism should be further studied. On the other hand, the other selected metabolites were not associated with cardiometabolic risk factors.

The median value of total ACN intake in the study was 1.6 mg/day, and such intake may not have been high enough to detect metabolome biomarkers found in randomized controlled trials (RCT) with ACN-rich foods [31–33]. Many short or long-term RCTs were conducted with capsulated ACNs or berries to discover biomarkers of ACNs intake. In these trials, daily intakes of ACNs typically varied from 100 to 300 mg as single dose intakes [33,34], or between 50–350 mg/day for four weeks [35–37]. In general, many parent ACNs and up to 70 phenolic compounds resultant of the gut microbial metabolism of ACNs have been identified [35,36]. Even though the majority of these metabolites were not identified in our study, 4-methylcatechol sulfate, 3,4-DHPHVA-3S, and 3,4-DHPA-3S had been previously associated with ACN intake. Maybe, longer half-lives of these metabolites vs. the others, or the competition of polyphenol substrates for bacteria able to metabolize them limited the production of ACNs metabolites under the low levels of ACN intake (exposure) in the study.

Among the strengths of this study are its observational nature and the fact that dietary data were assessed with 24-HDRs instead of food frequency questionnaires. This last characteristic allowed us to have exact intake data both in terms of amounts and specific food items compared to food frequency questionnaires. However, this also brings the limitation of measurement errors in estimating ACN intake and the short time period surveyed (one 24-HDR at each evaluation time). Another limitation was that the median consumption of dietary ACNs within the population of the DCH-NG MAX study was 1.6 mg/day, which was considerably lower than other studies in which the median intake varied between 9.3 to 52.6 mg/day [38–41]. This fact could have limited the number of plasma metabolites associated with dietary ACNs. Furthermore, the mean fasting time of the participants at the time the blood samples were drawn was 5 h and the impact of fasting on serum metabolome is uncertain. Nonetheless, this is the first study evaluating the impact of ACNs coming from different dietary sources on plasma metabolome and therefore our results cannot be contrasted with others. While berries contain other polyphenols in addition to ACNs, further research is needed to fully understand the individual and combined effects of different polyphenols of berries on health outcomes. Our approach to isolating the effects of ACNs from berries was conducted from a bioinformatic approach and a more precise study testing the effects of isolated ACNs from berries should corroborate our results. While berries contain other polyphenols in addition to ACNs, further research is needed to fully understand the individual and combined effects of different polyphenols of berries on health outcomes. Our approach to isolating the effects of ACNs from berries was conducted from a bioinformatic approach and a more precise study testing the effects of isolated ACNs from berries should corroborate our results. Last, it is not clear if the microbial metabolites were exclusively related to the ACNs from the dietary source pointed out in the MGM analysis or could have been also produced from ACNs coming from other foods, or even from food components other than ACNs (e.g., other polyphenols apart from ACNs). Although MGM models adjust every association for all the other variables included in the analysis, these sources of confounding cannot be ruled out.

In conclusion, this study shows that the metabolomic fingerprint of ACN consumption depended on its dietary sources. Metabolites associated with the consumption of berries' ACNs showed inverse associations with visceral adipose tissue. Future RCTs should validate the importance of these foods for cardiometabolic health and their potential mechanisms of action.

Supplementary Materials: The following supporting information can be downloaded at: https://www.mdpi.com/article/10.3390/nu15051208/s1. Table S1. ACN-containing food list in the DCH-NG MAX study, Table S2. Dietary characteristics of the MAX study population according to tertiles of dietary ACN intake.

Author Contributions: Conceptualization: H.M., T.M. and C.A.-L.; data curation: T.M., F.L., A.L.R.-H., J.H. and A.T.; methodology: T.M. and H.M.; resources: C.A.-L., R.L. and J.H.; formal analysis: T.M., H.M., A.M., A.S.-P. and M.C.-C.; software: T.M., A.M., A.S.-P. and M.C.-C.; writing—original draft preparation, H.M. and T.M.; writing—review and editing, N.E.-T., F.L., A.M., R.Z.-R., A.L.R.-H., J.H., A.T., R.L. and C.A.-L.; supervision, R.L., J.H. and C.A.-L. All authors have read and agreed to the published version of the manuscript.

Funding: This work was accomplished as part of the DiGuMet project "Diet × gut microbiome-based metabotypes to determine cardio-metabolic risk and tailor intervention strategies for improved health" supported within the European Joint Programming Initiative "A Healthy Diet for a Healthy Life" (http://www.healthydietforhealthylife.eu), granted by Ministerio de Economía, Industria y Competitividad (MINECO) (Spain, PCIN-2017-076). The study was supported by Grant PID2020-114921RB-C21 and PID2021-128542OA-I00 funded by MCIN/AEI/10.13039/501100011033. The Generalitat de Catalunya's Agency AGAUR of 2021SGR00687. Maria de Maeztu Unit of Excellence grant (CEX2021-001234-M) funded by (MICIN/AEI/FEDER, UE). The work was partially funded from CIBERFES, CB16/10/00269, funded by the Instituto de Salud Carlos III and co-funded by the European Regional Development Fund's "Away to make Europe". The DCH-NG cohort was supported by the Danish Cancer Society, Knæk Cancer 2012 and Den A.P. Møllerske Støttefond (grant nº 10619). The establishment of the DCH-NG MAX study was partly funded by FORMAS (DNR 2016-00314). HM would like to thank the scholarship ajuts de personal investigador predoctoral en formació (APIF) from the University of Barcelona. FL receives support from the Chilean government for doing his PhD through the National Agency for Research and Development (ANID)/Food and Nutrition Doctoral Program/DOCTORADO BECAS CHILE/2019 – 72200061. RZ-R was supported by the "Miguel Servet" program (CPII20/00009) from the Institute of Health Carlos III (co-funded by the European Social Fund (ESF) – ESF investing in your future). C.A-L. is grateful for the ICREA Academia Award 2018. The contributions by RL were supported by Swedish Research Council grant nr 2019-01264 (Discovery and validation of novel biomarkers of gut microbiota, diet and their interactions associated with type 2 diabetes risk).

Institutional Review Board Statement: The DCH-NG cohort study was approved by the Danish Data Protection Agency ((journal number 2013-41-2043/2014-231-0094) and by the Committee on Health Research Ethics for the Capital Region of Denmark (journal number H-15001257).

Informed Consent Statement: Informed consent was obtained from all subjects involved in the study.

Data Availability Statement: Data may be available upon request to the Danish Cancer Society (contact: dchdata@cancer.dk).

Acknowledgments: The authors would like to thank all the participants of the study and the staff involved in the DCH-NG MAX study. Myfood24 was developed through Medical Research Council Funding grant G110235. Myfood24 is now being supported by spinout company Dietary Assessment Ltd. Request to use myfood24 should be made to enquiries@myfood24.org. The software for the assessment of polyphenol intake from food surveys (UBTT0422) (http://diposit.ub.edu/dspace/handle/2445/187698) (7 February 2023) was developed through the institutional support of the University of Barcelona, the Bellvitge Biomedical Research Institute (IDIBELL) and the Centro de Investigación Biomédica en Red (CIBER—Instituto de Salud Carlos III). Requests to use this software should be made to candres@ub.edu or rzamora@idibell.cat.

Conflicts of Interest: The authors declare no conflict of interest.

References

1. Celli, G.B.; Tan, C.; Selig, M.J.; States, U. Anthocyanidins and Anthocyanins. *Encycl. Food Chem.* **2017**, 218–223. [CrossRef]
2. Fang, J. Bioavailability of Anthocyanins. *Drug Metab. Rev.* **2014**, *46*, 508–520. [CrossRef]
3. Vendrame, S.; Guglielmetti, S.; Riso, P.; Arioli, S.; Klimis-Zacas, D.; Porrini, M. Six-Week Consumption of a Wild Blueberry Powder Drink Increases Bifidobacteria in the Human Gut. *J. Agric. Food Chem.* **2011**, *59*, 12815–12820. [CrossRef]
4. Li, D.; Wang, P.; Luo, Y.; Zhao, M.; Chen, F.; Group, F. Health Benefits of Anthocyanins and Molecular Mechanisms: Update from Recent Decade. *Crit. Rev. Food Sci. Nutr.* **2017**, *57*, 1729–1741. [CrossRef]
5. Lee, S.; Keirsey, K.I.; Kirkland, R.; Grunewald, Z.I.; Fischer, J.G.; de La Serre, C.B. Blueberry Supplementation Influences the Gut Microbiota, Inflammation, and Insulin Resistance in High-Fat-Diet-Fed Rats. *J. Nutr.* **2018**, *148*, 209–219. [CrossRef]
6. Cassidy, A. Berry Anthocyanin Intake and Cardiovascular Health. *Mol. Aspects Med.* **2018**, *61*, 76–82. [CrossRef]
7. Stalmach, A.; Edwards, C.A.; Wightman, J.D.; Crozier, A. Gastrointestinal Stability and Bioavailability of (Poly) Phenolic Compounds Following Ingestion of Concord Grape Juice by Humans. *Mol. Nutr. Food Res.* **2012**, *56*, 497–509. [CrossRef]
8. Azzini, E.; Vitaglione, P.; Intorre, F.; Napolitano, A.; Durazzo, A.; Foddai, M.S.; Fumagalli, A.; Catasta, G.; Rossi, L.; Venneria, E.; et al. Bioavailability of Strawberry Antioxidants in Human Subjects. *Br. J. Nutr.* **2010**, *104*, 1165–1173. [CrossRef]
9. De Ferrars, R.M.; Czank, C.; Zhang, Q.; Botting, N.P.; Kroon, P.A. The Pharmacokinetics of Anthocyanins and Their Metabolites in Humans. *Br. J. Clin. Pharmacol.* **2014**, *171*, 3268–3282. [CrossRef]
10. Tjønneland, A.; Olsen, A.; Boll, K.; Stripp, C.; Christensen, J.; Engholm, G.; Overvad, K. Study Design, Exposure Variables, and Socioeconomic Determinants of Participation in Diet, Cancer and Health: A Population-Based Prospective Cohort Study of 57,053 Men and Women in Denmark. *Scand. J. Public Health* **2007**, *35*, 432–441. [CrossRef]
11. Petersen, K.E.N.; Halkjær, J.; Loft, S.; Tjønneland, A.; Olsen, A. Cohort Profile and Representativeness of Participants in the Diet, Cancer and Health-Next Generations Cohort Study. *Eur. J. Epidemiol.* **2022**, *37*, 117–127. [CrossRef] [PubMed]
12. Wark, P.A.; Hardie, L.J.; Frost, G.S.; Alwan, N.A.; Carter, M.; Elliott, P.; Ford, H.E.; Hancock, N.; Morris, M.A.; Mulla, U.Z.; et al. Validity of an Online 24-h Recall Tool (Myfood24) for Dietary Assessment in Population Studies: Comparison with Biomarkers and Standard Interviews. *BMC Med.* **2018**, *16*, 136. [CrossRef] [PubMed]
13. Finglas, P.M.; Roe, M.; Pinchen, H.M.; Berry, R.; Church, S.; Dodhia, S.K.; Farron-Wilson, M.; Swan, G. *McCance and Widdowson's The Composition of Foods, Seventh Summary Edition*; Royal Society of Chemistry: Cambridg, UK, 2015.
14. Tjønneland, A.; Overvad, K.; Haraldsdóttir, J.; Bang, S.; Ewertz, M.; Jensen, O.M. Validation of a Semiquantitative Food Frequency Questionnaire Developed in Denmark. *Int. J. Epidemiol.* **1991**, *20*, 906–912. [CrossRef] [PubMed]
15. Knaze, V.; Rothwell, J.A.; Zamora-Ros, R.; Moskal, A.; Kyrø, C.; Jakszyn, P.; Skeie, G.; Weiderpass, E.; De Magistris, M.S.; Agnoli, C.; et al. A New Food-Composition Database for 437 Polyphenols in 19,899 Raw and Prepared Foods Used to Estimate Polyphenol Intakes in Adults from 10 European Countries. *Am. J. Clin. Nutr.* **2018**, *108*, 517–524. [CrossRef] [PubMed]
16. Lanuza, F.; Zamora-Ros, R.; Rostgaard-Hansen, A.L.; Tjønneland, A.; Landberg, R.; Halkjær, J.; Andres-Lacueva, C. Descriptive Analysis of Dietary (Poly)Phenol Intake in the Subcohort MAX from DCH-NG: "Diet, Cancer and Health—Next Generations Cohort". *Eur. J. Nutr.* **2022**, *62*, 337–350. [CrossRef] [PubMed]
17. Guglielmetti, S.; Bernardi, S.; Del Bo', C.; Cherubini, A.; Porrini, M.; Gargari, G.; Hidalgo-Liberona, N.; Gonzalez-Dominguez, R.; Peron, G.; Zamora-Ros, R.; et al. Effect of a Polyphenol-Rich Dietary Pattern on Intestinal Permeability and Gut and Blood Microbiomics in Older Subjects: Study Protocol of the MaPLE Randomised Controlled Trial. *BMC Geriatr.* **2020**, *20*, 77. [CrossRef]
18. González-Domínguez, R.; Jáuregui, O.; Queipo-Ortuño, M.I.; Andrés-Lacueva, C. Characterization of the Human Exposome by a Comprehensive and Quantitative Large-Scale Multianalyte Metabolomics Platform. *Anal. Chem.* **2020**, *92*, 13767–13775. [CrossRef]
19. Andres-Lacueva, C.; Shukitt-Hale, B.; Galli, R.L.; Jauregui, O.; Lamuela-Raventos, R.M.; Joseph, J.A. Anthocyanins in Aged Blueberry-Fed Rats Are Found Centrally and May Enhance Memory. *Nutr. Neurosci.* **2005**, *8*, 111–120. [CrossRef]
20. Castellano-Escuder, P.; Gonzalez-Domnguez, R.; Carmona-Pontaque, F.; Andrés-Lacueva, C.; Sanchez-Pla, A. POMAShiny: A User-Friendly Web-Based Workflow for Metabolomics and Proteomics Data Analysis. *PLoS Comput. Biol.* **2021**, *17*, e1009148. [CrossRef]
21. Leek, J.T.; Johnson, W.E.; Parker, H.S.; Jaffe, A.E.; Storey, J.D. The sva package for removing batch effects and other unwanted variation in high-throughput experiments. *Bioinformatics* **2012**, *28*, 882–883. [CrossRef]

22. Croissant, Y.; Millo, G. Panel Data Econometrics in R: The Plm Package. *J. Stat. Softw.* **2008**, *27*, 1–43. [CrossRef]
23. Haslbeck, J.M.B.; Waldorp, L.J. MGM: Estimating Time-Varying Mixed Graphical Models in High-Dimensional Data. *J. Stat. Softw.* **2020**, *93*. [CrossRef]
24. Roura, E.; Andrés-Lacueva, C.; Estruch, R.; Bilbao, M.L.M.; Izquierdo-Pulido, M.; Lamuela-Raventós, R.M. The Effects of Milk as a Food Matrix for Polyphenols on the Excretion Profile of Cocoa (-)-Epicatechin Metabolites in Healthy Human Subjects. *Br. J. Nutr.* **2008**, *100*, 846–851. [CrossRef] [PubMed]
25. Muñoz-González, I.; Jiménez-Girón, A.; Martín-Álvarez, P.J.; Bartolomé, B.; Moreno-Arribas, M.V. Profiling of Microbial-Derived Phenolic Metabolites in Human Feces after Moderate Red Wine Intake. *J. Agric. Food Chem.* **2013**, *61*, 9470–9479. [CrossRef]
26. Ulaszewska, M.; Garcia-Aloy, M.; Vázquez-Manjarrez, N.; Soria-Florido, M.T.; Llorach, R.; Mattivi, F.; Manach, C. Food Intake Biomarkers for Berries and Grapes. *Genes Nutr.* **2020**, *15*, 17. [CrossRef]
27. Faraj, B.A.; Camp, V.M.; Davis, D.C.; Lenton, J.D.; Kutner, M. Elevation of Plasma Salsolinol Sulfate in Chronic Alcoholics as Compared to Nonalcoholics. *Alcohol. Clin. Exp. Res.* **1989**, *13*, 155–163. [CrossRef]
28. Rojkovicova, T.; Mechref, Y.; Starkey, J.A.; Wu, G.; Bell, R.L.; McBride, W.J.; Novotny, M.V. Quantitative Chiral Analysis of Salsolinol in Different Brain Regions of Rats Genetically Predisposed to Alcoholism. *J. Chromatogr. B Anal. Technol. Biomed. Life Sci.* **2008**, *863*, 206–214. [CrossRef]
29. Chen, P.S.; Yang, Y.K.; Yeh, T.L.; Lee, I.H.; Yao, W.J.; Chiu, N.T.; Lu, R.B. Correlation between Body Mass Index and Striatal Dopamine Transporter Availability in Healthy Volunteers-A SPECT Study. *Neuroimage* **2008**, *40*, 275–279. [CrossRef]
30. Meireles, M.; Rodríguez-Alcalá, L.M.; Marques, C.; Norberto, S.; Freitas, J.; Fernandes, I.; Mateus, N.; Gomes, A.; Faria, A.; Calhau, C. Effect of Chronic Consumption of Blackberry Extract on High-Fat Induced Obesity in Rats and Its Correlation with Metabolic and Brain Outcomes. *Food Funct.* **2016**, *7*, 127–139. [CrossRef]
31. Zhang, X.; Sandhu, A.; Edirisinghe, I.; Burton-Freeman, B.M. Plasma and Urinary (Poly)Phenolic Profiles after 4-Week Red Raspberry (Rubus Idaeus L.) Intake with or without Fructo-Oligosaccharide Supplementation. *Molecules* **2020**, *25*, 4777. [CrossRef]
32. Baron, G.; Altomare, A.; Regazzoni, L.; Fumagalli, L.; Artasensi, A.; Borghi, E.; Ottaviano, E.; Del Bo, C.; Riso, P.; Allegrini, P.; et al. Profiling Vaccinium Macrocarpon Components and Metabolites in Human Urine and the Urine Ex-Vivo Effect on Candida Albicans Adhesion and Biofilm-Formation. *Biochem. Pharmacol.* **2020**, *173*, 113726. [CrossRef] [PubMed]
33. Curtis, P.J.; Berends, L.; van der Velpen, V.; Jennings, A.; Haag, L.; Chandra, P.; Kay, C.D.; Rimm, E.B.; Cassidy, A. Blueberry Anthocyanin Intake Attenuates the Postprandial Cardiometabolic Effect of an Energy-Dense Food Challenge: Results from a Double Blind, Randomized Controlled Trial in Metabolic Syndrome Participants. *Clin. Nutr.* **2022**, *41*, 165–176. [CrossRef] [PubMed]
34. Straßmann, S.; Passon, M.; Schieber, A. Chemical Hemisynthesis of Sulfated Cyanidin-3-o-Glucoside and Cyanidin Metabolites. *Molecules* **2021**, *26*, 2146. [CrossRef] [PubMed]
35. Huang, L.; Xiao, D.; Zhang, X.; Sandhu, A.K.; Chandra, P.; Kay, C.; Edirisinghe, I.; Burton-Freeman, B. Strawberry Consumption, Cardiometabolic Risk Factors, and Vascular Function: A Randomized Controlled Trial in Adults with Moderate Hypercholesterolemia. *J. Nutr.* **2021**, *151*, 1517–1526. [CrossRef]
36. Heiss, C.; Istas, G.; Feliciano, R.P.; Weber, T.; Wang, B.; Favari, C.; Mena, P.; Del Rio, D.; Rodriguez-Mateos, A. Daily Consumption of Cranberry Improves Endothelial Function in Healthy Adults: A Double Blind Randomized Controlled Trial. *Food Funct.* **2022**, *13*, 3812–3824. [CrossRef]
37. Behrendt, I.; Röder, I.; Will, F.; Mostafa, H.; Gonzalez-Dominguez, R.; Meroño, T.; Andres-Lacueva, C.; Fasshauer, M.; Rudloff, S.; Kuntz, S. Influence of Plasma-Isolated Anthocyanins and Their Metabolites on Cancer Cell Migration (HT-29 and Caco-2) In Vitro: Results of the ATTACH Study. *Antioxidants* **2022**, *11*, 1341. [CrossRef]
38. Zamora-Ros, R.; Knaze, V.; Luján-Barroso, L.; Slimani, N.; Romieu, I.; Touillaud, M.; Kaaks, R.; Teucher, B.; Mattiello, A.; Grioni, S.; et al. Estimation of the Intake of Anthocyanidins and Their Food Sources in the European Prospective Investigation into Cancer and Nutrition (EPIC) Study. *Br. J. Nutr.* **2011**, *106*, 1090–1099. [CrossRef]
39. Lagiou, P.; Rossi, M.; Lagiou, A.; Tzonou, A.; La Vecchia, C.; Trichopoulos, D. Flavonoid Intake and Liver Cancer: A Case-Control Study in Greece. *Cancer Causes Control* **2008**, *19*, 813–818. [CrossRef]
40. Lagiou, P.; Samoli, E.; Lagiou, A.; Tzonou, A.; Kalandidi, A.; Peterson, J.; Dwyer, J.; Trichpoulos, D. Intake of Specific Flavonoid Classes and Coronary Heart Disease—A Case-Control Study in Greece. *Eur. J. Clin. Nutr.* **2004**, *58*, 1643–1648. [CrossRef]
41. Rossi, M.; Garavello, W.; Talamini, R.; Negri, E.; Bosetti, C.; Dal Maso, L.; Lagiou, P.; Tavani, A.; Polesel, J.; Barzan, L.; et al. Flavonoids and the Risk of Oral and Pharyngeal Cancer: A Case-Control Study from Italy. *Cancer Epidemiol. Biomark. Prev.* **2007**, *16*, 1621–1625. [CrossRef]

nutrients

MDPI

Article

Grape-Seed Proanthocyanidins Modulate Adipose Tissue Adaptations to Obesity in a Photoperiod-Dependent Manner in Fischer 344 Rats

Èlia Navarro-Masip, Marina Colom-Pellicer, Francesca Manocchio, Anna Arola-Arnal, Francisca Isabel Bravo, Begoña Muguerza and Gerard Aragonès *

Nutrigenomics Research Group, Department of Biochemistry and Biotechnology, Universitat Rovira i Virgili, Marcel·lí Domingo 1, 43007 Tarragona, Spain
* Correspondence: gerard.aragones@urv.cat; Tel.: +34-977-558-188

Abstract: Seasonal rhythms drive metabolic adaptations that influence body weight and adiposity. Adipose tissue is a key regulator of energy homeostasis in the organism, and its healthiness is needed to prevent the major consequences of overweight and obesity. In this context, supplementation with proanthocyanidins has been postulated as a potential strategy to prevent the alterations caused by obesity. Moreover, the effects of these (poly)phenols on metabolism are photoperiod dependent. In order to describe the impact of grape-seed proanthocyanidins extract (GSPE) on important markers of adipose tissue functionality under an obesogenic environment, we exposed Fischer 344 rats to three different photoperiods and fed them a cafeteria diet for five weeks. Afterwards, we supplemented them with 25 mg GSPE/kg/day for four weeks. Our results revealed that GSPE supplementation prevented excessive body weight gain under a long photoperiod, which could be explained by increased lipolysis in the adipose tissue. Moreover, cholesterol and non-esterified fatty acids (NEFAs) serum concentrations were restored by GSPE under standard photoperiod. GSPE consumption slightly helped combat the obesity-induced hypertrophy in adipocytes, and adiponectin mRNA levels were upregulated under all photoperiods. Overall, the administration of GSPE helped reduce the impact of obesity in the adipose tissue, depending on the photoperiod at which GSPE was consumed and on the type of adipose depots.

Keywords: adiponectin; BAT; flavonoids; GSPE; seasonal rhythms

Citation: Navarro-Masip, È.; Colom-Pellicer, M.; Manocchio, F.; Arola-Arnal, A.; Bravo, F.I.; Muguerza, B.; Aragonès, G. Grape-Seed Proanthocyanidins Modulate Adipose Tissue Adaptations to Obesity in a Photoperiod-Dependent Manner in Fischer 344 Rats. *Nutrients* **2023**, *15*, 1037. https://doi.org/10.3390/nu15041037

Academic Editors: Luis Goya and Sonia de Pascual-Teresa

Received: 7 February 2023
Revised: 16 February 2023
Accepted: 17 February 2023
Published: 19 February 2023

1. Introduction

Obesity is a public health issue worldwide that has nearly tripled since 1975, and in 2016, 39% of adults were overweight and 13% were obese [1]. The principal cause of obesity is an energy imbalance between food intake and energy expenditure [2–4]. The World Health Organization (WHO) and other entities report that obesity is preventable by consuming healthier foods, such as vegetables, legumes, whole grains, and nuts, and increasing physical activity by up to 150 min per week for adults. Moreover, seasonal changes affect the metabolic features that control energy balance, such as food intake, adiposity, or energy metabolism [5,6]. In this sense, humans have a higher biological susceptibility to gaining weight in the summer than in the winter, meaning that adopting unhealthy habits in the summer can contribute to weight gain and obesity. Indeed, children gained more weight in the summer holidays rather than in the Christmas holidays [7,8]. Despite the recommendations, the incidence of obesity is still increasing every year; therefore, the need for the development of new strategies to ameliorate obesity and obesity-related disorders becomes greater.

A healthy functionality of the white adipose tissue (WAT) is crucial to prevent metabolic disorders since alterations in WAT metabolism can lead to an inflammatory response and systemic complications [9]. In an obesogenic environment, the healthiness of WAT is seriously threatened, and the general consequences that can arise from this situation depend on WAT location. In fact, the WAT is largely distributed throughout the organism, being classified into subcutaneous WAT and visceral WAT [10]. The subcutaneous WAT (sWAT) has a better adaptation to an obesogenic environment, yet it is capable of better modulating the inflammatory response and preventing lipotoxicity. However, the visceral WAT (vWAT) is less tolerant to a calory-excess situation: its development under obesity has a stronger impact on systemic inflammation, as it produces a considerable amount of pro-inflammatory cytokines and promotes an exacerbated immune response, followed by lipotoxicity [11].

The brown adipose tissue (BAT) appears as an important thermogenic tissue in mammals that strongly contributes to energy metabolism. Its role and characteristics are different from WAT: briefly, brown adipocytes contain high amounts of mitochondria and low amounts of small lipid droplets [12]. Moreover, they count on the presence of Uncoupling Protein 1 (UCP1) in the membrane of their mitochondria, which is a protein capable of dissipating the electron respiratory chain and using the energy to produce heat [12]. Consequently, its adaptation to distinct environments such as obesity or seasonality is different from the WAT and can help the organism face metabolic complications [6].

The consumption of specific bioactive compounds, such as (poly)phenols, is highly associated with health benefits [13]. These molecules have been related to a reduced incidence of metabolic disorders such as obesity, glucose intolerance, or cardiovascular disease [14,15]. In addition, phenolic compounds have shown synchronizing characteristics that can ease the metabolic adaptations of the organism to its environment [16,17]. In fact, an influence of photoperiod on the metabolic effects of grape-seed proanthocyanidin extract (GSPE) was recently demonstrated in healthy and obese animals. More specifically, GSPE modulated hepatic glucose and lipid metabolism in a photoperiod-dependent manner [18]. Moreover, GSPE effects on microbiota were dependent on photoperiod in obese animals [19]. GSPE contributed to the adaptation to new photoperiods by regulating body weight and energy expenditure [20]. However, the seasonal effects of GSPE consumption on adipose tissue adaptations to obesity have not been reported to date. Therefore, the aim of this study was to explore the effects of GSPE seasonal consumption on the functionality of different adipose tissue depots (BAT, visceral, and subcutaneous WAT) in animals fed a cafeteria (CAF) diet.

2. Materials and Methods

2.1. Grape-Seed Proanthocyanidin Extract

GSPE was kindly provided by Les Dérivés Résiniques et Terpéniques (Dax, France). According to the manufacturer, the GSPE composition used in this study contained monomers (21.3%), dimers (17.4%), trimers (16.3%), tetramers (13.3%), and oligomers (5–13 units; 31.7%) of proanthocyanidins. The exact phenolic composition of GSPE was determined by HPLC-MS/MS (TOF 6210, Agilent) [21], according to what was described by Quiñones et al. [22], and can be found in Supplementary Table S1.

2.2. Animal Experimental Procedure

Eight-week-old male Fischer 344 rats (n = 72) were purchased from Charles River Laboratories (Barcelona, Spain) and pair-housed in animal quarters at 22 °C with a light/dark period of 12 h. Animals had an adaptation period of 1 week fed with standard diets (STD) (Panlab, Barcelona, Spain) and tap water ad libitum. After the adaptation period, animals were submitted to three different light schedules for 9 weeks to mimic seasonal day lengths: L12 photoperiod (12 h light–12 h darkness); L18 photoperiod (18 h light–6 h darkness); and L6 photoperiod (6 h light–18 h darkness). Animals in each photoperiod group were further divided into three more groups (n = 8); one of them was fed an STD diet *ad libitum*, and the other two were fed cafeteria (CAF) diet *ad libitum*. The CAF diet consisted of biscuits with pâté, biscuits with cheese, *ensaïmada* (sweetened pastry), bacon, carrots, and sweetened milk (20% sucrose *w/v*) in addition to the standard diet (STD Panlab A04, Panlab, Barcelona,

Spain). The CAF diet is a highly palatable diet that is able to induce voluntary hyperphagia. Its composition was 10% protein, 31.9% fat, and 58.1% carbohydrates. Each component of the CAF diet was freshly provided to the animals daily, and they could choose and eat ad libitum. At week 6, the 4-week treatment period started while the animals kept eating the same type of diet. The three STD diet groups received the vehicle (VH) treatment, consisting of an oral dose of water and sweetened milk. Three of the CAF groups (one from each photoperiod) received VH, and the other three received 25 mg/kg/day of GSPE diluted in water and sweetened milk. At the end of the experiment, animals were fasted for 3 h and then sacrificed by live decapitation. Total blood was collected from the neck and then centrifuged (1500× *g*, 15 min, 4 °C) to obtain serum. All adipose tissue depots were excised, weighted and immediately frozen into liquid nitrogen, one piece of epididymal white adipose tissue (eWAT) and inguinal white adipose tissue (iWAT) were also kept in formol for histological analysis. Both serum and tissues were stored at −80 °C until further use. A schematic representation of the experimental design can be found in Supplementary Figure S1. The Ethics Review Committee for Animal Experimentation of the University Rovira i Virgili (Tarragona, Spain) and the Generalitat de Catalunya approved all the procedures of the investigation (reference number 9495), which was carried out in accordance with the ethical standards and the Declaration of Helsinki.

2.3. Serum Analysis

Serum glucose, total cholesterol, and triglycerides (TG) were measured with enzymatic colorimetric kits (QCA, Barcelona, Spain). Serum non-esterified fatty acids (NEFAs) were analyzed with the enzymatic colorimetric HR NEFA series kit (Wako, CA, USA).

2.4. Histology of Adipose Tissues

Frozen iWAT and eWAT samples were thawed and fixed in 4% formaldehyde. Tissues underwent successive dehydration and paraffin infiltration immersion (Citadel 2000, HistoStar, Thermo Scientific, Madrid, Spain) and the paraffin blocks were cut into 2-µm-thick sections using a microtome (Microm HM 355S, ThermoScientific). The sections were subjected to automated hematoxylin–eosin staining (Varistain Gemini, Shandom, Thermo Scientific). Sections were observed and acquired at ×10 magnification using AxioVision ZeissImaging software (Carl Zeiss Iberia, S.L., Madrid, Spain). The area of adipocytes was measured using the Adiposoft open-source software (CIMA, University of Navarra, Pamplona, Spain). Four fields per sample were measured, and six samples per group were analyzed. The adipocyte area was calculated from the average value of the area of cells in all measured fields for each sample. The total adipocyte volume was calculated using the formula $\left[\frac{\pi}{6}\right] \times \left[3\sigma^2 \times \overline{d} + \overline{d}^3\right]$, where $\overline{d}$ is the mean diameter and σ is the standard deviation of the diameter. Afterwards, this value was converted to the adipocyte weight using the adipocyte density (0.92 g/mL). In order to obtain the total adipocyte number in each depot, the weights of iWAT and eWAT depots were divided by the adipocyte weight. The frequencies of adipocytes were calculated by dividing all counted cells per sample into two groups according to their area, <3000 µm^2 or >3000 µm^2; then, the total number of counted adipocytes to calculate the percentage of adipocytes in both categories.

2.5. Gene Expression Analysis

Total RNA from iWAT, eWAT and BAT was extracted using TRIzol reagent (Thermo Fisher Scientific, Barcelona, Spain) following the manufacturer's protocol. RNA was quantified in NanoDrop ND-1000 spectrophotometer (Thermo Scientific, Wilmington, DE, USA). The integrity of the RNA was evaluated by the RNA integrity number (RIN) trough 2100 Bioanalyzer Instrument (Agilent Technologies). A RIN higher than six was accepted for total RNA samples. cDNA was synthesized using the High-Capacity cDNA Reverse Transcription Kit (Applied Biosystems, Barcelona, Spain) in a Multigene ThermalCycler (Labnet, Madrid, Spain). The cDNA was subjected to a quantitative reverse transcriptase polymerase chain reaction amplification using iTaq™Universal SYBR Green Supermix

(Bio-Rad, Madrid, Spain) in the 7900HT Fast Real-Time PCR System (Applied Biosystems). The primers used for the different genes are described in Supplementary Table S2 and were obtained from Biomers.net (Ulm, Germany). The relative expression of each gene was calculated according to *cyclophilin peptidylprolyl isomerase A* (*Ppia*) mRNA levels and normalized to the levels measured in the corresponding control group. The ΔΔCt method was used and corrected for primer efficiency [23].

2.6. Statistical Analysis

The effects of both CAF diet administration and seasonal GSPE supplementation on biometric, biochemical, and serum variables as well as gene expression were evaluated by a two-way ANOVA with Tukey's *post hoc* test for multiple comparisons. In addition, for biometric, biochemical, and serum variables, a one-way ANOVA followed by Tukey's *post hoc* test was conducted to detect significant differences between groups in the same photoperiod. GraphPad Prism 9 (GraphPad Software, La Jolla, CA, USA) was used for all statistical analysis. The values are expressed as means ± SEM. $p< 0.05$ was considered significant.

3. Results

3.1. GSPE Consumption Restored Cholesterol and NEFAs Serum Concentrations in a Photoperiod-Dependent Manner in Obese Animals

After nine weeks of study, animals fed the CAF diet significantly increased their body weight, body weight gain, and food intake (Table 1). However, the four-week GSPE supplementation was able to significantly reduce body weight gain and food intake in all groups. When we conducted one-way ANOVA tests to assess the individual effects of GSPE on each photoperiod, we detected that the greater effects on reducing body weight gain were in animals exposed to L18. Furthermore, the CAF diet significantly increased total fat mass and the mass of individual adipose tissue depots (iWAT, eWAT, and BAT) in all animals. In this case, GSPE supplementation did not significantly affect these values.

A diet effect was also observed in all circulating metabolic parameters, where CAF consumption significantly increased serum concentrations of glucose, cholesterol, TG, and NEFAs. However, GSPE supplementation was able to restore cholesterol and NEFAs levels in a photoperiod-dependent manner.

3.2. GSPE Consumption Restored Adipocyte Morphology in iWAT in a Photoperiod-Independent Manner

The histological analysis of iWAT showed a significant effect of the CAF diet on both area and volume of adipocytes in iWAT, which were increased in obese animals (Figure 1A,B, Supplementary Figure S2). Consequently, the adipocyte area distribution also changed, and the number of larger adipocytes increased with respect to the smaller ones (Figure 1D). However, no differences were detected in adipocyte number (Figure 1C), and, when we applied one-way ANOVA analyses to assess the significant effects within each photoperiod, we only observed statistical significance on L12 animals. Noteworthy, as indicated by two-way ANOVA analyses, GSPE consumption reduced the CAF diet impact because the cellular profile of CAF-fed animals supplemented with GSPE was more similar to the STD diet fed groups, showing reduced adipocyte area and volume. No interaction between GSPE and photoperiod was detected, but the GSPE effect only reached significance in L12 groups after applying one-way ANOVA tests.

3.3. Gene Expression of iWAT Was Affected by GSPE in a Photoperiod-Dependent Manner

However, after exploring the expression of the key metabolic genes in iWAT, we detected an interaction effect between GSPE and photoperiod in the expression levels of the adipogenic genes *Cebpα* and *Pparγ* (Figure 2A). The effect was only observed in animals exposed to the L6 photoperiod, in which the expression of these genes was significantly restored in response to GSPE consumption. A similar effect was detected in the lipid transport-related genes *Cd36* and *Fabp4* (Figure 2B), the lipolysis-related genes *Hsl* and *Atgl* (Figure 2C), and the adipokine genes *Lep* and *Adipoq* (Figure 2D). Differently, no significant effects in response to GSPE consumption were detected in the gene expression of the lipogenic genes *Acaca* and *Fasn* (Supplementary Figure S3).

3.4. GSPE Consumption Restored Adipocyte Morphology in eWAT in a Photoperiod-Independent Manner

Similar to iWAT, a significant effect of both the CAF diet and GSPE consumption was detected in the histological analysis of eWAT. Again, the CAF diet increased adipocyte area and volume, whereas GSPE consumption attenuated this significant increase regardless of the photoperiod (Figure 3A,B, Supplementary Figure S4). However, these changes in response to proanthocyanidins were not fully reflected by adipocyte number or adipocyte area frequency (Figure 3C,D), and one-way ANOVA tests did not show significancy between CAF-VH and CAF-GSPE groups on area and volume values under any photoperiod.

Table 1. Biometric and circulating parameters.

	P	STD-VH	CAF-VH	CAF-GSPE	*p* Diet [1]	*p* Diet × P [1]	*p* GSPE [1]	*p* GSPE × P [1]
Body weight (g)	L12	479.8 ± 12.1 ^{a}	584.5 ± 11.9 ^{b}	555.5 ± 14.2 ^{b}	<0.0001	*ns*	*ns*	*ns*
	L18	491.9 ± 12.9 ^{a}	598.3 ± 11.4 ^{b}	559.1 ± 15.9 ^{b}				
	L6	491.4 ± 12.0 ^{a}	554.9 ± 19.9 ^{b}	554.8 ± 18.5 ^{b}				
Body weight gain (g)	L12	96.5 ± 8.9 ^{a}	180. ± 10.5 ^{b}	156.5 ± 9.7 ^{b}	<0.0001	*ns*	0.0403	*ns*
	L18	104.8 ± 9.1 ^{a}	197.3 ± 8.2 ^{c}	159.4 ± 12.8 ^{b}				
	L6	106.1 ± 8.1 ^{a}	158.3 ± 14.9 ^{b}	155.8 ± 16.1 ^{b}				
Food intake (g)	L12	201.3 ± 3.4 ^{a}	308.9 ± 14.2 ^{c}	256.8 ± 16.4 ^{b}	<0.0001	*ns*	0.0015	*ns*
	L18	212.1 ± 6.9 ^{a}	301.8 ± 13.4 ^{b}	292.1 ± 12.8 ^{b}				
	L6	219.3 ± 2.2 ^{a}	311.9 ± 7.8 ^{c}	264.3 ± 12.7 ^{b}				
Fat mass (g)	L12	35.3 ± 3.4 ^{a}	83.5 ± 5.3 ^{b}	82.2 ± 4.8 ^{b}	<0.0001	*ns*	*ns*	*ns*
	L18	46.5 ± 3.5 ^{a}	95.1 ± 5.4 ^{b}	82.5 ± 6.8 ^{b}				
	L6	40.1 ± 3.7 ^{a}	74.3 ± 9.4 ^{b}	76.1 ± 5.9 ^{b}				
iWAT mass (g)	L12	4.1 ± 0.3 ^{a}	8.6 ± 0.7 ^{b}	9.4 ± 1.1 ^{b}	<0.0001	*ns*	*ns*	*ns*
	L18	4.8 ± 0.1 ^{a}	9.2 ± 1.3 ^{b}	7.1 ± 1.1 ^{ab}				
	L6	4.2 ± 0.6 ^{a}	8.2 ± 1.5 ^{b}	7.2 ± 0.8 ^{ab}				
eWAT mass (g)	L12	7.4 ± 0.8 ^{a}	18.9 ± 1.1 ^{b}	18.1 ± 1.1 ^{b}	<0.0001	*ns*	*ns*	*ns*
	L18	10.9 ± 0.5 ^{a}	19.6 ± 1.5 ^{b}	16.4 ± 1.7 ^{b}				
	L6	9.3 ± 0.9 ^{a}	16.0 ± 1.7 ^{b}	18.6 ± 1.1 ^{b}				
BAT mass (g)	L12	0.54 ± 0.05 ^{a}	0.96 ± 0.07 ^{b}	0.94 ± 0.09 ^{b}	<0.0001	*ns*	*ns*	*ns*
	L18	0.49 ± 0.05 ^{a}	0.92 ± 0.16 ^{ab}	0.99 ± 0.14 ^{b}				
	L6	0.64 ± 0.06 ^{a}	0.93 ± 0.12 ^{ab}	1.01 ± 0.10 ^{b}				
Glucose (mmol/L)	L12	5.7 ± 0.3 ^{a}	8.1 ± 0.3 ^{b}	7.1 ± 0.4 ^{ab}	<0.0001	*ns*	*ns*	*ns*
	L18	5.7 ± 0.2 ^{a}	7.1 ± 0.2 ^{b}	7.6 ± 0.1 ^{b}				
	L6	6.3 ± 0.1	7.4 ± 0.4	6.9 ± 0.4				
Cholesterol (mmol/L)	L12	1.23 ± 0.07 ^{a}	1.95 ± 0.28 ^{b}	1.23 ± 0.09 ^{a}	*ns*	*ns*	*ns*	0.0056
	L18	1.28 ± 0.11 ^{ab}	1.18 ± 0.10 ^{a}	1.57 ± 0.08 ^{b}				
	L6	1.20 ± 0.06	1.36 ± 0.19	1.15 ± 0.14				
NEFAs (mmol/L)	L12	0.57 ± 0.06 ^{a}	1.17 ± 0.12 ^{b}	0.82 ± 0.01 ^{a}	0.0003	0.0024	*ns*	0.0005
	L18	0.69 ± 0.07 ^{a}	0.64 ± 0.09 ^{a}	1.03 ± 0.10 ^{b}				
	L6	0.60 ± 0.06 ^{a}	0.93 ± 0.10 ^{b}	0.68 ± 0.05 ^{ab}				
TG (mmol/L)	L12	0.79 ± 0.04 ^{a}	2.79 ± 0.36 ^{c}	1.82 ± 0.16 ^{b}	<0.0001	*ns*	*ns*	*ns*
	L18	0.97 ± 0.13 ^{a}	1.88 ± 0.25 ^{b}	2.18 ± 0.23 ^{b}				
	L6	0.88 ± 0.11 ^{a}	2.12 ± 0.39 ^{b}	1.49 ± 0.20 ^{ab}				

Values are presented as the mean ± SEM of eight animals per group. [1] Denotes two-way ANOVA analysis: Diet: diet increasing effect within VH groups; Diet × P: interaction increasing effect between diet and photoperiod within VH groups; GSPE: GSPE consumption decreasing effect within CAF groups; GSPE × P: interaction decreasing effect between GSPE consumption and photoperiod within CAF groups; *ns:* no significant effects. Different letters denote significant differences within each photoperiod group (assessed with one-way ANOVA followed by Tukey's *post hoc* test, $p < 0.05$). BAT: brown adipose tissue; CAF: cafeteria diet; eWAT: epididymal white adipose tissue; GSPE: grape-seed proanthocyanidins extract; iWAT: inguinal white adipose tissue; NEFAs: non-esterified fatty acids; P: photoperiod; STD: standard diet; TG: triglycerides; VH: vehicle

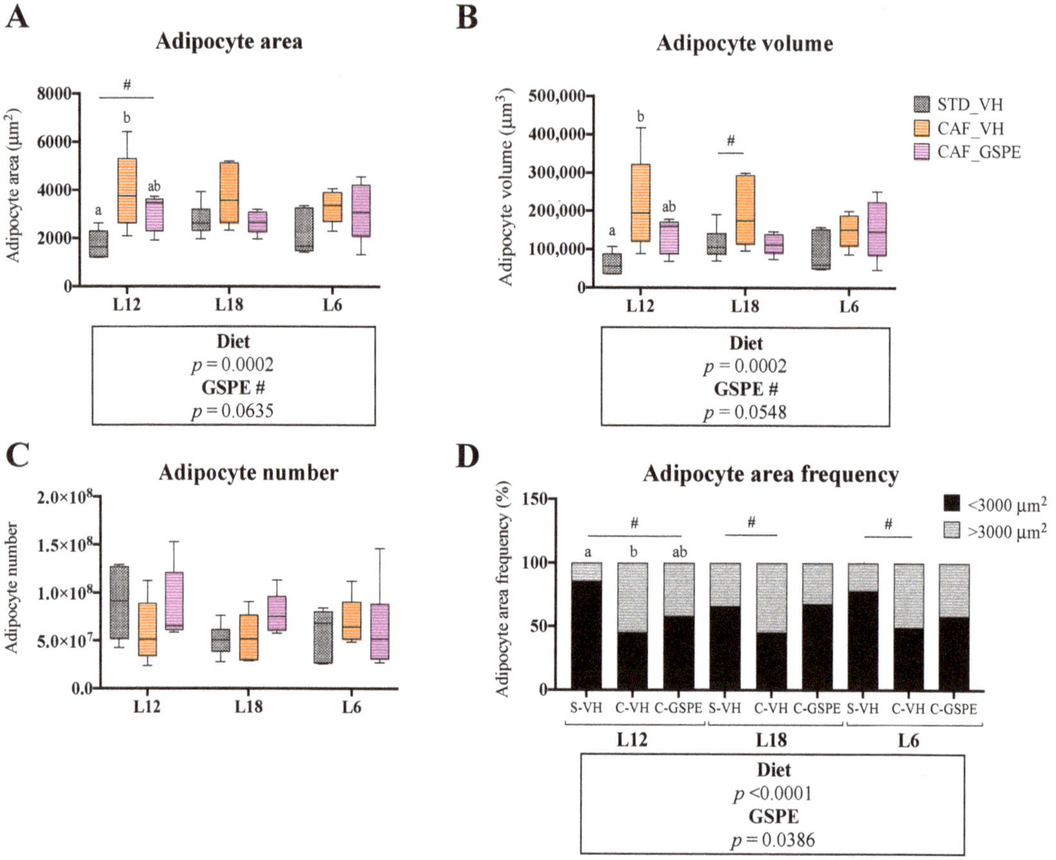

Figure 1. Adipocyte area (**A**), adipocyte volume (**B**), adipocyte number (**C**) and adipocyte area frequencies (**D**) of inguinal white adipose tissue (iWAT) of rats treated with grape seed proanthocyanidin extract (GSPE) at different photoperiods, 12 h of light:12 h of darkness (L12); 18 h of light:6 h of darkness (L18); 6 h of light:18 h of darkness (L6). Histological study of iWAT morphology of Fisher 344 rats fed with standard diet (STD) or cafeteria diet (CAF) for 9 weeks and treated with GSPE (25 mg/kg body weight) or vehicle (VH) the last 4 weeks of the study; STD-VH group; CAF-VH group; CAF-GSPE group, under different photoperiods; L12, L18 or L6. For adipocyte area frequency, adipocytes were distributed in two groups according to their areas (<3000 or >3000 μm^2). The results are presented as the mean ± SEM ($n = 6$). Significant differences were assessed through two-way ANOVA analysis ($p < 0.05$). Diet: diet effect within VH groups; GSPE: GSPE consumption effect within CAF groups; #: tendency ($0.05 < p < 1$). Different letters denote significant differences within each photoperiod group (assessed with one-way ANOVA followed by Tukey's *post hoc* test, $p < 0.05$; #: tendency: $0.05 < p < 1$).

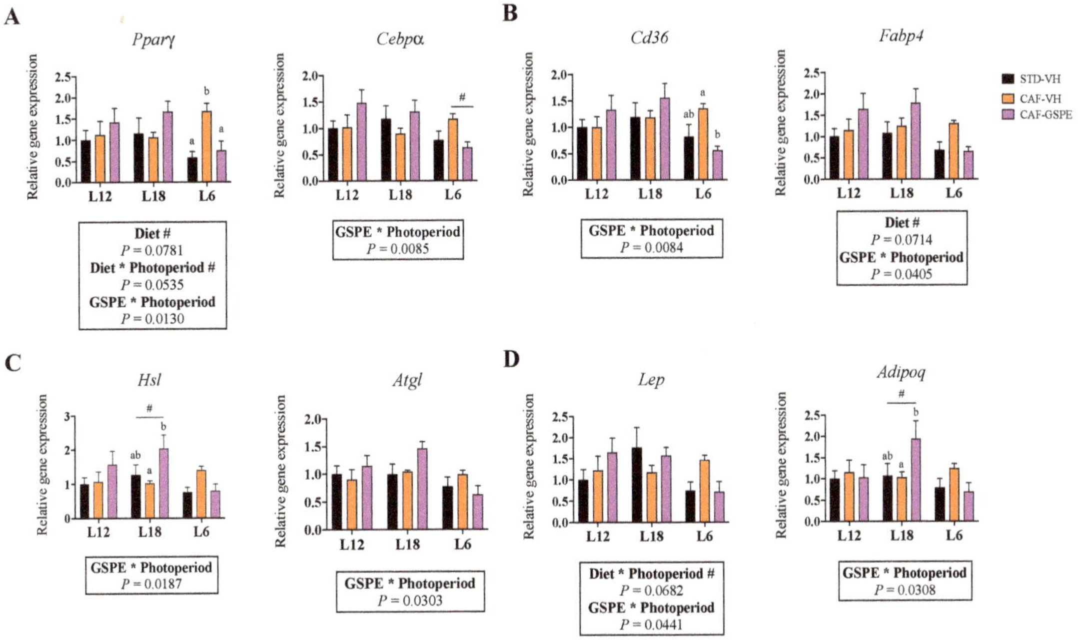

Figure 2. Effect of grape seed proanthocyanidin extract (GSPE) consumption at different photoperiods, 12 h of light:12 h of darkness (L12); 18 h of light:6 h of darkness (L18); 6 h of light:18 h of darkness (L6) on the expression of gens related to adipogenesis, lipogenesis, lipolysis and adipokines in inguinal white adipose tissue (iWAT) of diet-induced obese rats. Fisher 344 rats were fed standard diet (STD) or cafeteria diet (CAF) for 9 weeks and treated with GSPE (25 mg/kg body weight) or vehicle (VH) the last 4 weeks of the study; STD-VH group; CAF-VH group; CAF-GSPE group, under different photoperiods; L12, L18 or L6. The expression of genes involved in adipogenesis (**A**), lipogenesis (**B**), lipolysis (**C**) and adipokines (**D**) was measured by qPCR and normalized by *Ppia*. The relative expression (presented as fold-change) of CAF-VH and CAF-GSPE was normalized to the corresponding STD-VH control group (L12 photoperiod). Significant differences were assessed through two-way ANOVA analysis ($p < 0.05$). Diet: diet effect within VH groups; Diet * Photoperiod: interaction effect between diet and photoperiod within VH groups; GSPE * Photoperiod: interaction effect between GSPE consumption and photoperiod within CAF groups. Different letters denote significant differences within each photoperiod group (assessed with two-way ANOVA followed by Tukey's *post hoc* test, $p < 0.05$; #: tendency: $0.05 < p < 1$).

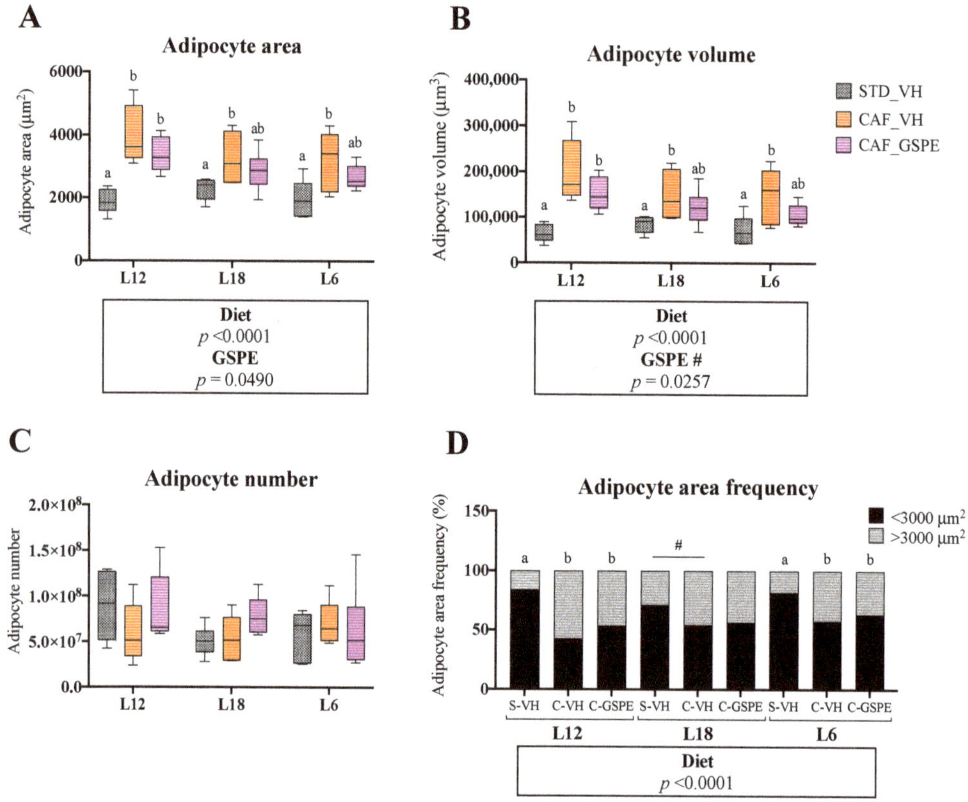

Figure 3. Adipocyte area (**A**), adipocyte volume (**B**), adipocyte number (**C**) and adipocyte area frequencies (**D**) of epididymal white adipose tissue (eWAT) of rats treated with grape seed proanthocyanidin extract (GSPE) at different photoperiods, 12 h of light:12 h of darkness (L12); 18 h of light:6 h of darkness (L18); 6 h of light:18 h of darkness (L6). Histological study of eWAT morphology of Fisher 344 rats fed with standard diet (STD) or cafeteria diet (CAF) for 9 weeks and treated with GSPE (25 mg/kg body weight) or vehicle (VH) the last 4 weeks of the study; STD-VH group; CAF-VH group; CAF-GSPE group, under different photoperiods; L12, L18 or L6. For adipocyte area frequency, adipocytes were distributed in two groups according to their areas (<3000 or >3000 µm^2). The results are presented as the mean ± SEM (n = 6). Significant differences were assessed through two-way ANOVA analysis ($p < 0.05$). Diet: diet effect within VH groups; GSPE: GSPE consumption effect within CAF groups; #: tendency ($0.05 < p < 1$). Different letters denote significant differences within each photoperiod group (assessed with one-way ANOVA followed by Tukey's *post hoc* test, $p < 0.05$; #: tendency: $0.05 < p < 1$).

3.5. GSPE Consumption Influenced the Expression of Adipogenic Genes in eWAT in a Photoperiod-Dependent Manner

As observed in iWAT, when we explored the gene expression profile of metabolic genes in eWAT, an interaction between GSPE and photoperiod was significantly detected in adipogenic genes (Figure 4A). Moreover, GSPE consumption significantly upregulated the gene expression of *Fabp4* (in L18 and L6 animals) and of *Cd36* (Figure 4B). Similarly, an interaction effect between GSPE consumption and photoperiod was observed in the lipolytic gene *Hsl* (Figure 4C). As expected, the CAF diet significantly downregulated the expression of the lipogenic genes *Acacα* and *Fasn* in all animal groups (Supplementary Figure S5).

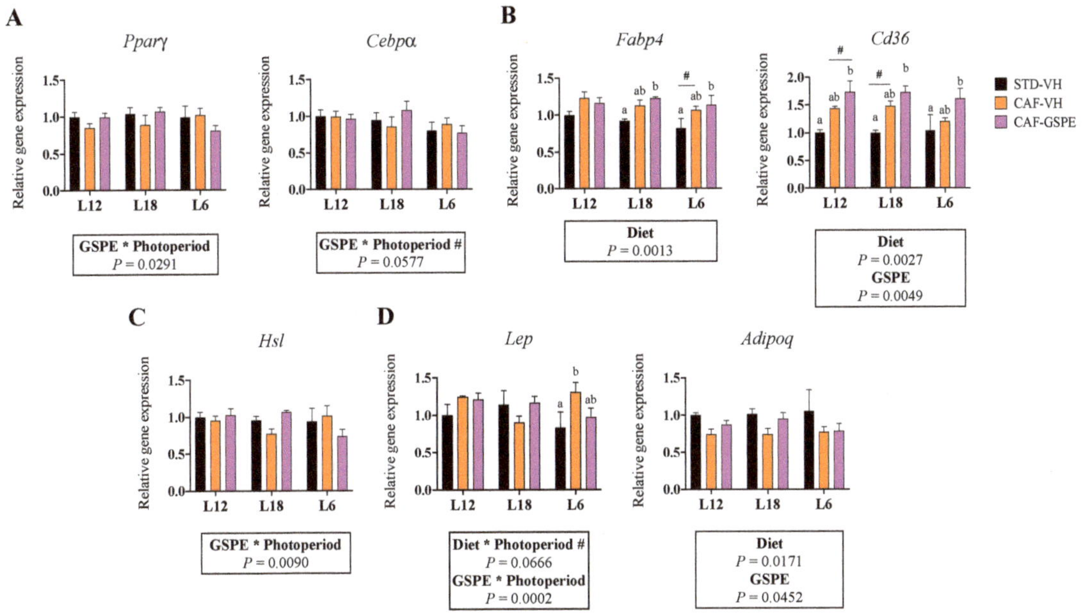

Figure 4. Effect of grape seed proanthocyanidin extract (GSPE) consumption at different photoperiods, 12 h of light:12 h of darkness (L12); 18 h of light:6 h of darkness (L18); 6 h of light:18 h of darkness (L6) on the expression of gens related to adipogenesis, lipid transport, lipolysis and adipokines in epididymal white adipose tissue (eWAT) of diet-induced obese rats. Fisher 344 rats were fed standard diet (STD) or cafeteria diet (CAF) for 9 weeks and treated with GSPE (25 mg/kg body weight) or vehicle (VH) the last 4 weeks of the study; STD-VH group; CAF-VH group; CAF-GSPE group, under different photoperiods; L12, L18 or L6. The expression of genes involved in adipogenesis (**A**), lipid transport (**B**), lipolysis (**C**) and adipokines (**D**) was measured by qPCR and normalized by *Ppia*. The relative expression (presented as fold-change) of CAF-VH and CAF-GSPE was normalized to the corresponding STD-VH control group (L12 photoperiod). Significant differences were assessed through two-way ANOVA analysis ($p < 0.05$). Diet: diet effect within VH groups; GSPE: GSPE consumption effect within CAF groups; Diet * Photoperiod: interaction effect between diet and photoperiod within VH groups; GSPE * Photoperiod: interaction effect between GSPE consumption and photoperiod within CAF groups; #: tendency: $0.05 < p < 1$. Different letters denote significant differences within each photoperiod group (assessed with two-way ANOVA followed by Tukey's *post hoc* test, $p < 0.05$; #: tendency: $0.05 < p < 1$).

Interestingly, an interaction effect between GSPE and photoperiod was also detected in the expression of *Lep*, which was reduced in response to GSPE consumption in the L12 and L6 groups but significantly upregulated in the L18 animals (Figure 4D). In addition, *Adipoq* gene expression was reduced by the CAF diet in all groups, but a tendency for GSPE to reverse this decrease was observed in animals subjected to the L12 and L18 photoperiods. Finally, a significant effect of the CAF diet on the expression of the inflammatory genes *Il6* and *Tnfα* was also observed (Supplementary Figure S5).

3.6. GSPE Consumption Reverted the Obesity-Induced Downregulation of Pgc1α in BAT in a Photoperiod-Dependent Manner

Being BAT such an important tissue in energy homeostasis, especially in rodents, we analyzed the most important components that drive the main metabolic pathways of brown adipocytes' activity. Despite the fact that no significant effects of CAF diet or GSPE administration on the gene expression of *Ucp1* were detected (Supplementary Figure S6), an interaction effect between GSPE and photoperiod was observed in the gene expression of *Pgc1α*, *Cpt1β*, and *Dio2* (Figure 5A–C). Particularly, GSPE supplementation subtly upregulated the expression of these genes under L12 and L6 photoperiods, while it downregulated its expression under L18. Moreover, the CAF diet downregulated the expression of *Pgc1α* in all photoperiods and the expression of *Dio2* in the L18 and L6 groups. Finally, mRNA levels of *Cpt1β* were influenced by the CAF diet in a photoperiod-dependent manner.

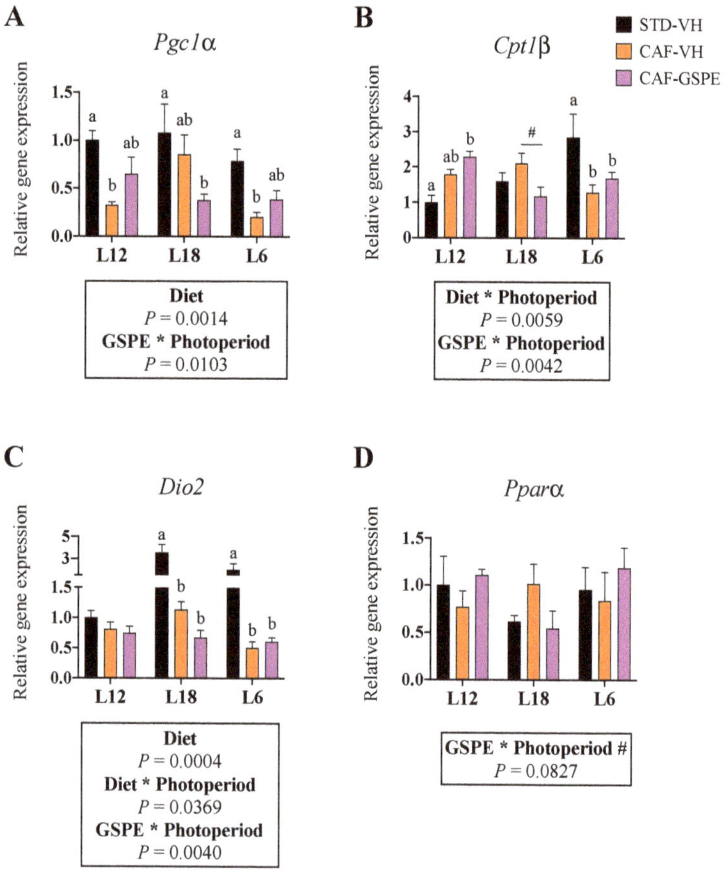

Figure 5. Effect of grape seed proanthocyanidin extract (GSPE) consumption at different photoperiods, 12 h of light:12 h of darkness (L12); 18 h of light:6 h of darkness (L18); 6 h of light:18 h of darkness (L6) on the expression of key metabolic genes in brown adipose tissue (BAT) of diet-induced obese rats. Fisher 344 rats were fed standard diet (STD) or cafeteria diet (CAF) for 9 weeks and treated with GSPE (25 mg/kg body weight) or vehicle (VH) the last 4 weeks of the study; STD-VH group; CAF-VH group; CAF-GSPE group, under different photoperiods; L12, L18 or L6. The gene expression of *Pgc1α* (**A**), *Cpt1β* (**B**), *Dio2* (**C**) and *Pparα* (**D**) was measured by qPCR and normalized by *Ppia*. The relative expression (presented as fold-change) of CAF-VH and CAF-GSPE was normalized to the corresponding STD-VH control group (L12 photoperiod). Significant differences were assessed through two-way ANOVA analysis ($p < 0.05$). Diet: diet effect within VH groups; Diet * Photoperiod: interaction effect between diet and photoperiod within VH groups; GSPE * Photoperiod: interaction effect between GSPE consumption and photoperiod within CAF groups; #: tendency: $0.05 < p < 1$. Different letters denote significant differences within each photoperiod group (assessed with two-way ANOVA followed by Tukey's post hoc test, $p < 0.05$; #: tendency: $0.05 < p < 1$).

Moreover, *Cd36*, *Lpl*, *Ppar*γ, and *Prdm16* gene expression was significantly downregulated by the CAF diet in all animal groups (Supplementary Figure S6). However, an interaction effect between GSPE and photoperiod was observed in the gene expression of *Ppar*α, which tended to be upregulated in the L12 and L6 groups but not in the L18 animals (Figure 5D).

4. Discussion

The beneficial effects of (poly)phenols on health and their natural presence in vegetable foods make these molecules of high interest in research. Additionally, a (poly)phenol-rich diet or the development of natural-based extracts as an alternative for preventing or treating metabolic disorders is more likely to be accepted by society than synthetic options [24]. In this context, (poly)phenol consumption can help reduce the negative effects of a disrupted metabolism in the adipose tissue and, consequently, in the whole organism [25]. Moreover, changes in the atmosphere can drive metabolic adaptations in mammals [26], meaning that seasonal rhythms can impact the effects of (poly)phenol consumption in adipose tissue. However, these effects have not been reported yet, even if evidence exists about the impact of seasonal consumption of (poly)phenols in other metabolic tissues such as the liver or the gut microbiota [18,19]. Our study aimed to elucidate the effects of proanthocyanidin seasonal consumption on the adaptations of the adipose tissue to an obesogenic environment. For this purpose, Fischer 344 rats were exposed to different photoperiod schedules in order to mimic the seasons of the year. Then, animals were fed a CAF diet and supplemented with either VH or 25 mg GSPE/ kg/day for four weeks. The dose of GSPE was chosen based on previous studies that showed metabolic effects with the same dose [27].

After nine weeks of study, body weight gain was reduced in response to proanthocyanidin consumption in CAF-fed animals. These results were particularly marked when GSPE was consumed at L18 than when it was at L12 and L6 photoperiods. According to the seasonal adaptations to photoperiod, increased weight gain in long photoperiods appears as a natural metabolic modulation in mammals, which, contrarily, lose more weight in short photoperiods [28]. Nevertheless, GSPE consumption resulted in a significant reduction in food intake only in animals subjected to L12 and L6 photoperiods but not in L18 animals. The effects of proanthocyanidins on reducing food intake were previously reported in Wistar rats [29] and were attributed to the astringency of these (poly)phenols [30]. Other mechanisms that could explain the GSPE-derived inhibition of food intake involve the central nervous system. On the one hand, enhanced leptin signaling was detected in obese rats supplemented with GSPE [31]. Furthermore, proanthocyanidins were shown to induce glucagon-like peptide (GLP-1) secretion, which is capable of increasing satiety and inhibiting feeding [32]. Nevertheless, considering the strong seasonal impact of long photoperiods on feeding behavior, which remains unaltered even with increased leptin levels, reflecting reduced leptin sensitivity under summer-like conditions, it would be possible that the GSPE lowering effects on food intake were not adopted by animals under the L18 photoperiod [33–35]. Similarly, even if previous studies observed increased energy expenditure in animals supplemented with GSPE under standard photoperiods [29,36], a previous study from our group showed that the energy expenditure of CAF-fed rats exposed to L18 and supplemented with GSPE for one week tended to be reduced [20], which suggests that the observed decrease in body weight gain in L18-CAF animals supplemented with proanthocyanidins in our study should be explained by other mechanisms. For instance, it was recently reported that obese rats exposed to L18 showed greater changes in the gut microbiota composition in response to proanthocyanidin consumption compared to animals exposed to L12 and L6 photoperiods. Particularly, GSPE consumption under L18 resulted in increased levels of *Bifidobacterium* and *Coprobacillus*, while decreased *Klebsiella* to levels associated with reduced body weight gain and obesity [19]. Moreover, other studies indicated that proanthocyanidins can modulate nutrient absorption in the gastrointestinal tract, through inhibiting digestive enzymes such as α-Amylase and α-Glucosidase, resulting in a 20% decrease in the amount of absorbed energy [29,37]. In future studies, it would be interesting to calculate energy/food efficiency in these animals in order to understand the proanthocyanidin-induced mechanism driving the reduction in body weight gain under the L18 photoperiod.

Given that WAT functionality responds to changes in photoperiod and that obesity drives major changes to the structure and activity of this tissue, histological analyses of WAT depots were performed. As expected, we observed a strong impact of the CAF diet on the area and volume of adipocytes, which were larger compared to animals fed a chow diet. These changes are common in obesity and contribute to the altered functionality of the WAT. In fact, under an obesogenic environment, the WAT needs to store the excess calories, which can be done through two processes: hyperplasia or hypertrophy [38]. The hyperplasia expansion is considered the healthy expansion, where the number of adipocytes increases via adipogenesis and they maintain their functionality and insulin sensitivity [9,38]. However, when hypertrophy occurs, existing adipocytes need to incorporate the excess of fat, increasing their volume and disrupting their functionality [9]. These dysfunctional adipocytes are less glucose tolerant and become pro-inflammatory, producing cytokines such as IL-6 or TNFα, which can induce macrophage infiltration in the adipose tissue and provoke a systemic immune response [39]. Adipose tissue expansion partly depends on the adipose depot; indeed, under healthy conditions, hyperplasia is more commonly developed in subcutaneous depots, while visceral depots are more susceptible to hypertrophy [40]. Additionally, it has been studied that hypertrophic visceral adipose tissue has a stronger negative impact on the whole metabolism of the organism, being highly associated with the metabolic syndrome [39]. Therefore, it is important to maintain the integrity and healthiness of this adipose depot. Furthermore, in our study, proanthocyanidin consumption reversed the effect induced by the CAF diet in iWAT regardless of the photoperiod, as adipocyte areas and volumes were at similar levels to their healthy counterparts in all photoperiod conditions. Similar results have been reported with GSPE supplementation in iWAT [25]. Further, the administration of blueberry polyphenol extract resulted in a reduced adipocyte area in the eWAT [41]. Nonetheless, in our study, this effect was less clear in eWAT, where a significant effect of GSPE was observed but the individual differences within groups in each photoperiod were not significant. In addition, when we studied the gene expression of inflammatory markers in eWAT (*Il6* and *Tnfα*), no effects were observed in response to GSPE consumption.

One of the main roles of WAT is to regulate energy balance in the organism. Therefore, lipid turnover is constantly active, involving distinct metabolic pathways that control fat storage and release from the WAT. The process that drives the formation of new adipocytes is named adipogenesis and is mainly driven by the master adipogenic molecules PPARγ and C/EBPs [42]. In our study, gene expression of *Pparγ* and *Cebpα* in both iWAT and eWAT showed an interaction effect between GSPE administration and photoperiod. In iWAT, we observed that GSPE increased its gene expression only in the L12 and L18 groups. Similar effects were previously observed in response to cyanidin-3-glucoside administration [43] and could explain the differences observed in the histological analyses, where the presence of larger adipocytes provoked by the CAF diet was partly reversed in response to GSPE consumption in L12 and L18 animals. Hence, increased adipogenesis would favor hyperplasia expansion rather than hypertrophy. Furthermore, the lipid transport-related genes (*Fabp4* and *Cd36*) were also affected by GSPE consumption in iWAT (L12 and L18 groups) and in eWAT (all groups), which could suggest a beneficial role for proanthocyanidins in helping release circulating TG. In fact, animals consuming GSPE at L12 and L18 photoperiods showed reduced serum TG concentrations compared to VH groups. However, another interaction effect between GSPE and photoperiod on the expression of lipolysis-related genes was detected in both iWAT and eWAT, which were upregulated in response to GSPE consumption only in the L12 and L18 groups. This activation of lipolysis could be involved in reducing the size of existing adipocytes, contributing to β-oxidation and global energy homeostasis. Moreover, the reduction in body weight gain observed in L18 animals could be partly explained by the increased lipolysis detected in these animals. In fact, supplementation with blueberry (poly)phenols suggested increased lipolysis in the WAT [41], and the lipolytic effects of other types of (poly)phenols such as epigallocatechin-gallate (EGCG) or resveratrol are already established [44].

Following these interaction effects between GSPE consumption and photoperiod, we observed that when proanthocyanidins were consumed at L6 photoperiod, the obesity-driven changes on the gene expression of iWAT were reverted. Thus, while the CAF diet induced the mRNA expression of genes related to adipogenesis (*Ppar*γ, *Cebp*α), lipid transport (*Fabp4*, *Cd36*), lipolysis (*Atgl*, *Hsl*), and adipokines (*Lep*, *Adipoq*), GSPE was capable of returning the expression of these genes to a healthy situation. In this context, (poly)phenol-induced reduction of lipid turnover has been reported previously in human and animal studies [45–47], which partly agrees with our results in the L6 group. However, a significant effect was also observed only under L18, where GSPE consumption upregulated the expression of *Adipoq*. High adiponectin serum concentrations are associated with anti-inflammatory effects that help prevent the metabolic consequences of an obesogenic environment [48]. In this context, a previous study reported that GSPE administration lowered inflammatory levels and increased adiponectin mRNA levels in the mesenteric adipose tissue of obese rats, which agrees with our findings [49]. Importantly, in our study, we could not observe the same effects in animals exposed to L12 and L6, which reinforce the photoperiod-dependent effects of proanthocyanidins on the modulation of gene expression in WAT.

Importantly, we also detected a browning effect in iWAT only in animals consuming proanthocyanidins under short photoperiods. In fact, GSPE was associated with increased thermogenesis in the adipose tissue of obese mice [36], but these results were obtained after supplementing with noticeable higher doses of GSPE. Therefore, in our study, it is possible that the dose of GSPE was too low to fully induce browning in the iWAT. In fact, proanthocyanidins did not seem to promote enhanced BAT activity in our study, as *Ucp1* gene expression was unchanged after both the CAF diet and (poly)phenols consumption. However, other molecules involved in BAT thermogenesis were affected by GSPE in a photoperiod-dependent manner. Thus, proanthocyanidins upregulated the expression of *Pgc1*α, *Cpt1*β and *Ppar*α in animals exposed to L12 and L6, while these genes were downregulated in L18 animals. Our results agree with previous findings [27], where GSPE reversed the obesity-induced mitochondrial dysfunction in BAT by increasing *Pgc1*α expression. However, in our study, we demonstrated that these effects are limited to L6 and L12 photoperiods.

In conclusion, our results indicate that the beneficial effects associated with GSPE consumption on diet-induced obese animals are strongly influenced by the photoperiod of exposure in a tissue-specific manner. Particularly, our findings suggest that L18 photoperiod is highly sensitive to GSPE with respect to body weight adaptation and modulation of metabolic genes, especially adiponectin, in the iWAT. Meanwhile, animals under the L12 and L6 photoperiods showed reduced food intake in response to GSPE and enhanced BAT activity. Therefore, our study reinforces the relevance of considering seasonal rhythms when investigating the metabolic effects of (poly)phenols in the organism in order to properly potentiate the metabolic response of the adipose tissue.

Supplementary Materials: The following supporting information can be downloaded at: https://www.mdpi.com/article/10.3390/nu15041037/s1, Figure S1: Experimental design used in this study. Figure S2: Representative images of iWAT histology; Figure S3: Effect of grape seed proanthocyanidin extract (GSPE) consumption at different photoperiods, 12 h of light:12 h of darkness (L12); 18 h of light:6 h of darkness (L18); 6 h of light:18 h of darkness (L6), on the gene expression of *Acac*α and *Fasn* in inguinal white adipose tissue (iWAT) of diet-induced obese rats; Figure S4: Representative images of eWAT histology; Figure S5: Effect of grape seed proanthocyanidin extract (GSPE) consumption at different photoperiods, 12 h of light:12 h of darkness (L12); 18 h of light:6 h of darkness (L18); 6 h of light:18 h of darkness (L6), on the expression of lipogenesis and inflammation-related genes in epididymal white adipose tissue (eWAT) of diet-induced obese rats; Figure S6. Effect of grape seed proanthocyanidin extract (GSPE) consumption at different photoperiods, 12 h of light:12 h of darkness (L12); 18 h of light:6 h of darkness (L18); 6 h of light:18 h of darkness (L6), on the expression of key metabolic genes in brown adipose tissue (BAT) of diet-induced obese rats; Table S1: Phenolic Composition of GSPE; Table S2: Primers for the Q-PCR analysis.

Author Contributions: Conceptualization, A.A.-A., F.I.B., B.M. and G.A.; methodology, È.N.-M. and M.C.-P.; formal analysis, È.N.-M. and M.C.-P.; investigation, È.N.-M., M.C.-P. and F.M.; resources, A.A.-A., F.I.B., B.M. and G.A.; data curation, È.N.-M. and M.C.-P.; writing—original draft preparation, È.N.-M.; writing—review and editing, È.N.-M. and G.A.; supervision, G.A.; project administration, A.A.-A., F.I.B., B.M. and G.A.; funding acquisition, A.A.-A., F.I.B., B.M. and G.A. All authors have read and agreed to the published version of the manuscript.

Funding: This research was funded by the Ministerio de Ciencia e Innovación MCIN/AEI/10.13039/501100011033/FEDER "Una manera de hacer Europa" (Grant number: AGL2016-77105-R). È.N.-M. and F.M. are recipients of a predoctoral fellowship from Universitat Rovira i Virgili—Martí i Franquès (2019PMF-PIPF-73 and 2019PMF-PIPF-19, respectively). M.C.-P. is the recipient of a predoctoral fellowship from Generalitat de Catalunya FI (2021 FI_B2 00150). F.I.B. and G.A. are Serra-Hunter fellows at the University of Rovira i Virgili.

Institutional Review Board Statement: All animal care and experimental protocols involving animals were approved by the Ethics Review Committee for Animal Experimentation of the Universitat Rovira i Virgili (reference number 9495, 18 September 2019) and were carried out in accordance with Directive 86/609EEC of the Council of the European Union and the procedure established by the Departament d'Agricultura, Ramaderia i Pesca of the Generalitat de Catalunya.

Informed Consent Statement: Not applicable.

Data Availability Statement: The data presented in this study are available on request from the corresponding author. The data are not publicly available due to a lack of a platform to publish them.

Acknowledgments: We deeply thank Niurka López and Rosa M. Pastor for their technical help and advice.

Conflicts of Interest: The authors declare no conflict of interest.

References

1. Abarca-Gómez, L.; Abdeen, Z.A.; Hamid, Z.A.; Abu-Rmeileh, N.M.; Acosta-Cazares, B.; Acuin, C.; Adams, R.J.; Aekplakorn, W.; Afsana, K.; Aguilar-Salinas, C.A.; et al. Worldwide Trends in Body-Mass Index, Underweight, Overweight, and Obesity from 1975 to 2016: A Pooled Analysis of 2416 Population-Based Measurement Studies in 128.9 Million Children, Adolescents, and Adults. *Lancet* **2017**, *390*, 2627–2642. [CrossRef] [PubMed]
2. World Health Organization (WHO). Global Health Observatory Data. In *Overweight and Obesity*; WHO: Geneva, Switzerland, 2016.
3. Nicolaidis, S. Environment and Obesity. *Metabolism* **2019**, *100*, 153942. [CrossRef] [PubMed]
4. Rohde, K.; Keller, M.; la Cour Poulsen, L.; Blüher, M.; Kovacs, P.; Böttcher, Y. Genetics and Epigenetics in Obesity. *Metabolism* **2019**, *92*, 37–50. [CrossRef] [PubMed]
5. Yoshimura, E.; Tajiri, E.; Hatamoto, Y.; Tanaka, S. Changes in Season Affect Body Weight, Physical Activity, Food Intake, and Sleep in Female College Students: A Preliminary Study. *Int. J. Environ. Res. Public Health* **2020**, *17*, 8713. [CrossRef]
6. Au-Yong, I.T.H.; Thorn, N.; Ganatra, R.; Perkins, A.C.; Symonds, M.E. Brown Adipose Tissue and Seasonal Variation in Humans. *Diabetes* **2009**, *58*, 2583–2587. [CrossRef] [PubMed]
7. Turicchi, J.; O'Driscoll, R.; Horgan, G.; Duarte, C.; Palmeira, A.L.; Larsen, S.C.; Heitmann, B.L.; Stubbs, J. Weekly, Seasonal and Holiday Body Weight Fluctuation Patterns among Individuals Engaged in a European Multi-Centre Behavioural Weight Loss Maintenance Intervention. *PLoS ONE* **2020**, *15*, e0232152. [CrossRef]
8. Yokoya, M.; Higuchi, Y. Day Length May Make Geographical Difference in Body Size and Proportions: An Ecological Analysis of Japanese Children and Adolescents. *PLoS ONE* **2019**, *14*, e0210265. [CrossRef]
9. Longo, M.; Zatterale, F.; Naderi, J.; Parrillo, L.; Formisano, P.; Raciti, G.A.; Beguinot, F.; Miele, C. Adipose Tissue Dysfunction as Determinant of Obesity-Associated Metabolic Complications. *Int. J. Mol. Sci.* **2019**, *20*, 2358. [CrossRef]
10. Frigolet, M.E.; Gutiérrez-Aguilar, R. The Colors of Adipose Tissue. *Gac. Med. Mex.* **2020**, *156*, 142–149. [CrossRef]
11. Pellegrinelli, V.; Carobbio, S.; Vidal-Puig, A. Adipose Tissue Plasticity: How Fat Depots Respond Differently to Pathophysiological Cues. *Diabetologia* **2016**, *59*, 1075–1088. [CrossRef]
12. Bargut, T.C.L.; Aguila, M.B.; Mandarim-de-Lacerda, C.A. Brown Adipose Tissue: Updates in Cellular and Molecular Biology. *Tissue Cell* **2016**, *48*, 452–460. [CrossRef] [PubMed]
13. Cardona, F.; Andrés-Lacueva, C.; Tulipani, S.; Tinahones, F.J.; Queipo-Ortuño, M.I. Benefits of Polyphenols on Gut Microbiota and Implications in Human Health. *J. Nutr. Biochem.* **2013**, *24*, 1415–1422. [CrossRef] [PubMed]

14. Mezhibovsky, E.; Knowles, K.A.; He, Q.; Sui, K.; Tveter, K.M.; Duran, R.M.; Roopchand, D.E. Grape Polyphenols Attenuate Diet-Induced Obesity and Hepatic Steatosis in Mice in Association With Reduced Butyrate and Increased Markers of Intestinal Carbohydrate Oxidation. *Front. Nutr.* **2021**, *8*, 675267. [CrossRef] [PubMed]
15. Tresserra-Rimbau, A.; Medina-Remón, A.; Pérez-Jiménez, J.; Martínez-González, M.A.; Covas, M.I.; Corella, D.; Salas-Salvadó, J.; Gómez-Gracia, E.; Lapetra, J.; Arós, F.; et al. Dietary Intake and Major Food Sources of Polyphenols in a Spanish Population at High Cardiovascular Risk: The PREDIMED Study. *Nutr. Metab. Cardiovasc. Dis.* **2013**, *23*, 953–959. [CrossRef] [PubMed]
16. Ávila-Román, J.; Soliz-Rueda, J.R.; Bravo, F.I.; Aragonès, G.; Suárez, M.; Arola-Arnal, A.; Mulero, M.; Salvadó, M.J.; Arola, L.; Torres-Fuentes, C.; et al. Phenolic Compounds and Biological Rhythms: Who Takes the Lead? *Trends Food Sci. Technol.* **2021**, *113*, 77–85. [CrossRef]
17. Arola-Arnal, A.; Cruz-Carrión, Á.; Torres-Fuentes, C.; Ávila-Román, J.; Aragonès, G.; Mulero, M.; Bravo, F.I.; Muguerza, B.; Arola, L.; Suárez, M. Chrononutrition and Polyphenols: Roles and Diseases. *Nutrients* **2019**, *11*, 2602. [CrossRef] [PubMed]
18. Rodríguez, R.M.; Colom-Pellicer, M.; Blanco, J.; Calvo, E.; Aragonès, G.; Mulero, M. Grape-Seed Procyanidin Extract (GSPE) Seasonal-Dependent Modulation of Glucose and Lipid Metabolism in the Liver of Healthy F344 Rats. *Biomolecules* **2022**, *12*, 839. [CrossRef]
19. Arreaza-Gil, V.; Escobar-Martínez, I.; Muguerza, B.; Aragonès, G.; Suárez, M.; Torres-Fuentes, C.; Arola-Arnal, A. The Effects of Grape Seed Proanthocyanidins in Cafeteria Diet-Induced Obese Fischer 344 Rats Are Influenced by Faecal Microbiota in a Photoperiod Dependent Manner. *Food Funct.* **2022**, *13*, 8363–8374. [CrossRef]
20. Soliz-Rueda, J.R.; López-Fernández-sobrino, R.; Bravo, F.I.; Aragonès, G.; Suarez, M.; Muguerza, B. Grape Seed Proanthocyanidins Mitigate the Disturbances Caused by an Abrupt Photoperiod Change in Healthy and Obese Rats. *Nutrients* **2022**, *14*, 1834. [CrossRef]
21. Serra, A.; MacI, A.; Romero, M.P.; Valls, J.; Bladé, C.; Arola, L.; Motilva, M.J. Bioavailability of Procyanidin Dimers and Trimers and Matrix Food Effects in in Vitro and in Vivo Models. *Br. J. Nutr.* **2010**, *103*, 944–952. [CrossRef]
22. Quiñones, M.; Guerrero, L.; Suarez, M.; Pons, Z.; Aleixandre, A.; Arola, L.; Muguerza, B. Low-Molecular Procyanidin Rich Grape Seed Extract Exerts Antihypertensive Effect in Males Spontaneously Hypertensive Rats. *Food Res. Int.* **2013**, *51*, 587–595. [CrossRef]
23. Pfaffl, M.W. Relative Quantification. In *Real-Time PCR*; Dorak, R., Ed.; International University Line: La Jolla, CA, USA, 2004; pp. 64–82, ISBN 9780203967317.
24. Sachdeva, V.; Roy, A.; Bharadvaja, N. Current Prospects of Nutraceuticals: A Review. *Curr. Pharm. Biotechnol.* **2020**, *21*, 884–896. [CrossRef]
25. Colom-Pellicer, M.; Rodríguez, R.M.; Navarro-Masip, È.; Bravo, F.I.; Mulero, M.; Arola, L.; Aragonès, G. Time-of-Day Dependent Effect of Proanthocyanidins on Adipose Tissue Metabolism in Rats with Diet-Induced Obesity. *Int. J. Obes.* **2022**, *46*, 1394–1402. [CrossRef] [PubMed]
26. Helfer, G.; Barrett, P.; Morgan, P.J. A Unifying Hypothesis for Control of Body Weight and Reproduction in Seasonally Breeding Mammals. *J. Neuroendocrinol.* **2019**, *31*, e12680. [CrossRef] [PubMed]
27. Pajuelo, D.; Quesada, H.; Díaz, S.; Fernández-Iglesias, A.; Arola-Arnal, A.; Bladé, C.; Salvadó, J.; Arola, L. Chronic Dietary Supplementation of Proanthocyanidins Corrects the Mitochondrial Dysfunction of Brown Adipose Tissue Caused by Diet-Induced Obesity in Wistar Rats. *Br. J. Nutr.* **2012**, *107*, 170–178. [CrossRef] [PubMed]
28. Ebling, F.J.P. On the Value of Seasonal Mammals for Identifying Mechanisms Underlying the Control of Food Intake and Body Weight. *Horm. Behav.* **2014**, *66*, 56–65. [CrossRef]
29. Serrano, J.; Casanova-Martí, À.; Gual, A.; Pérez-Vendrell, A.M.; Blay, M.T.; Terra, X.; Ardévol, A.; Pinent, M. A Specific Dose of Grape Seed-Derived Proanthocyanidins to Inhibit Body Weight Gain Limits Food Intake and Increases Energy Expenditure in Rats. *Eur. J. Nutr.* **2017**, *56*, 1629–1636. [CrossRef]
30. Soares, S.; Brandão, E.; Mateus, N.; de Freitas, V. Sensorial Properties of Red Wine Polyphenols: Astringency and Bitterness. *Crit. Rev. Food Sci. Nutr.* **2017**, *57*, 937–948. [CrossRef] [PubMed]
31. Ibars, M.; Ardid-Ruiz, A.; Suárez, M.; Muguerza, B.; Bladé, C.; Aragonès, G. Proanthocyanidins Potentiate Hypothalamic Leptin/STAT3 Signalling and Pomc Gene Expression in Rats with Diet-Induced Obesity. *Int. J. Obes.* **2017**, *41*, 129–136. [CrossRef] [PubMed]
32. Serrano, J.; Casanova-Martí, À.; Blay, M.; Terra, X.; Ardévol, A.; Pinent, M. Defining Conditions for Optimal Inhibition of Food Intake in Rats by a Grape-Seed Derived Proanthocyanidin Extract. *Nutrients* **2016**, *8*, 652. [CrossRef]
33. Togo, Y.; Otsuka, T.; Goto, M.; Furuse, M.; Yasuo, S. Photoperiod Regulates Dietary Preferences and Energy Metabolism in Young Developing Fischer 344 Rats but Not in Same-Age Wistar Rats. *Am. J. Physiol. Endocrinol. Metab.* **2012**, *303*, 777–786. [CrossRef] [PubMed]
34. Ibars, M.; Aragonès, G.; Ardid-Ruiz, A.; Gibert-Ramos, A.; Arola-Arnal, A.; Suárez, M.; Bladé, C. Seasonal Consumption of Polyphenol-Rich Fruits Affects the Hypothalamic Leptin Signaling System in a Photoperiod-Dependent Mode. *Sci. Rep.* **2018**, *8*, 13572. [CrossRef]
35. Mariné-Casadó, R.; Domenech-Coca, C.; del Bas, J.M.; Bladé, C.; Arola, L.; Caimari, A. The Exposure to Different Photoperiods Strongly Modulates the Glucose and Lipid Metabolisms of Normoweight Fischer 344 Rats. *Front. Physiol.* **2018**, *9*, 416. [CrossRef] [PubMed]

36. Du, H.; Wang, Q.; Li, T.; Ren, D.; Yang, X. Grape Seed Proanthocyanidins Reduced the Overweight of C57BL/6J Mice through Modulating Adipose Thermogenesis and Gut Microbiota. *Food Funct.* **2021**, *12*, 8467–8477. [CrossRef]
37. Yilmazer-Musa, M.; Griffith, A.M.; Michels, A.J.; Schneider, E.; Frei, B. Grape Seed and Tea Extracts and Catechin 3-Gallates Are Potent Inhibitors of α-Amylase and α-Glucosidase Activity. *J. Agric. Food Chem.* **2012**, *60*, 8924–8929. [CrossRef] [PubMed]
38. Ursula, W.; Eric, R. Dynamics of Adipose Tissue Turnover in Human Metabolic Health and Disease HHS Public Access. *Physiol. Behav.* **2019**, *176*, 139–148. [CrossRef]
39. Despres, J.P.; Lemieux, I. Abdominal Obesity and Metabolic Syndrome. *Nature* **2006**, *444*, 881–887. [CrossRef] [PubMed]
40. Drolet, R.; Richard, C.; Sniderman, A.D.; Mailloux, J.; Fortier, M.; Huot, C.; Rhéaume, C.; Tchernof, A. Hypertrophy and Hyperplasia of Abdominal Adipose Tissues in Women. *Int. J. Obes.* **2008**, *32*, 283–291. [CrossRef]
41. Jiao, X.; Wang, Y.; Lin, Y.; Lang, Y.; Li, E.; Zhang, X.; Zhang, Q.; Feng, Y.; Meng, X.; Li, B. Blueberry Polyphenols Extract as a Potential Prebiotic with Anti-Obesity Effects on C57BL/6 J Mice by Modulating the Gut Microbiota. *J. Nutr. Biochem.* **2019**, *64*, 88–100. [CrossRef]
42. Ali, A.T.; Hochfeld, W.E.; Myburgh, R.; Pepper, M.S. Adipocyte and Adipogenesis. *Eur. J. Cell Biol.* **2013**, *92*, 229–236. [CrossRef]
43. Matsukawa, T.; Inaguma, T.; Han, J.; Villareal, M.O.; Isoda, H. Cyanidin-3-Glucoside Derived from Black Soybeans Ameliorate Type 2 Diabetes through the Induction of Differentiation of Preadipocytes into Smaller and Insulin-Sensitive Adipocytes. *J. Nutr. Biochem.* **2015**, *26*, 860–867. [CrossRef]
44. Wang, S.; Moustaid-Moussa, N.; Chen, L.; Mo, H.; Shastri, A.; Su, R.; Bapat, P.; Kwun, I.S.; Shen, C.L. Novel Insights of Dietary Polyphenols and Obesity. *J. Nutr. Biochem.* **2014**, *25*, 1–18. [CrossRef] [PubMed]
45. Khalilpourfarshbafi, M.; Gholami, K.; Murugan, D.D.; Abdul Sattar, M.Z.; Abdullah, N.A. Differential Effects of Dietary Flavonoids on Adipogenesis. *Eur. J. Nutr.* **2019**, *58*, 5–25. [CrossRef] [PubMed]
46. Most, J.; Warnke, I.; Boekschoten, M.v.; Jocken, J.W.E.; de Groot, P.; Friedel, A.; Bendik, I.; Goossens, G.H.; Blaak, E.E. The Effects of Polyphenol Supplementation on Adipose Tissue Morphology and Gene Expression in Overweight and Obese Humans. *Adipocyte* **2018**, *7*, 190–196. [CrossRef]
47. de Lima, L.P.; de Paula Barbosa, A. A Review of the Lipolytic Effects and the Reduction of Abdominal Fat from Bioactive Compounds and Moro Orange Extracts. *Heliyon* **2021**, *7*, e07695. [CrossRef]
48. Fang, H.; Judd, R.L. Adiponectin Regulation and Function. *Compr. Physiol.* **2018**, *8*, 1031–1063. [CrossRef] [PubMed]
49. Terra, X.; Montagut, G.; Bustos, M.; Llopiz, N.; Ardèvol, A.; Bladé, C.; Fernández-Larrea, J.; Pujadas, G.; Salvadó, J.; Arola, L.; et al. Grape-Seed Procyanidins Prevent Low-Grade Inflammation by Modulating Cytokine Expression in Rats Fed a High-Fat Diet. *J. Nutr. Biochem.* **2009**, *20*, 210–218. [CrossRef]

nutrients

MDPI

Article

Chemical Characterization, Antioxidant Capacity and Anti-Oxidative Stress Potential of South American Fabaceae *Desmodium tortuosum*

José-Luis Rodríguez [1,2,*], Paola Berrios [1], Zoyla-Mirella Clavo [1], Manuel Marin-Bravo [3], Luis Inostroza-Ruiz [4], Mariella Ramos-Gonzalez [1], Miguel Quispe-Solano [5], Maria S. Fernández-Alfonso [6], Olga Palomino [6] and Luis Goya [7]

1 Faculty of Veterinary Medicine, Universidad Nacional Mayor de San Marcos, Lima 15021, Peru
2 Faculty of Veterinary, Universidad Complutense de Madrid, 28040 Madrid, Spain
3 Faculty of Biological Sciences, Universidad Nacional Mayor de San Marcos, Lima 15021, Peru
4 Faculty of Pharmacy, Universidad Nacional Mayor de San Marcos, Lima 15021, Peru
5 Faculty of Engineering in Food Industries, Universidad Nacional del Centro del Perú, Huancayo 12006, Peru
6 Faculty of Pharmacy, Universidad Complutense de Madrid, 28040 Madrid, Spain
7 Department of Metabolism and Nutrition, Spanish National Research Council (CSIC), Institute of Food Science, Technology and Nutrition (ICTAN), Jose Antonio Novais 10, 28040 Madrid, Spain
* Correspondence: josero05@ucm.es

Abstract: It has been proposed that oxidative stress is a pathogenic mechanism to induce cytotoxicity and to cause cardiovascular and neuronal diseases. At present, natural compounds such as plant extracts have been used to reduce the cytotoxic effects produced by agents that induce oxidative stress. Our study aimed to evaluate the antioxidant and cytoprotective capacity of *Desmodium tortuosum* (*D. tortuosum*) extract in the co- and pre-treatment in EA.hy926 and SH-SY5Y cell lines subjected to oxidative stress induced by tert-butylhydroperoxide (t-BOOH). Cell viability, reactive oxygen species (ROS), nitric oxide (NO), caspase 3/7 activity, reduced glutathione (GSH), glutathione peroxidase (GPx), glutathione reductase (GR), and molecular expression of oxidative stress biomarkers (SOD2, NRF2 and NFκB1) and cell death (APAF1, BAX, Caspase3) were all evaluated. It was observed that the *D. tortuosum* extract, in a dose-dependent manner, was able to reduce the oxidative and cytotoxicity effects induced by t-BOOH, even normalized to a dose of 200 μg/mL, which would be due to the high content of phenolic compounds mainly phenolic acids, flavonoids, carotenoids and other antioxidant compounds. Finally, these results are indicators that the extract of *D. tortuosum* could be a natural alternative against the cytotoxic exposure to stressful and cytotoxic chemical agents.

Keywords: plant antioxidants; phytochemicals; polyphenols; medicinal plants; vascular endothelium; neuroprotection

Citation: Rodríguez, J.-L.; Berrios, P.; Clavo, Z.-M.; Marin-Bravo, M.; Inostroza-Ruiz, L.; Ramos-Gonzalez, M.; Quispe-Solano, M.; Fernández-Alfonso, M.S.; Palomino, O.; Goya, L. Chemical Characterization, Antioxidant Capacity and Anti-Oxidative Stress Potential of South American Fabaceae *Desmodium tortuosum*. *Nutrients* **2023**, *15*, 746. https://doi.org/10.3390/nu15030746

Academic Editor: Paul Prenzler

Received: 18 January 2023
Revised: 28 January 2023
Accepted: 30 January 2023
Published: 1 February 2023

1. Introduction

The chemical products as xenobiotics can cause the production of ROS and other free radical that may result in inflammatory and fibrotic processes [1]. After absorption, the first tissue affected is the vascular system, especially the endothelia, this being the first step towards the development of vascular diseases [2]. Oxidative stress is capable of inducing vascular endothelial damage, which would produce a change in the vascular structure, this being one of the causes of diseases such as diabetes or nephropathy. Therefore, it is important to reduce this stressful effect on the vascular endothelium, with medicinal plants being a good therapeutic alternative [3–5]. During the last decade, studies have been published indicating that the antioxidant compounds of medicinal plants have the function of reducing endothelial damage and this would make them effective against cardiovascular diseases [6,7].

Neurodegenerative diseases, characterized by progressive dysfunction and cellular senescence of specific neuronal systems, involving structural and functional damage of neurons, may be caused by oxidative stress [1,8,9]. Recent mounting evidence suggests that oxidative stress in neuronal cells contributes to neuroinflammation, facilitated by the constant activation of microglia, hence inducing neuronal necrosis and apoptosis [8,9]. Since the onset of oxidative stress seems to be a main contributive cause of cardiovascular and neurodegenerative pathologies, protection of cells, tissues and organs against this challenging condition is the major goal of many studies dealing with nutritional and pharmacological prevention of pathologies.

During the last two decades, extended research on plant antioxidants, especially polyphenols, has unequivocally demonstrated their role as bioactive compounds that protect against oxidative stress and prevent or delay the onset of many pathologies [5,9,10]. For example, metabolic diseases, such as type 2 diabetes, or cardiovascular complications have endothelial damage as a common factor and, also, are characterized by an excess of free radical, oxidative stress and pro-inflammatory cytokines creating an oxidative and pro-inflammatory environment [11]. Some alternatives to counteract these effects are medicinal plants, such as *D. tortuosum*, which has antioxidant activity due to its high concentration in polyphenols, polyterpenes and flavonoids and their metabolites. In addition, continuing with the vascular endothelium, cocoa catechins are known to have a positive effect on healthy vascular function [10,12] which has been proclaimed since the pioneer studies from two decades ago [13–15], up to recent reviews [12,16,17]. However, the beneficial effect on cardiovascular function is not privative of flavanols or flavonoids in general [16] as we have already described before, also many other phenolic compounds or plant extracts have shown this potential [18,19]. Indeed, by using EA.hy926 cells, we have shown the protective effect of an extract from *Silybum marianum*, rich in flavonol derivatives known as flavonolignans, on cultured endothelial cells subjected to high glucose concentrations [20]. In the same model, *Vochysia rufa* stem bark extract, rich in reducing sugars and flavonoids, also showed significant protection against high glucose damage [21]. Similarly, in yerba mate and green coffee extracts, their main hydroxycinnamic acids and microbial metabolites prevented Tumor Necrosis Factor-alpha (TNF-α)-induced inflammation [22]. More recently, cocoa flavanols were proven to protect the same EA.hy926 cells against chemically induced oxidative stress [23]. All these studies confirm the protective effect on endothelial function of plant extracts and pure antioxidant compounds, as well as endorse the reliability of the cell culture model.

On the other hand, the preventive effects of polyphenolic antioxidants in aging and neurodegeneration associated with oxidative stress have been largely reported [1,9,24,25]. Since the culture of primary neurons is rather difficult, established cell lines have been widely used to test the neuro-regulatory effect of different bioactive compounds and their specific effects at the cellular and molecular levels. Neuroblastoma SH-SY5Y cells, derived from the SK-N-SH cell line, is one of the most commonly used neuronal-like cell cultures, and it has been recently validated as a simple reliable model of neuronal-like cells that is amenable to biological, biochemical and electrophysiological investigation [26]. In fact, using this human cell line as a neuronal cell culture model, the chemo-protective effect of an aqueous extract of cocoa phenolic compounds (mainly flavanols) against oxidative stress-induced neurodegeneration has been recently reported [27], as well as that of an extract from *Sambucus nigra* (elderflower), rich in flavonoids and hydroxycinnamic acids [28].

Thus, in this study, the polyphenolic chemical composition of the *D. tortuosum* extract was determined and to delineate the potential protective mechanisms through which *D. tortuosum* extracts protect endothelial and neuron cell function, two human cell lines, EA.hy926 and SH-SY5Y cells, were treated with t-BOOH a strong pro-oxidant used to induce oxidative stress in cell cultures.

2. Materials and Methods

2.1. Reagents

t-BOOH, glutathione reductase (GR), reduced (GSH) and oxidized (GSSG) glutathione, o-phthaldialdehyde, nicotine adenine dinucleotide phosphate reduced salt (NADPH), 2,4-dinitrophenylhydrazine, gentamicin, penicillin G and streptomycin, 2′,7′-dichlorofluorescin-diacetate (DCFH-DA), 4-amino-5-methylamino-2,7-difluorofluorescein-diacetate (DAF-FM-DA), 3-(4,5-dimethylthiazol-2-yl)-2,5-diphenyl-tetrazolium bromide (MTT), Dulbecco's phosphate-buffered saline (DPBS, D8537) were obtained from Sigma-Aldrich (Madrid, Spain). Acetonitrile, methanol of high performance liquid chromatography (HPLC) grade, dimethyl sulfoxide (DMSO) of analytical grade and all other usual laboratory reagents were acquired from Panreac (Barcelona, Spain). Nucleo-spin-RNA, qPCRBIO-cDNA-synthesis, ICgreen-amplification-PCR, DMEM-culture-media and fetal bovine serum (FBS) were from Cultek (Madrid, Spain). The Apo-ONE® Homogeneous Caspase-3/7 Assay kit was acquired from Promega (Madison, WI, USA). Bradford reagent was from BioRad Laboratories S.A (Hercules, CA, USA). All other chemical reagents used were of high-purity for cell and molecular biology and were available in the laboratory.

2.2. Plant Selection and Extract Preparation

The biological material was collected in the Monte Alegre district, Padre Abad province, Ucayali region, Peru coordinates 0498477LS-9030950LW, at an altitude of 194 m.a.s.l.; 10 kg of stems and leaves were collected (voucher number RS978). In the Ucayali Veterinary Institute-Pucallpa Regional Herbarium (HRUIP), the plants were dried, herbalized, and assembled, and the taxonomic verification of the species was carried out by comparison with existing samples and the use of a specialized bibliography, voucher number RS978. The obtained plants belonged to the species *D. tortuosum* of the Fabaceae family and were entered in the Herbarium under the registration number 12208.

The stems and leaves were collected and washed-and-dried in open air, then completely dried in an oven and reduced to a fine powder. The decoction was made with distilled water and the powdered plant material (10:1) in a beaker, heating until boiling and maintaining for twenty minutes. Plant material was filtered off and the aqueous extract was concentrated and lyophilized.

2.3. Chemical Characterization of Extract

Briefly, the sample (20 mg) was diluted with methanol (20 mL). The mixture was ultrasonicated for 10 min. Then, it was filtered through a 0.25 µm filter, and 3 µL were injected into a Dionex Ultimate 3000 (Thermo Scientific, Waltham, MA, USA) UHPLC system. Column was a Luna© Omega (Phenomenex Inc., Torrance, CA, USA) C18 100 Å, Phenomenex (150 × 2.1 mm, 1.6 µm), temperature 40 °C, flow rate 0.25 mL/min, with distilled water 1% formic acid and acetonitrile 1% formic acid eluents. The UHPLC system was coupled to a QExactive PLus mass spectrometer (Thermo Scientific, Waltham, MA, USA). Full Mass Spectrometry (MS) scan parameters were in the range of 120–1500 m/z, resolution 70,000, microscans 1, Automatic Gain Control (AGC) target of 1×10^6, and maximum intensity (IT) of 100 ms. The parameters of the Resolution Type MS2 (MS2) resolution were 17,500, an AGC target of 2×10^5, and a maximum IT of 50 ms. The ionization source parameters were Electrospray ionization (ESI) (negative/positive), spray voltage 2.5/3.0, temperature 280 °C, N_2 (sheath gas flow rate: 40, aux gas flow rate: 10), gas heater temp of 350 °C, S-lens radio frequency (RF) level of 100, and a normalized collision energy of 20, 40, 60. The m/z values of the ions were detected in full ESI-MS (positive and/or negative) and the main fragments observed in the MS/MS spectra, the error in ppm is also indicated for the calculation of the molecular formula ($\leq$5 ppm).

2.4. Cell Culture

Human EA.hy926 cells were a gift from Prof. Patricio Aller (CSIC, Madrid, Spain), and the SH-SY5Y cell line was a gift from Prof. Ignacio Torres Alemán (Instituto Cajal,

Madrid, Spain), and later (same depositary batch) from Prof. Carlos Guillén, School of Pharmacy, University Complutense, Madrid, Spain. The cells were cultured and passaged in DMEM-F12 with FBS (10%) and 50 mg/L of gentamicin, penicillin and streptomycin. Cells were incubated in humid conditions with 5% CO_2 and 95% air, at 37 °C; the culture medium was changed every other day.

Different concentrations of the *D. tortuosum* extract (1, 10, 25, 50, 100 and 200 μg/mL) dissolved in DMEM-F12 were added to microwell plates. To evaluate the protective effect of the *D. tortuosum* extract against oxidative stress, co- and pre-treatment were carried out. In the co-treatment assay, EA.hy926 and SH-SY5Y cells were simultaneously treated for 22 h with 100 μM t-BOOH plus the different concentrations of *D. tortuosum*; whereas in the pre-treatment assay cells were first treated with noted doses of extract for 18 h, then washed and submitted to a new media containing 200 μM t-BOOH for 4 h, after which the assay was performed [29].

2.5. Cell Viability Evaluation (MTT)

In this assay, it is observed if the mitochondria is active and is capable of reducing tetrazolium-MTT [30]. Briefly, after treatments, 0.5 mg/mL MTT as the final concentration in was added to each well for 2 h, during this time, metabolically active EA.hy926 and SH-SY5Y cells reduced the tetrazolium-MTT to a formazan-salt. Absorbance was measured at 540 nm (SPECTROstar BMG microplate reader (BMG Labtech, Ortenberg, Baden-Wurttemberg, Germany)). Cell viability is represented as % of control.

2.6. Intracellular ROS Production

Oxidative stress was assessed by the ROS intracellular production according to standardized protocols using the DCFH-DA-fluorescence assay [31]. DCFH-DA enters the cell and is hydrolyzed by esterases to allow the release of DCFH and its reaction with ROS to generate a fluorescent compound. Briefly, 10 μM of DCFH-DA was added to each well (2×10^5 cells/well under incubation conditions) in a black multi-well plate for 30 min, and immediately measured in a fluorescent microplate reader (FLx800 Fluorimeter, BioTek, Winooski, VT, USA) at 485 nm/530 nm (λ excitation/λ emission).

2.7. Determination of Nitric Oxide (NO) Levels

NO levels were determined by direct measurement using the DAF-FM-DA assay. The cells were seeded in black 96-well plates at a rate of 8×10^4 cells. After treatment, 1 mM DAF-FM-DA stock solution was added to each assay well to obtain a final concentration of 10 μM for 30 min. Then, the intensity of the fluorescent signal was measured in a microplate fluorescence reader (FLx800 Fluorimeter, BioTek, Winooski, VT, USA) at a λ excitation of 495 nm and λ emission of 515 nm [32].

2.8. Apoptotic Assay with Caspase 3/7 Activity

EA.hy926 and SH-SY5Y cells (15×10^3 cells/well) were seeded in black 96-well plates. After treatment, Apo-ONE® (Promega (Madison, WI, USA)) Caspase-3/7 was prepared and used according to the manufacturer's instructions, for 60 min in the dark. Fluorescence (λ excitation/λ emission, 485/528 nm) was measured using a plate reader (FLx800, BioTek, Winooski, VT, USA). Data were evaluated as % of the control [32].

2.9. Antioxidant Defenses

2.9.1. Reduced Glutathione (GSH)

GSH content was evaluated by a fluorometric assay [29]. The method takes advantage of the reaction of GSH with o-phthalaldehyde at pH 8.0. Fluorescence was measured at 340 nm/460 nm (λ excitation/λ emission). Fluorescence data were interpolated from a standard curve of pure GSH (5–1000 ng).

2.9.2. Antioxidant Enzymes

Determination of GPx activity was based on the oxidation of GSH by GPx, using t-BOOH as a substrate, coupled to the disappearance rate of NADPH by GR [29]. GR activity was determined by following the decrease in absorbance due to the oxidation of NADPH utilized in the reduction in oxidized glutathione [29]. Protein concentration in the samples was measured by the Bradford reagent.

2.10. Molecular Assay by Real-Time PCR

After treatment, total RNA was obtained using the NucleoSpin®-RNA-Plus Kit (Macherey-Nagel, Germany) according to the manufacturer's instructions. The total RNA was quantified using a Nano-Spectrophotometer (Microdigital, Seoul, Korea), obtaining A260/A280 ratios > 1.9–2.1 < in all the samples. cDNA synthesis was obtained from 1 µg of total RNA by retro-transcription using the qPCRBIO cDNA Synthesis Kit (PCRBiosystems, Wayne, PA, USA). Finally, the cDNA was diluted in nuclease-free water (*v:v*, 1:10) and stored at −80 °C. Real-time PCR (qPCR) assays for SOD2, NRF2, NFκB1 (genes from oxidative stress-antioxidant), APAF1, BAX, and Caspase 3 (genes from cell death) were performed using a real-time PCR system (BioRad CFX, Hercules, CA, USA), using the ICgreen Mastermix (Nippon Genetics, Duren, Germany) according to the manufacturer's instructions. For qPCR, it was necessary to use primers with concentrations of 400 nM, and the thermocycling protocol was: 95 °C for 2 min, 40 cycles of 5 s at 95 °C and 30 s at 60 °C.

Forward and reverse primers were SOD2: 'CCACTGCTGGGGATTGATGT' 'CGTGGTTTACTTTTTGCAAGCC'; NRF2: 'CTGGTCATCGGAAAACCCCA' 'TCTGCAATTCTGAGCAGCCA'; NFκB1: 'TTTTCGACTACGCGGTGACA' 'GTTACCCAAGCGGTCCAGAA'; APAF1: 'TCTTCCAGTGGTAAAGATTCAGTT' 'CGGAGACGGTCTTTAGCCTC'; BAX: 'CCCCCGAGAGGTCTTTTTCC' 'CCTTGAGCACCAGTTTGCTG'; Caspase3: 'GTGGAGGCCGACTTCTTGTA' 'TTTCAGCATGGCACAAAGCG'. GAPDH was used as a housekeeping gene and extracting the efficiencies from the raw data using the LinRegPCR software [33].

2.11. Statistics

Analysis of the data obtained from the cell culture studies was performed with one-way ANOVA followed by Tukey's post hoc test, and the level of significance was $p < 0.05$ using the GraphPad Prism version 7.0 program (Boston, MA, USA).

3. Results

3.1. Chemical Analysis of D. Tortuosom Extract

Thirty compounds were identified in *D. tortuosum* based on their relative retention time, mass spectra and commercial standards. Table 1 shows the retention time (RT), molecular formula, accurate mass of the molecular ion $(M-H)^-$ after negative–positive ionization, and MS^2 fragments of the main compounds identified in *D. tortuosum* by Ultra High Perfomance Liquid Chromatography (UHPLC) UHPLC/MS (Supplementary Figure S1). The largest number of compounds identified were phenolic acids and flavonoids, and to a lesser extent compounds derived from linoleic acid, and others.

3.2. Cell Viability

The first parameter to ascertain the potential cyto-protection of the *D. tortuosum* extracts was cell viability after oxidative stress. *D. tortuosum* extracts at 1, 10, 25, 50, 100 and 200 µg/mL were tested for their cyto-protective capacity. As depicted in Figure 1, exposure to t-BOOH produced a significant decrease in cell viability of around 62% in EA.hy926 cells and 50% in SH-SY5Y cells. Increasing the concentrations of extract evoked a partial but significant dose-dependent recovery of cell viability in the two cell lines for both co- and pre-treatment. *D. tortuosum* extracts at 100 and 200 µg/mL significantly increased SH-SY5Y cell viability for both co- and pre-treatment (Figure 1B), and in the pre-treatment of EA.hy926, whereas concentrations above 50 µg/mL were necessary to significantly recover EA.hy926 cell viability with co-treatment (Figure 1A). The highest recovery of cell viability from the oxidative stress was observed at co- and pre-treatment with 200 μg/mL extract in SH-SY5Y cells (Figure 1B). Once the cyto-protective effect of some of the tested concentrations of the extract was ensured, the study of the redox status and antioxidant response was carried out.

Table 1. Identification of bioactive compounds detected by UHPLC/MS in the *D. tortuosum* extract.

Identified Compound	Retention Time (min)	Molecular Formula	MS-ESI−	MS2	MS-ESI+	MS2	Nominal Mass
HYDROXYCINNAMIC ACIDS AND HYDROXYCINNATES							
4-Coumaric acid	8.79	$C_9H_8O_3$	163	119, 93			164
CAROTENOIDS							
Loliolide	9.60	$C_{11}H_{16}O_3$			197	179, 161, 135, 133	196
FLAVONES							
6-C-xylosyl-8-C-galactosylapigenin	8.16	$C_{26}H_{28}O_{14}$	563	473, 443, 383	565	547, 529, 511	564
Vitexin-2″-O-rhamnoside	8.66	$C_{27}H_{32}O_{15}$	577	457, 413, 341, 323	579	433, 415, 367, 337	578
Vitexin	9.55	$C_{21}H_{20}O_{10}$	431	341, 311, 283, 268			432
Isovitexin	8.85	$C_{21}H_{20}O_{10}$	431	341,323, 311, 295	433	415, 397, 379, 367	432
Saponarin	8.10	$C_{27}H_{30}O_{15}$	593	473, 431, 311, 297	595	433, 415, 367, 337	594
luteolin-7-glucoside	9.03	$C_{21}H_{20}O_{11}$	447	357,285,256			448
Luteolin-6-C-glucoside	8.33	$C_{21}H_{20}O_{11}$	447	429, 357, 327	449	431, 383, 353, 329	448
6-C-arabinosyl-8-C-β-D-xylosylapigenin	8.66	$C_{25}H_{26}O_{13}$	533	443, 413, 383, 353	535	517, 499, 481, 469	534
FLAVANONES							
Naringenin	11.62	$C_{15}H_{12}O_5$	271	177, 151, 119			272
Prunin	9.53	$C_{21}H_{22}O_{10}$	433	271, 177, 151			434
8-Prenylnaringenin	13.79	$C_{20}H_{20}O_5$	339	245, 233, 219	341	285,183,165	340
2′,4′,5,7-Tetrahydroxy-8-prenylflavanone	13.32	$C_{20}H_{20}O_6$	355	193,161,149	357	301, 283	356
FLAVONOLS							
Hyperoside	8.95	$C_{21}H_{20}O_{12}$	463	300, 271, 255	465	303, 229	464
Isorhamnetin-3-O-glucoside	9.46	$C_{21}H_{20}O_{11}$	447	314, 285, 271, 243			448
PHENOLIC ACIDS							
3,4-Dihydroxybenzoic acid	3.69	$C_7H_6O_4$	153	109,108			154
2,5-Dihydroxybenzoic acid	5.88	$C_7H_6O_4$	153	123, 108, 95			154
6,8-di-C-glucosylapigenin	7.70	$C_{27}H_{30}O_{15}$	593	503, 473, 383	595	577, 559, 511	594
4-hydroxybenzaldehyde	7.92	$C_7H_6O_2$	121	108, 95, 93			122
12-hydroxyjasmonic acid glucoside	7.93	$C_{18}H_{28}O_9$	387	207, 163, 119			388
Uralenneoside	4.00	$C_{12}H_{14}O_8$	285	152, 108			286
[2-hydroxy-3-[3,4,5-trihydroxy-6-[[3,4,5-trihydroxy-6-(hydroxymethyl)oxan-2-yl]oxymethyl]oxan-2-yl]oxypropyl] hexadecanoate	14.87	$C_{31}H_{58}O_{14}$	699	653, 415			654
p-hydroxybenzoic acid	10.00	$C_7H_6O_3$	137	93			138
GLYCOSYLGLYCEROLS							
[2-hydroxy-3-[3,4,5-trihydroxy-6-[[3,4,5-trihydroxy-6-(hydroxymethyl)oxan-2-yl]oxymethyl]oxan-2-yl]oxypropyl] (9E,12E,15E)-octadeca-9,12,15-trienoate	13.90	$C_{33}H_{56}O_{14}$	721	675, 415			722
GLYCEROPHOSPHOCHOLINES							
1-Palmitoyl-sn-glycero-3-phosphocholine	16.10	$C_{24}H_{50}NO_7$	540	480, 255,152, 78	496	184, 125, 86	495
LINOLEIC ACIDS AND DERIVATIVES							
9,12,13-Trihydroxy-10,15-octadecadienoic acid	11.44	$C_{18}H_{32}O_5$	327	291, 229			328
9,12,13-Trihydroxy-10-octadecenoic acid	11.85	$C_{18}H_{34}O_5$	329	229, 211			330
9,10,13-Trihydroxy-10-octadecenoic acid	12.47	$C_{18}H_{34}O_5$	329	293, 211			330
9,10-DHOME or Leukotoxin Diol	14.38	$C_{18}H_{34}O_4$	313	277, 201			314

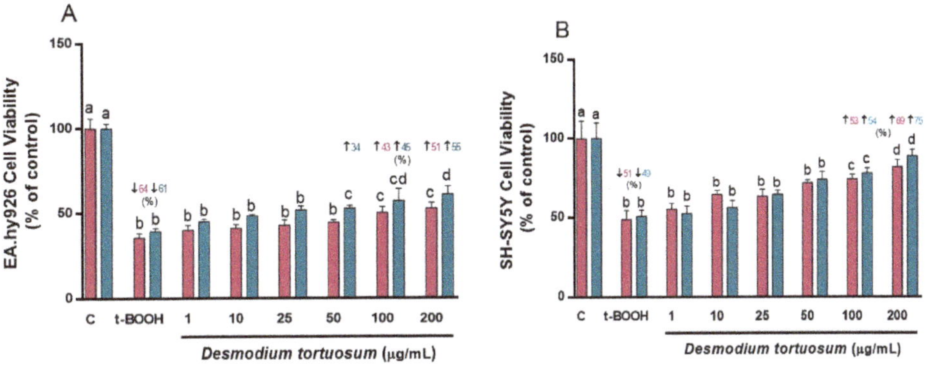

Figure 1. Cytoprotective effects by *D. tortuosum* extract on EA.hy926 (**A**) and SH-SY5Y (**B**) cells after co-treatment (red bars, ■) and pre-treatment (blue bars, ■) periods. Data are presented as % control, and as mean ± SEM of six independent experiments. [a,b,c,d] Different letters show significance between groups at $p < 0.05$. ↓ represents percentage decrease with respect to control, ↑ represents percentage increase with respect to t-BOOH.

3.3. Intracellular ROS Production

The addition of t-BOOH to cell cultures induced a remarkable increase in ROS generation of around 100% in EA.hy926 and around 40% in SH-SY5Y cells, very similar for both types of treatments, ensuring the reliability of the model for oxidative damage (Figure 2). The same extract concentrations (1–200 μg/mL) tested for cell viability were also assayed for their ROS-quenching capacity. Similar to the assay of cell viability, a significant dose-dependent reduction in ROS production was observed with increasing doses of the extract and, in the case of EA.hy926 cells, a decline in ROS almost to control pre-stress values was reached with the highest tested concentration of 200 μg/mL of *D. tortuosum* extract (Figure 2A). The two highest extract concentrations, 100 and 200 μg/mL, were also efficient in preventing ROS overproduction induced by t-BOOH in SH-SY5Y cells. The results clearly indicate that both co- and pre-treatment with extracts from *D. tortuosum* in the range of 50–200 μg/mL, significantly reduced ROS production induced by oxidative stress in these two cell lines (Figure 2).

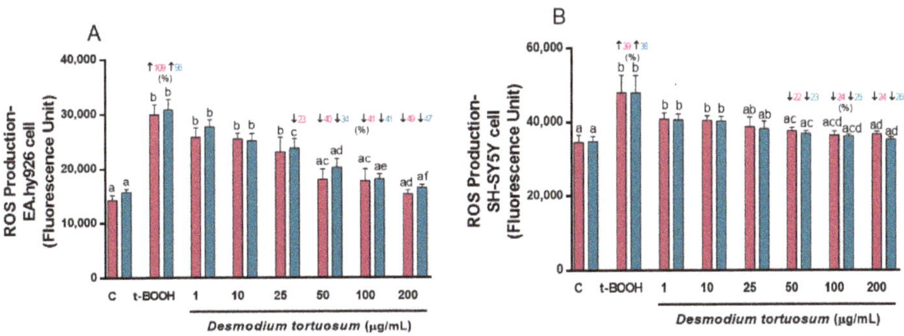

Figure 2. Antioxidant effects by *D. tortuosum* extract on ROS production induced by t-BOOH in EA.hy926 (**A**) and SH-SY5Y (**B**) cells after co-treatment (red bars, ■) and pre-treatment (blue bars, ■) periods. ROS production was measured as fluorescence units. Data are presented as mean ± SEM of six independent experiments. [a,b,c,d,e,f] Different letters show significance between groups at $p < 0.05$. ↑ represents percentage increase with respect to control, ↓ represents percentage decrease with respect to t-BOOH.

Since the highest protection against an oxidative challenge for both co- and pre-treatment approaches was obtained with the three highest concentrations of the extract, 50, 100 and 200 µg/mL, these three doses were tested for the rest of the oxidative stress biomarkers.

3.4. Determination NO Levels

When t-BOOH was added to cell cultures, an increase in NO levels was observed in EA.hy926 (around 50%) and SH-SY5Y cells (around 40%) (Figure 3), while the extracts of *D. tortuosum* in the highest doses prevented the effect of t-BOOH. In EA.hy296 cells, doses of 50, 100, and 200 µg/mL of *D. toruosum* reduced NO levels that were previously induced by t-BOOH; and in SH-SY5Y cells this effect could only be observed with the 100 and 200 µg/mL doses of *D. tortuosum* (Figure 3).

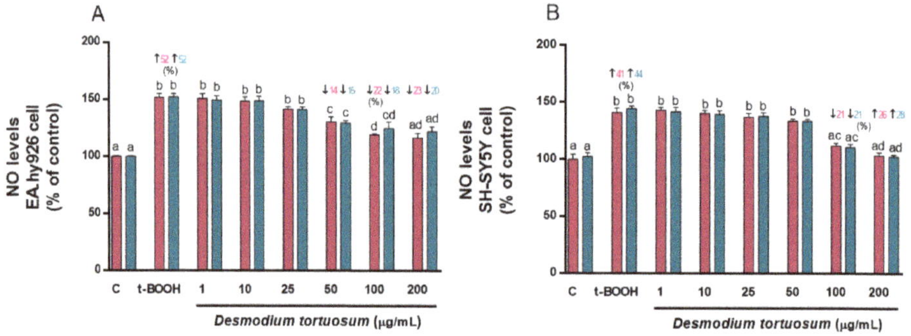

Figure 3. *D. tortuosum* extract effect on NO levels induced by t-BOOH in EA.hy926 (**A**) and SH-SY5Y (**B**) cells after co-treatment (red bars,) and pre-treatment (blue bars,) periods. NO levels are presented as % of control-change. Data represent the mean ± SEM of six independent experiments. [a,b,c,d] Different letters show significance between groups at $p < 0.05$. ↑ represents percentage increase with respect to control, ↓ represents percentage decrease with respect to t-BOOH.

3.5. Apoptotic Assay: Caspase 3/7 Activity

The capacity of *D. tortuosum* extract to reduce the apoptotic effects produced by t-BOOH was evaluated. The activity of the caspase 3/7 enzyme was increased by the effect of t-BOOH in EA.hy926 cells by 70% (co- and pre-treatment), and in SH-SY5Y cells by 95% (co- and pre-treatment) (Figure 4). *D. tortuosum* was able to reduce the apoptotic activity induced by t-BOOH in both cell lines from doses of 25, 50, 100, to 200 µg/mL; this effect was similar for both co- and pre-treatment with *D. tortuosum* (Figure 4).

3.6. Antioxidant Defenses

3.6.1. GSH Concentration

The concentration of GSH was determined as a reliable biomarker of the intracellular non-enzymatic antioxidant defenses. Since an acute treatment with pure compounds or extracts rich in natural antioxidants might evoke changes in the steady-state level of GSH that might affect its response to induced oxidative stress, both cell lines were subjected to a direct treatment for 22 h with the three extracts. Figure 5 shows that 50–200 µg/mL of extract did not induce any change in the basal concentration of GSH of EA.hy926 cells, whereas 100–200 µg/mL evoked a significant decrease in SH-SY5Y cells, indicating that neuronal-like cells were more sensitive to the presence of the extract concentrations for 22 h.

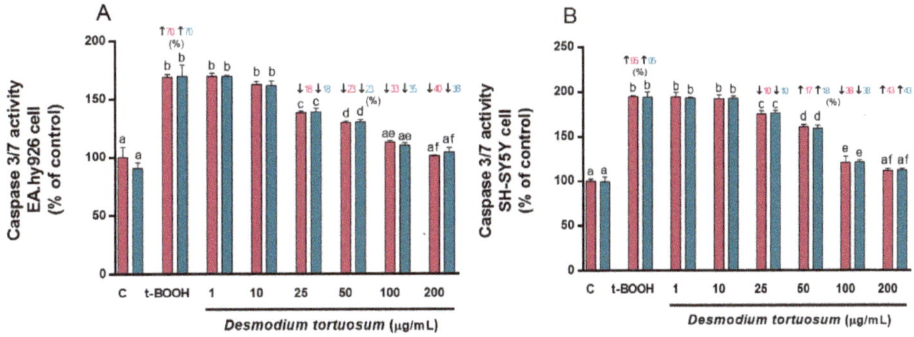

Figure 4. *D. tortuosum* extract effect on Caspase 3/7 activity induced by t-BOOH in EA.hy926 (**A**) and SH-SY5Y (**B**) cells after co-treatment (red bars,) and pre-treatment (blue bars,). Caspase 3/7 activity normalized as % of control-change, and the data are presented as mean ± SEM of six independent experiments. [a,b,c,d,e,f] Different letters show significance between groups at $p < 0.05$. ↑ represents percentage increase with respect to control, ↓ represents percentage decrease with respect to t-BOOH.

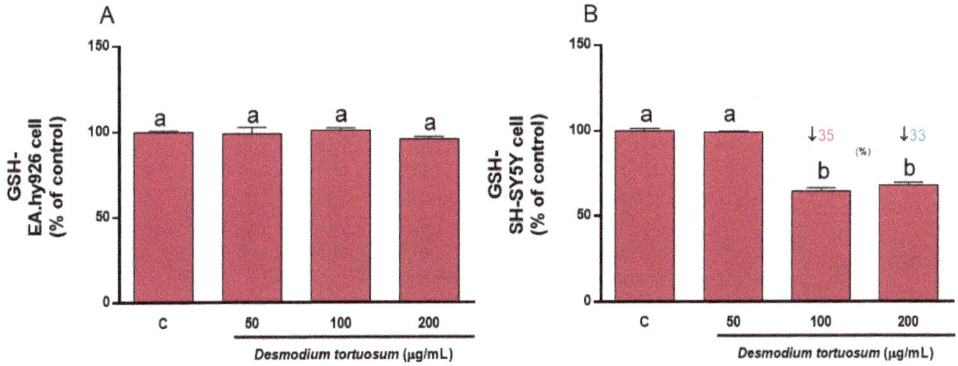

Figure 5. Direct effect of *D. tortuosum* extract on GSH levels in EA.hy926 (**A**) and SH-SY5Y (**B**) cells after a 22 h of treatment period. GSH levels was determined as % of control-change, and represent the mean ± SEM of four independent experiments. [a,b,c] Different letters show significance between groups at $p < 0.05$. ↓ represents percentage decrease with respect to control.

Treatment of EA.hy926 and SH-SY5Y cells with 100 µM t-BOOH for 22 h (co-treatment) or with 200 µM t-BOOH for 4 h (pre-treatment) significantly reduced the cell GSH concentration (Figure 6A,B). There was a slight but significant increase in GSH in EA.hy926 cells subjected to co-treatment with the extracts; however, a significant dose-dependent recovery was observed when endothelial cells were pre-treated with the extract concentrations prior to exposure to the potent pro-oxidant (Figure 6A, pre-treatment). In SH-SY5Y cells, none of the three tested doses of extract recovered the depleted GSH when they were added simultaneously (co-treatment) to the pro-oxidant (Figure 6B). Nevertheless, when neuroblastoma cells were pre-treated with the extract prior to the stress, all three concentrations were capable of fully preventing the GSH decrease (Figure 6B), indicating a chemo-preventive effect of *D. tortuosum* on neuronal-like damage.

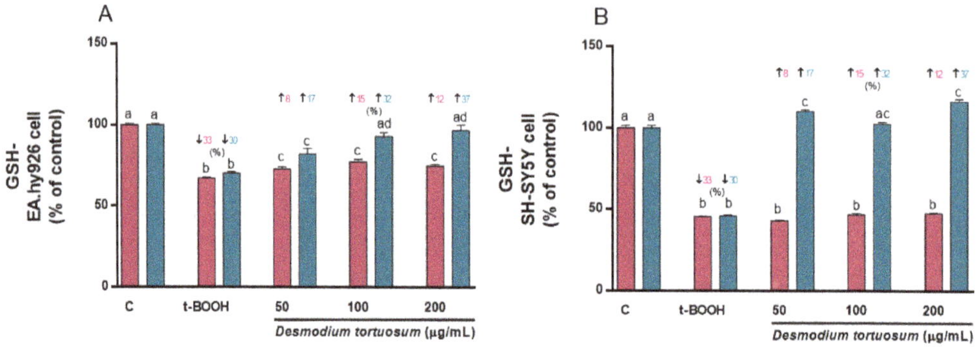

Figure 6. Effect of *D. tortuosum* extract on GSH levels altered by t-BOOH in EA.hy926 (**A**) and SH-SY5Y (**B**) cells after co-treatment (red bars, ■) and pre-treatment (blue bars, ■). GSH levels were determined as % of control-change, and represent the mean ± SEM of four independent experiments. [a,b,c,d] Different letters show significance between groups at $p < 0.05$. ↓ represents percentage decrease with respect to control, ↑ represents percentage increase with respect to t-BOOH.

The results indicate that a pre-treatment with the extracts of *D. tortuosum* was effective in reducing cell death and quenching ROS and NO over-production, as well as diminishing caspase 3/7 activity, and significantly preventing the intense depletion of GSH induced by oxidative stress in both EA.hy926 and SH-SY5Y cells.

3.6.2. Antioxidant Enzymes

As the main antioxidant enzymes, evaluation of GPx and GR activity guarantees an archetypal response of the antioxidant system to a stressful challenge. As in the case of GSH, acute treatment with pure phytochemicals or natural extracts rich in antioxidants might evoke changes in the basal pre-stress activity of GPx and GR that might affect its further response to induced oxidative stress; thus, both cell lines were firstly subjected to a direct treatment for 22 h with the three extract doses. A slight but significant decrease in GPx activity was observed after treatment with 50–100 µg/mL extract in EA.hy926 cells (Figure 7A); similarly, a reduced activity of GR was found when EA.hy926 cells were treated with 100–200 µg/mL (Figure 7B). Direct treatment of SH-SY5Y cells with *D. tortuosum* extract for 22 h did not evoke any change in GPx activity (Figure 7C), whereas all three tested doses (50–200 µg/mL) induced a significant decrease in GR activity (Figure 7D), similar to what was observed in endothelial cells.

Treatment of EA.hy926 cells with 100 µM t-BOOH for 22 h evoked a 100% increase in GPx activity (Figure 8A), whereas treatment with 200 µM t-BOOH for 4 h provoked a 50% enhancement of the enzyme's activity (Figure 8A). This result confirms the expected response of GPx to face the over-production of ROS induced by t-BOOH in EA.hy926 cells. In agreement with the GPx increase, GR activity was also stimulated by 100 µM of the pro-oxidant for 22 h to more than 100% (Figure 8B), and around two-fold by 200 µM of t-BOOH for 4 h (Figure 8B). This result of GR ensures appropriate recycling of GSSG to GSH for re-utilization. Remarkably, both co- and pre-treatment of EA.hy926 cells with 100–200 µg/mL of the extract evoked a significant reduction in the enhanced GPx activity, that was dose-dependent in the co-treatment, and reverted to basal activity at the end of the stress period (Figure 8A). Similarly, a dose-dependent rescue of the altered GR activity was observed when EA.hy926 cells were co- or pre-treated with the three extract concentrations (Figure 8B).

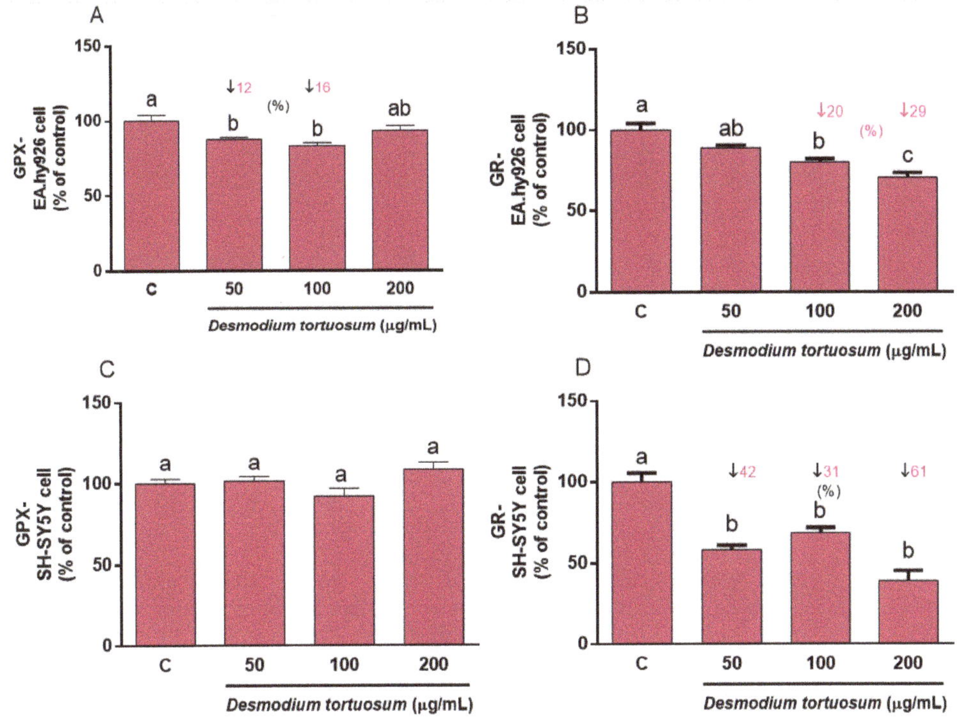

Figure 7. Direct effect of *D. tortuosum* extract on GPX and GR activities in EA.hy926 (**A**,**B**) and SH-SY5Y (**C**,**D**) cells after a 22 h of treatment period. GPX and GR activities were determined as % of control-change, and represent the mean ± SEM of four independent experiments. [a,b,c] Different letters show significance between groups at $p < 0.05$. ↓ represents percentage decrease with respect to control.

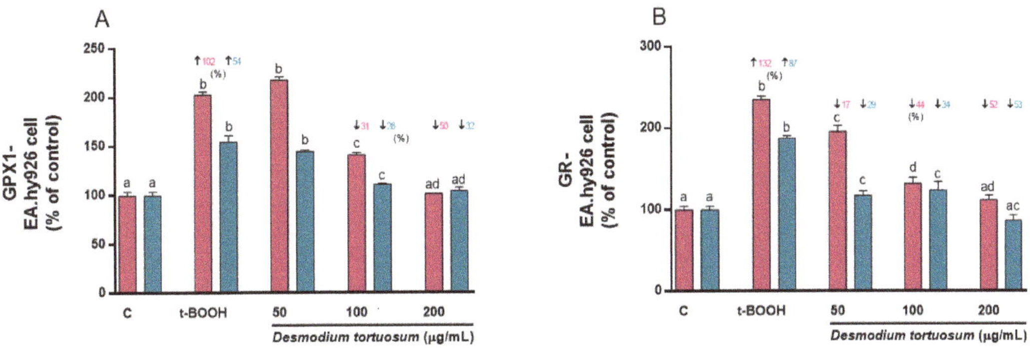

Figure 8. Effect of *D. tortuosum* extract on GPx (**A**) and GR (**B**) activity altered by t-BOOH in EA.hy926 cells after co-treatment (red bars,) and pre-treatment (blue bars,) periods. Data represent the mean ± SEM of three independent experiments. [a,b,c,d] Different letters show significance between groups at $p < 0.05$. ↑ represents percentage increase with respect to control, ↓ represents percentage decrease with respect to t-BOOH.

Treatment of SH-SY5Y cells with 100 μM t-BOOH for 22 h or with 200 μM t-BOOH for 4 h provoked a significant enhancement of GPx and GR activity (Figure 9), confirming the predictable response of both enzymes to face the over-production of ROS induced by t-BOOH and the suitable reprocessing of GSSG to GSH for re-utilization in SH-SY5Y cells. Co-treatment with the extract did not evoke a significant rescue of the enhanced GPx activity, whereas pre-treatment of the SH-SY5Y cells with the three doses of *D. tortuosum* extract remarkably reverted the stimulated GPx activity to the control pre-stress values (Figure 9). Unexpectedly, no significant changes in GR activity were found in the SH-SY5Y cells treated with 100 μM t-BOOH (co-treatment) or 200 μM t-BOOH (pre-treatment). As GR activity was very low in all conditions, the assay was not sensitive enough to detect any measurable changes in enzyme activity (data not shown). Overall, EA.hy926 cells were more robust and responsive to stressful conditions than SH-SY5Y cells but, in general, both co- and pre-treatment of endothelial and neuronal-like cells with the *D. tortuosum* extract significantly prevented the permanent enhancement of both antioxidant enzyme activities, especially GPx (Figures 8 and 9).

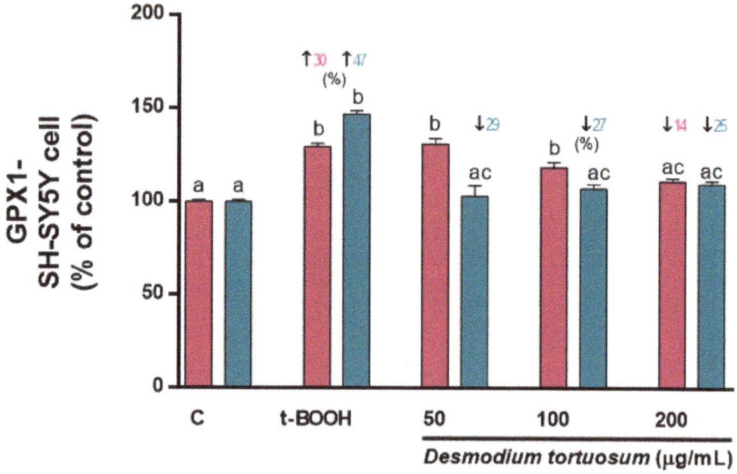

Figure 9. Effect of *D. tortuosum* extract on GPx altered by t-BOOH on SH-SY5Y cells after co-treatment (red bars, ■) and pre-treatment (blue bars, ■) periods. Data represent the mean ± SEM of three independent experiments. [a,b,c] Different letters show significance between groups at $p < 0.05$. ↑ represents percentage increase with respect to control, ↓ represents percentage decrease with respect to t-BOOH.

3.7. Gene Expression of Oxidative–Antioxidative and Cell Death Biomarkers

Genes related to oxidative–antioxidative (SOD2, NRF2 and NFκB1) and cell death (APAF1, BAX and Caspase3) proteins were evaluated for the effect of t-BOOH or *D. tortuosum* extract (co- and pre-treatment) on EA.hy926 and SH -SY5Y cells.

Molecular expression of SOD2 was decreased by around 45% due to the effect of t-BOOH in both cell types, but this effect was reversed by the effect of the highest concentration of the *D. tortuosum* extract (200 μg/mL) in EA.hy926 and in SH-SY5Y cells (Figure 10A,D). Likewise, NRF2 gene levels decreased by around 50% and 55% due to the effect of t-BOOH in EA.hy926 and SH-SY5Y cells, respectively; while the 100 and 200 μg/mL doses, in both cell types, had the ability to significantly reduce the effect of t-BOOH (Figure 10B,E). Moreover, t-BOOH was able to increase NFκB1 expression levels above 250% and 175% in EA.hy926 and SH-SY5Y cells, respectively; and the *D. tortuosum* extract at all its doses reduced this effect in EA.hy926 cells, while only at 100 and 200 μg/mL of *D. tortuosum* in SH-SY5Y cells (Figure 10C,F).

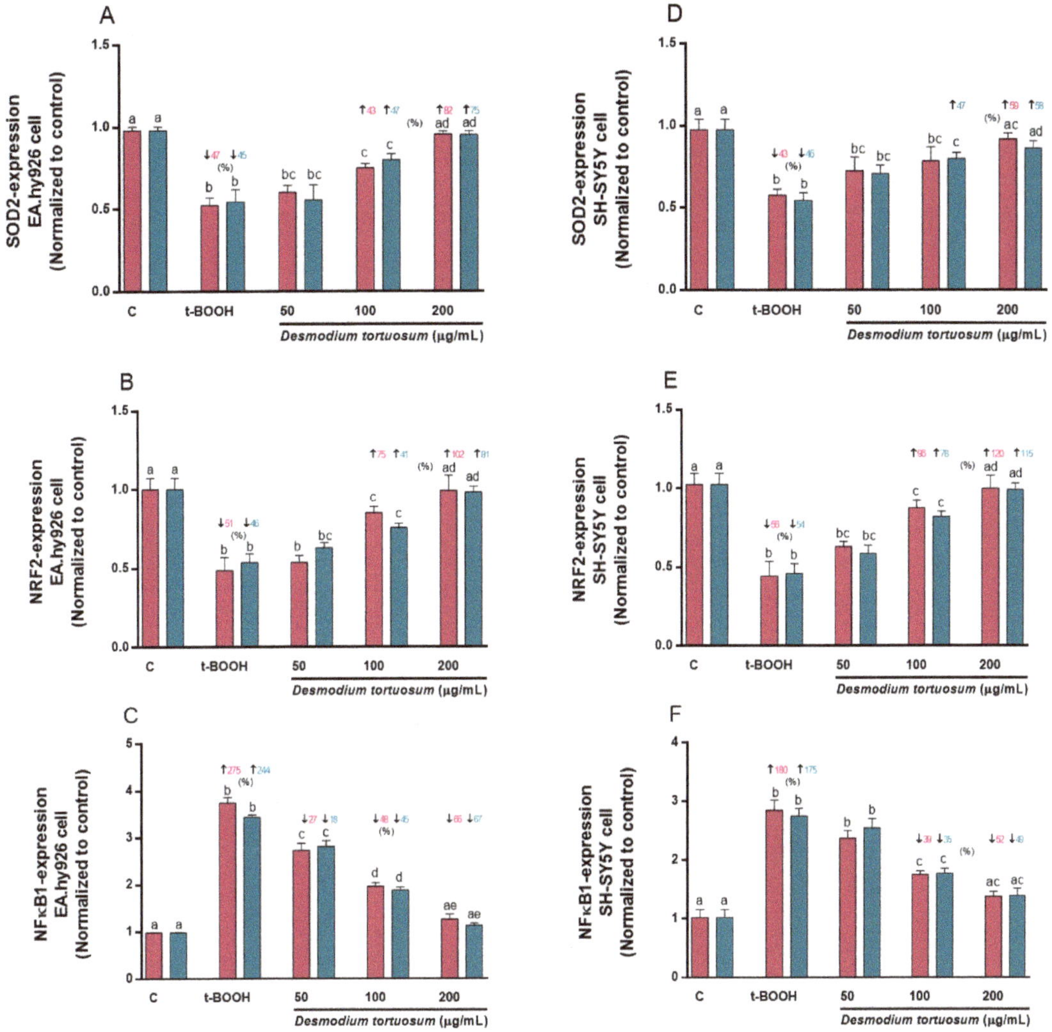

Figure 10. Effect of *D. tortuosum* extract (50, 100, and 200 µM) on the expression of oxidative-antioxidative genes (SOD2, NRF2, NFκB1) in EA.hy926 (**A–C**) and SH-SY5Y (**D–F**) cells after 24 h co-treatment (red bars, ■) and pre-treatment (blue bars, ■) periods. The gene expression was determined as control-normalized value, and represented as the mean ± SEM of four independent experiments. [a,b,c,d] Different letters show significance between groups at $p < 0.05$. ↓ represents percentage decrease or ↑ represents percentage increase with respect to control or with respect to t-BOOH.

The molecular expression of APAF1 was significantly increased above 140% and 90% by the effect of t-BOOH in EA.hy926 and SH-SY5Y cells, respectively; but this effect was reversed by the effect of the concentrations of 50 (only pre-treatment in EA.hy926 cells), 100 and 200 µg/mL of *D. tortuosum* in EA.hy926 and SH-SY5Y cells, (Figure 11A,D). Likewise, BAX mRNA expression was significantly increased by the effect of t-BOOH by 215% (co-treatment) and 179% (pre-treatment) in the EA.hy926 cells, and around 170% in the SH-SY5Y cells; while all doses of *D. tortuosum*, in both cell types had the ability to significantly and dose-dependently reduce the effect of t-BOOH (Figure 11B,E). It was also observed that t-BOOH induced Caspase3 expression above 89% in both cell types, which was reduced by the effect of the *D. tortuosum* extract (50, 100 and 200 µg/mL) in both cell types (Figure 11C,F).

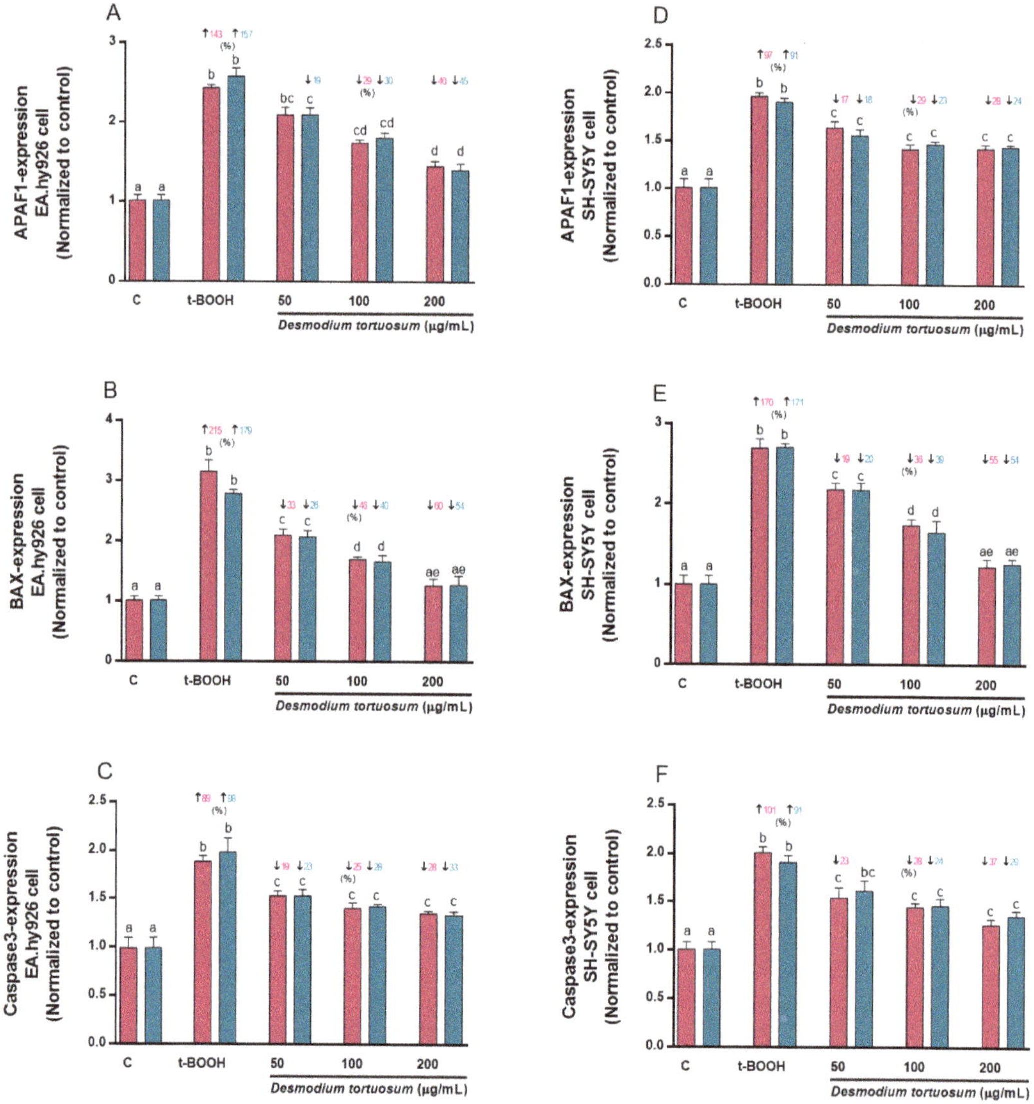

Figure 11. Effect of *D. tortuosum* extract (50, 100, and 200 µM) on the expression of cell death genes (APAF-1, BAX, Caspase3) in EA.hy926 (**A–C**) and SH-SY5Y (**D–F**) cells after 24 h co-treatment (red bars, ■) and pre-treatment (blue bars, ■) periods. The gene expression was determined as control-normalized values, and is represented the mean ± SEM of four independent experiments. [a,b,c,d,e] Different letters show significance between groups at $p < 0.05$. ↓ represents percentage decrease or ↑ represent percentage increase with respect to control or with respect to t-BOOH.

4. Discussion

In this study, an aqueous extract of *D. tortuosum* was prepared and its main phenolic compounds were characterized by UHPLC/MS, showing phenolic acids, flavonoids such as flavones, flavanones and flavanols, carotenoids and others antioxidant compounds (Table 1) as the main compounds with bioactive potential. The extract showed significant antioxidant capacity in vitro and effects in endothelial and neuronal-like cell culture that include the regulation on ROS production and NO concentration, caspase 3/7 activity and a remarkable anti-oxidative stress protection and molecular regulation of biomarkers of oxidative stress and cell death. All these effects support the use of the plant since ancient times in traditional medicine.

The range of doses of the D. *tortuosum* extract to test the anti-oxidative stress potential was selected according to previous data from other studies working with plant extracts, foodstuff and juices. A concentration of 35 μM of flavanol epicatechin was found in rat serum 1 h after oral administration of 172 μmol epicatechin per Kg of body weight [34]. Similarly, levels of 30–40 μM of cranberry phytochemicals have been detected in plasma after the intake of cranberry juice [35,36]. Hence, the range of *D. tortuosum* extract concentrations tested is not far from realistic; in fact, in previous works we have report the protective activity of *Vochysia rufa* (0.5–100 μg/mL) [21], *Silybum marianum* (5–25 μg/mL) [20], and cocoa extract (2.5–20 μg/mL) [23] in EA.hy926 cells. Additionally, we have recently reported that doses of 5–25 μg/mL of a *Sambucus nigra* extract [28] and 25–200 μg/mL of an aqueous extract of cocoa phenolic compounds [27] protect SH-SY5Y cells from oxidative stress.

Previous results have indicated that the treatment of EA.hy926 cells with t-BOOH is an excellent oxidative model in cell culture [20,23,37]. Similarly, very recently we have also established a oxidative model in SH-SY5Y cells by a comparable treatment with the same pro-oxidant, t-BOOH [27,28]. As most organic peroxides, t-BOOH decompose to other alkoxyl and peroxyl radicals in a reaction assisted by metal ions that can generate ROS, including H_2O_2 [29]. If the over-production of ROS is long-lasting, damage to macromolecules, proteins, lipids and DNA, might be excessive and irreversibly endanger cell viability, as observed in t-BOOH-treated cells. However, under these stressful conditions, significant inhibition of t-BOOH-induced cytotoxicity when both EA.hy926 and SH-SY5Y cells were pre- or co-treated with plant extracts at realistic doses for 20 h indicated that the integrity of the stressed cells was remarkably protected against the potent oxidative challenge. The relevant amount of the bioactive phenolics in the extract was effective enough for partial but significant dose-dependent cell protection, although the different responses to co- and pre-treatment suggest a differential sensitivity of the two cell types to the extracts in stressful conditions; pre-treatment being more effective in both cell lines. As reported above, a similar cytoprotective capacity has been reported with other phenolic extracts in both cell lines, EA.hy926 [20,21,23] and SH-SY5Y cells [27,28].

Assessment of ROS generation is a reliable index of the redox status as well as the oxidative damage to living cells [29]. The addition of t-BOOH to either cell line in both, co- and pre-treatment conditions, evoked a significant increase in ROS generation that might be the main cause for the increased cell death. The significant dose-dependent reduction in ROS induced by t-BOOH observed with co- and pre-treatment with extract in both cell lines unequivocally support the antioxidant nature of the phenolic components and could be a primary explanation for the reduced oxidative stress and subsequent cell protection. Interestingly, a comparable ROS-quenching capacity has been reported not only in both EA.hy926 and SH-SY5Y cells as referred above [20,21,23,27,28], but also in cultured hepatic cells [38–42], clearly indicating that this chemo-protective effect is not specific of a particular cell type or tissue but an systemic anti-oxidative stress capacity of natural antioxidants.

After administering t-BOOH to both cell cultures, NO levels increased significantly. This similar effect of increasing NO in cell cultures by cytotoxic substances has been previously reported [32,43]. The NO concentrations were reduced in a dose-dependent manner by the effect of *D. tortuosum* from 50 μg/mL in both co- and pre-treatment in EA.hy926 cells, and from 100 μg/mL in co- and pre-treatment in SH-SY5Y cells. NO is a signaling molecule that plays an important role in prolonging inflammation and immune responses. *D. tortuosum* and other extracts could act as NO scavengers or inhibitors of its production, through the inhibition of NO activity, inducible nitric oxide synthase (iNO) or through free radical scavenging activities [44,45].

Many forms of cellular stress can lead to cell death, through intracellular stress or mitochondrial dysfunction [46]. In this study it was observed that t-BOOH is capable of inducing oxidative stress by increasing the levels of ROS and NO. This effect may have produced an increase in the caspase 3/7 activity, enzymes involved in cell death and evaluated in this study. It is also known that under conditions of oxidative stress, high levels of ROS (superoxide, hydroxyl radical and hydrogen peroxide) are generated, which induce cell damage and cell death [47]. This cell death often involves the induction of apoptosis through the activation of caspase enzymes [48]. In this study, the high caspase 3/7 activity found in EA.hy926 and SH-SY5Y cells was reduced in a dose-dependent manner by the effect of the *D. tortuosum* extract starting at 25 μg/mL for both pre- and co-treatment. Similar effects have been reported in other studies, where they observed that natural compounds with a high phenolic content reduced the activity of caspase enzymes in EA.hy925 [49,50] and SH-SY5Y cells [51,52].

The best biomarker of the cell redox status is the concentration of GSH; thus, GSH depletion or reduction indicates increased intracellular oxidation and precarious redox status, whereas a balanced GSH concentration positions the cell in an advantageous situation to face potential oxidative stress [53]. Concentration of GSH is tightly regulated within the cell and direct exposure to plant extracts at non-toxic concentrations does not usually evoke significant changes in basal GSH levels [20–22,29]. Thus, the decline in GSH concentrations observed in SH-SY5Y cells treated with 100–200 μg/mL extract may be a consequence from the direct conjugation of some extract compounds to GSH, a fact previously reported for flavonoids, such as catechin [54] and epigallocatechin-3-gallate [55]. This direct conjugation might be only relevant in the case of direct treatment with the highest concentrations tested because of the larger amount of flavonoids and other antioxidants in the extract that are not consumed to face the oxidative stress. On the other hand, the decreased GSH concentration induced by t-BOOH suggests a state of oxidative stress that might result in irreparable oxidative damage to macromolecules: lipids, proteins and nucleic acids. This hazardous situation was dose-dependently prevented by the pre-treatment in EA.hy926 cells and completely prevented by pre-treatment with all three doses in the SH-SY5Y cells. The results suggest that the condition of co-treatment, with the continuous presence of a strong pro-oxidant during the whole assay, was too severe for the extract compounds to recover the consumed GSH to suitable levels in the cells. In any case, the response in the pre-treatment assay is in agreement with reports of other plant extracts rich in phenolic antioxidants such as *Silybum marianum* [20], *Vochysia rufa* [21] and green coffee [22] in EA.hy926 cells, as well as cocoa [27] and *Sambucus nigra* [28] in SH-SY5Y cells. This outcome is essential since preserving GSH concentrations above an appropriate threshold while struggling against a stressful situation represents a crucial advantage for cell survival.

Activities of GPx and GR enzymes are essential to balance the cellular redox state. GPx induces the reduction in cell-damaging peroxide species, along with the conversion of GSH to oxidized glutathione [29,53], whereas GR recycles oxidized glutathione back to GSH [29,53], recovering the steady state of cellular GSH. The increase in GPx and GR activities observed after the noted treatments with t-BOOH unambiguously indicates a positive response of the cell's defense system to face oxidative stress [29,38–42]. Consequently, during or after induced oxidative stress the antioxidant defense system of the cells pre-treated with *D. tortuosum* extract rapidly returned to a steady-state condition minimizing cell damage and, thus allowing the cell to deal with further oxidative insults in conditions that are more favorable. We have previously demonstrated a similar chemo-protective response of antioxidant defense enzymes by other antioxidant extracts in the same two cell lines [20,21,23,27,28].

Oxidative stress is one mechanisms through which cells respond by activating cell survival or cell death pathways. Initially, cell tries to respond to oxidative damage in a positive way, but if the damage too extensive the cell death mechanism will be activated. In other words, the alteration of the oxidant–antioxidant mechanisms will activate cell survival or death mechanisms in the cells [56]. In this study, we observed that the molecular expression of the antioxidant biomarkers SOD2 and NRF2, and the oxidant biomarker NFκB1 are altered by the effect of t-BOOH and are completely restored with a 200 μg/mL dose of *D. tortuosum*. Likewise, it was observed that molecular biomarkers of cell death, such as APAF1, BAX and Caspase3, were overexpressed by the effect of t-BOOH, and this effect was reversed by the various concentrations of *D. tortuosum* in the EA.hy926 and SH-SY5Y cells. Therefore, we can conclude that the *D. tortuosum* extract has an antioxidant cytoprotective effect and a direct or indirect antiapoptotic effect (through antioxidant mechanisms). In other studies where natural extracts have been used, this antioxidant–antiapoptotic association effect has also been observed. This is the case of the increase in SOD2 levels and the decrease in the expression of BAX and Caspase3 proteins due to the effect of the *Scrophularia buergeriana* extract in the SH-SY5Y cells [57]. Furthermore, the genus *Astragalus* was reported to increase SOD levels and decrease NFκB activity in EA.hy926 cells [58]. Likewise, in cell cultures, the ability of natural extracts to increase the levels of SOD and NRF2, two molecules involved in cell antioxidant activity, has been observed [59,60].

In general, the response of EA.hy926 cells to stress was more constant and robust than that of the SH-SY5Y cells and the protective effect of the *D. tortuosum* extract was more efficient as a pre-treatment versus co-treatment. Overall data indicate that, under chemically induced oxidative stress, treatment of endothelial and neuronal-like cells with the *D. tortuosum* extract rich in antioxidant compounds reduces ROS production, NO generation, caspase 3/7 activity, and limits GSH depletion resulting in a restricted requirement for antioxidant enzyme activity. Likewise, it was observed that there is an important association between the expression of antioxidant molecules and the decrease in molecules that induce cell death. This inclusive biochemical and molecular response evoked by the bioactive extract could systematically explain the observed endothelial and neuronal-like cyto-protection.

5. Conclusions

The extract of *D. tortuosum* is rich in phenolic compounds with antioxidant capacity. This work demonstrates that the doses (50, 100 and 200 μg/mL) of the extract contributed to the cytoprotection of EA.hy926 endothelial and SH-SY5Y neuronal cells subjected to oxidative damage by t-BOOH, through the regulation of ROS, NO, GSH, antioxidant enzyme activity, caspase3/7 activity, and molecular biomarkers from oxidative stress and cell death. Taking into account all the data, it can be concluded that the treatment of EA.hy926 and SH-SY5Y cells with the *D. tortuosum* extract (from 50 μg/mL) practically normalizes (200 μg/mL) the antioxidant defense system of the cells after oxidative stress. More studies are needed to evaluate the mechanism of action and biological activity in vivo of *D. tortuosum*, before categorically concluding the potential protective effect of this botanic extract in animals and humans.

Supplementary Materials: The following supporting information can be downloaded at: https://www.mdpi.com/article/10.3390/nu15030746/s1, Figure S1: TIC chromatograms of the *D. tortuosum* sample analyzed in negative (top) ESI mode and in positive (bottom) ESI mode.

Author Contributions: Conceptualization, J.-L.R., Z.-M.C. and L.G.; methodology, J.-L.R., M.M.-B., L.I.-R., M.Q.-S. and L.G.; software, M.R.-G., M.M.-B. and O.P.; validation, J.-L.R., and L.G.; formal analysis, M.Q.-S. and J.-L.R.; investigation, P.B., L.I.-R. and M.S.F.-A.; resources, Z.-M.C.; data curation, J.-L.R., M.R.-G. and L.G.; writing—original draft preparation, L.G. and J.-L.R.; writing—review and editing, L.G.; visualization, P.B., J.-L.R. and L.G.; supervision, L.G., Z.-M.C. and J.-L.R.; project administration, Z.-M.C. and L.G.; funding acquisition, Z.-M.C. and L.G. All authors have read and agreed to the published version of the manuscript.

Funding: The project of Funding number "PCONFIGI A21081221" was funded by the Universidad Nacional Mayor de San Marcos, Lima-Perú.

Institutional Review Board Statement: Not applicable.

Informed Consent Statement: Not applicable.

Data Availability Statement: Not applicable.

Acknowledgments: Patricio Aller, CIB, CSIC, Spain; Ignacio Torres Alemán, Instituto Cajal, CSIC, Madrid, Spain, and Carlos Guillén, School of Pharmacy, University Complutense, Madrid, Spain.

Conflicts of Interest: The authors declare no conflict of interest.

References

1. Ionescu-Tucker, A.; Cotman, C.W. Emerging roles of oxidative stress in brain aging and Alzheimer's disease. *Neurobiol. Aging.* **2021**, *107*, 86–95. [CrossRef] [PubMed]
2. Favero, G.; Paganelli, C.; Buffoli, B.; Rodella, L.F.; Rezzani, R. Endothelium and its alterations in cardiovascular diseases: Life style intervention. *BioMed Res. Int.* **2014**, *2014*, 801896. [CrossRef] [PubMed]
3. Polovina, M.M.; Potpara, T.S. Endothelial dysfunction in metabolic and vascular disorders. *Postgrad. Med.* **2014**, *126*, 38–53. [CrossRef] [PubMed]
4. Paneni, F.; Beckman, J.A.; Creager, M.A.; Cosentino, F. Diabetes and vascular disease: Pathophysiology, clinical consequences and medical therapy: Part I. *Eur. Heart J.* **2013**, *34*, 2436–2443. [CrossRef] [PubMed]
5. Ríos, J.L.; Francini, F.; Schinella, G.R. Natural products for the treatment of type 2 Diabetes mellitus. *Planta Med.* **2015**, *81*, 975–994. [CrossRef] [PubMed]
6. González, J.; Valls, N.; Brito, R.; Rodrigo, R. Essential hypertension and oxidative stress: New insights. *World J. Cardiol.* **2014**, *6*, 353–366. [CrossRef] [PubMed]
7. Song, P.; Zou, M.H. Redox regulation of endothelial cell fate. *Cell. Mol. Life Sci.* **2014**, *71*, 3219–3239. [CrossRef]
8. Niedzielska, E.; Smaga, I.; Gawlik, M.; Moniczewski, A.; Stankowicz, P.; Pera, J.; Filip, M. Oxidative stress in neurodegenerative diseases. *Mol. Neurobiol.* **2016**, *53*, 4094–4125. [CrossRef]
9. Rivas, F.; Poblete-Aro, C.; Pando, M.E.; Allel, M.J.; Fernandez, V.; Soto, A.; Nova, P.; Garcia-Diaz, D. Effects of polyphenols in aging and neurodegeneration associated with oxidative stress. *Curr. Med. Chem.* **2022**, *29*, 1045–1060. [CrossRef]
10. Martín, M.A.; Goya, L.; Ramos, S. Protective effects of tea, red wine and cocoa in diabetes. Evidence from human studies. *Food Chem. Toxicol.* **2017**, *109*, 302–314. [CrossRef]
11. Romacho, T.; Valencia, I.; Ramos-González, M.; Vallejo, S.; López-Esteban, M.; Lorenzo, O.; Cannata, P.; Romero, A.; San Hipólito-Luengo, A.; Gómez-Cerezo, J.F.; et al. Visfatin/eNampt induces endothelial dysfunction in vivo: A role for Toll-Like Receptor 4 and NLRP3 inflammasome. *Sci. Rep.* **2020**, *10*, 5386. [CrossRef] [PubMed]
12. Martin, M.A.; Ramos, S. Impact of cocoa flavanols on human health. *Food Chem. Toxicol.* **2021**, *151*, 112121. [CrossRef] [PubMed]
13. Kris-Etherton, P.M.; Keen, C.L. Evidence that the antioxidant flavonoids in tea and cocoa are beneficial for cardiovascular health. *Curr. Opin. Lipidol.* **2002**, *13*, 41–49. [CrossRef]
14. Fisher, N.D.; Hughes, M.; Gerhard-Herman, M.; Hollenberg, N.K. Flavanol rich cocoa induces nitric oxide-dependent vasodilation in healthy humans. *J. Hypertens.* **2003**, *1*, 2281–2286. [CrossRef]
15. Heiss, C.; Dejam, A.; Kleinbongard, P.; Schewe, T.; Sies, H.; Kelm, M. Vascular effects of cocoa rich in flavan-3-ols. *J. Am. Med. Assoc.* **2003**, *290*, 1030–1031. [CrossRef] [PubMed]
16. Ciumarnean, L.; Milaciu, M.V.; Runcan, O.; Vesa, S.C.; Rachis, A.L.; Negrean, V.; Perné, M.-G.; Donca, V.I.; Alexescu, T.-G.; Para, I. The effects of flavonoids in cardiovascular diseases. *Molecules* **2020**, *25*, 4320. [CrossRef]
17. Ebaditabar, M.; Djafarian, K.; Saeidifard, N.; Shab-Bidar, S. Effect of dark chocolate on flow-mediated dilatation: Systematic review, meta-analysis, and dose–response analysis of randomized controlled trials. *Clin. Nutr. ESPEN* **2020**, *36*, 17–27. [CrossRef]
18. Bravo, L.; Mateos, R.; Sarriá, B.; Baeza, G.; Lecumberri, E.; Ramos, S.; Goya, L. Hypocholesterolaemic and antioxidant effects of yerba mate (Ilex paraguariensis) in high-cholesterol fed rats. *Fitoterapia* **2014**, *92*, 219–229. [CrossRef]
19. Gutiérrez-Del-Río, I.; López-Ibáñez, S.; Magadán-Corpas, P.; Fernández-Calleja, L.; Pérez-Valero, Á.; Tuñón-Granda, M.; Miguélez, E.M.; Villar, C.J.; Lombó, F. Terpenoids and polyphenols as natural antioxidant agents in food preservation. *Antioxidants* **2021**, *10*, 1264. [CrossRef]
20. Palomino, O.M.; Gouveia, N.M.; Ramos, S.; Martín, M.A.; Goya, L. Protective effect of Silybum marianum on endothelial cells submitted to high glucose concentration. *Planta Med.* **2017**, *83*, 97–103. [CrossRef]
21. de Gouveia, N.M.; Ramos, S.; Martín, M.A.; Spindola, F.; Goya, L.; Palomino, O.M. Vochysia rufa stem bark extract protects endothelial cells against high glucose damage. *Medicines* **2017**, *4*, 9. [CrossRef]

22. Wang, S.-L.; Sarriá, B.; Mateos, R.; Goya, L.; Bravo, L. TNF-α induced inflammation in human EA.hy926 endothelial cells is prevented by yerba mate and green coffee extracts, their main hydroxycinnamic acids, and microbial metabolites. *Int. J. Food Sci. Nutr.* **2019**, *70*, 267–284. [CrossRef]
23. Martins, T.F.; Palomino, O.M.; Álvarez-Cilleros, D.; Ramos, S.; Goya, L. Cocoa flavanols protect human endothelial cells from oxidative stress. *Plant. Food Hum. Nutr.* **2020**, *75*, 161–168. [CrossRef]
24. Martín, M.A.; Goya, L.; de Pascual-Teresa, S. Effect of Cocoa and Cocoa Products on Cognitive Performance in Young Adults. *Nutrients* **2020**, *12*, 3691. [CrossRef]
25. Goya, L.; Román, R.S.; de Pascual-Teresa, S. Polyphenols' effect on cerebrovascular health. *Curr. Med. Chem.* **2022**, *29*, 1029–1044. [CrossRef]
26. Strother, L.; Miles, G.B.; Holiday, A.R.; Cheng, Y.; Doherty, G.H. Long-term culture of SH-SY5Y neuroblastoma cells in the absence of neurotrophins: A novel model of neuronal ageing. *J. Neurosci. Methods* **2021**, *362*, 109301. [CrossRef]
27. Carballeda Sangiao, N.; Chamorro, S.; de Pascual-Teresa, S.; Goya, L. Aqueous extract of cocoa phenolic compounds protects differentiated neuroblastoma SH-SY5Y cells from oxidative stress. *Biomolecules* **2021**, *11*, 1266. [CrossRef]
28. Palomino, O.; García-Aguilar, A.; González, A.; Guillén, C.; Benito, M.; Goya, L. Biological actions and molecular mechanisms of Sambucus nigra L. in neurodegeneration: A cell culture approach. *Molecules* **2021**, *26*, 4829. [CrossRef]
29. Alía, M.; Ramos, S.; Mateos, R.; Bravo, L.; Goya, L. Quercetin protects human hepatoma cell line (HepG2) against oxidative stress induced by tertbutyl hydroperoxide. *Toxicol. Appl. Pharmacol.* **2006**, *212*, 110–118. [CrossRef]
30. Denizot, F.; Lang, R. Rapid colorimetric assay for cell growth and survival. Modifications to the tetrazolium dye procedure giving improved sensitivity and reliability. *J. Immunol. Methods* **1986**, *89*, 271–277. [CrossRef]
31. Wang, H.; Joseph, J.A. Quantifying cellular oxidative stress by dichlorofluorescein assay using microplate reader. *Free Radic. Biol. Med.* **1999**, *27*, 612–616. [CrossRef] [PubMed]
32. Barrios-Arpi, L.; Arias, Y.; Lopez-Torres, B.; Ramos-Gonzalez, M.; Ticli, G.; Prosperi, E.; Rodríguez, J.L. In vitro neurotoxicity of flumethrin pyrethroid on SH-SY5Y neuroblastoma cells: Apoptosis associated with oxidative stress. *Toxics* **2022**, *10*, 131. [CrossRef]
33. Ramakers, C.; Ruijter, J.M.; Deprez, R.H.L.; Moorman, A.F.M. Assumption-free analysis of quantitative real-time polymerase chain reaction (PCR) data. *Neurosci. Lett.* **2003**, *339*, 62–66. [CrossRef]
34. Baba, S.; Osakabe, N.; Natsume, N.; Muto, Y.; Takizawa, T.; Terao, J. In vivo comparison of the bioavailability of catechin, epicatechin and their mixture in orally administered rats. *J. Nutr.* **2001**, *131*, 2885–2891. [CrossRef] [PubMed]
35. Pappas, E.; Schaich, K.M. Phytochemicals of cranberries and cranberry products: Characterization, potential health effects and processing stability. *Crit. Rev. Food Sci. Nutr.* **2009**, *49*, 741–781. [CrossRef] [PubMed]
36. Pedersen, C.B.; Kyle, J.; Jenkinson, A.M.; Gardner, P.T.; PcPhail, D.B.; Duthie, G.G. Effects of cranberry and blueberry juice consumption on the plasma antioxidant capacity of healthy female volunteers. *Eur. J. Clin. Nutr.* **2000**, *54*, 405–408. [CrossRef] [PubMed]
37. Palomino, O.; Giordani, V.; Chowen, J.A.; Fernández Alonso, S.; Goya, L. Physiological doses of oleic and palmitic acids protect human endothelial cells from oxidative stress. *Molecules* **2022**, *27*, 5217. [CrossRef]
38. Martín, M.A.; Ramos, S.; Mateos, R.; Serrano, A.B.G.; Izquierdo-Pulido, M.; Bravo, L.; Goya, L. Protection of Human HepG2 Cells against Oxidative Stress by Cocoa Phenolic Extract. *J. Agric. Food Chem.* **2008**, *56*, 7765–7772. [CrossRef]
39. Martín, M.A.; Ramos, S.; Cordero-Herrera, I.; Bravo, L.; Goya, L. Cocoa phenolic extract protects pancreatic beta cell viability and function against oxidative stress. *Nutrients* **2013**, *5*, 2955–2968. [CrossRef]
40. León-González, A.; Mateos, R.; Ramos, S.; Martín, M.A.; Sarriá, B.; Martín-Cordero, C.; López-Lázaro, M.; Bravo, L.; Goya, L. Chemo-protective activity and characterization of phenolic extracts from Corema album. *Food Res. Int.* **2012**, *49*, 728–738. [CrossRef]
41. Baeza, G.; Amigo-Benavent, M.; Sarriá, B.; Goya, L.; Mateos, R.; Bravo, L. Green coffee hydroxycinnamic acids but not caffeine protect human HepG2 cells against oxidative stress. *Food Res. Int.* **2014**, *62*, 1038–1046. [CrossRef]
42. Martín, M.A.; Ramos, S.; Mateos, R.; Marais, J.; Bravo, L.; Khoo, C.; Goya, L. Chemical characterization and chemo-protective activity of cranberry phenolic extracts in a model cell culture. Response of the antioxidant defences and regulation of signaling pathways. *Food Res. Int.* **2015**, *71*, 68–82. [CrossRef]
43. Castillo, G.; Barrios-Arpi, L.; Ramos-Gonzalez, M.; Vidal, P.; Gonzales-Irribarren, A.; Ramos-Cevallos, N.; Rodríguez, J.L. Neurotoxicity associated with oxidative stress and inflammasome gene expression induced by allethrin in SH-SY5Y cells. *Toxicol. Ind. Health* **2022**, *38*, 777–788. [CrossRef]
44. Lee, M.H.; Lee, J.M.; Jun, S.H.; Lee, S.H.; Kim, N.W.; Lee, J.H.; Ho, N.Y.; Mun, S.H.; Kim, B.K.; Lim, B.O.; et al. The anti-inflammatory effects of Pyrolae herba extract through the inhibition of the expression of inducible nitric oxide synthase (iNOS) and NO production. *J. Ethnopharmacol.* **2007**, *112*, 49–54. [CrossRef]
45. Adebayo, S.A.; Ondua, M.; Shai, L.J.; Lebelo, S.L. Inhibition of nitric oxide production and free radical scavenging activities of four South African medicinal plants. *J. Inflamm. Res.* **2019**, *12*, 195–203. [CrossRef]
46. Marino, G.; López-Otín, C. Autophagy: Molecular mechanisms, physiological functions and relevance in human pathology. *Cell Mol. Life Sci. CMLS* **2004**, *61*, 1439–1454. [CrossRef]
47. Pelicano, H.; Carney, D.; Huang, P. ROS stress in cancer cells and therapeutic implications. *Drug Resist. Updat.* **2004**, *7*, 97–110. [CrossRef]

48. Chen, Y.; McMillan-Ward, E.; Kong, J.; Israels, S.J.; Gibson, S.B. Oxidative stress induces autophagic cell death independent of apoptosis in transformed and cancer cells. *Cell Death Differ.* **2008**, *15*, 171–182. [CrossRef]
49. Li, J.K.; Ge, R.; Tang, L.; Li, Q.S. Protective effects of farrerol against hydrogen-peroxide-induced apoptosis in human endothelium-derived EA.hy926 cells. *Can J. Physiol. Pharmacol.* **2013**, *91*, 733–740. [CrossRef]
50. Guo, S.; Long, M.; Li, X.; Zhu, S.; Zhang, M.; Yang, Z. Curcumin activates autophagy and attenuates oxidative damage in EA.hy926 cells via the Akt/mTOR pathway. *Mol. Med. Rep.* **2016**, *13*, 2187–2193. [CrossRef]
51. Morán-Santibañez, K.; Vasquez, A.H.; Varela-Ramirez, A.; Henderson, V.; Sweeney, J.; Odero-Marah, V.; Fenelon, K.; Skouta, R. Larrea tridentata extract mitigates oxidative stress-induced cytotoxicity in human neuroblastoma SH-SY5Y cells. *Antioxidants* **2019**, *8*, 427. [CrossRef] [PubMed]
52. Jantas, D.; Malarz, J.; Le, T.N.; Stojakowska, A. Neuroprotective properties of kempferol derivatives from maesa membranacea against oxidative stress-induced cell damage: An association with cathepsin D inhibition and PI3K/Akt Activation. *Int. J. Mol. Sci.* **2021**, *22*, 10363. [CrossRef]
53. Myhrstad, M.C.; Carlsen, H.; Nordström, O.; Blomhoff, R.; Moskaug, J.O. Flavonoids increase the intracellular glutathione level by transactivation of the gamma-glutamylcysteine synthetase catalytical subunit promoter. *Free Rad. Biol. Med.* **2002**, *32*, 386–393. [CrossRef] [PubMed]
54. Moridani, M.Y.; Scobie, H.; Salehi, P.; O'Brien, P.J. Catechin metabolism: Glutathione conjugate formation catalyzed by tyrosinase, peroxidase, and cytochrome p450. *Chem. Res. Toxicol.* **2001**, *14*, 841–848. [CrossRef] [PubMed]
55. Galati, G.; Lin, A.; Sultan, A.M.; O'Brien, P.J. Cellular and in vivo hepatotoxicity caused by green tea phenolic acids and catechins. *Free Radic. Biol. Med.* **2006**, *40*, 570–580. [CrossRef]
56. Battistelli, M.; Malatesta, M.; Meschini, S. Oxidative stress to promote cell death or survival. *Oxid. Med. Cell. Longev.* **2016**, *2016*, 2054650. [CrossRef] [PubMed]
57. Lee, H.J.; Spandidos, D.A.; Tsatsakis, A.; Margina, D.; Izotov, B.N.; Yang, S.H. Neuroprotective effects of Scrophularia buergeriana extract against glutamate-induced toxicity in SH-SY5Y cells. *Int. J. Mol. Med.* **2019**, *43*, 2144–2152. [CrossRef]
58. Huang, W.M.; Liang, Y.Q.; Tang, L.J.; Ding, Y.U.E.; Wang, X.H. Antioxidant and anti-inflammatory effects of Astragalus polysaccharide on EA.hy926 cells. *Exp. Ther. Med.* **2013**, *6*, 199–203. [CrossRef]
59. Ma, D.; Wang, Z.; He, Z.; Wang, Z.; Chen, Q.; Qin, F.; Zeng, M.; Chen, J. Pine pollen extract alleviates ethanol-induced oxidative stress and apoptosis in HepG2 cells via MAPK signaling. *Food Chem. Toxicol.* **2023**, *171*, 113550. [CrossRef]
60. Chu, W.L.; Lim, Y.W.; Radhakrishnan, A.K.; Lim, P.E. Protective effect of aqueous extract from Spirulina platensisagainst cell death induced by free radicals. *BMC Complement. Altern. Med.* **2010**, *10*, 53. [CrossRef]

nutrients

Article

The Influence of *APOE* Genotype, DHA, and Flavanol Intervention on Brain DHA and Lipidomics Profile in Aged Transgenic Mice

Anneloes Martinsen [1], Rasha N. M. Saleh [1,2,*], Raphael Chouinard-Watkins [3], Richard Bazinet [3], Glenn Harden [1], James Dick [4], Noemi Tejera [1], Matthew G. Pontifex [1], David Vauzour [1] and Anne-Marie Minihane [1]

1 Norwich Medical School, University of East Anglia, Norwich NR4 7TJ, UK
2 Clinical Pathology Department, Faculty of Medicine, Alexandria University, Alexandria 21526, Egypt
3 Department of Nutritional Sciences, Faculty of Medicine, University of Toronto, Toronto, ON M5S 1A8, Canada
4 Nutrition Analytical Service, Institute of Aquaculture, University of Stirling, Stirling FK9 4LA, UK
* Correspondence: r.saleh@uea.ac.uk

Abstract: The apolipoprotein E4 (*APOE4*) genotype is predictive of Alzheimer's disease (AD). The brain is highly enriched with the omega-3 polyunsaturated fatty acid (n3-PUFA), docosahexaenoic acid (DHA). DHA's metabolism is defective in *APOE4* carriers. Flavanol intake can play a role in modulating DHA levels. However, the impact of flavanol co-supplementation with fish oil on brain DHA uptake, status and partitioning, and according to *APOE* genotype is currently unknown. Here, using a humanised *APOE3* and *APOE4* targeted replacement transgenic mouse model, the interactive influence of cocoa flavanols (FLAV) and *APOE* genotype on the blood and subcortical brain PUFA status following the supplementation of a high fat (HF) enriched with DHA from fish oil (FO) was investigated. DHA levels increased in the blood ($p < 0.001$) and brain ($p = 0.001$) following supplementation. Compared to *APOE3*, a higher red blood cell (RBC) DHA ($p < 0.001$) was evident in *APOE4* mice following FO and FLAV supplementation. Although FO did not increase the percentage of brain DHA in *APOE4*, a 17.1% ($p < 0.05$) and 20.0% ($p < 0.001$) higher DHA level in the phosphatidylcholine (PC) fraction in the HF FO and HF FO FLAV groups, and a 14.5% ($p < 0.05$) higher DHA level in the phosphatidylethanolamine (PE) fraction in the HF FO FLAV group was evident in these animals relative to the HF controls. The addition of FLAV (+/− FO) did not significantly increase the percentage of brain DHA in the group as a whole. However, a higher brain: RBC DHA ratio was evident in *APOE3* only ($p < 0.05$) for HF FLAV versus HF. In conclusion, our data shows only modest effects of FLAV on the brain DHA status, which is limited to *APOE3*.

Keywords: Alzheimer's disease; apolipoprotein E; docosahexaenoic acid; brain; flavonoids; phospholipids; PUFAs

Citation: Martinsen, A.; Saleh, R.N.M.; Chouinard-Watkins, R.; Bazinet, R.; Harden, G.; Dick, J.; Tejera, N.; Pontifex, M.G.; Vauzour, D.; Minihane, A.-M. The Influence of *APOE* Genotype, DHA, and Flavanol Intervention on Brain DHA and Lipidomics Profile in Aged Transgenic Mice. *Nutrients* **2023**, *15*, 2032. https://doi.org/10.3390/nu15092032

Academic Editors: Asim K. Duttaroy and Maria Annunziata Carluccio

Received: 13 January 2023
Revised: 11 April 2023
Accepted: 17 April 2023
Published: 23 April 2023

1. Introduction

The apolipoprotein E4 (*APOE4*) genotype is the strongest prevalent genetic determinant of Alzheimer's disease (AD) risk with heterozygotes (*APOE3/E4*) and homozygotes (*APOE4/E4*) carriers at 3–4- and 12–15-fold increased risk, respectively, relative to the "wild-type" *APOE3/E3* genotype [1,2]. Despite this, the *APOE4* allele remains predictive rather than deterministic, with half of *APOE4/E4* never progressing to the development of AD [3]. This observed phenomenon suggests that environmental factors/circumstances may play a crucial role in *APOE4*-mediated AD development. Behavioural changes, such as dietary interventions, may modify or even prevent *APOE4*-mediated disease pathophysiology and ultimately cognitive decline.

The grey matter of the brain, particularly synaptic phospholipids, are docosahexaenoic acid (DHA)-rich, with several-fold higher DHA levels than in systemic tissue. The cognitive benefits associated with a higher DHA intake and status have been widely acknowledged in experimental animals [4] and in prospective cohort studies [5]. DHA is a crucial molecule for the neuronal structure and function [6,7]. It has anti-inflammatory and pro-resolving properties [8], improves nerve signalling [9], and reduces the amyloid burden [10]. More than 80% of brain DHA is esterified in the phospholipid pool [11]. DHA is highly concentrated in phosphatidylethanolamine (PE) followed by phosphatidylserine (PS), then phosphatidylcholine (PC), and finally in the phosphatidylinositol (PI) pool [12]. The distribution of DHA across these subclasses has been shown to be affected by the *APOE* genotype. Lower PE-DHA and PS-DHA levels were observed in older *APOE4* mice [13,14], which were corrected by DHA supplementation [14].

In fish oil (FO) intervention trials, the *APOE* genotype has been shown to affect DHA metabolism [15], including brain uptake and utilisation [7,16,17]. While responses were more favourable in *APOE3* carriers, *APOE4* carriers showed inconsistent and variable DHA levels in response to supplementation [18]. In some preclinical and clinical studies, the brain and blood DHA status was lower in *APOE4* carriers [18–20]. On the contrary, a higher uptake of DHA was seen in *APOE4* carriers in less advanced AD pathology compared to more advanced pathology, highlighting the importance of the timing of n-3 fatty acid supplementation in at-risk *APOE4* carriers. We have previously demonstrated an age-related decline in habitual cortex and hippocampal DHA levels, accompanied by diminished DHA-derived specialised anti-inflammatory and pro-resolving mediators (SPMs), which was more prominent in *APOE4* animals [21]. This was corrected by DHA supplementation with the complete restoration of brain DHA levels in *APOE4* animals [4].

There is growing evidence that the co-supplementation of FO with antioxidants can increase DHA status. The addition of selenium, tocopherols, vitamin A, vitamin B, vitamin C, and vitamin D all increased DHA blood levels [22–25]. Flavonoids, such as cocoa/tea-derived flavanols (FLAV) possess neuroprotective properties, with their anti-inflammatory properties often cited as mechanistically responsible [26–28]. Indeed, flavonoids have been shown to improve the hallmarks of AD pathology-mitigating abnormal glial cell activation, amyloid β deposition, and tau phosphorylation. Intriguingly, these effects may be more evident in non-*APOE4* carriers [29]. However, in the COSMOS-mind study, flavanol-rich cocoa extract alone did not benefit cognition [30]. Flavonoids may increase DHA uptake by preventing polyunsaturated fatty acid peroxidation, whilst the n-3 polyunsaturated fatty carbon chain facilitates the penetration of dietary flavonoids through cell membranes [31]. Therefore, the combined consumption of flavonoids and n-3 polyunsaturated fatty acids may enhance their therapeutic benefit. Despite this promising synergistic potential, no study has comprehensively investigated the impact of FLAV supplementation on habitual brain DHA and lipidomic profiles, their response to DHA intervention, and according to APOE genotype status.

We hypothesize that the co-supplementation of fish oil with a flavanol-rich cocoa extract will modulate brain DHA in an *APOE* dependent fashion. A full lipidomic analysis was conducted to provide a more granular and holistic overview of DHA derivatives in different brain lipid pools.

2. Materials and Methods

2.1. Study Approval

All the experiments were approved by the Animal Welfare and Ethical Review Body (AWERB); approval date: 17th February 2016 under the project licence code: PPL 70/8710). The experiments were conducted within the provisions of the Home Office Animals (Scientific Procedures) Act 1986. The reporting of the results complies with the guidance of the Animal Research: Reporting of In Vivo Experiments (ARRIVE) guidelines.

2.2. Animals Experimental Design and Dietary Treatments

120 male humanised *APOE3* (B6.129P2-Apoe$^{tm2(APOE*3)Mae}$ N8) and *APOE4* (B6.129P2-Apoe$^{tm2(APOE*4)Mae}$ N8) targeted replacement mice homozygous for the human *APOE3*

or *APOE4* gene (Taconic, Germantown, NY, USA) were used in the experiments [32]. The mice were 10-months-old at the start of the experiments. They were maintained in a controlled environment as described previously [4]. In week 1, the animals were on a standard chow diet (RM3-P, Special Diets Services, Essex, UK). At week 2, the mice were assigned to five experimental groups and provided with one of the following diets (Research Diets Inc., New Brunswick, USA) for 22 weeks: a low fat diet (LF, 10 kcal%), a high fat diet (HF, 45 kcal%), a HF diet supplemented with fish oil (HF FO, 45 kcal% + 5.1 mg/g EPA/DHA; EPAX TGN, Oslo, Norway), a HF diet supplemented with cocoa flavanols (HF FLAV, 45 kcal% + 892 µg/g total cocoa flavanols from CHD-Q65ACTICOA-558; Barry Callebaut, Lebbeke-Wieze, Belgium) or a HF diet supplemented with a combination of both (HF FO FLAV). A full dietary composition is provided in the supplementary data (Supplementary Table S1). The doses of the n-3 polyunsaturated fatty acids and flavanols were chosen to be equivalent to a human diet. Using allometric scaling, the animals fed with FO received 9.2 mg/d EPA/DHA, equivalent to a 1.5 g/d human equivalent dose (HED), such as that found in a half-to-full portion of oily fish [33]. Similarly, the animals fed the FLAV diet received 3.1 mg/d of total cocoa flavanols of which 0.7 mg/d comprised catechin/epicatechin, which corresponds to a physiologically relevant HED of 508 mg/d of which 107 mg/d comprised catechin/epicatechin, which are the levels found in 33 g of FLAV-rich cocoa. All the diets were adjusted for caffeine (20 mg/d HED) and theobromine (157 mg/d HED), which are naturally present in cocoa.

Food intake and body weight were recorded twice a week, three weeks prior to the intervention and for the duration of the intervention. Food pellets were replaced twice a week to avoid the oxidation of the bioactive compounds. At week 22, the mice were anaesthetised and blood was collected by cardiac puncture followed by the transcardiac perfusion of an ice-cold saline solution containing 10 IU/mL heparin (Sigma, Hertfordshire, UK). The plasma samples were isolated by centrifugation at 2000× *g* for 10 min and the samples were snap-frozen and stored at −80 °C. The brains were rapidly removed, bisected, snap-frozen, and stored at −80 °C. The half-brains were further dissected into corteces, hippocampi, subcortical regions, and cerebella, snap-frozen in liquid nitrogen and stored at −80 °C until further analysis.

2.3. *Lipid Extraction and Fatty Acid Analysis in Red Blood Cells and Feeds*

The total lipids were extracted according to the method of Folch et al. [34] as described before [35]. Fatty acid methyl esters (FAME) were prepared by the acid-catalyzed transesterification of the total lipids [36,37]. The samples were dried overnight in a desiccator then methylation was carried out using 1.25M HCL in methanol. The FAME were separated using 50% saturated KCL then purified by elusion through SPE silica cartridges (Clean-up 203 Cusil 156, UCT) using isohexane: diethylether (95:5). The FAME were evaporated under oxygen-free nitrogen and re-suspended in 500 µL of iso-hexane. The purified FAME were then separated by gas-liquid chromatography using a ThermoFisher Trace GC 2000 (ThermoFisher, Hemel Hempstead, UK). On-column injection was carried out using a fused silica capillary column (ZBWax, 60 m × 0.32 × 0.25 mm i.d.; Phenomenex, Macclesfield, UK) and hydrogen as a carrier gas. The temperature gradient was from 50 to 150 °C at 40 °C/min, and then to 195 °C at 1.5 °C/min, and finally to 220 °C at 2 °C/min. Individual methyl esters were identified by comparison to known standards (Supelco 37-FAME mix; Sigma-Aldrich Ltd., Poole, UK) and by reference to published data [38]. The data were analysed using the Chromcard software package (Thermoquest Italia, Milan, Italy). Individual fatty acids were reported as percentages (% total fatty acids). This was calculated by measuring the relative area % of the chromatogram peak generated by the individual fatty acid and dividing it by the sum of all the relative area % of all the fatty acids in the chromatogram. The fatty acid content per g of tissue (nmol/g) was calculated using heptadecanoic acid (17:0) as the internal standard. Both measurements are reported as it is important to consider brain fatty acid content as a percentage of the total fatty acids and relative to the weight of the total lipids when interpreting the physiological meaning of the altered status [39].

2.4. Lipid Extraction and Fatty Acid Analysis in Plasma

Fatty acids in the plasma were extracted by a method modified from Folch et al. [34]. Briefly, the samples and the internal standard docosatrienoic ethyl ester (22:3n-3) were homogenised by vortex in a solution of 2:1 chloroform: methanol. The mixtures were kept at 4 °C overnight and brought to room temperature the next day. Potassium chloride (0.88% (w/v)) was added to the separate phases. The bottom organic phase containing the total lipid extract was transferred into new test tubes and dried down under a stream of nitrogen. Fatty acids were transmethylated at 100 °C for one hour with 14% (v/v) boron trifluoride methanol. The FAME were quantified on a Varian 430 gas chromatograph (Bruker, Billerica, MA, USA) equipped with a SP-2560 biscyanopropyl siloxane capillary column (100 m length × 0.25 mm diameter × 0.2 µm film thickness; Supelco, Belle-fonte, PA, USA). Details of the chromatography set-up method have been described previously [40].

2.5. Brain Lipidomics Profile

Lipids were extracted from the subcortical region samples (n = 10) via a modified Bligh–Dyer extraction [41], using methanol/water/dichloromethane in the presence of deuterated internal standards, using the Metabolon TrueMass® Complex Lipid Panel (Rowarth, UK).

The extracts were dried under nitrogen and reconstituted in ammonium acetate dichloromethane: methanol. The extracts were transferred to vials for infusion-MS analysis, performed on a Shimadzu LC with nano PEEK tubing and the Sciex SelexIon-5500 QTRAP. The samples were analysed via both a positive and negative mode electrospray. The 5500 QTRAP was operated in MRM mode with a total of more than 1100 MRMs. Individual lipid species were quantified by taking the ratio of the signal intensity of each target compound to that of its assigned internal standard, then multiplying by the concentration of the internal standard added to the sample. The lipid class concentrations were calculated from the sum of all the molecular species within a class, and the fatty acid compositions were determined by calculating the proportion of each class comprised of individual fatty acids.

2.6. Statistical Analysis

Body weight and fatty acids data are presented as means ± SEM. The data analysis was performed using one-way and two-way ANOVAs. Standard diagnostic tests (e.g., the normality of residuals assessed using quantile–quantile plots and Shapiro–Wilk tests, outlier tests, high-leverage/influence data points tested using Cook's distance) were used to verify that the data were appropriate for ANOVA analysis. Where necessary, transformations were applied to the response variance to ensure that the data complied with the ANOVA assumptions. If transformations could not be applied to ensure the normality of data, appropriate non-parametric statistical techniques (e.g., the Wilcoxon rank sum test) were used instead. Post hoc tests were carried out using the Tukey's honest significant differences test. A multiple-test correction was performed. Along with a p-diet*genotype interaction, the analysis considered the effect of the genotype (independent of dietary group. p-genotype) and the effect of the diet (independent of genotype. p-diet) on the response to intervention. A statistical analysis was performed using R statistical software v.3.5.1 (R Foundation for Statistical Computing, Vienna, Austria). We utilised the brain: blood ratio of enrichment as a *pseudo* measure of brain DHA uptake [42].

3. Results

To determine whether the impact of a high fat diet on DHA concentrations in the brain can be modulated by a diet enriched with a combination of fish oil with flavanols and is influenced by *APOE* genotype, we analysed fatty acids' profiles in the brain and blood pools of male *APOE3* and *APOE4* mice after a 22-week dietary intervention.

3.1. Effect of Diet on Brain, Plasma, and Red Blood Cell DHA Content

First, the impact of FO, FLAV, and FO + FLAV on brain and blood DHA levels was analysed independently of *APOE* genotype. The brain and plasma DHA concentrations were higher for HF FO ($p < 0.001$) (Figure 1A,B). No additional effect of FLAV on brain DHA was observed (Figure 1A), while FLAV addition led to a higher plasma DHA concentration, $p < 0.05$ (Figure 1B). FO intervention reduced the brain: plasma DHA concentration ratio (Figure 1C), with the degree of enrichment in the plasma not reflected by a comparable increase in brain tissue (Figure 1A,B). Comparable trends were observed when the brain DHA (%) was compared to RBC (Figure 1D–F). A high-fat (HF) diet was not associated with changes in the DHA compared to a low-fat (LF) diet in the brain or the blood (Figure 1A,B,D,E); however, LF was associated with a higher brain: plasma DHA ratio.

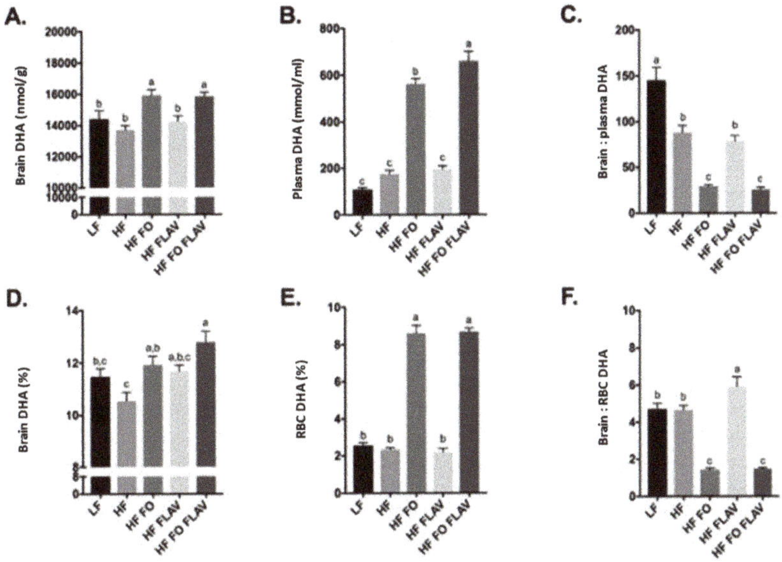

Figure 1. Effect of diet on brain, plasma, and red blood cells DHA content in mice, expressed in nmol/g of brain tissue (**A**), mmol/mL of plasma (**B**), and as % of total fatty acids (**D**,**E**). (**C**,**F**) are ratios of brain to plasma and brain to RBCs DHA concentrations respectively. Values are expressed as means ± SEM. One-way ANOVA Tukey's post-hoc test adjusted *p*-values. a, b, c: columns (groups) with different letters are significantly different at $p = 0.05$, while columns sharing the same letters are not significantly different. Data of each dietary group was obtained from *APOE3* and *APOE4* mice combined. Number of mice per dietary group: LF n = 20, HF n = 21, HF FO n = 19, HF FLAV n = 19, and HF FO + FLAV n = 19.

3.2. Effect of APOE Genotype and Diet on Brain, Plasma, and RBC DHA Levels

There was no effect of *APOE* genotype on the brain DHA status (Figure 2A). The *APOE4* mice had a higher DHA percentage in the plasma ($p = 0.025$ Figure 2B) and RBC ($p < 0.001$ Figure 2C) than the *APOE3* mice, independent of the dietary intervention. A diet–genotype interaction was evident for the RBC DHA ($p < 0.001$) with a higher percentage enrichment in the *APOE4* mice following the FO and FLAV intervention (Figure 2C), with comparable trends evident for the plasma DHA. We observed a genotype and diet-genotype interaction for the brain: RBC DHA ratio, with the higher ratio in *APOE3* versus *APOE4* animals ($p < 0.001$, Figure 2E), largely reflected by FLAV supplementation ($p < 0.001$, Figure 2E).

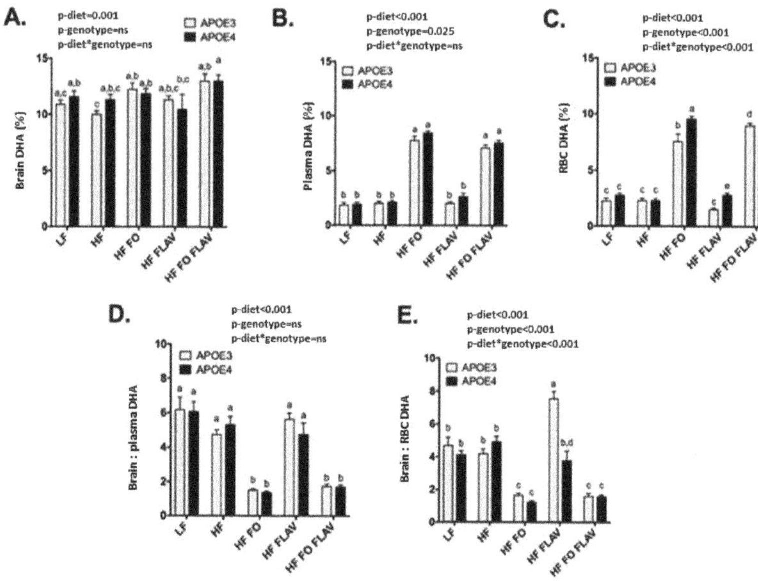

Figure 2. Effect of diet and genotype in the brain, plasma, and red blood cells DHA content of *APOE3* and *APOE4-TR* mice, expressed in % of total fatty acids (**A**–**C**) and ratios of brain: plasma (**D**) and brain: RBC DHA (**E**). Values are expressed as means ± SEM. a, b, c, d: columns (groups) with different letters are significantly different at $p = 0.05$, while columns sharing the same letters are not significantly different. Number of mice in each dietary and genotype group are: LF *APOE3* n = 11, LF *APOE4* n = 9, HF *APOE3* n = 13, HF *APOE4* n = 8, HF FO *APOE3* n = 11, HF FO *APOE4* n = 8, HF FLAV *APOE3* n = 10, HF FLAV *APOE4* n = 9, HF FO + FLAV *APOE3* n = 10, and HF FO + FLAV *APOE4* n = 8. P-diet*genotype: significance of diet and genotype interactive effect.

3.3. *Impact of Diet and Genotype on DHA: AA Is More Evident in APOE4 Mice*

Due to its impact on systemic and neuro-inflammatory status, the DHA: AA ratio was calculated in the blood and the brain. A higher RBC DHA: AA ratio was evident in *APOE4* compared to *APOE3* ($p < 0.0001$, Table 1). A diet–genotype interaction was observed in the DHA: AA ratio in the blood pools only (plasma and RBC, $p = 0.021$ and $p < 0.000$, respectively), with a significantly higher ratio in *APOE4* carriers in the HF FO, HF FLAV, and HF FO + FLAV diet groups compared to their corresponding *APOE3* dietary group.

A lipidomic analysis in the brain of *APOE3* and *APOE4* mice revealed no overall diet–genotype interaction on lipid classes or fatty acid concentrations. The total lipids DHA and arachidonic acids (AA) values did not differ between the genotypes (Supplemental Table S2).

Table 1. Effect of diet and genotype on arachidonic acid (AA), DHA, and DHA: AA in plasma, red blood cells, and brain of *APOE3* and *APOE4-TR* mice after 22 weeks of intervention.

	APOE3					*APOE4*							
	LF	**HF**	**HF FO**	**HF FLAV**	**HF FO FLAV**	**LF**	**HF**	**HF FO**	**HF FLAV**	**HF FO FLAV**	**Diet**	**Genotype**	**Interaction**
Plasma (nmol/mL)													
20:4n-6 (AA)	857 ± 188	1595 ± 161	878 ± 95.7	2101 ± 137	1133 ± 144	850 ± 123	2224 ± 253	1008 ± 97.5	1887 ± 216	1268 ± 93.5	**<0.000**	NS	NS
22:6n-3 (DHA)	109 ± 12.8	149 ± 14.1	542 ± 43.3	183 ± 10	596 ± 62.5	108 ± 11.9	199 ± 27.2	582 ± 20	208 ± 26.1	724 ± 42.3	**<0.000**	**0.025**	NS
DHA:AA	0.14 ± 0.01	0.09 ± 00	0.64 ± 0.04	0.09 ± 00	0.54 ± 0.02	0.13 ± 0.01	0.09 ± 00	0.58 ± 0.06	0.11 ± 0.01	0.57 ± 0.06	**<0.000**	NS	**0.021**
Red Blood Cells (%)													
20:4n-6 (AA)	17 ± 0.74	19.1 ± 0.46	11.4 ± 0.57	15.2 ± 0.26	13.3 ± 0.37	18.3 ± 0.57	18.1 ± 0.65	13.7 ± 0.58	19.7 ± 0.46	11.8 ± 0.58	**<0.000**	**0.001**	**<0.000**
22:6n-3 (DHA)	2.48 ± 0.17	2.45 ± 0.12	7.74 ± 0.61	1.53 ± 0.07	8.42 ± 0.45	2.84 ± 0.14	2.34 ± 0.12	9.63 ± 0.2	2.84 ± 0.14	8.33 ± 0.26	**<0.000**	**<0.000**	**<0.000**
DHA:AA	0.14 ± 0	0.13 ± 0.01	0.67 ± 0.04	0.10 ± 00	0.63 ± 0.02	0.16 ± 0.01	0.13 ± 0.01	0.71 ± 0.04 a	0.14 ± 0.01 b	0.72 ± 0.05 c	**<0.000**	**<0.000**	**<0.000**
Brain (%)													
20:4n-6 (AA)	8.48 ± 0.4	8.64 ± 0.28	7.55 ± 0.32	9.68 ± 0.37	8.45 ± 0.42	9.21 ± 0.32	9.53 ± 0.33	8.03 ± 0.33	8.95 ± 0.72	8.41 ± 0.39	**0.010**	NS	NS
22:6n-3 (DHA)	10.9 ± 0.33	10.0 ± 0.3	12.3 ± 0.54	11.3 ± 0.33	13.0 ± 0.63	11.6 ± 0.45	11.3 ± 0.43	11.9 ± 0.42	10.5 ± 1.3	13.0 ± 0.53	**0.001**	NS	NS
DHA: AA	1.3 ± 0.03	1.16 ± 0.03	1.63 ± 0.06	1.18 ± 0.04	1.55 ± 0.05	1.26 ± 0.03	1.19 ± 0.03	1.49 ± 0.04	1.11 ± 0.13	1.56 ± 0.04	**<0.000**	NS	NS

Values are expressed as means ± SEM. a, b, c significantly higher in *APOE4* compared to *APOE3*. Number of mice in each dietary and genotype group are: LF *APOE3* n = 11, LF *APOE4* n = 9, HF *APOE3* n = 13, HF *APOE4* n = 8, HF FO *APOE3* n = 11, HF FO *APOE4* n = 8, HF FLAV *APOE3* n = 10, HF FLAV *APOE4* n = 9, HF FO + FLAV *APOE3* n = 10, and HF FO + FLAV *APOE4* n = 8. Bold numbers show significant *p*-values.

3.4. Impact of Diet and APOE Genotype on Brain Phospholipid Subclasses' DHA Content

The DHA concentration in the different phospholipid fractions was not different between the LF and the HF diet. In the *APOE3* mice, DHA supplementation (HF FO and HF FO + FLAV) increased the brain DHA in total phospholipids compared to the high-fat diet group (HF), being mostly reflected in the PC fraction (Table 2). Similar results were observed in the *APOE4* mice. Interestingly, a 17.1% ($p < 0.05$) and 20.0% ($p < 0.001$) higher DHA level in the PC fraction of the HF FO and HF FO + FLAV groups, and a 14.5% ($p <$ 0.05) higher DHA level in the PE fraction of the HF FO + FLAV group was evident in the *APOE4* animals relative to the HF controls (Table 2).

Table 2. Diet effect on brain DHA phospholipid fractions in *APOE3* and *APOE4-TR* mice after 22 weeks of treatment.

Diet Group	LPC	LPE	PC	PE	PI	Total
APOE3						
LF	3.65 ± 0.66	4.55 ± 0.35	1049 ± 43.2	13,376 ± 802	9.66 ± 0.91	14,628 ± 802
HF	3.79 ± 0.23	4.35 ± 0.21	955 ± 41.2	12,355 ± 305	6.94 ± 0.97	13,519 ± 316
HF FO	3.53 ± 0.57	4.88 ± 0.34	1151 ± 29.8 **	14,732 ± 469	9.43 ± 1.04	16,142 ± 461 **
HF FLAV	2.50 ± 0.64	4.32 ± 0.30	932 ± 43.2	13,169 ± 451	7.71 ± 1.76	14,305 ± 473
HF FO FLAV	3.96 ± 0.42	5.29 ± 0.62	1089 ± 50.0	14,495 ± 455	8.57 ± 1.30	15,851 ± 448 *
APOE4						
LF	2.06 ± 0.53	4.59 ± 0.35	927 ± 32.4	13,090 ± 186	6.10 ± 1.29	14,199 ± 186
HF	2.43 ± 0.41	4.70 ± 0.37	941 ± 28.1	13,025 ± 295	7.10 ± 0.99	14,157 ± 310
HF FO	4.12 ± 0.71	5.09 ± 0.36	1102 ± 52.9 *	14,295 ± 508	7.82 ± 1.52	15,643 ± 468
HF FLAV	1.82 ± 0.50	4.32 ± 0.41	810 ± 106	11,801 ± 1536	5.11 ± 1.84	12,799 ± 1631
HF FO FLAV	3.47 ± 0.56	5.42 ± 0.77	1129 ± 34.2 **	14,896 ± 418 *	5.22 ± 0.97	16,269 ± 386

Values expressed in nmol/g of brain tissue. LF: low fat; HF: high fat; HF FO: high fat enriched in fish oil; HF FLAV: high fat supplemented in cocoa flavanols; HF FO FLAV: high fat combined with both fish oil and cocoa flavanols; LPC, lysophosphatidylcholine; LPE, lysophosphatidylethanolamine; PC, phosphatidylcholine; PE, phosphatidylethanolamine; PI, phosphatidylinositol. Total, total lipid in sample. Values are expressed as means ± SEM. * $p < 0.05$; ** $p < 0.01$ compared to HF. Number of mice in each dietary and genotype group are: LF *APOE3* n = 11, LF *APOE4* n = 9, HF *APOE3* n = 13, HF *APOE4* n = 8, HF FO *APOE3* n = 11, HF FO *APOE4* n = 8, HF FLAV *APOE3* n = 10, HF FLAV *APOE4* n = 9, HF FO + FLAV *APOE3* n = 10, and HF FO + FLAV *APOE4* n = 8.

The PC or PE fractions of arachidonic acid (AA) were not significantly different between the diet groups in the *APOE4* mice (Supplemental Table S3). In the *APOE3* mice, only the AA-PE fraction was reduced in the DHA-enriched groups in the *APOE3* animals (Supplemental Table S3).

4. Discussion

The present study provides the first comprehensive investigation of the impact of supplementation with FLAV with or without DHA on the brain lipidomic profiles and according to *APOE* genotype status, with a complex picture of DHA*FLAV**APOE* interactions. Although the addition of FLAV did not significantly increase brain DHA in the group as a whole, FLAV led to an increase in the brain: RBC DHA ratio, which was limited to the *APOE3* animals. As expected, the brain DHA levels increased in the DHA-enriched dietary groups; however this was not influenced by the addition of FLAV. Following supplementation, higher blood (RBC and trend for plasma) DHA levels were observed in the *APOE4* mice relative to *APOE3*. This was accompanied by a lower brain: RBC ratio, which together highlights deficits in either DHA transport, storage, use, or turnover associated with the *APOE4* carrier status. However, in the *APOE4* animals a significant increase in DHA in both the brain PC and PE fractions was evident following DHA supplementation.

Although the size effect was modest, our research does highlight the potential of dietary flavanols as an approach to increase the brain DHA status independent of intake. Such strategies are of wide public health relevance, given concerns over fish and EPA/DHA sustainability [43], and given that the majority of the global population fail to meet dietary oily fish or EPA+DHA recommendations [44,45] and have a suboptimal EPA and DHA status [46]. It is particularly relevant to the estimated 25% of the world population who follow a vegetarian or vegan diet by choice or necessity, with negligible dietary EPA and DHA intake.

The number of people living with dementia worldwide is around 50 million [47]. With expanding and ageing populations, the incidence is predicted to almost triple by 2050. *APOE* genotype and obesity have been independently and interactively linked to AD risk, in humans and rodents [48]. In the present study we compared fatty acid and lipidomic profiles in the wild-type *APOE3* and with at-risk *APOE4* transgenic mice [2]. Animals were exposed to a high-fat diet to mimic human Western dietary practices, and in order to induce a human-like overweight phenotype with the associated insulin resistance and accelerated cognitive decline [48].

Observational studies have shown an association between DHA intake and reduced AD risk [49,50]. However, RCTs have shown inconsistent results, likely in part explained by the *APOE* genotype and AD stage. There is now relatively consistent evidence that the cognitive benefits associated with DHA supplementation are reduced in *APOE4* carriers [18,19,51]. Mechanistically, impaired BBB-mediated uptake and increased DHA oxidation and turnover are purported to be responsible. While there was an increase in the brain DHA level following fish oil supplementation in *APOE3* only, no difference between genotypes was evident for habitual DHA in the control (HF) group. We recently demonstrated a reduced habitual brain DHA level in 18-month-old female *APOE4* mice relative to younger females or age-matched *APOE3* males [21]. In addition, rodent studies performed on males and females have successfully demonstrated a genotype difference in response to DHA [42,52]. In the present study only male mice were used, which might explain some of the discrepancy. The cognitive deficits associated with *APOE4* are emerging as being more evident in females, potentially due to the additive effect of menopause and *APOE4*-carrier status on neurocognitive processes, with females also shown to benefit more from fish intake and DHA intake in old age [7,50]. Future research on DHA-*APOE* genotype-neurocognitive associations should consider sex as a potentially important mediating factor.

The capacity for DHA production in the brain is limited, with DHA predominantly derived from the systemic circulation, through dietary and systemic tissue pools. It has been shown that the plasma DHA metabolism is impaired in *APOE4* carriers [53]. We observed higher plasma and RBC DHA levels in the *APOE4* mice, in agreement with previous findings [42]. These genotype-mediated differences were reflected in the brain: RBC DHA values, with a higher ratio in *APOE3*. This may suggest a lower tissue uptake of DHA in *APOE4*, although a higher brain turnover cannot be precluded [17]. Furthermore, it indicates that supplementation with 1.5g DHA (HED) was not sufficient to raise DHA in the brain in the *APOE4* genotype. A higher dose of DHA (up to 3 g/d) may be needed in *APOE4* carriers, as has been suggested [52].

The combination of fish oil/DHA with other dietary compounds, such as flavonoids, is of particular interest, as it could help mitigate the negative impact of an *APOE4* genotype and potentially increase DHA bioavailability. In a 2010 analysis, green tea flavonoids (catechins) and fish oil had an additive effect on the inhibition of cerebral Aβ deposits in a mouse model of AD [54]. This combination of DHA and catechin also had an effect on cognitive performance (maze behaviour) in old mice, with the addition of catechin to the diet increasing brain DHA, possibly through the antioxidant properties of catechin [55]. In the present study the addition of cocoa FLAV did not significantly increase brain DHA levels. The lack of effect of catechin/epicatechin (FLAV) supplementation in our current study relative to the observations of Shirai and Suzuki [55] could be partly due to our use of a high-fat regime background diet (45% versus 6.5%). Interestingly, we observed that relative to LF feeding, the HF regime reduced the DHA concentration (in nmol/g) in the brain, as reflected in the lower brain: plasma DHA ratio, which could potentially buffer any physiological impact of FLAV on brain DHA levels. Furthermore, the brain region may have an impact on the size effect, with our analysis conducted in the subcortical region versus the whole brain in the previous analysis [55]. Somewhat consistent with a previous finding [55], we did observe a higher brain: RBC (%) ratio of DHA in the *APOE3*, but not the *APOE4* animals (Figure 2D), attributable to lower RBC DHA and a trend towards higher brain DHA (%). The aetiology of this finding is unknown, but is suggestive of an impact of FLAV on DHA biosynthesis, or retention/metabolism within the brain in the common *APOE3* genotype. This is worthy of further investigation, as it represents a potential means of increasing brain DHA without the need for increased intake.

The balance between DHA and AA is tightly regulated in the phospholipids of neuronal membranes and is crucial to ensure membrane fluidity, plasticity, and neurotransmission [52]. The DHA: AA ratio is also used in the diagnosis of MCI/AD pathology [56]. We recently demonstrated that old age and carrying an *APOE4* genotype decreased the brain DHA: AA ratio [21]. Interestingly, we did not observe any significant genotype effect in the brain DHA: AA ratio in the present study, which is likely attributable to the sex of the animals, with the effect more evident in females [4,7].

Less than 2% of fatty acids are present in the brain as free fatty acids. The majority are present as membrane phospholipids: PC, PE, PS, and PI. PS and PE are quantitatively the major brain DHA pools [39]. As ApoE is involved in brain phospholipid transport and metabolism, carrying an *APOE4* genotype may impair the structure of the neuronal membrane affecting DHA transport and downstream signalling pathways [57]. We observed that the higher total brain DHA in the DHA-enriched diet group in *APOE3* mice was reflected in the PC fraction. In *APOE4*, DHA supplementation increased DHA in both the PC and PE fraction, with the size effect for LPC not reaching significance. The addition of FLAV to the HF diet (with or without DHA) did not result in phospholipid-fraction enrichment. Although FO did not increase the percentage of brain DHA in *APOE4*, a 17.1% ($p < 0.05$) and 20.0% ($p < 0.001$) higher DHA level in the PC fraction of the HF FO and HF FO + FLAV groups, and a 14.5% ($p < 0.05$) higher DHA level in the PE fraction of the HF FO + FLAV group was evident in these animals relative to the HF controls. Flavonoids have been reported to increase the level of different phospholipids. For example, quercetin and naringenin increased PC and to a lesser extent PE in LPS-induced macrophages, which showed higher levels of longer chain and more polyunsaturated fatty acids (36–40 carbons) in PCs while no effect was observed in the PI and PS fractions. [58].

A limitation was that our analysis did not include the PS fraction, which is known to be a major anionic phospholipid class in neuronal membranes and particularly sensitive to DHA deprivation [59]. However, PE, which is a precursor of PS, was captured.

5. Conclusions

The current analysis shows an *APOE*-dependent change in brain and blood DHA levels, in response to fish oil and cocoa flavanols interventions. The *APOE4* genotype was associated with lower brain-DHA enrichment despite having higher blood DHA, compared to *APOE3*. An investigation into human bio-banked samples from human males and females is needed to confirm these findings and to help establish if *APOE4s* should be targeted with recommendations to optimise the DHA brain status. There was some evidence of a positive impact of flavanols on the brain-DHA status in *APOE3* only, which merits further investigation to confirm the findings and establish underpinning mechanisms.

Supplementary Materials: The following supporting information can be downloaded at: https://www.mdpi.com/article/10.3390/nu15092032/s1, Table S1: Fatty acid content in feeds; Table S2: Impact of diet and *APOE* genotype on brain neutral lipid AA and DHA content; Table S3: Diet effect on arachidonic acid phospholipid fractions in *APOE3* and *APOE4*-TR mice after 22 weeks of treatment.

Author Contributions: A.M., D.V. and A.-M.M. conceptualised and designed the experiments and analytical approaches; D.V. provided the Home Office Animal License. D.V., N.T., M.G.P., G.H. and A.M. performed the research and subsequent sample processing; J.D. performed the fatty acids analysis on feeds. R.C.-W. performed the fatty acid analysis on blood samples. A.M. performed all other analyses and analysed the data. A.M. and R.N.M.S. wrote the manuscript with contributions from all authors. R.C.-W. and R.B. critically revised the manuscript. G.H. contributed to the animal husbandry. All authors have read and agreed to the published version of the manuscript.

Funding: The research was funded as part of a Biotechnology and Biological Sciences Research Council (BBSRC, BB/J004545/1) Institute Programme Grant.

Institutional Review Board Statement: Experimental procedures and protocols were reviewed and approved by the Animal Welfare and Ethical Review Board (AWERB) and were conducted within the provisions of the Home Office Animals (Scientific Procedures) Act 1986.

Informed Consent Statement: Not applicable.

Data Availability Statement: The data presented in this study are available on request from the corresponding author.

Acknowledgments: The authors thank the staff of the Disease Modelling Unit at the University of East Anglia for expertise and help conducting the rodent studies. We would also like to thank Jack Dainty for his statistical support.

Conflicts of Interest: The authors declare no conflict of interest.

References

1. Davidson, Y.; Gibbons, L.; Pritchard, A.; Hardicre, J.; Wren, J.; Stopford, C.; Julien, C.; Thompson, J.; Payton, A.; Pickering-Brown, S.M.; et al. Apolipoprotein E ε4 Allele Frequency and Age at Onset of Alzheimer's Disease. *Dement. Geriatr. Cogn. Disord.* **2006**, *23*, 60–66. [CrossRef] [PubMed]
2. Heffernan, A.L.; Chidgey, C.; Peng, P.; Masters, C.L.; Roberts, B.R. The Neurobiology and Age-Related Prevalence of the epsilon4 Allele of Apolipoprotein E in Alzheimer's Disease Cohorts. *J. Mol. Neurosci.* **2016**, *60*, 316–324. [CrossRef] [PubMed]
3. Michaelson, D.M. APOE ε4: The most prevalent yet understudied risk factor for Alzheimer's disease. *Alzheimer's Dement.* **2014**, *10*, 861–868. [CrossRef] [PubMed]
4. Pontifex, M.G.; Martinsen, A.; Saleh, R.N.M.; Harden, G.; Fox, C.; Muller, M.; Vauzour, D.; Minihane, A.-M. DHA-Enriched Fish Oil Ameliorates Deficits in Cognition Associated with Menopause and the APOE4 Genotype in Rodents. *Nutrients* **2022**, *14*, 1698. [CrossRef]
5. Jennings, A.; Cunnane, S.C.; Minihane, A.M. Can nutrition support healthy cognitive ageing and reduce dementia risk? *BMJ* **2020**, *369*, m2269. [CrossRef] [PubMed]
6. Arterburn, L.M.; Hall, E.B.; Oken, H. Distribution, interconversion, and dose response of n−3 fatty acids in humans. *Am. J. Clin. Nutr.* **2006**, *83*, 1467S–1476S. [CrossRef]
7. Pontifex, M.; Vauzour, D.; Minihane, A.-M. The effect of APOE genotype on Alzheimer's disease risk is influenced by sex and docosahexaenoic acid status. *Neurobiol. Aging* **2018**, *69*, 209–220. [CrossRef]
8. López-Vicario, C.; Rius, B.; Alcaraz-Quiles, J.; García-Alonso, V.; Lopategi, A.; Titos, E.; Clària, J. Pro-resolving mediators produced from EPA and DHA: Overview of the pathways involved and their mechanisms in metabolic syndrome and related liver diseases. *Eur. J. Pharmacol.* **2016**, *785*, 133–143. [CrossRef]
9. Tanaka, K.; Farooqui, A.A.; Siddiqi, N.J.; Alhomida, A.S.; Ong, W.-Y. Effects of Docosahexaenoic Acid on Neurotransmission. *Biomol. Ther.* **2012**, *20*, 152–157. [CrossRef]
10. Lim, G.P.; Calon, F.; Morihara, T.; Yang, F.; Teter, B.; Ubeda, O.; Salem, N., Jr.; Frautschy, S.A.; Cole, G.M. A Diet Enriched with the Omega-3 Fatty Acid Docosahexaenoic Acid Reduces Amyloid Burden in an Aged Alzheimer Mouse Model. *J. Neurosci.* **2005**, *25*, 3032–3040. [CrossRef]
11. Taha, A.Y.; Cheon, Y.; Ma, K.; Rapoport, S.I.; Rao, J.S. Altered fatty acid concentrations in prefrontal cortex of schizophrenic patients. *J. Psychiatr. Res.* **2013**, *47*, 636–643. [CrossRef] [PubMed]
12. Norris, C.; Fong, B.; MacGibbon, A.; McJarrow, P. Analysis of Phospholipids in Rat Brain Using Liquid Chromatography—Mass Spectrometry. *Lipids* **2009**, *44*, 1047–1054. [CrossRef] [PubMed]
13. Sharman, M.J.; Shui, G.; Fernandis, A.Z.; Lim, W.L.F.; Berger, T.; Hone, E.; Taddei, K.; Martins, I.J.; Ghiso, J.; Buxbaum, J.D.; et al. Profiling Brain and Plasma Lipids in Human APOE ε2, ε3, and ε4 Knock-in Mice Using Electrospray Ionization Mass Spectrometry. *J. Alzheimer's Dis.* **2010**, *20*, 105–111. [CrossRef]
14. Kariv-Inbal, Z.; Yacobson, S.; Berkecz, R.; Peter, M.; Janaky, T.; Lütjohann, D.; Broersen, L.M.; Hartmann, T.; Michaelson, D.M. The Isoform-Specific Pathological Effects of ApoE4 in vivo are Prevented by a Fish Oil (DHA) Diet and are Modified by Cholesterol. *J. Alzheimer's Dis.* **2012**, *28*, 667–683. [CrossRef]
15. Chouinard-Watkins, R.; Rioux-Perreault, C.; Fortier, M.; Tremblay-Mercier, J.; Zhang, Y.; Lawrence, P.; Vohl, M.C.; Perron, P.; Lorrain, D.; Brenna, J.T.; et al. Disturbance in uniformly ^{13}C-labelled DHA metabolism in elderly human subjects carrying the apoE ε4 allele. *Br. J. Nutr.* **2013**, *110*, 1751–1759. [CrossRef] [PubMed]
16. Yassine, H.N.; Rawat, V.; Mack, W.J.; Quinn, J.F.; Yurko-Mauro, K.; Bailey-Hall, E.; Aisen, P.S.; Chui, H.C.; Schneider, L.S. The effect of APOE genotype on the delivery of DHA to cerebrospinal fluid in Alzheimer's disease. *Alzheimer's Res. Ther.* **2016**, *8*, 25. [CrossRef] [PubMed]

17. Ebright, B.; Assante, I.; Poblete, R.A.; Wang, S.; Duro, M.V.; Bennett, D.A.; Arvanitakis, Z.; Louie, S.G.; Yassine, H.N. Eicosanoid lipidome activation in post-mortem brain tissues of individuals with APOE4 and Alzheimer's dementia. *Alzheimer's Res. Ther.* **2022**, *14*, 152. [CrossRef]
18. Yassine, H.N.; Braskie, M.N.; Mack, W.J.; Castor, K.J.; Fonteh, A.N.; Schneider, L.S.; Harrington, M.; Chui, H.C. Association of Docosahexaenoic Acid Supplementation With Alzheimer Disease Stage in Apolipoprotein E ε4 Carriers. *JAMA Neurol.* **2017**, *74*, 339–347. [CrossRef]
19. Quinn, J.F.; Raman, R.; Thomas, R.G.; Yurko-Mauro, K.; Nelson, E.B.; Van Dyck, C.; Galvin, J.E.; Emond, J.; Jack, C.R.; Weiner, M.; et al. Docosahexaenoic Acid Supplementation and Cognitive Decline in Alzheimer Disease: A randomized trial. *JAMA* **2010**, *304*, 1903–1911. [CrossRef]
20. Chouinard-Watkins, R.; Conway, V.; Minihane, A.M.; Jackson, K.G.; Lovegrove, J.A.; Plourde, M. Interaction between BMI and APOE genotype is associated with changes in the plasma long-chain–PUFA response to a fish-oil supplement in healthy participants. *Am. J. Clin. Nutr.* **2015**, *102*, 505–513. [CrossRef]
21. Martinsen, A.; Tejera Hernandez, N.; Vauzour, D.; Harden, G.; Dick, J.; Shinde, S.; Barden, A.; Mori, T.; Minihane, A.M. Altered SPMs and age-associated decrease in brain DHA in APOE4 female mice. *FASEB J.* **2019**, *33*, 10315–10326. [CrossRef] [PubMed]
22. Trebble, T.; Arden, N.K.; Stroud, M.A.; Wootton, S.A.; Burdge, G.C.; Miles, E.A.; Ballinger, A.B.; Thompson, R.L.; Calder, P.C. Inhibition of tumour necrosis factor-α and interleukin 6 production by mononuclear cells following dietary fish-oil supplementation in healthy men and response to antioxidant co-supplementation. *Br. J. Nutr.* **2003**, *90*, 405–412. [CrossRef] [PubMed]
23. De Cosmi, V.; Mazzocchi, A.; D'oria, V.; Re, A.; Spolidoro, G.C.I.; Milani, G.P.; Berti, C.; Scaglioni, S.; Giavoli, C.; Bergamaschi, S.; et al. Effect of Vitamin D and Docosahexaenoic Acid Co-Supplementation on Vitamin D Status, Body Composition, and Metabolic Markers in Obese Children: A Randomized, Double Blind, Controlled Study. *Nutrients* **2022**, *14*, 1397. [CrossRef]
24. Song, L.; Zhou, H.; Yu, W.; Ding, X.; Yang, L.; Wu, J.; Song, C. Effects of Phytosterol Ester on the Fatty Acid Profiles in Rats with Nonalcoholic Fatty Liver Disease. *J. Med. Food* **2020**, *23*, 161–172. [CrossRef]
25. van Soest, A.P.M.; van de Rest, O.; Witkamp, R.F.; Cederholm, T.; de Groot, L.C.P.G.M. DHA status influences effects of B-vitamin supplementation on cognitive ageing: A post-hoc analysis of the B-proof trial. *Eur. J. Nutr.* **2022**, *61*, 3731–3739. [CrossRef] [PubMed]
26. Vauzour, D.; Martinsen, A.; Layé, S. Neuroinflammatory processes in cognitive disorders: Is there a role for flavonoids and n-3 polyunsaturated fatty acids in counteracting their detrimental effects? *Neurochem. Int.* **2015**, *89*, 63–74. [CrossRef] [PubMed]
27. Jin, H.; Kim, H.S.; Yu, S.T.; Shin, S.R.; Lee, S.H.; Seo, G.S. Synergistic anticancer effect of docosahexaenoic acid and isoliquiritigenin on human colorectal cancer cells through ROS-mediated regulation of the JNK and cytochrome c release. *Mol. Biol. Rep.* **2021**, *48*, 1171–1180. [CrossRef]
28. Pontifex, M.G.; Malik, M.M.A.H.; Connell, E.; Müller, M.; Vauzour, D. Citrus Polyphenols in Brain Health and Disease: Current Perspectives. *Front. Neurosci.* **2021**, *15*, 640648. [CrossRef]
29. Agarwal, P.; Holland, T.M.; James, B.D.; Cherian, L.J.; Aggarwal, N.T.; Leurgans, S.E.; Bennett, D.A.; Schneider, J.A. Pelargonidin and Berry Intake Association with Alzheimer's Disease Neuropathology: A Community-Based Study. *J. Alzheimer's Dis.* **2022**, *88*, 653–661. [CrossRef]
30. Baker, L.D.; Manson, J.E.; Rapp, S.R.; Sesso, H.D.; Gaussoin, S.A.; Shumaker, S.A.; Espeland, M.A. Effects of cocoa extract and a multivitamin on cognitive function: A randomized clinical trial. *Alzheimer's Dement.* **2022**, *19*, 1308–1319. [CrossRef]
31. Sun, C.Q.; Johnson, K.D.; Wong, H.; Foo, L.Y. Biotransformation of Flavonoid Conjugates with Fatty Acids and Evaluations of Their Functionalities. *Front. Pharmacol.* **2017**, *8*, 759. [CrossRef]
32. Zhu, Y.; Nwabuisi-Heath, E.; Dumanis, S.B.; Tai, L.M.; Yu, C.; Rebeck, G.W.; Ladu, M.J. APOE genotype alters glial activation and loss of synaptic markers in mice. *Glia* **2012**, *60*, 559–569. [CrossRef] [PubMed]
33. Minihane, A.M. Fish oil omega-3 fatty acids and cardio-metabolic health, alone or with statins. *Eur. J. Clin. Nutr.* **2013**, *67*, 536–540. [CrossRef] [PubMed]
34. Folch, J.; Lees, M.; Sloane Stanley, G.H. A simple method for the isolation and purification of total lipides from animal tissues. *J. Biol. Chem.* **1957**, *226*, 497–509. [CrossRef] [PubMed]
35. Pontifex, M.G.; Martinsen, A.; Saleh, R.N.M.; Harden, G.; Tejera, N.; Müller, M.; Fox, C.; Vauzour, D.; Minihane, A. APOE4 genotype exacerbates the impact of menopause on cognition and synaptic plasticity in APOE-TR mice. *FASEB J.* **2021**, *35*, e21583. [CrossRef]
36. Ghioni, C.; Bell, J.; Sargent, J. Polyunsaturated fatty acids in neutral lipids and phospholipids of some freshwater insects. *Comp. Biochem. Physiol. Part B Biochem. Mol. Biol.* **1996**, *114*, 161–170. [CrossRef]
37. Christie, W.W. *Lipid Analysis*, 3rd ed.; The Oily Press: Bridgewater, UK, 2003; pp. 205–224.
38. Ackman, R.G. Fish lipids. In *Advances in Fish Science and Technology*; Connell, J.J., Ed.; Fishing News Books Ltd.: Farnham, UK, 1980; pp. 83–103.
39. Lacombe, R.; Chouinard-Watkins, R.; Bazinet, R.P. Brain docosahexaenoic acid uptake and metabolism. *Mol. Asp. Med.* **2018**, *64*, 109–134. [CrossRef]
40. Chouinard-Watkins, R.; Chen, C.T.; Metherel, A.H.; Lacombe, R.S.; Thies, F.; Masoodi, M.; Bazinet, R.P. Phospholipid class-specific brain enrichment in response to lysophosphatidylcholine docosahexaenoic acid infusion. *Biochim. Biophys. Acta (BBA)-Mol. Cell Biol. Lipids* **2017**, *1862*, 1092–1098. [CrossRef]
41. Jensen, S.K. Improved Bligh and Dyer extraction procedure. *Lipid Technol.* **2008**, *20*, 280–281. [CrossRef]

42. Vandal, M.; Alata, W.; Tremblay, C.; Rioux-Perreault, C.; Salem, N.; Calon, F.; Plourde, M. Reduction in DHA transport to the brain of mice expressing human APOE4 compared to APOE2. *J. Neurochem.* **2014**, *129*, 516–526. [CrossRef]
43. Culliford, A.; Bradbury, J. A cross-sectional survey of the readiness of consumers to adopt an environmentally sustainable diet. *Nutr. J.* **2020**, *19*, 138. [CrossRef] [PubMed]
44. Derbyshire, E. Oily Fish and Omega-3s Across the Life Stages: A Focus on Intakes and Future Directions. *Front. Nutr.* **2019**, *6*, 165. [CrossRef] [PubMed]
45. Micha, R.; Khatibzadeh, S.; Shi, P.; Fahimi, S.; Lim, S.; Andrews, K.G.; Engell, R.E.; Powles, J.; Ezzati, M.; Mozaffarian, D.; et al. Global, regional, and national consumption levels of dietary fats and oils in 1990 and 2010: A systematic analysis including 266 country-specific nutrition surveys. *BMJ* **2014**, *348*, g2272. [CrossRef] [PubMed]
46. Stark, K.D.; Van Elswyk, M.E.; Higgins, M.R.; Weatherford, C.A.; Salem, N., Jr. Global survey of the omega-3 fatty acids, docosahexaenoic acid and eicosapentaenoic acid in the blood stream of healthy adults. *Prog. Lipid Res.* **2016**, *63*, 132–152. [CrossRef]
47. GBD 2016 Dementia Collaborators. Global, regional, and national burden of Alzheimer's disease and other dementias, 1990–2016: A systematic analysis for the Global Burden of Disease Study 2016. *Lancet Neurol.* **2019**, *18*, 88–106. [CrossRef]
48. Jones, N.S.; Rebeck, G.W. The Synergistic Effects of APOE Genotype and Obesity on Alzheimer's Disease Risk. *Int. J. Mol. Sci.* **2018**, *20*, 63. [CrossRef]
49. I Kosti, R.; I Kasdagli, M.; Kyrozis, A.; Orsini, N.; Lagiou, P.; Taiganidou, F.; Naska, A. Fish intake, n-3 fatty acid body status, and risk of cognitive decline: A systematic review and a dose–response meta-analysis of observational and experimental studies. *Nutr. Rev.* **2022**, *80*, 1445–1458. [CrossRef]
50. Zhang, Y.; Zhuang, P.; He, W.; Chen, J.N.; Wang, W.Q.; Freedman, N.D.; Abnet, C.; Wang, J.B.; Jiao, J.J. Association of fish and long-chain omega-3 fatty acids intakes with total and cause-specific mortality: Prospective analysis of 421 309 individuals. *J. Intern. Med.* **2018**, *284*, 399–417. [CrossRef]
51. Huang, T.L.; Zandi, P.P.; Tucker, K.L.; Fitzpatrick, A.L.; Kuller, L.H.; Fried, L.P.; Burke, G.L.; Carlson, M.C. Benefits of fatty fish on dementia risk are stronger for those without APOEε4. *Neurology* **2005**, *65*, 1409–1414. [CrossRef]
52. Chouinard-Watkins, R.; Vandal, M.; Léveillé, P.; Pinçon, A.; Calon, F.; Plourde, M. Docosahexaenoic acid prevents cognitive deficits in human apolipoprotein E epsilon 4-targeted replacement mice. *Neurobiol. Aging* **2017**, *57*, 28–35. [CrossRef]
53. Nock, T.G.; Chouinard-Watkins, R.; Plourde, M. Carriers of an apolipoprotein E epsilon 4 allele are more vulnerable to a dietary deficiency in omega-3 fatty acids and cognitive decline. *Biochim. Biophys. Acta (BBA)-Mol. Cell Biol. Lipids* **2017**, *1862*, 1068–1078. [CrossRef] [PubMed]
54. Giunta, B.; Hou, H.; Zhu, Y.; Salemi, J.; Ruscin, A.; Shytle, R.D.; Tan, J. Fish oil enhances anti-amyloidogenic properties of green tea EGCG in Tg2576 mice. *Neurosci. Lett.* **2010**, *471*, 134–138. [CrossRef] [PubMed]
55. Shirai, N.; Suzuki, H. Effect of Dietary Docosahexaenoic Acid and Catechins on Maze Behavior in Mice. *Ann. Nutr. Metab.* **2004**, *48*, 51–58. [CrossRef] [PubMed]
56. Abdullah, L.; Evans, J.E.; Emmerich, T.; Crynen, G.; Shackleton, B.; Keegan, A.P.; Luis, C.; Tai, L.; LaDu, M.J.; Mullan, M.; et al. APOE ε4 specific imbalance of arachidonic acid and docosahexaenoic acid in serum phospholipids identifies individuals with preclinical Mild Cognitive Impairment/Alzheimer's Disease. *Aging* **2017**, *9*, 964–985. [CrossRef]
57. Bazan, N.G.; Molina, M.F.; Gordon, W.C. Docosahexaenoic Acid Signalolipidomics in Nutrition: Significance in Aging, Neuroinflammation, Macular Degeneration, Alzheimer's, and Other Neurodegenerative Diseases. *Annu. Rev. Nutr.* **2011**, *31*, 321–351. [CrossRef]
58. Conde, T.A.; Mendes, L.; Gaspar, V.M.; Mano, J.F.; Melo, T.; Domingues, M.R.; Duarte, I.F. Differential Modulation of the Phospholipidome of Proinflammatory Human Macrophages by the Flavonoids Quercetin, Naringin and Naringenin. *Molecules* **2020**, *25*, 3460. [CrossRef]
59. Hamilton, J.; Greiner, R.; Salem, N.; Kim, H.-Y. n−3 Fatty acid deficiency decreases phosphatidylserine accumulation selectively in neuronal tissues. *Lipids* **2000**, *35*, 863–869. [CrossRef]

nutrients

MDPI

Article

Soy Extract, Rich in Hydroxylated Isoflavones, Exhibits Antidiabetic Properties In Vitro and in *Drosophila melanogaster* In Vivo

Kai Lüersen [1,*], Alexandra Fischer [1], Ilka Bauer [1], Patricia Huebbe [1], Yukiko Uekaji [2], Keita Chikamoto [2], Daisuke Nakata [2], Naoto Hiramatsu [3], Keiji Terao [2] and Gerald Rimbach [1]

1 Institute of Human Nutrition and Food Science, University of Kiel, 24118 Kiel, Germany
2 CycloChem Bio Co., Ltd., 7-4-5 Minatojima-Minamimachi, Chuo-ku, Kobe 650-0047, Hyogo, Japan
3 Toyo Hakko Co., Ltd., 1-39-1 Yoshikawa-cho, Obu-shi 474-0046, Aichi, Japan
* Correspondence: luersen@foodsci.uni-kiel.de; Tel.: +49-431-880-5312

Abstract: In the context of the growing prevalence of type 2 diabetes (T2DM), control of postprandial hyperglycemia is crucial for its prevention. Blood glucose levels are determined by various factors including carbohydrate hydrolyzing enzymes, the incretin system and glucose transporters. Furthermore, inflammatory markers are recognized predictors of diabetes outcome. Although there is some evidence that isoflavones may exhibit anti-diabetic properties, little is known about to what extent their corresponding hydroxylated metabolites may affect glucose metabolism. We evaluated the ability of a soy extract before (pre-) and after (post-) fermentation to counteract hyperglycemia in vitro and in *Drosophila melanogaster* in vivo. Fermentation with *Aspergillus* sp. *JCM22299* led to an enrichment of hydroxy-isoflavones (HI), including 8-hydroxygenistein, 8-hydroxyglycitein and 8-hydroxydaidzein, accompanied by an enhanced free radical scavenging activity. This HI-rich extract demonstrated inhibitory activity towards α-glucosidase and a reduction of dipeptidyl peptidase-4 enzyme activity. Both the pre- and post-fermented extracts significantly inhibited the glucose transport via sodium-dependent glucose transporter 1. Furthermore, the soy extracts reduced c-reactive protein mRNA and secreted protein levels in interleukin-stimulated Hep B3 cells. Finally, supplementation of a high-starch *D. melanogaster* diet with post-fermented HI-rich extract decreased the triacylglyceride content of female fruit flies, confirming its anti-diabetic properties in an in vivo model.

Keywords: soy; isoflavones; hydroxy isoflavones; bioactivity; glucose metabolism

Citation: Lüersen, K.; Fischer, A.; Bauer, I.; Huebbe, P.; Uekaji, Y.; Chikamoto, K.; Nakata, D.; Hiramatsu, N.; Terao, K.; Rimbach, G. Soy Extract, Rich in Hydroxylated Isoflavones, Exhibits Antidiabetic Properties In Vitro and in *Drosophila melanogaster* In Vivo. *Nutrients* **2023**, *15*, 1392. https://doi.org/10.3390/nu15061392

Academic Editor: Adam Matkowski

Received: 10 January 2023
Revised: 3 March 2023
Accepted: 7 March 2023
Published: 14 March 2023

1. Introduction

The prevalence of diabetes, especially type 2 diabetes mellitus (T2DM), is increasing globally [1]. This metabolic disease is characterized by hyperglycaemia, induced by a progressive insulin secretory defect or a diminished or missing response of insulin receptors [2]. Continuously high blood sugar levels may result in long-term complications, including renal and cardiovascular diseases, retinopathy or an impaired blood flow [3]. Thus, controlling postprandial hyperglycemia through dietary means is crucial for the prevention of T2DM. Blood glucose levels are determined by various factors including food constituents and food matrix, glucose transporters, carbohydrate hydrolyzing enzymes (e.g., α-glucosidase, α-amylase) and hormones such as the incretin system. The incretin system relates to the gut hormones glucagon-like peptide-1 (GLP-1) and glucose-dependent insulin polypeptide (GIP), which increase postprandial insulin production by acting on pancreatic beta cells. Dipeptidyl peptidase-4 (DPP-4) degrades circulating GLP-1 and GIP and reduces the circulating postprandial glucagon level [4]; hence, DPP4 inhibitors are regarded as a novel means for extending the action of insulin and treating T2DM. In addition, intestinal glucose absorption is largely achieved by sodium/glucose symporter 1 (SGLT1) [5]. SGLT1 expression is regulated by the diet, e.g., it is strongly elevated in response to intraluminal glucose and by compounds that activate sweet taste receptors [6]. Moreover, it is induced in patients with T2DM [7]. The development of T2DM is usually associated with chronic inflammation. Accordingly, C-reactive protein (CRP), which is considered a sensitive systemic marker of low-grade inflammation, has been found to be a proper biomarker for T2DM [8]. CRP predominantly synthesized and secreted by hepatocytes is elicited by the dual activity of interleukin (IL)-6 and IL-1ß, and enhances inflammatory pathways by inducing IL-6 secretion [9,10]. A recent meta-analysis investigated the impact of soy intake on inflammatory markers and revealed significantly decreased CRP levels in women; however, the underlying mechanisms by which soy foods and their ingredients influence inflammatory biomarkers has not yet been elucidated [11].

Importantly, legumes, specifically soybeans, are a widespread dietary source of isoflavones, with genistein, daidzein and glycitein forming the major fraction [12]. In contrast, their hydroxylated counterparts, such as 8-hydroxygenistein (8OHGen), 8-hydroxydaidzein (8OHDai) and 8-hydroxyglycitein (8OHGly), are barely found in plants. Hydroxylation at either the C6 or C8 carbon position of the isoflavone backbone is not prioritized during isoflavone biosynthesis in plants, as cytochrome P450 (CYP)-dependent enzymes from plants seem to not catalyze the ortho-hydroxylation. However, food processing can affect the isoflavone concentration and isomer composition of soy products. Thus, most hydroxylated isoflavones (HI) are derived from fermented soybean foods, such as miso, natto, soy sauce and tempeh, where microorganisms, mainly fungi (e.g., *Aspergillus*) and bacteria (e.g., *Rhizopus, Streptomyces*), incorporate the hydroxyl-group into the isoflavone molecule during fermentation in a CYP-dependent manner [12,13]. Furthermore, the production of HIs is also feasible via microbial production using genetic engineering [14]. In addition, fermentation increases the cleavage of glycoside bonds of isoflavones, thereby enhancing their bioavailability [15,16].

The basic chemical structure of the isoflavones consists of two benzene rings (1 and 2) linked via a heterocyclic pyrone ring (3) (Figure 1).

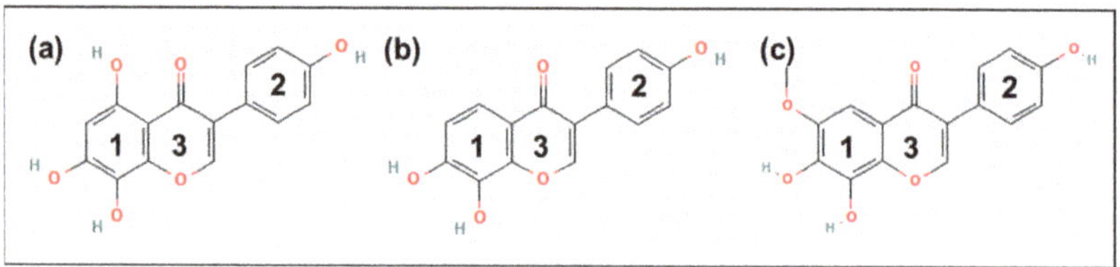

Figure 1. Chemical structure of (**a**) 8-hydroxygenistein (8OHGen; 4′,5,7,8-Tetrahydroxyisoflavone; $C_{15}H_{10}O_6$, PubChem CID: 5492944), (**b**) 8-hydroxydaidzein (8OHDai; 7,8,4′-Trihydroxyisoflavone; $C_{15}H_{10}O_5$, PubChem CID: 5466139) and (**c**) 8-hydroxyglycitein (8OHGly; 7,8,4′-Trihydroxy-6-methoxyisoflavone; $C_{16}H_{12}O_6$, PubChem CID: 10870296). Structures were taken from PubChem [17]. **1:** Benzene ring 1; **2:** Benzene ring 2; **3:** Heterocyclic pyrone ring.

Isoflavones have been shown to induce endogenous antioxidant defense mechanisms, such as glutathione peroxidase, catalase and superoxide dismutase [18], presumably via an Nrf2-dependent signal transduction pathway [19]. Furthermore, an inhibition of lipoxygenase due to isoflavones has been described [20]. In addition, the dietary intake of genistein and daidzein improved the resistance of LDL against oxidation [21].

Genistein and daidzein are relatively weak scavengers of hydroxyl, superoxide, and nitric oxide free radicals, as determined using spin trapping and electron spin resonance spectroscopy [22]. Interestingly, it has been shown in vitro that the corresponding hydroxylated metabolites of genistein and daidzein exhibited significantly stronger bioactivity in terms of prevention of lipid peroxidation as compared to the parent compounds per se [23,24]. Hirota et al. [25] found that, among various isoflavones isolated from soybean miso, 8OHGen represented the most potent antimutagenic and antiproliferative activity. Likewise, in contrast to daidzein, 8OHDai has been found to be a potent aldose reductase inhibitor in vitro [26] and may, therefore, represent a potential substance for the treatment of diabetic complications.

However, fermented isoflavones remain an "understudied" group of soy compounds and little is known about the bioactivity of isoflavones regarding their antidiabetic properties [15], especially to what extent their corresponding hydroxylated metabolites, including 8OHGen, 8OHDai and 8OHGly, may affect glucose metabolism and biomarkers of inflammation. The aim of our study was to investigate a soybean extract before (pre-) fermentation and, in particular, a HI-rich extract after (post-) fermentation, regarding their antidiabetic and anti-inflammatory properties in vitro. *Aspergillus* sp. JCM 22,299 was applied for fermentation, which resulted in a substantial increase in HI, where 8OHGen represented the highest fraction. We tested the in vitro impact of the HI extract on α-glucosidase inhibition, starch digestion via α-amylase and the incretin system regarding DPP4 inhibition. The influence of soy bean pre-extract and HI extract was further evaluated based on the activity of glucose transport by SGLT1 using the Caco-2 cell culture model and CRP expression (mRNA and secreted protein level) in cultured hepatocytes.

In order to take into account pharmacological aspects, such as bioavailability, the influence of the gut microbiota and biotransformation, the verification of in vitro bioactivities in animal models usually represents the next important step in the development of new therapeutic approaches. In recent years, the fruit fly *Drosophila melanogaster* has been acknowledged as a valuable model in food science [27]. In particular, owing to the remarkable similarity between *D. melanogaster* and humans in terms of the metabolic pathways involved in energy metabolism and its hormonal regulation, several genetic and diet-based diabetes models have been established in fruit flies [28–30], which can serve as helpful tools to test putative antidiabetic substances [31,32]. Accordingly, we employed a high-sugar diet *D. melanogaster* obesity model, which we have previously used to validate the efficacy of plant extracts with α-amylase and α-glucosidase inhibitor activities [32] to examine whether the post-fermented HI extract exerted antidiabetic properties in vivo.

2. Materials and Methods

2.1. Preparation of Pre-Fermented and Hydroxy-Isoflavone (HI)-Enriched Post-Fermented Soybean Extract and Isoflavone Analysis Using HPLC

Soy isoflavone extract was supplied by Toyo Hakko Co., Ltd. (Aichi, Japan). To obtain an extract enriched in HI, soy isoflavone extracts (pre-fermented samples) were sterilized at 121 °C for 20 min and, subsequently, fermented with *Aspergillus* sp. JCM 22,299 for 8 d at 30 °C with aeration and continuous stirring. After ethanol extraction and two separation steps, the extract was concentrated using evaporation and filtrated through a 30-mesh filter. The obtained post-fermented samples as well as the pre-fermented samples were analyzed in terms of their isoflavone content through HPLC (Prominence UFLC, Shimadzu Corporation, Kyoto, Japan) using a Phenomenex, Kinetex C18 column, 100A (5 μm, 4.6 mm I.D. × 250 mm, Phenomenex Inc., Torrance, CA, USA). Mobile phase A consisted of 0.1% acetic acid and mobile phase B was made of acetonitrile while using a gradient profile (0~50 min (15~35% B), 50~55 min (100% B), 55~60 min (15% B)) at a flow rate of 1.0 mL/min and a temperature of 35 °C. The injection volume was 10 μL and detection was carried out at 254 nm. An external standard curve was applied to calculate the concentration using the peak area (Figure 2). Hence, the concentrations of 8OHGen, 8OHGly and 8OHDai were determined to be 35%, 9% and 8% of total isoflavones content in the post-fermented extract, respectively.

2.2. Antioxidant Capacity Assays

The free radical scavenging properties of pre- and post-fermented soy extracts (10 μg/mL) were determined with the ferric-reducing ability of plasma (FRAP) assay and the trolox equivalent antioxidant capacity (TEAC) assay, as previously described [33,34]. For the FRAP assay, which measures how well a test compound reduces ferric ions (Fe^{3+}) to ferrous ions (Fe^{2+}), pre- or post-fermented soy extracts were added to a 2 mM iron (III) chloride solution with 2,4,6-tris(2-pyridyl)-s-triazine (TPTZ, 1 mM) in acetate buffer (228 mM, pH 3.6). After 15 min of incubation, absorbance was measured at 620 nm in an iEMS reader MF (Labsystems, Helsinki, Finland). Results are given in µmol ascorbic acid equivalents per mg extract. The TEAC assay is related to the reduction of ABTS (2,20-azino-bis-(3-ethylbenzthiazoline-6-sulfonic acid) radical cation by antioxidants. The increase in reduction by pre- or post-fermented extracts was photometrically measured at 750 nm in a Tecan Infinite 200 (Tecan Group Ltd., Crailsheim, Germany) microplate reader and compared to trolox as external standard. TEAC values are given in µmol trolox equivalents per mg extract.

(a)

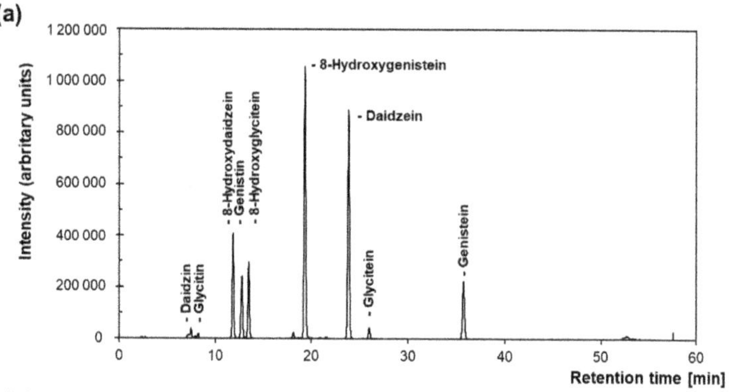

(b)

Peak number	Isoflavone	Retention time t_R (min)	Isoflavon conc. pre-fermented (% of total isoflavones)	Isoflavon conc. post-fermented (HI) (% of total isoflavones)
1	Daidzin	7.5	35%	3%
2	Glycitin	8.3	13%	1%
3	8-Hydroxydaidzein	11.9	-	8%
4	Genistin	12.8	49%	9%
5	8-Hydroxyglycitein	13.6	-	9%
6	8-Hydroxygenistein	19.3	-	35%
7	Daidzein	23.9	1%	28%
8	Glycitein	26.0	-	1%
9	Genistein	35.7	1%	7%

Figure 2. (**a**) Representative HPLC chromatogram of hydroxy-isoflavones (HI)-enriched post-fermented soybean extract at 254 nm and (**b**) the corresponding relative concentrations of identified isoflavones in pre- and post-fermented soybean extract.

2.3. *Enzymatic Assays*

2.3.1. In Vitro α-Glucosidase Inhibition Assay

A total of 15 µL of diluted (0.05–10 mg/mL) HI extracts was added to 105 µL of 0.1 M phosphate buffer, pH 6.8 and 15 µL of 0.5 U/mL baker yeast α-glucosidase (Sigma-Aldrich, Taufkirchen, Germany). Acarbose was used as a reference inhibitor. After 5 min of pre-incubation at 37 °C, 15 µL of the substrate p-nitrophenyl-α-D-glucopyranoside (10 mM, Sigma-Aldrich, Taufkirchen, Germany) was added and incubated for 20 min at 37 °C in a 96-well microtest plate (VWR, Darmstadt, Germany). The reaction was stopped by adding 50 µL 2 M Na_2CO_3 (VWR, Darmstadt, Germany) and the absorbance of samples was measured photometrically at 405 nm (iEMS Reader MF).

2.3.2. In Vitro α-Amylase Inhibition (Disc) Assay

Four filter discs with a diameter of 0.5 cm were placed in a 92 × 16 mm Petri dish (Sarstedt, Nuernbrecht, Germany) filled with medium comprising 1% agar–agar (Carl Roth GmbH & Co. KG, Karlsruhe, Germany) and 1% starch (VWR, Darmstadt, Germany). Then,

80 µL of HI extract at concentrations of 0–10 mg/mL was mixed with 20 µL α-amylase (derived from porcine pancreas, Sigma-Aldrich, Taufkirchen, Germany). Acarbose was utilized as a reference inhibitor. A total of 20 µL of each sample was pipetted onto filter discs and left at 37 °C overnight. After removing the filter discs, plates were incubated with 5 mM iodine in 3% potassium iodide solution (Merck, Darmstadt, Germany). After 15 min, the diameters of the cleared zones were evaluated and the percentage inhibition of α-amylase was calculated. The disc assay was performed on two independent testing days.

2.3.3. In Vitro Dipeptidyl Peptidase-4 (DPP4) Inhibition Assay

The DPP4 inhibitor activities of HI extracts were determined using the DPP4 inhibitor screening kit according to the manufacturer's instructions (MAK203, Sigma-Aldrich, Taufkirchen, Germany). The HI extract was dissolved in DMSO to a concentration of 100 mg/mL and further diluted to a final concentration of 1 mg/mL, 250 µg/mL and 100 µg/mL with assay buffer. Then, 18 nM of the established DPP4 inhibitor sitagliptin (representing its IC_{50} concentration) served as the positive inhibitor control, whereas assay buffer only was used as the control for DPP4 enzyme activity and was set to 100%. Subsequently, 49 µL assay buffer and 1 µL DPP4 enzyme were mixed with 25 µL of HI extract and 18 nM sitagliptin or assay buffer. After 10 min pre-incubation at 37 °C, a reaction mix of 23 µL assay buffer and 2 µL substrate was given to each well. The fluorescence signal (excitation: 360 nm, emission: 465 nm) was measured in black 96-well microtiter plates at 37 °C over a period of 30 min in 1 min intervals (Tecan Infinite 200 microplate reader).

2.4. Testing for Mycoplasma Contamination

All cell lines were regularly tested for mycoplasma contamination via the Mycoplasma Detection Kit for conventional PCR (Venor®GeM Classic, Minerva Biolabs, Berlin, Germany) using MB Taq Polymerase (5 Unit/µL, 50 Units). All tested cell lines were found to be mycoplasma-negative.

2.5. Sodium-Dependent Glucose Transporter 1 (SGLT1) Assay Using Ussing Chambers in Caco-2/PD7 Cells

SGLT1 was determined in Caco-2/PD7 cells by employing the Ussing chambers methodology [32]. Cells were provided by Edith Brot-Laroche, Unité de Recherches sur la Différenciation Cellulaire Intestinale (Villejuif Cedex, France) and seeded at a density of 1×10^6 cells/well into 6-well Corning® Costar® Snapwell cell culture inserts (0.4 µm pore size, 1.12 cm^2surface area, Merck, Darmstadt, Germany). Following 21 days of culturing, 0.5 mL of the cell-containing medium was given to the apical side (upper compartment) and 2.5 mL of cell-free medium was seeded into the basolateral side (lower compartment). After 7 days, the apical medium was withdrawn FBS. Only monolayers with a transepithelial electrical resistance (TEER) value exceeding 300 Ω cm^2, measured via a Millicell ERS-2 Volt-Ohm Meter, equipped with a STX01 planar electrode (Merck, Darmstadt, Germany), were regarded as functional barriers and used in the transport studies. Before starting the experiments, Hank's balanced salt solution (HBSS, pH 7.2) was heated to 37 °C and oxygenated using an influx of carbogen-gas (95% oxygen, 5% carbon dioxide). HBSS was used to fill half-chambers and wash Caco-2/PD7 monolayers before mounting the Snapwell inserts in Ussing chamber slides. Subsequently, both half-chambers were replenished with HBSS solution containing mannitol (10 mmol/L) apically and glucose (10 mmol/L) basolaterally. The measurement of the transepithelial potential difference was performed at 37 °C under open-circuit conditions using a DVC 1000 amplifier (WPI) and continuous carbogen bubbling. The potential difference was continuously monitored and recorded through Ag–AgCl electrodes and KBR agarose bridges. The short-circuit current (I_{SC}; µA cm^{-2}) was measured via an automatic VCCMC8 MultiChannel Voltage Current Clamp (Physiologic Instruments) and data were stored using the Acquire & Analyze Data II acquisition software (Physiologic Instruments). The potential difference was allowed to stabilize for 20 min. Then, 10 mM glucose solution was given apically and 10 mM mannitol solution basolaterally. The glucose-stimulated I_{SC} was challenged by applying either pre-fermented extract (1 mg/mL), post-fermented HI-rich extract (1 mg/mL) or phlorizin (0.1 mM) as a positive control for inhibition of SGLT1 activity. The decline in the glucose-induced I_{SC} was assessed.

2.6. Induction of CRP in Hep 3B Cells and Measurement of CRP mRNA and Secreted Protein Level

Induction of CRP in Hep 3B cells and measurement of CRP mRNA and secreted protein levels were conducted according to [35,36]. Hep 3B cells were kindly gifted by Claudia Geismann (Laboratory of Molecular Gastroenterology & Hepatology, Department of Internal Medicine I, UKSH-Campus Kiel, 24,105 Kiel, Germany). Cells were cultivated for 5 days in MEM with Earle's balanced salt solution (EBSS), L-glutamine (PAN Biotech, Aidenbach, Germany) and 2.2 g/L $NaHCO_3$; and supplemented with 10% (vol/vol) heat inactivated fetal bovine serum (Gibco™ by Thermo Fisher Scientific GmbH, Life Technologies™, Darmstadt, Germany) and 1% penicillin/streptomycin (PAN Biotech, Aidenbach, Germany). For CRP induction, Hep 3B cells were incubated with 10 µg/mL isoflavone extract in DMSO in serum-free media containing 1 µM of dexamethasone (Dex) and stimulated with interleukin-1ß (IL1ß, 400 U/mL) and interleukin-6 (IL6, 200 U/mL) (both from ImmunoTools GmbH, Friesoythe, Germany) for 18 h for mRNA isolation or 48 h for ELISA analyses. DMSO served as the solvent control at a final dilution of 0.1%.

RNA isolation and quantitative RT-PCR were performed as previously described [37]. In brief, cells were harvested and RNA isolated with peqGOLD TriFast (VWR International, Radnor, PA, USA). The RNA isolation procedure is based on phenol and guanidinium thiocyanate extraction and on separation of RNA, protein and DNA into three phases upon centrifugation after adding chloroform. RNA concentrations and purity (260/280 nm) were determined with a Nanodrop 2000 (Thermo Fisher Scientific GmbH, Life Technologies, Darmstadt, Germany). Gene expression was determined using quantitative RT-PCR with the SensiFAST™ SYBR® No-ROX One-Step Kit (Bioline, Luckenwalde, Germany) via Rotorgene 6000 cycler (Corbett Life Science, Sydney, Australia). Gene expression levels were analyzed using a standard curve and normalized to the expression level of GAPDH. Primers were as follows: CRP forward primer: 5′-CCCTGAACTTTCAGCCGAATACA-3′; CRP reverse primer 5′-CGTCCTGCTGCCAGTGATACA-3′; GAPDH forward primer: 5′-CAATGACCCCTTCATTG ACC-3′; and GAPDH reverse primer: 5′-GATCTCGCTCCTGGAAGATG-3′.

To determine the secreted CRP protein levels, the cell culture medium was diluted 1:250 before being used for the ELISA according to the manufacturer's instructions (Hu-man C-reactive Protein ELISA Kit, Sigma-Aldrich, Taufkirchen, Germany).

2.7. Drosophila Melanogaster Feeding Assay Using a High-Starch Diet

The *D. melanogaster* wild-type strain w^{1118} (#5905, Bloomington Drosophila Stock Center, Indiana University, Bloomington, United States) was maintained under standard conditions in climate cabinets HPP750 or HPP110 (Memmert, Schwabach, Germany) at 25 °C, 60% humidity, and a 12/12 h light/dark cycle [32]. The fruit flies were cultured on Caltech medium (6.0% cornmeal, 5.5% dextrose, 3.0% sucrose, 2.5% inactive dry yeast, 1.0% agar Type II, Kisker, Steinfurt, Germany). Propionic acid (0.3%, Carl Roth, Karlsruhe, Germany) and Tegosept (0.15%, Genesee Scientific, San Diego, SC, USA) were added to the medium as preservatives. The feeding assay was started by transferring freshly eclosed adult animals to CT medium for mating. On day 3, female fruit flies were sorted and transferred onto a starch-based control diet (20% soluble starch (Carl Roth), 5% yeast, 2% agar, 0.18% nipagin, 0.3% propionic acid) or experimental diets that were supplemented with 0.8%, 1.6% or 2.4% of the post-fermented HI-rich extract. A medium containing 1.8 µg/mL acarbose was used as positive control [32]. The mated female flies were then transferred to the respective fresh experimental medium every other day. On day 10, the animals were harvested and ten flies per vial were homogenized for 10 min at 4 °C and 25 Hz in 0.05% Triton X100-containing PBS using a tissue lyser (Qiagen TissueLyser II, Hilden, Germany). The protein and triglyceride content of the fly lysates were measured using a Pierce BCA Protein Assay Kit (Pierce Biotechnology, Rockford, IL, USA) and colorimetric assay reagent (GPO-PAP Kit, Dialab, Wiener Neudorf, Austria), respectively.

2.8. Statistics

Statistical analyses were performed using the software GraphPad Prism (Ver. 7.05). The IC_{50} value of glucosidase inhibition by the post-fermented HI-rich extract was calculated using nonlinear regression. Prior to statistical tests, normal distribution of data was approved using the Shapiro–Wilk normality test. An analysis of variance (ANOVA) was conducted for α-amylase inhibition and in vivo fly data, followed by a post-hoc multiple comparison test of Dunnett to compare means of treatment with the post-fermented HI extract to the controls. Results from Ussing chamber experiments, CRP, FRAP and TEAC measurements were analyzed with two-sided unpaired Student's *t*-tests. In cases without normally distributed data, non-parametric tests were applied. Data from the DPP4 inhibiting assay were tested using the Kruskal–Wallis test and the Dunn's multiple comparison test. For secreted CRP protein level, the Mann–Whitney test was conducted. *p*-values less than 0.05 were considered significantly different.

3. Results

3.1. Post-Fermented Hydroxy-Isoflavone (HI)-Rich Soybean Extract Exhibited Significant Inhibitory Activity towards α-Glucosidase and DPP4 In Vitro, but Not towards α-Amylase

In order to test the ability of HI extracts to modulate carbohydrate-hydrolyzing enzymes in vitro, we first examined the influence on α-amylase activity. However, we did not observe a significant modulation of α-amylase enzyme activity by the post-fermented HI-rich extract up to a concentration of 10 mg/mL (Figure 3a).

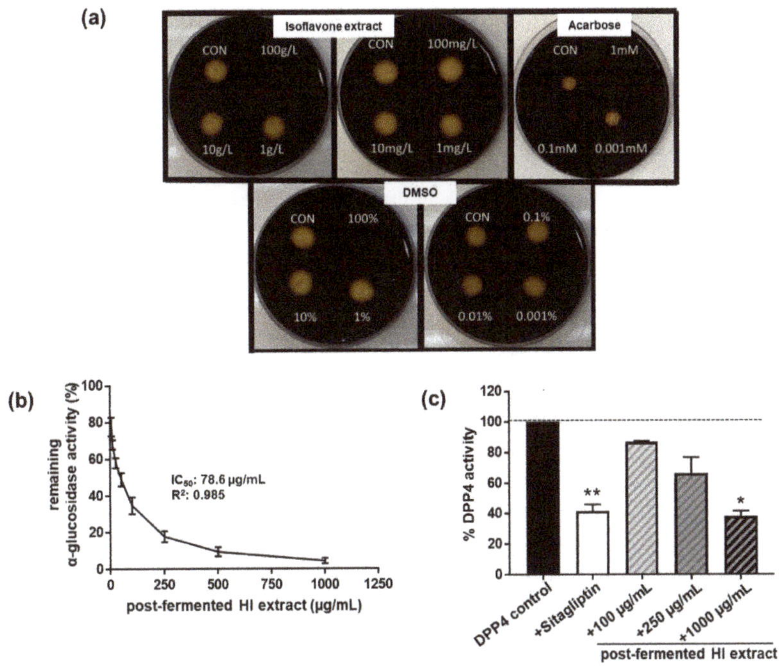

Figure 3. The post-fermented HI-rich extract did not affect (**a**) in vitro α-amylase activity, but inhibited (**b**) α-glucosidase activity as well as (**c**) dipeptidyl peptidase-4 (DPP4) enzyme activity in a dose-dependent manner. For measuring in vitro α-amylase inhibition, the disc diffusion assay was applied. Porcine pancreatic α-amylase was mixed with increasing concentrations of HI extract (final

concentrations as indicated) and administered onto the filter discs filled with medium comprising 1% agar–agar and 1% starch. After overnight incubation at 37 °C, starch/agar plates were iodide-stained. α-Amylase inhibition in samples was determined by matching the diameter of the cleared zones. HI extract-, DMSO- (solvent control for the HI extract) and acarbose- (positive control) treated filter discs were compared to control filter discs (CON, α-amylase alone). The assay was conducted on two independent testing days. Exemplary results are shown. α-Glucosidase activity was determined spectrophotometrically. IC_{50} was calculated using GraphPad Prism. DPP4 assay was performed in the presence of the following substances: DPP4 control: assay buffer; sitagliptin: 18 nM; post-fermented HI-rich extract as indicated. Values of remaining DPP4 enzyme activity (in %) compared to the DPP4 control are displayed. Results are mean values of n = 2–4 independent experiments. Error bars indicate standard deviation. ** $p < 0.01$; * $p < 0.05$, Kruskal–Wallis test ($p < 0.001$), followed by Dunn's multiple comparison test, compared to DPP4 control enzyme activity.

When we looked at the in vitro inhibition of α- glucosidase activity, we discovered a concentration-dependent inhibitory effect of the soy HI extract (Figure 3b). The IC_{50} value of the extract was estimated to be 78.6 μg/mL (R^2: 0.985; 95% confidence interval: 60.6–102 μg/mL). The soy HI extract was six times more potent than the positive control acarbose (IC_{50} = 493 μg/mL, R^2: 0.973; 95% confidence interval: 348–697 μg/mL) at inhibiting α-glucosidase activity.

Furthermore, the HI extract inhibited the dipeptidyl peptidase activity of DPP4 in a dose-dependent manner (ANOVA: $p < 0.001$), leading to an approximately 60% inhibition of enzyme activity at the highest concentration of 1 mg/mL when compared to controls (Figure 3c). A similar inhibition of DPP4 activity was achieved by the inhibitor control sitagliptin, but at a much lower concentration of 18 nM (this equates to 7.33 ng/mL). Thus, HI extract might serve only as a moderate inhibitor of DPP4 enzyme activity.

3.2. *Pre- and Post-Fermented Soy Isoflavone Extracts Were Moderate Inhibitors of SGLT1-Mediated Glucose Transport*

To examine whether soy isoflavone extracts before (pre-) fermentation and HI-enriched after (post-) fermentation affect SGLT1-mediated glucose transport, we employed Ussing chamber experiments using the Caco-2/PD7 cell monolayer model. Representative runs are given in Figure 4a, c. Adding either pre- or post-fermented extract at a concentration of 1 mg/mL to the Ussing chamber system substantially lowered the glucose-induced short-circuit current from 6.90 ± 0.36 to 4.05 ± 0.66 $\mu A/cm^2$ (pre-fermented extract) and from 7.45 ± 0.69 to 4.51 ± 0.63 $\mu A/cm^2$ (post-fermented HI-rich extract), respectively (Figure 4d). This represents a SGLT1 inhibition of approximately 60% for both extracts. In comparison, glucose uptake was almost completely blocked by the established SGLT1 inhibitor phlorizin at a concentration of 0.1 mM (Figure 4b).

3.3. *Expression of C-Reactive Protein (CRP)-Coding mRNA and CRP Protein Secretion Were Reduced in Hep 3B Cells by Pre- and HI-Enriched Post-Fermented Soy Extract*

Incubation of Hep 3B cells with 10 μg/mL pre-fermented soy extract significantly inhibited the mRNA expression of the inflammatory marker CRP after IL1ß plus IL6 stimulation by ca. 30% (Figure 5). An even more potent inhibition of ca. 60% was observed by incubating the cells with post-fermented HI-rich extract (10 μg/mL). However, when analyzing the impact on the level of secreted protein, both pre- and post-fermented extracts significantly reduced the CRP concentration similarly by about 50%.

3.4. *Post-Fermented HI-Rich Soy Extract Exhibited Higher Antioxidative Capacity Than Pre-Fermented Soy Extract*

We next tested the free radical scavenging properties of pre- or post-fermented soy extracts (10 μg/mL) by employing the ferric-reducing ability of plasma (FRAP) assay as well as the Trolox equivalent antioxidant capacity (TEAC) assay, respectively. As shown in Figure 6, in both cases the fermented soy extract with the increased HI content exhibited a significantly higher antioxidative capacity than the pre-fermented soy extract.

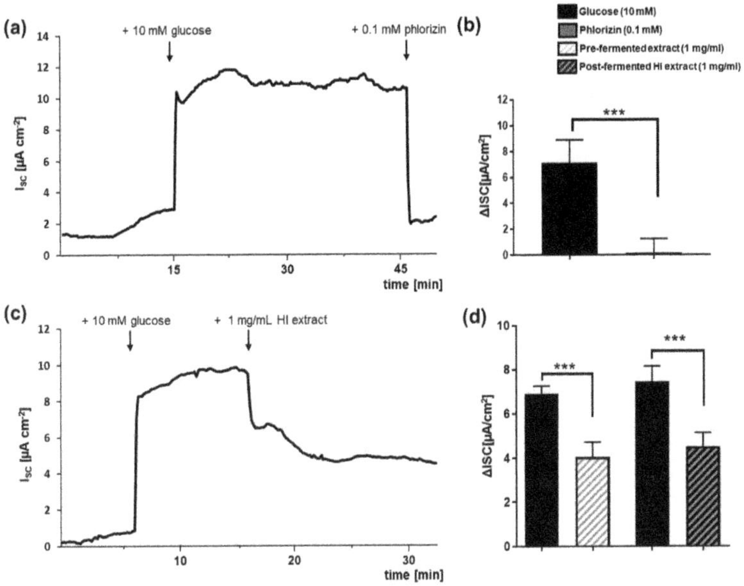

Figure 4. Influence of pre-fermented and HI-enriched post-fermented soy isoflavone extracts on sodium-dependent glucose transporter 1 (SGLT1). SGLT1-dependent glucose transport was measured in Caco-2/PD7 cell monolayers via Ussing chambers. Short-circuit current (I_{SC}) was followed over time (exemplary runs depicted in (**a**,**c**)). After addition of 10 mM glucose to the apical side at the indicated time points, the I_{SC} reached a stable plateau within approximately 10 min, before 0.1 mM phlorizin (**a**) as positive control or 1 mg/mL soy isoflavone extract (**c**) was added. The corresponding I_{SC} values are shown in (**b**,**d**) and the I_{SC} values for the pre-fermented extract are shown in (**d**). Error bars indicate standard deviation of n = 4–7 independent experiments. *** $p < 0.001$, unpaired t-test. HI: Hydroxy-isoflavone extract.

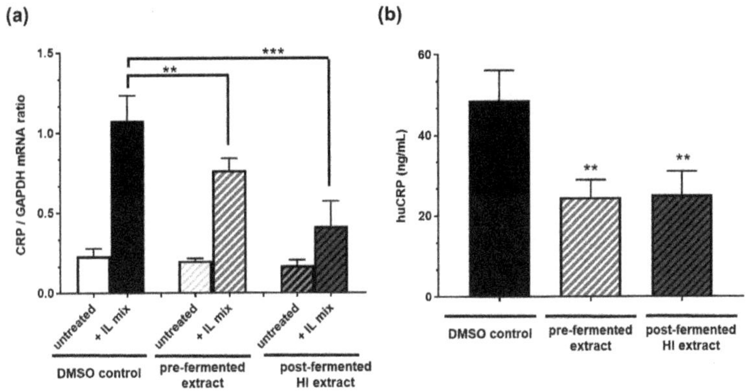

Figure 5. Soy isoflavone extracts before (pre-) and after (post-) fermentation (HI) reduced (**a**) CRP mRNA expression and (**b**) the amount of secreted CRP protein in human hepatoma (Hep B3) cells. Cells were incubated with either pre- or post-fermented extract (10 µg/mL) solved in DMSO. CRP expression was stimulated with 400 U/mL interleukin-1ß and 200 U/mL interleukin-6 (IL Mix). DMSO only served as solvent control. n = 6 independent experiments. Results are mean values ± SD. Error bars indicate standard deviation. *** $p < 0.001$; ** $p < 0.01$: unpaired t-test (mRNA) or Mann–Whitney test (secreted protein level) between treatment compared to DMSO solvent control after stimulation.

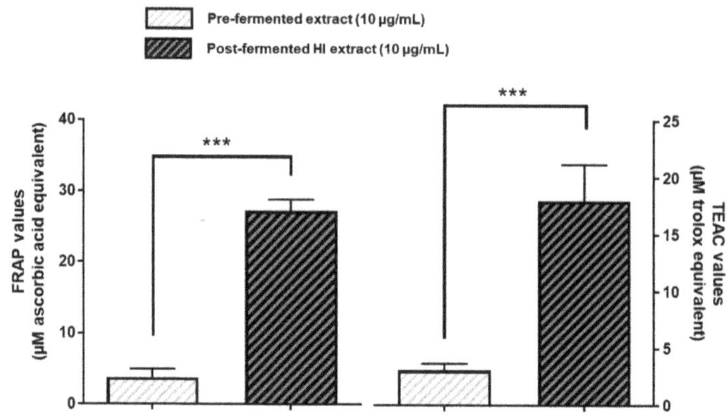

Figure 6. Post-fermented HI-rich soy extract exhibited higher antioxidative capacity than pre-fermented soy extract. For the ferric-reducing ability of plasma (FRAP) assay, pre- and post-fermented soy extracts were incubated with 2 mM iron (III) chloride in the presence of TPTZ for 15 min. The resulting complex between TPTZ and reduced iron (II) was measured photometrically at 620 nm. Results are given in µM ascorbic acid equivalents. The ability to reduce ABTS by pre- or post-fermented soy extracts was investigated in the Trolox equivalent antioxidant capacity (TEAC) assay and measured spectrophotometrically at 750 nm. TEAC values are given in µM trolox equivalents. Results are mean values ± SD of $n = 6$–10 independent experiments. Error bars indicate standard deviation. *** $p < 0.001$: unpaired *t*-test between pre- and post-fermented HI extract.

3.5. Supplementation of a High-Starch Drosophila Melanogaster Diet with Post-Fermented HI-Rich Extract Decreased the Triacylglyceride (TAG) Content of Female Fruit Flies

Dietary supplementation of the 20% starch-based diet with increasing concentrations of the post-fermented HI-rich extract led to a dose-dependent reduction of the triglyceride content in 10-day-old female flies (Figure 7). In flies fed 2.4% of the post-fermented HI-rich extract, the TAG to protein ratio was found to be 0.36 compared to control animals which had a value of 0.51. The treatment with the positive control acarbose induced an even more drastic decline in lipid storage to a TAG to protein ratio of 0.08.

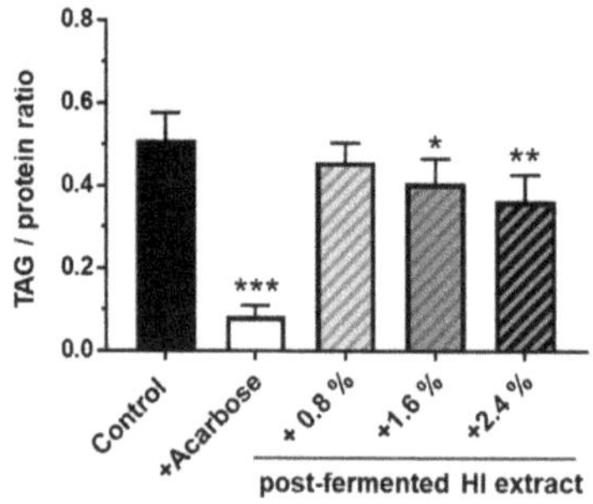

Figure 7. Supplementation of post-fermented HI-rich soy extract decreased the triacylglyceride (TAG) content in *D. melanogaster* fed a high-starch diet. Freshly eclosed male and female fruit flies were mated for 2 days. On day 3, females were allocated to control medium (20% starch, 5% yeast extract) or different experimental diets supplemented with the indicated amount of post-fermented HI-rich soy extract. A diet containing 1.8 mg/L acarbose was used as positive control. After 1 week, the flies were harvested and the TAG to protein ratios were determined. Bars represent the mean values ± SD of n = 6 independent experiments. Error bars show standard deviation * $p < 0.05$, ** $p < 0.01$, and *** $p < 0.001$, ANOVA ($p < 0.001$) with post-hoc multiple comparison test of Dunnett, compared to controls. HI: Hydroxy-isoflavone.

4. Discussion

The prevalence of T2DM is growing globally; hence, controlling postprandial hyperglycemia and inflammation is central for halting disease progression. Soy isoflavones are believed to play a role in diabetes prevention [38]. The dietary intake of soy products has consistently been inversely associated with the risk of T2DM among women [39]. However, the underlying mechanisms and, in particular, the role of soy-derived HI in diabetes prevention remain unclear. By applying a portfolio of numerous in vitro assays related to various important steps within the glucose metabolism (from intestinal digestion to glucose uptake), as well as assessing potential anti-inflammatory properties, we have addressed this research question in the present study. Furthermore, we have included adequate positive controls (e.g., acarbose, sitaglibtin, phlorizin) in the respective assays.

We have shown that a HI-enriched soy extract demonstrated inhibitory activity towards α-glucosidase, moderately reduced the DPP4 enzyme activity and significantly inhibited SGLT1-dependent glucose transport. Furthermore, the fermented HI-rich extract substantially decreased CRP mRNA and secreted protein levels in cultured Hep B3 hepatocytes. Thus, the HI-rich soy extract mediated antidiabetic properties by addressing multiple targets. Since we observed an inhibition of α-glucosidase but not α-amylase, HI may, nevertheless, exhibit a certain specificity as far as carbohydrate digesting enzymes are concerned. A shortcoming of our present experimental approach may be that we studied only the soy isoflavone-rich extracts (although analytically well characterized) but not their purified constituents, which should be taken into consideration in future studies.

A decrease in intestinal glucose uptake could be an important mechanism in counteracting hyperglycemia [40]. Interestingly, we observed a significant inhibition of SGLT1 due to a HI-rich soy extract, as previously reported for other extracts rich in secondary plant metabolites [32,41]. We did not investigate whether the decrease in glucose uptake was mediated via a competitive inhibition of SGLT1. Phlorizin was used as a positive control in our Ussing chamber experiments. Thus, it would also be interesting to study whether there is a synergistic interaction between phlorizin and isoflavones/HI in terms of SGLT1 inhibition. Furthermore, other glucose transporters, such as Glut4, as a potential target of flavonoids [42] could be considered in response to the treatment with HI in additional studies. In terms of the SGLT1 assay, the fermented isoflavones did not show higher bioactivity than the unfermented extract as far as sodium-dependent glucose uptake was concerned. However, regarding anti-inflammatory properties, we observed a stronger inhibition of CRP gene expression in interleukin 1ß- and interleukin 6-stimulated hepatocytes in response to fermented versus unfermented isoflavones, whereas both extracts reduced the amount of secreted CRP protein to the same extent. Thus, fermentation may affect bioactivity in some but not all assays. Furthermore, it was unclear whether the inhibition of CRP gene expression was via a nuclear factor kappa B-controlled signal transduction pathway, as previously reported for the flavone quercetin in cultured hepatocytes [43]. We further observed a moderate inhibition of DPP4 activity due to HI in vitro. Our data were in line with previous studies indicating that prenyl isoflavones improve glucose homeostasis by inhibiting DPP4 in hyperglycemic rats in vivo [44]. Accordingly, genistein has been shown to inhibit DPP4 in diabetic laboratory mice. This bioactivity was accompanied by an enhanced GLP1 concentration [45], which was not monitored in the present study.

However, we have validated the antidiabetic activity of the fermented HI-rich extract in a starch-based high-sugar diet model of *D. melanogaster*. High-sugar diets have been frequently demonstrated to lead to enhanced triglyceride levels in fruit flies [46–48]. By choosing starch as the sole carbohydrate source, we addressed all steps of the carbohydrate degradation pathway including the intestinal enzymes α-amylase and α-glucosidase. Therefore, we cannot currently assess which target molecule(s) is/are responsible for the triglyceride-lowering effect of the post-fermented HI-rich extract. Accordingly, further studies are necessary to clarify the precise in vivo mechanism of action. Overall, data from the present study and literature suggest that structural modifications of isoflavones, either through fermentation or endogenous metabolism, affect their pharmacological properties in terms of their bioactivity and possibly also their bioavailability [16]. The inclusion of additional hydroxyl groups into isoflavone molecules due to fermentation often enhances their bioactivity [23]. In contrast, sulfation [49] or glucuronidation [50], which mainly occur in the small intestine as well as in the liver, are associated with the loss of hydroxyl groups, and thereby decrease the bioactivity of isoflavones. Changes in the bioactivity through structural modifications were also evident in the case of the free radical scavenging activity of the post-fermented HI versus the pre-fermented soy extract. Thus, hydroxylation of isoflavones was accompanied with improved free radical scavenging properties, which has also been reported elsewhere for 8OHGen [25], 3OHDai [51], 6-hydroxyequol [52] and 8OHDai [53]. Accordingly, fermentation of soybean residues with *R. oligosporus* and *L. plantarum* resulted in an improved yield of isoflavone aglycones and gamma amino butyric acid, which led to lowered ROS levels and an increased antioxidative capacity, better blood glucose homeostasis and improved blood biochemistry in STZ-induced hyperglycemic mice [54]. Hence, we cannot fully exclude the possibility that beside HI, other ingredients could have contributed to the antidiabetic and anti-inflammatory effect, seen in our study. Improved free radical scavenging activity due to fermentation could also impact the food quality and shelf life of HI-rich soy derived food. Several efforts have been made to increase the bioavailability of isoflavones from soy beans, including the functional cloning of a soy isoflavone conjugate hydrolyzing β-glucosidase as a potential candidate for soy isoflavone bioavailability enhancement [16]. Although the bioavailability of genistein and daidzein has been studied in laboratory rodents [55,56], as well as in humans [57,58], little is known in terms of the bioavailability (e.g., plasma and tissue concentration) of HI including 8OHGen and 8OHGly. Nevertheless, it has been suggested that 8-OHDai is relatively easily absorbed in rats and distributed to peripheral tissues [59]. Such studies are necessary to evaluate whether the isoflavone concentrations used in in vitro studies are physiologically achievable under in vivo conditions. On the other hand, bioavailability was not an issue when isoflavones and HI inhibited intestinal targets, such as α-glucosidase and SGLT1,96 identified here.

5. Conclusions

A soy isoflavone extract rich in 8-hydroxygenistein, 8-hydroxyglycitein and 8-hydroxydaidzein exhibited antidiabetic properties in vitro and in an in vivo diabetes model of *Drosophila melanogaster*. Such an extract may have the capability to serve as a dietary natural plant bioactive for prevention strategies in terms of T2DM. However, in the future, the potential antidiabetic and anti-inflammatory properties of HI need to be validated in laboratory rodents, as well as in human intervention studies, also taking their bioavailability into account.

Author Contributions: Conceptualization, K.L., A.F., I.B., K.T. and G.R.; methodology, K.L., I.B., P.H., Y.U., K.C., D.N. and N.H.; data curation, K.L., A.F., P.H., Y.U. and G.R.; writing—original draft preparation, K.L., A.F. and G.R.; writing—review and editing, K.L., A.F., P.H., Y.U., K.C., D.N., N.H. and G.R.; supervision, K.L., K.T. and G.R. All authors have read and agreed to the published version of the manuscript.

Funding: This study was partly funded by CycloChem Bio Co., Ltd., 7-4-5 Minatojima-minamimachi, Chuo-ku, Kobe, 650-0047, Japan.

Institutional Review Board Statement: Not applicable.

Informed Consent Statement: Not applicable.

Data Availability Statement: The data presented in this study are available upon reasonable request from the corresponding author.

Acknowledgments: We thank Vivien Schmuck for their excellent technical assistance.

Conflicts of Interest: K.T. and D.N. are board members of CycloChem Bio Co., Ltd. K.C. and Y.U. are employees of CycloChem Bio Co., Ltd. N.H. is an employee of Toyo Hakko Co., Ltd.

References

1. Khan, M.A.B.; Hashim, M.J.; King, J.K.; Govender, R.D.; Mustafa, H.; Al Kaabi, J. Epidemiology of Type 2 Diabetes–Global Burden of Disease and Forecasted Trends. *J. Epidemiol. Glob. Health* **2020**, *10*, 107–111. [CrossRef] [PubMed]
2. ElSayed, N.A.; Aleppo, G.; Aroda, V.R.; Bannuru, R.R.; Brown, F.M.; Bruemmer, D.; Collins, B.S.; Hilliard, M.E.; Isaacs, D.; Johnson, E.L.; et al. 2. Classification and Diagnosis of Diabetes: Standards of Care in Diabetes-2023. *Diabetes Care* **2023**, *46*, S19–S40. [CrossRef]
3. Francini, F.; Schinella, G.R.; Ríos, J.L. Activation of AMPK by Medicinal Plants and Natural Products: Its Role in Type 2 Diabetes Mellitus. *Mini Rev. Med. Chem.* **2019**, *19*, 880–901. [CrossRef]
4. Chai, S.; Zhang, R.; Zhang, Y.; Carr, R.D.; Zheng, Y.; Rajpathak, S.; Ji, L. Effect of dipeptidyl peptidase-4 inhibitors on postprandial glucagon level in patients with type 2 diabetes mellitus: A systemic review and meta-analysis. *Front. Endocrinol.* **2022**, *13*, 994944. [CrossRef]
5. Tyagi, N.K.; Kumar, A.; Goyal, P.; Pandey, D.; Siess, W.; Kinne, R.K. D-Glucose-recognition and phlorizin-binding sites in human sodium/D-glucose cotransporter 1 (hSGLT1): A tryptophan scanning study. *Biochemistry* **2007**, *46*, 13616–13628. [CrossRef]
6. Stearns, A.T.; Balakrishnan, A.; Rhoads, D.B.; Tavakkolizadeh, A. Rapid upregulation of sodium-glucose transporter SGLT1 in response to intestinal sweet taste stimulation. *Ann. Surg.* **2010**, *251*, 865–871. [CrossRef]
7. Dyer, J.; Wood, I.S.; Palejwala, A.; Ellis, A.; Shirazi-Beechey, S.P. Expression of monosaccharide transporters in intestine of diabetic humans. *Am. J. Physiol. Gastrointest. Liver Physiol.* **2002**, *282*, G241–G248. [CrossRef]
8. Yang, X.; Tao, S.; Peng, J.; Zhao, J.; Li, S.; Wu, N.; Wen, Y.; Xue, Q.; Yang, C.X.; Pan, X.F. High-sensitivity C-reactive protein and risk of type 2 diabetes: A nationwide cohort study and updated meta-analysis. *Diabetes/Metab. Res. Rev.* **2021**, *37*, e3446. [CrossRef]
9. Ridker, P.M.; Rane, M. Interleukin-6 signaling and anti-interleukin-6 therapeutics in cardiovascular disease. *Circ. Res.* **2021**, *128*, 1728–1746. [CrossRef]
10. Serban, C.; Sahebkar, A.; Antal, D.; Ursoniu, S.; Banach, M. Effects of supplementation with green tea catechins on plasma C-reactive protein concentrations: A systematic review and meta-analysis of randomized controlled trials. *Nutrition* **2015**, *31*, 1061–1071. [CrossRef]
11. Bajerska, J.; Łagowska, K.; Mori, M.; Reguła, J.; Skoczek-Rubińska, A.; Toda, T.; Mizuno, N.; Yamori, Y. A Meta-Analysis of Randomized Controlled Trials of the Effects of Soy Intake on Inflammatory Markers in Postmenopausal Women. *J. Nutr.* **2022**, *152*, 5–15. [CrossRef] [PubMed]
12. Rüfer, C.E.; Maul, R.; Donauer, E.; Fabian, E.J.; Kulling, S.E. In vitro and in vivo metabolism of the soy isoflavone glycitein. *Mol. Nutr. Food Res.* **2007**, *51*, 813–823. [CrossRef]
13. Chang, T.S. Isolation, bioactivity, and production of ortho-hydroxydaidzein and ortho-hydroxygenistein. *Int. J. Mol. Sci.* **2014**, *15*, 5699–5716. [CrossRef] [PubMed]
14. Lee, H.; Kim, B.G.; Ahn, J.H. Production of bioactive hydroxyflavones by using monooxygenase from Saccharothrix espanaensis. *J. Biotechnol.* **2014**, *176*, 11–17. [CrossRef] [PubMed]
15. Zaheer, K.; Humayoun Akhtar, M. An updated review of dietary isoflavones: Nutrition, processing, bioavailability and impacts on human health. *Crit. Rev. Food Sci. Nutr.* **2017**, *57*, 1280–1293. [CrossRef]
16. Kumar, S.; Awana, M.; Rani, K.; Kumari, S.; Sasi, M.; Dahuja, A. Soybean (Glycine max) isoflavone conjugate hydrolysing β-glucosidase (GmICHG): A promising candidate for soy isoflavone bioavailability enhancement. *3 Biotech* **2023**, *13*, 52. [CrossRef]
17. Kim, S.; Chen, J.; Cheng, T.; Gindulyte, A.; He, J.; He, S.; Li, Q.; Shoemaker, B.A.; Thiessen, P.A.; Yu, B.; et al. PubChem 2023 update. *Nucleic Acids Res.* **2022**, *51*, D1373–D1380. [CrossRef]
18. Rimbach, G.; Boesch-Saadatmandi, C.; Frank, J.; Fuchs, D.; Wenzel, U.; Daniel, H.; Hall, W.L.; Weinberg, P.D. Dietary isoflavones in the prevention of cardiovascular disease—A molecular perspective. *Food Chem. Toxicol.* **2008**, *46*, 1308–1319. [CrossRef]
19. Li, Y.; Zhang, H. Soybean isoflavones ameliorate ischemic cardiomyopathy by activating Nrf2-mediated antioxidant responses. *Food Funct.* **2017**, *8*, 2935–2944. [CrossRef]
20. Voss, C.; Sepulveda-Boza, S.; Zilliken, F.W. New isoflavonoids as inhibitors of porcine 5-lipoxygenase. *Biochem. Pharmacol.* **1992**, *44*, 157–162. [CrossRef]
21. Zhang, H.; Pang, X.; Yu, H.; Zhou, H. Genistein suppresses ox-LDL-elicited oxidative stress and senescence in HUVECs through the SIRT1-p66shc-Foxo3a pathways. *J. Biochem. Mol. Toxicol.* **2022**, *36*, e22939. [CrossRef] [PubMed]
22. Guo, Q.; Rimbach, G.; Moini, H.; Weber, S.; Packer, L. ESR and cell culture studies on free radical-scavenging and antioxidant activities of isoflavonoids. *Toxicology* **2002**, *179*, 171–180. [CrossRef] [PubMed]
23. Esaki, H.; Onozaki, H.; Morimitsu, Y.; Kawakishi, S.; Osawa, T. Potent Antioxidative Isoflavones Isolated from Soybeans Fermented with Aspergillus saitoi. *Biosci. Biotechnol. Biochem.* **1998**, *62*, 740–746. [CrossRef]
24. Turner, R.; Baron, T.; Wolffram, S.; Minihane, A.M.; Cassidy, A.; Rimbach, G.; Weinberg, P.D. Effect of circulating forms of soy isoflavones on the oxidation of low density lipoprotein. *Free Radic. Res.* **2004**, *38*, 209–216. [CrossRef]
25. Hirota, A.; Taki, S.; Kawaii, S.; Yano, M.; Abe, N. 1,1-Diphenyl-2-picrylhydrazyl radical-scavenging compounds from soybean miso and antiproliferative activity of isoflavones from soybean miso toward the cancer cell lines. *Biosci. Biotechnol. Biochem.* **2000**, *64*, 1038–1040. [CrossRef]
26. Fujita, T.; Funako, T.; Hayashi, H. 8-Hydroxydaidzein, an aldose reductase inhibitor from okara fermented with *Aspergillus* sp. HK-388. *Biosci. Biotechnol. Biochem.* **2004**, *68*, 1588–1590. [CrossRef] [PubMed]

27. Staats, S.; Lüersen, K.; Wagner, A.E.; Rimbach, G. Drosophila melanogaster as a Versatile Model Organism in Food and Nutrition Research. *J. Agric. Food Chem.* **2018**, *66*, 3737–3753. [CrossRef] [PubMed]
28. Nayak, N.; Mishra, M. Simple techniques to study multifaceted diabesity in the fly model. *Toxicol. Mech. Methods* **2019**, *29*, 549–560. [CrossRef] [PubMed]
29. Chatterjee, N.; Perrimon, N. What fuels the fly: Energy metabolism in Drosophila and its application to the study of obesity and diabetes. *Sci. Adv.* **2021**, *7*, eabg4336. [CrossRef] [PubMed]
30. Baenas, N.; Wagner, A.E. Drosophila melanogaster as a Model Organism for Obesity and Type-2 Diabetes Mellitus by Applying High-Sugar and High-Fat Diets. *Biomolecules* **2022**, *12*, 307. [CrossRef]
31. Miao, Y.; Chen, R.; Wang, X.; Zhang, J.; Tang, W.; Zhang, Z.; Liu, Y.; Xu, Q. Drosophila melanogaster diabetes models and its usage in the research of anti-diabetes management with traditional Chinese medicines. *Front. Med.* **2022**, *9*, 953490. [CrossRef] [PubMed]
32. Günther, I.; Rimbach, G.; Nevermann, S.; Neuhauser, C.; Stadlbauer, V.; Schwarzinger, B.; Schwarzinger, C.; Ipharraguerre, I.R.; Weghuber, J.; Lüersen, K. Avens Root (*Geum Urbanum* L.) Extract Discovered by Target-Based Screening Exhibits Antidiabetic Activity in the Hen's Egg Test Model and Drosophila melanogaster. *Front. Pharmacol.* **2021**, *12*, 794404. [CrossRef] [PubMed]
33. Pallauf, K.; Duckstein, N.; Hasler, M.; Klotz, L.O.; Rimbach, G. Flavonoids as Putative Inducers of the Transcription Factors Nrf2, FoxO, and PPARgamma. *Oxidative Med. Cell. Longev.* **2017**, *2017*, 4397340. [CrossRef]
34. Bayram, B.; Esatbeyoglu, T.; Schulze, N.; Ozcelik, B.; Frank, J.; Rimbach, G. Comprehensive analysis of polyphenols in 55 extra virgin olive oils by HPLC-ECD and their correlation with antioxidant activities. *Plant Foods Hum. Nutr. (Dordr. Neth.)* **2012**, *67*, 326–336. [CrossRef]
35. Zhang, D.; Jiang, S.L.; Rzewnicki, D.; Samols, D.; Kushner, I. The effect of interleukin-1 on C-reactive protein expression in Hep3B cells is exerted at the transcriptional level. *Biochem. J.* **1995**, *310*, 143–148. [CrossRef] [PubMed]
36. Yoshida, T.; Yamagishi, S.; Nakamura, K.; Matsui, T.; Imaizumi, T.; Inoue, H.; Ueno, T.; Sata, M. Pigment epithelium-derived factor (PEDF) blocks the interleukin-6 signaling to C-reactive protein expression in Hep3B cells by suppressing Rac-1 activation. *Life Sci.* **2006**, *79*, 1981–1987. [CrossRef] [PubMed]
37. Kuhn, G.; Pallauf, K.; Schulz, C.; Rimbach, G. Flavonoids as putative modulators of Delta4-, Delta5-, and Delta6-desaturases: Studies in cultured hepatocytes, myocytes, and adipocytes. *BioFactors* **2018**, *44*, 485–495. [CrossRef]
38. Li, N.; Wu, X.; Zhuang, W.; Xia, L.; Chen, Y.; Zhao, R.; Yi, M.; Wan, Q.; Du, L.; Zhou, Y. Soy and Isoflavone Consumption and Multiple Health Outcomes: Umbrella Review of Systematic Reviews and Meta-Analyses of Observational Studies and Randomized Trials in Humans. *Mol. Nutr. Food Res.* **2020**, *64*, 1900751. [CrossRef]
39. Yan, F.; Eshak, E.S.; Shirai, K.; Dong, J.Y.; Muraki, I.; Tamakoshi, A.; Iso, H. Soy Intake and Risk of Type 2 Diabetes Among Japanese Men and Women: JACC Study. *Front. Nutr.* **2021**, *8*, 813742. [CrossRef]
40. Merino, B.; Fernández-Díaz, C.M.; Cózar-Castellano, I.; Perdomo, G. Intestinal fructose and glucose metabolism in health and disease. *Nutrients* **2019**, *12*, 94. [CrossRef]
41. Schloesser, A.; Esatbeyoglu, T.; Schultheiss, G.; Vollert, H.; Luersen, K.; Fischer, A.; Rimbach, G. Antidiabetic Properties of an Apple/Kale Extract In Vitro, In Situ, and in Mice Fed a Western-Type Diet. *J. Med. Food* **2017**, *20*, 846–854. [CrossRef] [PubMed]
42. Mishra, K.; Nath, M.; Halder, N.; Velpandian, T. Evaluation of the possibility of selective modulation of retinal glucose transporters in diabetic complications: An experimental study. *Indian J. Pharmacol.* **2020**, *52*, 495–504. [CrossRef] [PubMed]
43. García-Mediavilla, V.; Crespo, I.; Collado, P.S.; Esteller, A.; Sánchez-Campos, S.; Tuñón, M.J.; González-Gallego, J. The anti-inflammatory flavones quercetin and kaempferol cause inhibition of inducible nitric oxide synthase, cyclooxygenase-2 and reactive C-protein, and down-regulation of the nuclear factor kappaB pathway in Chang Liver cells. *Eur. J. Pharm.* **2007**, *557*, 221–229. [CrossRef] [PubMed]
44. Altenhofen, D.; da Luz, G.; Frederico, M.J.; Venzke, D.; Brich, M.; Vigil, S.; Fröde, T.S.; Linares, C.E.; Pizzolatti, M.G.; Silva, F.R. Bis-Pyrano Prenyl Isoflavone Improves Glucose Homeostasis by Inhibiting Dipeptidyl Peptidase-4 in Hyperglycemic Rats. *J. Cell. Biochem.* **2017**, *118*, 92–103. [CrossRef] [PubMed]
45. Rajput, M.S.; Sarkar, P.D.; Nirmal, N.P. Inhibition of DPP-4 Activity and Neuronal Atrophy with Genistein Attenuates Neurological Deficits Induced by Transient Global Cerebral Ischemia and Reperfusion in Streptozotocin-Induced Diabetic Mice. *Inflammation* **2017**, *40*, 623–635. [CrossRef]
46. Musselman, L.P.; Fink, J.L.; Narzinski, K.; Ramachandran, P.V.; Hathiramani, S.S.; Cagan, R.L.; Baranski, T.J. A high-sugar diet produces obesity and insulin resistance in wild-type Drosophila. *Dis. Model. Mech.* **2011**, *4*, 842–849. [CrossRef] [PubMed]
47. Abrat, O.B.; Storey, J.M.; Storey, K.B.; Lushchak, V.I. High amylose starch consumption induces obesity in Drosophila melanogaster and metformin partially prevents accumulation of storage lipids and shortens lifespan of the insects. *Comp. Biochem. Physiol. Part A Mol. Integr. Physiol.* **2018**, *215*, 55–62. [CrossRef]
48. Eickelberg, V.; Lüersen, K.; Staats, S.; Rimbach, G. Phenotyping of Drosophila Melanogaster-A Nutritional Perspective. *Biomolecules* **2022**, *12*, 221. [CrossRef]
49. Rimbach, G.; Weinberg, P.D.; de Pascual-Teresa, S.; Alonso, M.G.; Ewins, B.A.; Turner, R.; Minihane, A.M.; Botting, N.; Fairley, B.; Matsugo, S.; et al. Sulfation of genistein alters its antioxidant properties and its effect on platelet aggregation and monocyte and endothelial function. *Biochim. Biophys. Acta* **2004**, *1670*, 229–237. [CrossRef]

50. Schrader, C.; Ernst, I.M.; Sinnecker, H.; Soukup, S.T.; Kulling, S.E.; Rimbach, G. Genistein as a potential inducer of the anti-atherogenic enzyme paraoxonase-1: Studies in cultured hepatocytes in vitro and in rat liver in vivo. *J. Cell. Mol. Med.* **2012**, *16*, 2331–2341. [CrossRef]
51. Chen, Y.C.; Sugiyama, Y.; Abe, N.; Kuruto-Niwa, R.; Nozawa, R.; Hirota, A. DPPH radical-scavenging compounds from dou-chi, a soybean fermented food. *Biosci. Biotechnol. Biochem.* **2005**, *69*, 999–1006. [CrossRef] [PubMed]
52. Nozawa, D.; Matsuyama, A.; Furuya, T. Biocatalytic synthesis and evaluation of antioxidant and antibacterial activities of hydroxyequols. *Bioorg. Med. Chem. Lett.* **2022**, *73*, 128908. [CrossRef] [PubMed]
53. Rimbach, G.; De Pascual-Teresa, S.; Ewins, B.A.; Matsugo, S.; Uchida, Y.; Minihane, A.M.; Turner, R.; VafeiAdou, K.; Weinberg, P.D. Antioxidant and free radical scavenging activity of isoflavone metabolites. *Xenobiot. Fate Foreign Compd. Biol. Syst.* **2003**, *33*, 913–925. [CrossRef]
54. Hariyanto, I.; Hsieh, C.W.; Hsu, Y.H.; Chen, L.G.; Chu, C.; Weng, B.B. In Vitro and In Vivo Assessments of Anti-Hyperglycemic Properties of Soybean Residue Fermented with Rhizopus oligosporus and Lactiplantibacillus plantarum. *Life* **2022**, *12*, 1716. [CrossRef] [PubMed]
55. Andrade, J.E.; Twaddle, N.C.; Helferich, W.G.; Doerge, D.R. Absolute bioavailability of isoflavones from soy protein isolate-containing food in female BALB/c mice. *J. Agric. Food Chem.* **2010**, *58*, 4529–4536. [CrossRef] [PubMed]
56. Cohen, R.; Schwartz, B.; Peri, I.; Shimoni, E. Improving bioavailability and stability of genistein by complexation with high-amylose corn starch. *J. Agric. Food Chem.* **2011**, *59*, 7932–7938. [CrossRef]
57. Zubik, L.; Meydani, M. Bioavailability of soybean isoflavones from aglycone and glucoside forms in American women. *Am. J. Clin. Nutr.* **2003**, *77*, 1459–1465. [CrossRef]
58. Rodríguez-Morató, J.; Farré, M.; Pérez-Mañá, C.; Papaseit, E.; Martínez-Riera, R.; de la Torre, R.; Pizarro, N. Pharmacokinetic Comparison of Soy Isoflavone Extracts in Human Plasma. *J. Agric. Food Chem.* **2015**, *63*, 6946–6953. [CrossRef]
59. Esaki, H.; Shirasaki, T.; Yamashita, K.; Nakamura, Y.; Kawakishi, S.; Osawa, T. Absorption and excretion of the 8-hydroxydaidzein in rats after oral administration and its antioxidant effect. *J. Nutr. Sci. Vitaminol.* **2005**, *51*, 80–86. [CrossRef]

nutrients

MDPI

Article

Acute Dose–Response Effectiveness of Combined Catechins and Chlorogenic Acids on Postprandial Glycemic Responses in Healthy Men: Results from Two Randomized Studies

Aya Yanagimoto [1,*], Yuji Matsui [2], Tohru Yamaguchi [2], Shinichiro Saito [1], Ryuzo Hanada [3] and Masanobu Hibi [1]

[1] Biological Science Laboratories, Kao Corporation, 2-1-3 Bunka, Sumida-ku, Tokyo 131-8501, Japan
[2] Health & Wellness Products Research Laboratories, Kao Corporation, 2-1-3 Bunka, Sumida-ku, Tokyo 131-8501, Japan
[3] SOUSEIKAI Sumida Hospital, 1-29-1 Honjo, Sumida-ku, Tokyo 130-0004, Japan
* Correspondence: yanagimoto.aya@kao.com; Tel.: +81-3-5630-7476

Abstract: Epidemiologic studies show that the risk of diabetes can be reduced by ingesting green tea or coffee. Previous studies have shown that simultaneously taking green tea catechins (GTC) and coffee chlorogenic acid (CCA) alters postprandial gastrointestinal hormones secretion and improves insulin sensitivity. However, there is no evidence on the acute effects of GTC and CCA on incretin and blood glucose, and on the respective dose of polyphenols. In this randomized, double-blind, placebo-controlled crossover study, we examined the effective dose of GTC and CCA on postprandial glucose, insulin, and incretin responses to a high-fat and high-carbohydrate cookie meal containing 75 g of glucose in healthy men. Study 1 (n = 18) evaluated two doses of GTC (270 or 540 mg) containing a fixed dose of CCA (270 mg) with 113 mg of caffeine and a placebo (0 mg GTC and 0 mg CCA) with 112 mg of caffeine. Study 2 (n = 18) evaluated two doses of CCA (150 or 300 mg) containing a fixed dose of GTC (540 mg) and a placebo with 99 mg of caffeine. The single combined ingestion of GTC and CCA significantly altered the incretin response and suppressed glucose and insulin levels. These findings suggest that the effective minimum dose is 540 mg of GTC and 150 mg of CCA.

Keywords: coffee; cookie meal test; glucagon-like peptide-1; glucose-dependent insulinotropic polypeptide; green tea

Citation: Yanagimoto, A.; Matsui, Y.; Yamaguchi, T.; Saito, S.; Hanada, R.; Hibi, M. Acute Dose–Response Effectiveness of Combined Catechins and Chlorogenic Acids on Postprandial Glycemic Responses in Healthy Men: Results from Two Randomized Studies. *Nutrients* **2023**, *15*, 777. https://doi.org/10.3390/nu15030777

Academic Editors: Sonia de Pascual-Teresa and Luis Goya

Received: 21 December 2022
Revised: 23 January 2023
Accepted: 31 January 2023
Published: 2 February 2023

1. Introduction

Type 2 diabetes (T2D), a serious chronic disease caused by insulin resistance and impaired insulin secretion [1], is a global health and wellness issue [1–3]. Persistent hyperglycemia leads to insulin resistance. Blood glucose levels are strictly regulated by insulin, glucagon, cortisol, catecholamines (adrenal corticosteroids), growth hormone, cortisol, and aldosterone and are also influenced by stress. Incretins, such as glucose-dependent insulinotropic polypeptide (GIP) and glucagon-like peptide-1 (GLP-1) secreted from the small intestine in response to dietary glucose and lipids, are required for insulin secretion from pancreatic β-cells. GLP-1 enhances glucose uptake in the liver and muscle and improves systemic insulin sensitivity [4,5]. GIP promotes fat accumulation via glucose and free fatty acid uptake in adipocytes [6]. Healthy eating habits from childhood are a factor in preventing the development of T2D. Improving insulin resistance and controlling blood glucose levels through increased physical activity may also prevent T2D. The effectiveness of non-pharmacologic therapies, however, including food-derived functional ingredients, for improving insulin resistance has not been established.

Green tea and coffee, which contain abundant polyphenols such as green tea catechins (GTC) and coffee chlorogenic acids (CCA), respectively, are the most commonly consumed beverages in the world. Consumption of GTC or CCA may reduce the risk of developing

T2D and other metabolic diseases [7–11]. A previous study in healthy men demonstrated that consuming green tea extract containing GTC ameliorates insulin sensitivity in healthy individuals [12]. Chronic consumption of beverages containing 582 mg of GTC improved hemoglobin A_{1C} (HbA_{1c}) levels in participants taking medicines that stimulate insulin secretion [13]. In individuals with impaired glucose tolerance, consumption of 400 mg of CCA for 3 months significantly reduced fasting blood glucose levels and the insulinogenic index on the basis of fasting and postprandial blood glucose and insulin levels [14]. Acute ingestion of CCA suppresses postprandial blood glucose levels [15] and promotes GLP-1 secretion [16]. These findings suggest that ingesting a combination of GTC and CCA may further enhance their effects to improve glucose metabolism. In addition to the control of glucose and insulin responses, regulation of incretins is a potential therapeutic treatment option for insulin resistance [17,18]. Few studies, however, have evaluated the combined effects of GTC and CCA on postprandial glucose metabolism and gastrointestinal hormone secretion in humans [19]. In addition, the acute effects of simultaneous consumption of GTC and CCA have not been clarified, and the mechanisms by which these polyphenols affect glucose metabolism are unclear. To the best of our knowledge, there are no studies evaluating human glucose metabolism using different GTC and CCA doses. Clarifying the minimum effective doses of GTC and CCA to better understand the relationship could provide valuable information for future study designs.

In the present study, we evaluated the acute effects of four combinations of GTC and CCA concentrations on postprandial glucose metabolism to determine the dose–response relationship after ingestion in healthy, non-diabetic individuals. The primary endpoint was the minimum dose required for a postprandial GLP-1 response, and the secondary endpoints were postprandial blood glucose, insulin, and GIP responses.

2. Materials and Methods

2.1. Study Design

Effective combined doses of GTC and CCA were investigated in two human clinical trials conducted at Kao Corporation (Tokyo, Japan) and Sumida Hospital (Tokyo, Japan). Study 1 investigated the effective dose of GTC with a fixed dose of CCA, and Study 2 investigated the effective dose of CCA with a fixed dose of GTC.

Both acute ingestion studies were conducted, in which the three test beverages were each consumed immediately after ingestion of the cookie test meal, and the effects on postprandial glucose metabolism following 4 h were evaluated. A one-week washout was provided between the consumption of each of the three test beverages in a randomized crossover manner. In each study, the participants consumed test beverages containing two different dose combinations of GTC and CCA or a placebo. All of the participants in both studies were enrolled and randomly allocated a number using a computer-generated stratified randomization method to test the sequence groups. The sequence group allocation was concealed among the participants, physicians, and outcome assessors from screening to finalizing the dataset.

Study 1 and Study 2 participants were recruited separately, each with another 18 participants. Study 1 was conducted from October 2017 to February 2018. Each participant orally ingested a test beverage containing a fixed dose of CCA (270 mg) combined with either a minimum dose of GTC (540 mg: reported as the minimum effective dose for metabolic disorders [20]) or a half dose of GTC (270 mg) or a placebo beverage containing 0 mg GTC and 0 mg CCA. All of the test beverages contained 112–113 mg of caffeine. Study 2 was conducted from August 2019 to October 2019. Each participant orally ingested a beverage containing a fixed dose of GTC (540 mg) combined with either a minimum dose of CCA (300 mg: reported as the minimum effective dose for metabolic disorders [21]), a half dose of CCA (150 mg), or a placebo beverage containing 0 mg GTC and 0 mg CCA. All of the test beverages contained 99 mg caffeine. The two studies were conducted in accordance with the Declaration of Helsinki, and the protocols were reviewed and approved by the ethics committees of Kao Corporation (No. T232-190615, 30 July 2019; Tokyo, Japan) and Ueno-Asagao Clinic (No.2019-19; 7 August 2019; Tokyo, Japan). All of the participants received a full verbal and written explanation of the study and provided written informed consent prior to registration. The respective trials were registered with the University Hospital Medical Information Network (UMIN; http://www.umin.ac.jp/ (accessed on 2 January 2023); Registration No. UMIN000047544 for Study 1, UMIN000037738 for Study 2).

2.2. Participants

In Study 1, potential participants were recruited from among healthy male volunteers living in the Tokyo metropolitan area in October 2017. Eighteen healthy volunteers (mean ± SD: age, 41 ± 9 years; body mass index [BMI]: 22.6 ± 1.8 kg/m^2; fasting glucose, 86 ± 5 mg/dL) with normal blood glucose levels (fasting blood glucose, ≤125 mg/dL) were recruited to participate in this study. The exclusion criteria were a history of diabetes or cardiovascular disease, hypertension, hypercholesterolemia, hypertriglyceridemia, and hepatic, renal, or gastrointestinal diseases; smoking habit; excessive alcohol consumption (≥30 g/day); allergies to the ingredients in the test food; working as a shift worker; or otherwise determined to be unqualified by the physician in charge. In Study 2, potential participants were recruited from among healthy male volunteers living in the Tokyo metropolitan area in September 2019. Eighteen healthy male volunteers (mean ± SD; age: 47 ± 11 years; BMI: 23.2 ± 2.9 kg/m^2; fasting glucose: 103 ± 7 mg/dL) with blood glucose levels ranging from normal to impaired glucose tolerance (fasting glucose 95–125 mg/dL) were recruited according to the same criteria as in Study 1. The sample size for each study was estimated according to the results of a preliminary study and calculated to be 18, assuming a significance level of 5% and a statistical power of 80%. In the preliminary study, the effect size of the GLP-1 area under the curve (AUC) was 7.7 mg/dL·4 h with a standard deviation (SD) of 11.0 mg/dL·4 h.

2.3. Experimental Procedures

In both studies, after a 1-week run-in period, measurements for the acute ingestion test were obtained after the participants consumed one of the three test beverages. The day before the measurements, the participants were instructed to consume pre-packaged meals and to go to bed before midnight without consuming any food or beverages other than water after 9:00 p.m. In Study 1, the total energy content of the pre-packaged meals the day before the test was 2412 kcal/day, 14 E% from protein, 25 E% from fat, 61 E% from carbohydrates; 25 E% for breakfast, 24 E% for lunch, 21 E% for snacking, and 30 E% for dinner. In Study 2, the total energy content of the pre-packaged meals the day before the test was 2166 kcal/day, 14 E% from protein, 25 E% from fat, 61 E% from carbohydrates; 34 E% for breakfast, 32 E% for lunch, and 34 E% for dinner. The height, weight, body fat (body fat scale by bioelectrical impedance analysis, model TF-780 in Study 1 and model DC-320 in Study 2, Tanita Corp., Tokyo, Japan), waist circumference, and blood pressure (digital blood pressure monitor, model HEM-1000, Omron, Kyoto, Japan in Study 1 and model TM-2571, A&D Co., Ltd., Tokyo, Japan in Study 2) were measured for each participant. The participants consumed a high-fat and high-carbohydrate cookie meal which consisted of 12 test-specific cookies (meal test S, Saraya Co., Ltd., Osaka, Japan) over a 15-min period, and then the subjects consumed the test beverage over a 10-min period. The cookie meal contained flour, butter, maltose, chicken eggs, and baking powder and had an energy content of 592 kcal (6 E% from protein, 43 E% from fat, and 50 E% from carbohydrates). Venous blood samples were obtained during fasting (0 h) and at 0.5, 1, 1.5, 2, and 4 h after cookie meal consumption in Study 1 and during fasting (0 h) and at 0.5, 1, 1.5, 1.75, 2, 3, and 4 h after cookie meal consumption in Study 2. The primary outcome measurement was postprandial GLP-1 concentration, and the key secondary outcomes were postprandial blood glucose, insulin, and GIP concentrations.

2.4. Test Beverages

The test beverage compositions are summarized in Table 1. The test beverages were 350 mL tea-flavored beverages in both Studies 1 and 2. All of the test beverages were prepared by the Kao Corporation (Tokyo, Japan). The amount of GTC and caffeine in the test beverages was measured by high-performance liquid chromatography using an L-column ODS (4.6 mm diameter × 250 mm length; Chemicals Evaluation and Research Institute, Tokyo, Japan) and CCA using a Cadenza CD C18 column (4.6 mm diameter × 150 mm length; Intact, Kyoto, Japan). In Study 1, the GTC comprised catechin, epicatechin, gallocatechin, epigallocatechin, catechin gallate, epicatechin gallate, gallocatechin gallate, and epigallocatechin gallate in green tea extracts. The CCA comprised 3-, 4-, 5-caffeoylquinic acids and 3-, 4-, 5-feruloylquinic acids in coffee bean extracts. Each group in Study 1 is referred to hereafter as follows: Placebo (0 mg GTC, 0 mg CCA, with 112 mg), Dose A (270 mg GTC, 270 mg CCA, with 113 mg caffeine), and Dose B (540 mg GTC, 270 mg CCA, with 113 mg caffeine). Each group in Study 2 is referred to hereafter as follows: Placebo (0 mg GTC, 0 mg CCA, with 99 mg), Dose C (540 mg GTC, 150 mg CCA, with 99 mg caffeine), and Dose D (540 mg GTC, 300 mg CCA, with 99 mg caffeine). All of the test beverages, including the placebo, were prepared to be indistinguishable in appearance, taste, and smell.

Table 1. Test beverage compositions.

			Study 1			**Study 2**
	Placebo	**Dose A**	**Dose B**	**Placebo**	**Dose C**	**Dose D**
Catechin (mg)	0	14	27	0	30	30
Epicatechin (mg)	0	15	30	0	32	32
Gallocatechin (mg)	0	60	119	0	119	119
Epigallocatechin (mg)	0	54	108	0	103	103
Catechin gallate (mg)	0	11	23	0	23	23
Epicatechin gallate (mg)	0	17	33	0	37	37
Gallocatechin gallate (mg)	0	49	99	0	93	93
Epigallocatechin gallate (mg)	0	50	101	0	105	105
Total catechins (mg)	0	270	540	0	540	540
3-caffeoylquinic acid (mg)	0	82	83	0	46	92
4-caffeoylquinic acid (mg)	0	62	62	0	34	68
5-caffeoylquinic acid (mg)	0	73	74	0	41	82
3-feruloylquinic acid (mg)	0	19	19	0	11	23
4-feruloylquinic acid (mg)	0	15	15	0	16	16
5-feruloylquinic acid (mg)	0	18	17	0	10	19
Total chlorogenic acids (mg)	0	270	270	0	150	300
Caffeine (mg)	112	113	113	99	99	99

These values were measured by high-performance liquid chromatography. The volume of the tea-flavored test beverage was 350 mL. Abbreviations: CCA, coffee chlorogenic acid and GTC, green tea catechin.

2.5. Biochemical Analysis

The plasma and serum samples were obtained from the participants in a fasting state and after cookie meal ingestion, snap-frozen using liquid nitrogen, and immediately stored at −80 °C until the analysis was performed. Glucagon, active GLP-1, and total GIP levels were measured by enzyme-linked immunosorbent assay (glucagon: Mercodia, Uppsala, Sweden; active GLP-1: Immune Biology Laboratories, Gunma, Japan; total GIP: Merck Millipore, Darmstadt, Germany). All of the other items were analyzed at LSI Medience, Co., Ltd. (Tokyo, Japan). Blood glucose was measured by an enzymatic method and insulin was measured by chemiluminescence immunoassay (CLIA).

2.6. Statistical Analysis

Microsoft Excel 14.0, SPSS version 28.0 (IBM Corp., Armonk, NY, USA) and SAS version 9.4 (SAS Institute Inc., Cary, NC, USA) were used for the statistical analysis. Data are expressed as mean ± standard error of the mean (SEM). The AUCs were calculated using the trapezoidal rule, maximum concentration (C_{max}), and effect size, and the 95% confidence interval (CI) was calculated based on measurements at each time-point for the primary and secondary outcome measures. The following statistical analyses were performed among the three test beverage conditions. The values at each time-point were determined for each participant and entered into a linear mixed-effmixed-effect model with the treatment, time, and treatment-by-time interaction as the fixed effects. The significance of the total AUC (tAUC) and incremental AUC (iAUC) was analyzed using a mixed-effect model with adjustment of the effects of period and order by the "SAS GLIMMIX procedure" which allows statistical models to be fitted when responses are not necessarily normally distributed, and the effect of each test beverage was estimated with the "ESTIMATE" statement. Significance was estimated using empirical variance ("EMPIRICAL" option). An unstructured ("type = UN") variance-covariance matrix in the data with repeated measures was assumed. The Tukey–Kramer method was used to adjust for statistical multiplicity. In addition, stratified and correlation analyses were performed according to subject characteristics. In all of the analyses, the significance level was set at 5%.

3. Results

3.1. Participants

All of the participants completed Study 1. In Study 2, two of the participants declined to continue the test for personal reasons; therefore, 16 participants completed the study. No participants were excluded from the analysis according to the study protocols; data at the end of the study for the 18 and 16 participants who completed the studies were determined and an intention to treat analysis was performed for each study. The anthropometric measurements and the blood glucose and insulin levels at baseline in Studies 1 and 2 are shown in Table 2. No adverse events caused by the test beverages were noted. The demographic characteristics of the participants enrolled in Studies 1 and 2 were generally similar in terms of age, weight, height, and BMI.

Table 2. Physical characteristics, fasting blood glucose, and insulin at baseline.

		Study 1		Study 2	
		Mean ± SD	**Range**	**Mean ± SD**	**Range**
Age	(y)	41 ± 9	30–59	47 ± 11	33–63
Weight	(kg)	68.5 ± 1.7	55.8–88.8	69.0 ± 10.1	49.8–91.0
BMI	(kg/m^2)	22.6 ± 1.8	19.5–26.5	23.2 ± 2.9	18.5–27.4
Body Fat	(%)	19.5 ± 3.5	14.1–26.2	21.6 ± 5.1	13.1–31.0
HbA1c	(%)	5.3 ± 0.2	4.8–5.6	5.4 ± 0.3	4.9–5.9
FBG	(mg/dL)	86 ± 5	79–98	103 ± 7	95–115
Insulin	(μU/mL)	5.0 ± 1.7	2.8–7.9	3.7 ± 1.4	1.6–6.4

Data are expressed as mean ± SD (Study 1: n = 18; Study 2: n = 18). Abbreviations: SD: standard deviation, BMI: body mass index, HbA1c: hemoglobin A1c, and FBG: fasting blood glucose.

3.2. Blood Glucose and Insulin

The fasting glucose levels of all of the participants were in the normal to borderline range (fasting blood glucose level ≤ 125 mg/dL) during all of the periods in Studies 1 and 2. Figure 1 shows changes in blood glucose and insulin concentrations before and after the cookie meal, and Table 3 (Study 1) and Table 4 (Study 2) show estimated values of the tAUC adjusted for the effects of the order and period. In Study 1, comparisons of the three conditions using linear mixed-effect models revealed a treatment effect for the change in blood glucose and serum insulin but no treatment by time interaction effect. The results of the mixed-effect model analysis of the comparison of the three groups revealed a difference in the tAUC for blood glucose, but not for insulin. Furthermore, the tAUC for blood glucose was significantly lower in the Dose B group than in the Placebo group.

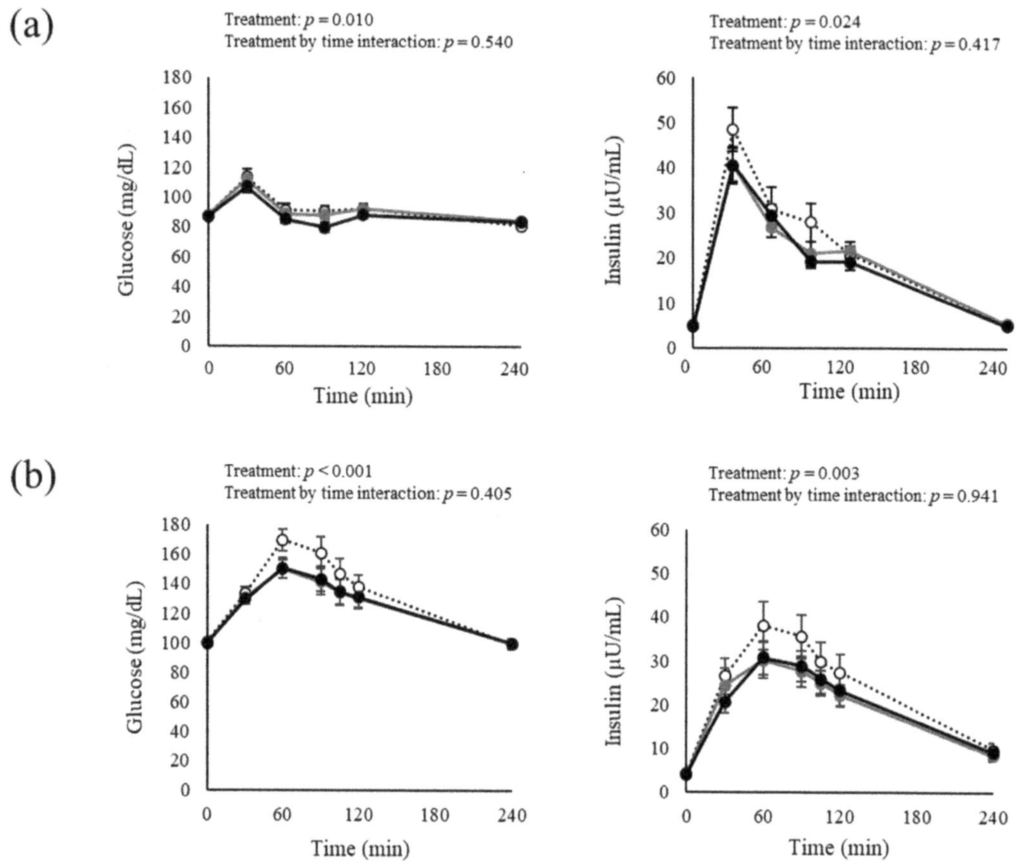

Figure 1. Changes in blood test results at fasting and after cookie meal consumption for Study 1 (**a**) and Study 2 (**b**). Data are expressed as mean ± SEM. (**a**) White circles indicate the Placebo. Gray and black circles indicate Dose A (270 mg GTC, 270 mg CCA, with 113 mg caffeine) and Dose B (540 mg GTC, 270 mg CCA, with 113 mg caffeine), respectively. Treatment main effect and treatment × time interactions were analyzed by linear mixed-effect models (n = 18). (**b**) White circles indicate the Placebo group. Gray and black circles indicate the Dose C group (540 mg GTC, 150 mg CCA, with 99 mg caffeine) and the Dose D group (540 mg GTC, 300 mg CCA, with 99 mg caffeine), respectively. Treatment main effects and treatment × time interactions were analyzed by linear mixed-effect models (n = 18; Placebo, n = 17; Dose C, Dose D). Abbreviations: CCA, coffee chlorogenic acid and GTC, green tea catechin.

In Study 2, comparisons of the three conditions using linear mixed-effect models revealed a treatment effect for the change in blood glucose and serum insulin. The comparison of the three groups analyzed by the mixed-effect model revealed a significant difference in the tAUC among the three groups in both blood glucose and serum insulin. Furthermore, the tAUC for blood glucose was significantly lower in the Dose C group and the Dose D group than in the Placebo group. Furthermore, the Dose D group had a significantly lower tAUC value than the Placebo group.

Table 3. AUC for blood-glucose-related indices and incretin up to 4 h after cookie meal ingestion in Study 1.

		Placebo	Dose A vs. Placebo	Dose B vs. Placebo	*p* Value (3 Group)
Glucose	tAUC mg/dL for 4 h	370 ± 7	363 ± 5	350 ± 6 *	0.058
Insulin	tAUC μU/mL for 4 h	83.1 ± 6.4	79.5 ± 5.8	76.3 ± 5.3	0.152
GLP-1	iAUC pmol/L for 4 h	17.7 ± 2.8	23.0 ± 2.5	26.6 ± 3.2 *	0.065
GIP	iAUC pg/mL for 4 h	2243.7 ± 107.3	1659.1 ± 82.7 ***	1518.8 ± 92.6 ***	<0.001 ***

Data are expressed as estimate ± SEM. Comparisons of the three groups and tests of the differences between the groups were analyzed by a mixed-effect model. Multiplicity was adjusted using Tukey–Kramer's test (n = 18, significant difference compared to the Placebo; * $p < 0.05$, *** $p < 0.001$). Dose A refers to the test beverage containing 270 mg of GTC, 270 mg of CCA, and 113 mg caffeine; Dose B refers to the test beverage containing 540 mg of GTC, 270 mg of CCA, and 113 mg caffeine in Study 1. Abbreviations: CCA, coffee chlorogenic acid; GIP, glucose-dependent insulinotropic polypeptide; GLP-1, glucagon-like peptide-1; and GTC, green tea catechin.

Table 4. AUC for blood-glucose-related indices and incretin up to 4 h after cookie meal ingestion in Study 2.

		PLA	Dose C vs. Placebo	Dose D vs. Placebo	*p* Value (3 Group)
Glucose	tAUC mg/dL for 4 h	535 ± 19	490 ± 15 **	497 ± 18 **	0.001 **
Insulin	tAUC μU/mL for 4 h	94.5 ± 9.5	78.8 ± 5.8	77.4 ± 7.5 *	0.048 *
GLP-1	iAUC pmol/L for 4 h	21.3 ± 2.1	26.1 ± 2.7 *	26.8 ± 2.8 *	0.011 *
GIP	iAUC pg/mL for 4 h	1970.2 ± 115.4	1818.2 ± 94.0	1800.3 ± 103.1 **	0.004 **

Data are expressed as estimate ± SEM. Comparison of the three groups and tests of the differences between the groups were analyzed by a mixed-effect model. Multiplicity was adjusted using Tukey–Kramer's test (n = 18; Placebo, n = 17; Dose C and Dose D, significant difference compared to the Placebo; * $p < 0.05$, ** $p < 0.01$). Dose C refers to the test beverage containing 540 mg GTC, 150 mg CCA, and 99 mg caffeine; Dose D refers to the test beverage containing 540 mg GTC, 300 mg CCA, and 99 mg caffeine in Study 2. Abbreviations: CCA, coffee chlorogenic acid; GIP, glucose-dependent insulinotropic polypeptide; GLP-1, glucagon-like peptide-1; GTC, and green tea catechin.

3.3. Incretins

Figure 2 shows changes in the plasma GLP-1 and GIP concentrations before and after the cookie meal, and Table 3 (Study 1) and Table 4 (Study 2) show estimated values of the iAUC adjusted for the effects of the order and period. We assessed the differences in the iAUC for incretin responses because the fasting values fluctuated between each test. The tAUC was used for the blood glucose and serum insulin levels because the fasting values between each test were strictly regulated within individuals, and thus, postprandial values could be less than 0.

In Study 1, the change in the postprandial GLP-1 and GIP levels revealed the main treatment effect, but no treatment by time interaction effect in a mixed-effect model of the measurement data. The results of the mixed-effect model analysis of comparison of the three groups showed a non-significant tendency toward a difference in the iAUC for plasma GLP-1 and a significant difference for plasma GIP. Furthermore, the iAUC for GLP-1 in the Dose A group exhibited an increasing trend, and the Dose B group had a significantly higher iAUC value than the Placebo group. On the other hand, the iAUC for plasma GIP was significantly lower in both the Dose A and Dose B groups than in the Placebo group.

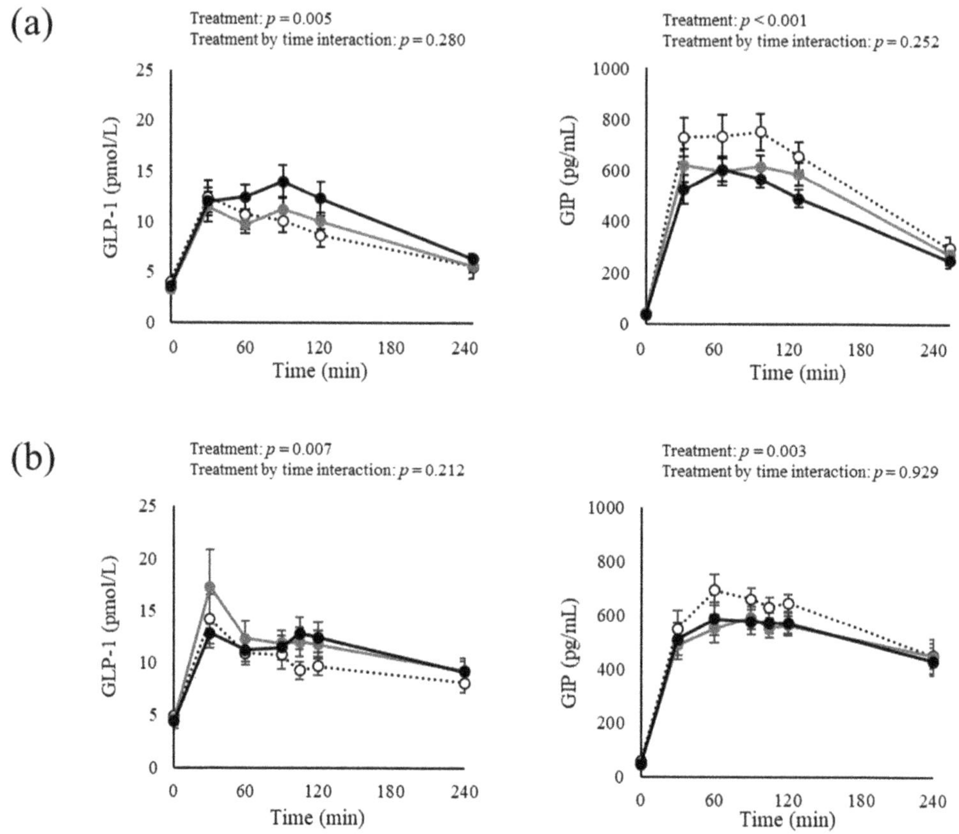

Figure 2. Changes in blood test results at fasting and after cookie meal consumption for Study 1 (**a**) and Study 2 (**b**). Data are expressed as mean ± SEM. (**a**) White circles indicate the Placebo. Gray and black circles indicate Dose A (270 mg GTC, 270 mg CCA, with 113 mg caffeine) and Dose B (540 mg GTC, 270 mg CCA, with 113 mg caffeine), respectively. Treatment main effects and treatment × time interactions were analyzed by linear mixed-effect models (n = 18). (**b**) White circles indicate the Placebo. Gray and black circles indicate Dose C (540 mg GTC, 150 mg CCA, with 99 mg caffeine), and Dose D (540 mg GTC, 300 mg CCA, with 99 mg caffeine), respectively. Treatment main effects and treatment × time interactions were analyzed by linear mixed-effect models (n = 18; Placebo, n = 17; Dose C and Dose D). Abbreviations: CCA, coffee chlorogenic acid; GIP, glucose-dependent insulinotropic polypeptide; GLP-1, glucagon-like peptide-1; and GTC, green tea catechin.

In Study 2, the change in the postprandial GLP-1 and GIP showed the main treatment effect, but no treatment-by-time interaction effect in the mixed-effect model. A comparison of the three groups analyzed by the mixed-effect model revealed a significant difference in the tAUC of both the plasma GLP-1 and GIP levels among the three groups. The tAUC for GLP-1 was significantly higher in both the Dose C and Dose D groups than in the Placebo group, while the tAUC for GIP was significantly lower in both the Dose C group and Dose D group than in the Placebo group.

4. Discussion

This study was performed to estimate the effective dose for a single ingestion of both GTC and CCA on postprandial glucose metabolism, including incretin responses. To our knowledge, this study is the first to assess these conditions, and our findings suggest that a single intake of a combination of varying doses of GTC and CCA significantly improved postprandial hyperglycemia, increased GLP-1, and decreased GIP secretion after consumption of a high-fat and high-carbohydrate cookie meal containing 75 g of glucose. In addition, a comparison of the test beverages with different concentrations of GTC or CCA revealed the minimum effective dose affecting the glycemic and incretin responses. Specifically, compared with the Placebo, the Dose B group (540 mg GTC, 270 mg CCA, with 113 mg caffeine), but not the Dose A group (270 mg GTC, 270 mg CCA, with 113 mg caffeine), maintained the treatment efficacy in Study 1, while both the Dose C group (540 mg GTC, 150 mg CCA, with 99 mg caffeine) and the Dose D group (540 mg GTC, 300 mg CCA, with 99 mg caffeine) maintained the treatment efficacy in Study 2. Based on the results of the present studies, the minimum effective doses of GTC and CCA for promoting the secretion of GLP-1 and GIP are 540 mg and 150 mg, respectively.

Previous studies in humans have not find postprandial enhancement of GLP-1 secretion by GTC intake alone. In contrast, studies investigating the effects of coffee or CCA alone revealed improvements in postprandial hyperglycemia and increased GLP-1 secretion [16,22]. Comparison of the present results with these studies is difficult, due to differences in the content of the test meals and the duration of the test beverage consumption period. Fujii et al. [23] reported that GLP-1 secretion is enhanced in a CCA dose-dependent manner in NCI-H716 cells. This may be related to the combined use with GTC as opposed to CCA alone. It should be noted that the CCA dose in the in vitro experiments cannot be directly compared to that in the human study, nor do the results of in vitro experiments reflect the effects of various dietary nutrients ingested simultaneously in vivo.

In the present study, acute incretin responses were observed following the ingestion of test beverages with high GTC and CCA concentrations, suggesting that GTC and CCA are involved in the mechanisms that promote GLP-1 and inhibit GIP secretions. Various phytochemicals, such as several types of flavonoids, black-tea polyphenols, oolong tea polymerized polyphenols, and potato polyphenolic compounds, inhibit the function of nutrient-digesting enzymes [24–27]. Blood incretin concentrations are affected by the intake of glycolytic and lipolytic enzyme inhibitors [28–30]. GLP-1 is rapidly secreted 15 to 30 min after a meal, mainly from L cells in the lower small intestine, and peaks at 90 to 120 min. GLP-1 secretion is presumed to be caused by stimulation of the vagus nerve or other hormones, whereas GIP secretion is thought to follow direct stimulation within the small intestinal lumen [4]. The nutrition factors reported to stimulate GLP-1 and GIP secretion include sodium–glucose cotransporter protein-mediated glucose and others and various fatty acids via G protein-coupled receptors [31–33]. It was suggested that the combined intake of GTC and CCA inhibits the digestion and absorption of nutrients such as carbohydrates and lipids, promotes increased secretion of GLP-1, and inhibits increased secretion of GIP, rather than directly stimulating K and L cells in the small intestine. In other words, the digestion of carbohydrates or lipids by amylase/glucosidase or lipase may be delayed by decreased stimulation in the upper part of the small intestine and increased stimulation in the lower small intestine. Of note, the consumption of the test beverages with high concentrations of both GTC and CCA decreased postprandial blood glucose levels, suggesting that polyphenols inhibit the absorption of dietary glucose into the blood. The difference in the effect on postprandial blood glucose levels between Studies 1 and 2 may be due to differences in the participants: in Study 1, no specific blood glucose levels were set as exclusion criteria, while in Study 2, blood glucose levels ranged from normal to impaired glucose tolerance (95–125 mg/dL); the mean fasting blood glucose level of the participants in Study 1 was 86 $\pm$ 5 mg/dL and that of the participants in Study 2 was 103 $\pm$ 7 mg/dL. Therefore, a more remarkable effect of the combined ingestion of GTC and CCA was observed in the participants of Study 2, who had higher blood glucose levels. Similarly, an increase in GLP-1 secretion and a decrease in GIP secretion is observed with α-glucosidase inhibitors [34] in drugs and apple juice [35], suggesting that the absorption site of carbohydrates is shifted to the lower part of the small intestine. Other studies have reported postprandial increases in GLP-1 following single and continuous

(90 days) ingestion of green-plant membranes [36]. The mechanism of this GLP-1 secretion enhancement is considered to be a direct effect on endocrine cells in the small intestine via prolonged digestion and absorption of dietary lipids or an indirect effect mediated by neural signals such as cholecystokinin, but this has not yet been clarified.

This study has some potential limitations. The participants were healthy (normal) or had impaired glucose tolerance but were non-diabetic individuals. Therefore, it is unclear whether a similar efficacy on GLP-1 secretion would occur in individuals with T2D. This study was conducted on men only, and caution should be exercised in making general inferences. We hypothesized that a woman's menstrual cycle could affect her blood glucose levels because of the 1-week washout period between each test. Caffeine is a major active component of green tea and coffee, but the synergistic effects of caffeine with catechins and chlorogenic acid were not considered in this study. While caffeine consumption may improve glucose tolerance, its effect on glucose levels varies among studies [37]. On the other hand, several studies report that caffeine consumption increases insulin secretion, but does not necessarily improve glucose levels in oral glucose tolerance tests [38–40], suggesting that caffeine influences insulin clearance. In the present study, the amount of caffeine was adjusted to be the same in the placebo and GTC plus CCA test beverages, but the possibility that caffeine contributed to the observed effects cannot be ruled out. Furthermore, due to differences in estimated energy intake the day before the test between Studies 1 and 2, fasting blood glucose levels on the day of the test may have been higher in Study 1. Therefore, although the same diet was consumed in each study, the findings should be cautiously interpreted.

In conclusion, the minimum dose for the effect of a single oral intake of GTC and CCA on incretin levels in healthy men after consuming the high-fat, high-carbohydrate cookie meal containing 75 g of glucose was found to be 540 mg of GTC and 150 mg of CCA. Furthermore, at the same time, 540 mg of GTC and 150 mg of CCA inhibited the increase in blood glucose levels after ingestion of the high-fat and high-carbohydrate cookie meal.

Author Contributions: A.Y., Y.M. and S.S. conceived and designed the study; A.Y. conducted the study; R.H. supervised the study; A.Y. and T.Y. analyzed the data; A.Y. wrote the manuscript and had primary responsibility for the final content; S.S. and M.H. critically reviewed the manuscript. All authors have read and agreed to the published version of the manuscript.

Funding: Kao Corporation (Tokyo, Japan) financially supported the study. The funder had no role in the study design, data collection, analysis, decision to publish, or the preparation of the manuscript.

Institutional Review Board Statement: The study was conducted in accordance with the Declaration of Helsinki. Study 1 was approved by the Ethics Committee of Kao Corporation (protocol code: T060-17926, date of approval: 17 October 2017) and Study 2 was approved by the Ethics Committee of Kao Corporation (protocol code; T232-190615, date of approval; 30 July 2019) and the Ethics Committee of Ueno-Asagao Clinic (protocol code: g2019004, date of approval: 7 August 2019).

Informed Consent Statement: Informed consent was obtained from all of the participants involved in the study.

Data Availability Statement: Not applicable.

Acknowledgments: We thank all of the volunteers for participating in this study and Haruo Nakamura of Mitsukoshi Health and Welfare Foundation for his instruction and advice regarding the study design, protocol, and discussion.

Conflicts of Interest: A.Y., Y.M., T.Y., S.S., and M.H. are employees of Kao Corporation, a chemical, cosmetic, and food company headquartered in Tokyo, Japan.

References

1. Akash, M.S.; Rehman, K.; Chen, S. Role of inflammatory mechanisms in pathogenesis of type 2 diabetes mellitus. *J. Cell. Biochem.* **2013**, *114*, 525–531. [CrossRef] [PubMed]
2. International Diabetes Federation. *IDF Diabetes Atlas*, 7th ed.; International Diabetes Federation: Brussels, Belgium, 2015.
3. Fujishima, M.; Kiyohara, Y.; Kato, I.; Ohmura, T.; Iwamoto, H.; Nakayama, K.; Ohmori, S.; Yoshitake, T. Diabetes and cardiovascular disease in a prospective population survey in Japan: The Hisayama Study. *Diabetes* **1996**, *45*, S14–S16. [CrossRef] [PubMed]
4. Baggio, L.L.; Drucker, D.J. Biology of incretins: GLP-1 and GIP. *Gastroenterology* **2007**, *132*, 2131–2157. [CrossRef] [PubMed]
5. Seino, Y.; Fukushima, M.; Yabe, D. GIP and GLP-1, the two incretin hormones: Similarities and differences. *J. Diabetes Investig.* **2010**, *1*, 8–23. [CrossRef] [PubMed]
6. Chen, S.; Okahara, F.; Osaki, N.; Shimotoyodome, A. Increased GIP signaling induces adipose inflammation via a HIF-1α-dependent pathway and impairs insulin sensitivity in mice. *Am. J. Physiol. Endocrinol. Metab.* **2014**, *308*, E414–E425. [CrossRef]
7. Legeay, S.; Rodier, M.; Fillon, L.; Faure, S.; Clere, N. Epigallocatechin gallate: A review of its beneficial properties to prevent metabolic syndrome. *Nutrients* **2015**, *7*, 5443–5468. [CrossRef]
8. Ferreira, M.; Silva, D.; de Morais, A., Jr.; Mota, J.; Botelho, P. Therapeutic potential of green tea on risk factors for type 2 diabetes in obese adults–a review. *Obes. Rev.* **2016**, *17*, 1316–1328. [CrossRef]
9. Thielecke, F.; Boschmann, M. The potential role of green tea catechins in the prevention of the metabolic syndrome—A review. *Phytochemistry* **2009**, *70*, 11–24. [CrossRef]
10. Tajik, N.; Tajik, M.; Mack, I.; Enck, P. The potential effects of chlorogenic acid, the main phenolic components in coffee, on health: A comprehensive review of the literature. *Eur. J. Nutr.* **2017**, *56*, 2215–2244. [CrossRef]
11. Meng, S.; Cao, J.; Feng, Q.; Peng, J.; Hu, Y. Roles of chlorogenic acid on regulating glucose and lipids metabolism: A review. *Evid. Based Complement. Alternat. Med.* **2013**, *2013*, 801457. [CrossRef]
12. Venables, M.C.; Hulston, C.J.; Cox, H.R.; Jeukendrup, A.E. Green tea extract ingestion, fat oxidation, and glucose tolerance in healthy humans. *Am. J. Clin. Nutr.* **2008**, *87*, 778–784. [CrossRef] [PubMed]
13. Nagao, T.; Meguro, S.; Hase, T.; Otsuka, K.; Komikado, M.; Tokimitsu, I.; Yamamoto, T.; Yamamoto, K. A catechin-rich beverage improves obesity and blood glucose control in patients with type 2 diabetes. *Obesity* **2009**, *17*, 310–317. [CrossRef] [PubMed]
14. Zuniga, L.Y.; Aceves-de la Mora, M.C.A.; Gonzalez-Ortiz, M.; Ramos-Nunez, J.L.; Martinez-Abundis, E. Effect of chlorogenic acid administration on glycemic control, insulin secretion, and insulin sensitivity in patients with impaired glucose tolerance. *J. Med. Food* **2018**, *21*, 469–473. [CrossRef] [PubMed]
15. Iwai, K.; Narita, Y.; Fukunaga, T.; Nakagiri, O.; Kamiya, T.; Ikeguchi, M.; Kikuchi, Y. Study on the Postprandial Glucose Responses to a Chlorogenic Acid-Rich Extract of Decaffeinated Green Coffee Beans in Rats and Healthy Human Subjects. *Food Sci. Technol. Res.* **2012**, *18*, 849–860. [CrossRef]
16. Jokura, H.; Watanabe, I.; Umeda, M.; Hase, T.; Shimotoyodome, A. Coffee polyphenol consumption improves postprandial hyperglycemia associated with impaired vascular endothelial function in healthy male adults. *Nutr. Res.* **2015**, *35*, 873–881. [CrossRef]
17. Nauck, M.A.; Meier, J.J. Incretin hormones: Their role in health and disease. *Diabetes Obes. Metab.* **2018**, *20* (Suppl. S1), 5–21. [CrossRef]
18. Abdulla, H.; Phillips, B.; Smith, K.; Wilkinson, D.; Atherton, P.J.; Idris, I. Physiological mechanisms of action of incretin and insulin in regulating skeletal muscle metabolism. *Curr. Diabetes Rev.* **2014**, *10*, 327–335. [CrossRef]
19. Yanagimoto, A.; Matsui, Y.; Yamaguchi, T.; Hibi, M.; Kobayashi, S.; Osaki, N. Effects of Ingesting Both Catechins and Chlorogenic Acids on Glucose, Incretin, and Insulin Sensitivity in Healthy Men: A Randomized, Double-Blinded, Placebo-Controlled Crossover Trial. *Nutrients* **2022**, *14*, 5063. [CrossRef]
20. Kozuma, K.; Mizuno, T.; Hibi, M. Effect of tea catechins in Japanese adult on visceral fat—A meta-analysis using individual participant data from seven randomized controlled trials. *Jpn. Pharmacol. Ther.* **2018**, *46*, 973–981.
21. Kozuma, K.; Watanabe, T.; Hibi, M. Effect of Coffee Chlorogenic Acid in Grade I Hypertension and High-normal blood pressure Japanese adults on blood pressure—A meta-analysis using individual participant data from randomized controlled trials—. *Jpn. Pharmacol. Ther.* **2018**, *46*, 1157–1166.
22. Johnston, K.L.; Clifford, M.N.; Morgan, L.M. Coffee acutely modifies gastrointestinal hormone secretion and glucose tolerance in humans: Glycemic effects of chlorogenic acid and caffeine. *Am. J. Clin. Nutr.* **2003**, *78*, 728–733. [CrossRef] [PubMed]
23. Fujii, Y.; Osaki, N.; Hase, T.; Shimotoyodome, A. Ingestion of coffee polyphenols increases postprandial release of the active glucagon-like peptide-1 (GLP-1(7-36)) amide in C57BL/6J mice. *J. Nutr. Sci.* **2015**, *4*, e9. [CrossRef] [PubMed]
24. Buchholz, T.; Melzig, M.F. Polyphenolic Compounds as Pancreatic Lipase Inhibitors. *Planta Med.* **2015**, *81*, 771–783. [CrossRef] [PubMed]
25. Kobayashi, M.; Ichitani, M.; Suzuki, Y.; Unno, T.; Sugawara, T.; Yamahira, T.; Kato, M.; Takihara, T.; Sagesaka, Y.; Kakuda, T.; et al. Black-tea polyphenols suppress postprandial hypertriacylglycerolemia by suppressing lymphatic transport of dietary fat in rats. *J. Agric. Food Chem.* **2009**, *57*, 7131–7136. [CrossRef]
26. Toyoda-Ono, Y.; Yoshimura, M.; Nakai, M.; Fukui, Y.; Asami, S.; Shibata, H.; Kiso, Y.; Ikeda, I. Suppression of postprandial hypertriglyceridemia in rats and mice by oolong tea polymerized polyphenols. *Biosci. Biotechnol. Biochem.* **2007**, *71*, 971–976. [CrossRef]

27. Kalita, D.; Holm, D.G.; LaBarbera, D.V.; Petrash, J.M.; Jayanty, S.S. Inhibition of alpha-glucosidase, alpha-amylase, and aldose reductase by potato polyphenolic compounds. *PLoS ONE* **2018**, *13*, e0191025. [CrossRef]
28. Mikada, A.; Narita, T.; Yokoyama, H.; Yamashita, R.; Horikawa, Y.; Tsukiyama, K.; Yamada, Y. Effects of miglitol, sitagliptin, and initial combination therapy with both on plasma incretin responses to a mixed meal and visceral fat in over-weight Japanese patients with type 2 diabetes. "the MASTER randomized, controlled trial". *Diabetes Res. Clin. Pract.* **2014**, *106*, 538–547. [CrossRef] [PubMed]
29. Damci, T.; Yalin, S.; Balci, H.; Osar, Z.; Korugan, U.; Ozyazar, M.; Ilkova, H. Orlistat augments postprandial increases in glucagon-like peptide 1 in obese type 2 diabetic patients. *Diabetes Care* **2004**, *27*, 1077–1080. [CrossRef]
30. Enc, F.Y.; Ones, T.; Akin, H.L.; Dede, F.; Turoglu, H.T.; Ulfer, G.; Bekiroglu, N.; Haklar, G.; Rehfeld, J.F.; Holst, J.J.; et al. Orlistat accelerates gastric emptying and attenuates GIP release in healthy subjects. *Am. J. Physiol. Gastrointest. Liver Physiol.* **2009**, *296*, G482–G489. [CrossRef]
31. Reimann, F.; Habib, A.M.; Tolhurst, G.; Parker, H.E.; Rogers, G.J.; Gribble, F.M. Glucose sensing in L cells: A primary cell study. *Cell Metab.* **2008**, *8*, 532–539. [CrossRef]
32. Rocca, A.S.; Brubaker, P.L. Stereospecific effects of fatty acids on proglucagon-derived peptide secretion in fetal rat intestinal cultures. *Endocrinology* **1995**, *136*, 5593–5599. [CrossRef] [PubMed]
33. Lauffer, L.; Iakoubov, R.; Brubaker, P.L. GPR119: "double-dipping" for better glycemic control. *Endocrinology* **2008**, *149*, 2035–2037. [CrossRef] [PubMed]
34. Narita, T.; Katsuura, Y.; Sato, T.; Hosoba, M.; Fujita, H.; Morii, T.; Yamada, Y. Miglitol induces prolonged and enhanced glucagon-like peptide-1 and reduced gastric inhibitory polypeptide responses after ingestion of a mixed meal in Japanese Type 2 diabetic patients. *Diabet. Med.* **2009**, *26*, 187–188. [CrossRef] [PubMed]
35. Johnston, K.L.; Michael, N.C.; Linda, M.M. Possible role for apple juice phenolic compounds in the acute modification of glucose tolerance and gastrointestinal hormone secretion in humans. *J. Sci. Food Agric.* **2002**, *82*, 1800–1805. [CrossRef]
36. Montelius, C.; Erlandsson, D.; Vitija, E.; Stenblom, E.L.; Egecioglu, E.; Erlanson-Albertsson, C. Body weight loss, reduced urge for palatable food and increased release of GLP-1 through daily supplementation with green-plant membranes for three months in overweight women. *Appetite* **2014**, *81*, 295–304. [CrossRef]
37. da Silva, L.A.; Wouk, J.; Weber, V.M.R.; da Luz Eltchechem, C.; de Almeida, P.; Martins, J.C.L.; Malfatti, C.R.M.; Osiecki, R. Mechanisms and biological effects of Caffeine on substrate metabolism homeostasis: A systematic review. *J. Appl. Pharm. Sci.* **2017**, *7*, 215–221. [CrossRef]
38. Graham, T.E.; Sathasivam, P.; Rowland, M.; Marko, N.; Greer, F.; Battram, D. Caffeine ingestion elevates plasma insulin response in humans during an oral glucose tolerance test. *Can. J. Physiol. Pharmacol.* **2001**, *79*, 559–565. [CrossRef]
39. Thong, F.S.; Graham, T.E. Caffeine-induced impairment of glucose tolerance is abolished by β-adrenergic receptor blockade in humans. *J. Appl. Physiol.* **2002**, *92*, 2347–2352. [CrossRef]
40. Robinson, L.E.; Savani, S.; Battram, D.S.; McLaren, D.H.; Sathasivam, P.; Graham, T.E. Caffeine ingestion before an oral glucose tolerance test impairs blood glucose management in men with type 2 diabetes. *J. Nutr.* **2004**, *134*, 2528–2533. [CrossRef]

nutrients

MDPI

Article

Rat Mucosal Immunity following an Intensive Chronic Training and an Exhausting Exercise: Effect of Hesperidin Supplementation

Patricia Ruiz-Iglesias [1,2], Sheila Estruel-Amades [1,2], Malén Massot-Cladera [1,2], Àngels Franch [1,2], Francisco J. Pérez-Cano [1,2,*] and Margarida Castell [1,2,3,*]

1 Secció de Fisiologia, Departament de Bioquímica i Fisiologia, Facultat de Farmàcia i Ciències de l'Alimentació, Universitat de Barcelona (UB), 08028 Barcelona, Spain
2 Institut de Recerca en Nutrició i Seguretat Alimentària (INSA-UB), UB, 08921 Santa Coloma de Gramenet, Spain
3 Centro de Investigación Biomédica en Red de Fisiopatología de la Obesidad y la Nutrición (CIBEROBN), Instituto de Salud Carlos III, 28029 Madrid, Spain
* Correspondence: franciscoperez@ub.edu (F.J.P.-C.); margaridacastell@ub.edu (M.C.); Tel.: +34-934-024-505 (F.J.P.-C. & M.C.)

Abstract: Stressful situations such as a high-intensity exercise or exhausting training programs can act as immune disruptors leading to transitory immunodepression status, which can be accompanied by alterations of the gastrointestinal functions. Hesperidin intake has demonstrated ergogenic activity and is able to influence the intestinal ecosystem and immunity. We aimed to investigate the effect of hesperidin consumption in rats submitted to an intense training and a final exhaustion test, focusing on the functionality of the intestinal immune system and on the cecal microbiota. Rats, supplemented or not with hesperidin, were intensively trained on a treadmill for 5 weeks. Samples were obtained 24 h after a regular training session, and immediately and 24 h after a final exhaustion test. Cecal microbiota and composition and function of mesenteric lymph node (MLN) lymphocytes and mucosal immunoglobulin A (IgA) were determined. Results showed that chronic intense exercise followed by an exhausting test induced changes in the intestinal immune compartment such as the distribution and function of MLN lymphocytes. Although the hesperidin supplementation did not prevent these alterations, it was able to enhance IgA synthesis in the intestinal compartment. This could be important in enhancing the immune intestinal barrier in this stressful situation.

Keywords: flavanone; flavonoid; IgA-coated bacteria; lymphocytes; mesenteric lymph node; microbiota; proliferation; secretory IgA

Citation: Ruiz-Iglesias, P.; Estruel-Amades, S.; Massot-Cladera, M.; Franch, À.; Pérez-Cano, F.J.; Castell, M. Rat Mucosal Immunity following an Intensive Chronic Training and an Exhausting Exercise: Effect of Hesperidin Supplementation. *Nutrients* **2023**, *15*, 133. https://doi.org/10.3390/nu15010133

Academic Editors: Sonia de Pascual-Teresa and Luis Goya

Received: 19 November 2022
Revised: 22 December 2022
Accepted: 23 December 2022
Published: 27 December 2022

1. Introduction

Immunonutrition focuses on the interface of nutrition and immunology and, in the last decades, it has become the focus of many studies. Within this field, physical exercise is a recent area of research that is fast developing due to the interest of athletes and also to the development of the nutrition market [1]. Nevertheless, high-intensity or exhausting training programs can cause stressful situations that can act as immune disruptors leading to a transitory immunodepression status. Consequently, elite endurance athletes have a higher frequency of upper respiratory tract infections following an intensive race such as a marathon or triathlon [2,3]. In particular, during and immediately after a high intensity exercise, a transient leukocytosis appears, which is later followed by leucopenia and the suppression of other immune players, producing a "window of opportunity" for pathogens [4,5]. Moreover, exhausting exercise changes the immunological profile, due to an increase in humoral immunity associated with a downregulated cellular immunity [4]. Therefore, vigorous exercise promotes the synthesis and release of T helper (Th)-2 (anti-inflammatory) cytokines in order to decrease the muscle damage resulting from inflammation, but increases susceptibility to infections [4–8]. Preclinical studies have demonstrated the alterations produced in systemic and intestinal lymphoid tissues after chronic and vigorous physical exercise. Thus, it has been reported that intensive training induces a decrease in the phagocytic activity, a differential pattern of cytokines, increased natural killer (NK) cytotoxic activity, and reduced lymphoproliferative activity [9,10].

Nevertheless, exercise does not only affect the immune system. Currently, the term of "exerkines" is used to define signaling moieties released in response to acute and/or chronic exercise that may affect a wide variety of organs [11]. The immune system, bone, nervous system, and cardiometabolic tissues (cardiovascular, adipose, endocrine systems, liver, and skeletal muscle) act as a source of exerkines and, at the same time, are directly affected by exercise [11]. Therefore, in addition to the disruption of the immune system, alterations of gastrointestinal functions can also appear. In fact athletes, especially endurance athletes, can suffer upper and lower gastrointestinal complaints, which can have negative effects on performance and also an impact on subsequent recovery [12,13]. In general, most gastrointestinal complaints are mild and include nausea, cramping, bloating, abdominal angina, and diarrhea [12,13], but mucosal erosions and ischemic colitis have also been observed after long-distance running [13]. The mechanisms underlying the prolonged exercise-induced gastrointestinal disorders likely include the redistribution of blood flow from the gut to the skin [12,13], but gut dysbiosis is also reported in some endurance runners [14,15]. In addition, preclinical studies have demonstrated that a high-intensity training reduces the salivary production of immunoglobulin (Ig) A, impairs the tight junction proteins' gene expression, and modifies the mesenteric lymph nodes (MLNs) lymphocyte composition and function, increasing the ratio between $T\alpha\beta+$ and B lymphocytes, reducing their proliferation capacity, and enhancing their interferon (IFN) γ secretion [16].

Hesperidin is a flavanone (hesperetin-7-rutinoside) found in citric fruits, such as mandarins, oranges, and lemons [17]. Many studies have demonstrated that hesperidin consumption offers beneficial health effects [18,19]. In this regard, in recent months, published reviews have summarized the effects of hesperidin in cancer [20–22], COVID-19 [23–26], Alzheimer's disease [21], cardiovascular disease [25,27,28], alcoholic liver disease [29], inflammatory disease [30], diabetes [31], and rheumatoid arthritis [31], among others.

Previous studies have reported the influence of hesperidin intake in exercise both in clinical [32–36], and preclinical studies [37–40]. In particular, clinical studies demonstrated that a customized citrus flavonoid extract taken for 4 weeks increased cycling time-trial performance in male athletes [32] and enhanced anaerobic capacity and peak power during high intensity exercise in moderately trained men and women [36]. A longer study demonstrated that the intake of 2S-hesperidin for 8 weeks increased endogenous antioxidant capacity and performance after exhausting exercise in amateur cyclists [34,35]. Furthermore, preclinical studies showed that the administration of 200 mg/kg/3 times per week of hesperidin was able to prevent the overproduction of reactive oxygen species and the decrease in thymic and splenic antioxidant activities after an exhaustion test in trained rats [39]. Moreover, the flavanone enhanced NK cell cytotoxic and monocyte phagocytic functions, whereas it attenuated the secretion of cytokines by macrophages [40]. Similarly, the intake of 100 mg/kg hesperidin for 4 weeks before a continuous swimming exercise in rats improved the biochemical profile and antioxidant biomarkers [37].

On the other hand, hesperidin supplement was able to influence the intestinal ecosystem and immunity [41–43]. The administration of the flavanone for 4 weeks in healthy rats influenced the cecal microbiota, increasing the number of bacteria and particularly the proportion of *Lactobacillus* population, and also raised the intestinal IgA content [42]. In healthy humans, the consumption of orange juice for two months also produced an increase in *Lactobacillus* spp. abundance and in the acetic acid production [44]. In another study, the intake of two orange juices per day for seven days resulted in microbiota composition shifts, with an increase in the abundance of Clostridia operational taxonomic units [44]. Moreover, hesperidin intake in rats was able to change the lymphocyte composition in the MLNs, intestinal epithelium and the lamina propria [41,42]. In addition, hesperidin intake increased the IFN-γ secretion by lymphocytes of MLNs in immunized rats [41]. In a model of physical exercise, the administration of 200 mg/kg of hesperidin (3 times per week, for 5 weeks) enhanced the NK function and the proportion of phagocytic monocytes, attenuated the secretion of macrophage-derived cytokines and prevented the leukocytosis induced by exhaustion [40]. These results and the intestinal influence of hesperidin prompted us to study the influence of hesperidin intake during an intensive training and an exhausting exercise, on the intestinal immune system. Thus, the aim of the present study was to investigate the effect of hesperidin consumption during 5 weeks in rats submitted to an intense training and a final exhaustion test, focusing on the functionality of the intestinal immune system and on the cecal microbiota.

2. Materials and Methods

2.1. Animals, Exercise Training Program, and Hesperidin Supplementation

The animals and exercise training program were the same as previously reported [40]. Briefly, 3-week-old female Wistar rats (Envigo, Huntingdon, UK) were used. Female animals were chosen because they showed better adaptability and higher performance than male rats [9]. Animals were fed, ad libitum, with Teklad Global 14% Protein Rodent Maintenance Diet (Teklad, Madison, WI, USA). The animal procedure was approved by the Ethical Committee for Animal Experimentation of the University of Barcelona and the Catalonia Government (CEEA/UB ref. 464/16 and DAAM 9257, respectively), in full compliance with national legislation following the European Union Directive 2010/63/EU for the protection of animals used for scientific purposes.

Two specialized treadmills for rodents—a LE8700 treadmill (Panlab, Harvard Apparatus, MA, USA) and an Exer3/6 treadmill (Columbus, OH, USA)—were used. Firstly, after a 7-day acclimation period, all rats (4-week-old animals) were adapted to the treadmill (10 days with increasing time and speed). Afterwards, animals were distributed into four groups with similar ability to run: non-supplemented runner animals (RUN, $n = 24$), hesperidin-supplemented runner animals (H-RUN, $n = 24$), non-supplemented sedentary animals (SED, $n = 8$), and hesperidin-supplemented sedentary animals (H-SED, $n = 8$). H-RUN and H-SED groups received 200 mg/kg of body weight (BW) of hesperidin (HealthTech BioActives, Murcia, Spain) by oral gavage 3 times per week for 5 weeks. RUN and SED receive a matched volume of vehicle. RUN and H-RUN groups were then submitted to a 5-week intensive training in which, every Monday and Friday, rats carried out an exhaustion test, whereas on Tuesday, Wednesday, and Thursday rats ran for a period of time according to the maximum speed achieved in the previous Monday's exhaustion test. At the end of the training program, each RUN and H-RUN groups were distributed into three subgroups to establish intestinal immunity at different time points. One subgroup (trained, T, and H-T groups, $n = 8$ each one) was euthanized 24 h after a regular training session. The other two subgroups carried out an additional final exhaustion test. One of these was euthanized immediately after carrying out an additional final exhaustion test (TE and H-TE groups, $n = 8$ each one), whereas the other one was euthanized 24 h after the final exhaustion test (TE24 and H-TE24 groups, $n = 8$ each one). Sedentary rats (both SED and H-SED groups) were euthanized randomly distributed over the three consecutive days. The experiment ended with 10-week-old animals.

2.2. Sample Collection and Processing

Rats were anesthetized with ketamine (90 mg/kg, Merial Laboratories S.A., Barcelona, Spain) and xylazine (10 mg/kg, Bayer A.G., Leverkusen, Germany) and exsanguinated. The MLNs, small intestine, submaxillary salivary glands (SMGs), and cecal content (CC) were collected.

Lymphocytes from MLNs were isolated by passing the tissue through a 40 µm sterile-mesh cell strainer (Thermo Fisher Scientific, Barcelona, Spain), as previously described [45].

The distal part of the small intestine was used to obtain gut wash (GW), as carried out previously [46]. Briefly, fecal content was removed from the intestine by flushing with cold phosphate buffered saline (PBS, pH 7.2), then the tissue was opened lengthwise, cut into 1–2 cm pieces, weighed, and incubated with PBS for 10 min in a shaker at 37 °C (55 shakings × min^{-1}). After centrifugation (538× *g*, 4 °C, 10 min), supernatants were collected and stored at −20 °C until IgA quantification.

SMGs, CC, and fecal (from feces obtained at the moment of anesthesia) homogenates were obtained using a tissue homogenizer (for SMG, Polytron, Kinematica, Lucerne, Switzerland) or a Pellet Pestle Cordless Motor (for CC and fecal samples, Kimble, Meiningen, Germany), as described in previous studies [47], and kept at −20 °C until IgA quantification.

2.3. Cecal Microbiota Analysis by Fluorescence In Situ Hybridization Coupled to Flow Cytometry

The previous effects observed by the polyphenols interventions on particular bacteria groups, such as *Clostridium coccoides/Eubacterium rectale, Bifidobacterium* and *Lactobacillus/Enterococcus,* led us to evaluate the changes induced by exercise only on these populations. For that, sequencing was initially discarded and targeted fluorescence in situ hybridization coupled to flow cytometry (FISH-FCM) as previously used in our group [48], was used. Group-specific fluorochrome-conjugated probes (Erec482 5′-GCTTCTTAGTCARGTACCG, Bif164 5′-CATCCGGCATTACCACCC, and Lab158 5′-GGTATTAGCAYCTGTTTCCA) (Sigma-Aldrich, Madrid, Spain), which hybridize the bacterial 16S RNA of each particular group were used. Data were acquired by a FacsAria SORP sorter (BD Biosciences, San Diego, CA, USA) in the Flow Cytometry Unit (FCU) of the Scientific and Technological Centers of the University of Barcelona (CCiTUB) and the analysis was performed with FlowJo v.10 software (Tree Star Inc., Ashland, OR, USA). The results were normalized by total bacteria, which was detected adding propidium iodide (PI; Sigma-Aldrich) prior to the FCM acquisition. Commercial Flow Check™ Fluorospheres (Beckman Coulter, Miami, FL, USA) were used to assess the total counts of bacteria.

2.4. Cecal Proportion of IgA-Coated Bacteria

The proportion of IgA-coated bacteria (IgA-CB) in CC was determined by FCM as previously described [48]. The CC homogenate was stained with rabbit anti-rat Ig polyclonal antibody conjugated to fluorescein isothiocyanate (FITC) (Abcam, Cambridge, UK). Bacteria were gated in a FacsAria SORP sorter (BD Biosciences) after PI staining (Sigma-Aldrich) and according to their forward (FSC) and side scatter (SSC) characteristics. Analysis was performed in the FCU of the CCiTUB using the FlowJo v.10 software.

2.5. Phenotypic Analysis of MLNs Lymphocytes

Lymphocytes from MLNs were extracellularly and intracellularly stained as previously reported [16]. Mouse anti-rat monoclonal antibodies (mAb) conjugated to fluorochromes—fluorescein isothiocyanate (FITC), phycoerythrin (PE), peridinin chlorophyll protein (PerCP), allophycocyanin (APC) or brilliant violet 421 (BV421): FITC-TCRαβ, FITC-CD8β, FITC-CD25, PE-CD161a, PE-TCRγδ, PE-CD4, PerCP-CD8α, APC-CD4, and BV421-CD45RA (BD Biosciences) and APC-FoxP3 (eBioscience, Frankfurt, Germany). For extracellular staining, cells were incubated with saturating amounts of mAb in PBS containing 2% Fetal Bovine Serum (FBS) and 0.1% NaN_3. For intracellular staining, cells were previously extracellularly labeled with anti-CD4-PE and anti-CD25-FITC mAb, then treated with Foxp3 fixation/permeabilization kit (eBioscience) and finally intracellularly stained with anti-Foxp3-APC mAb. Cells were fixed with 0.5% p-formaldehyde and stored at 4 °C in darkness until analysis by FCM. A negative control staining without any mAb and a staining control for each mAb were included.

Analyses were performed using a Gallios Cytometer (Beckman Coulter, Miami, FL, USA) in the FCU of CCiT-UB and by Flowjo v10 software.

2.6. Proliferative Activity of Lymphocytes from MLNs

Cells (10^5 cells/well) were incubated in quadruplicate in 96-well plates (TPP, Sigma-Aldrich) and stimulated or not with concanavalin (Con) A (5 µg × mL^{-1}, Sigma-Aldrich) for 48 h as previously described [16]. Cell proliferation was quantified using a BrdU Cell Proliferation Assay kit (MerckMillipore, Darmstadt, Germany), according to manufacturer's instructions. The proliferation rate was calculated by dividing the optical density of ConA stimulated cells with the optical density of non-stimulated cells.

2.7. IgA Quantification

The concentrations of IgA in GW, SMGs, and in CC and fecal homogenates were quantified by a sandwich ELISA (Bethyl Laboratories Inc., Montgomery, AL, USA) as previously described [49]. The IgA content in SMGs and CC was normalized by total protein concentration which was measured using the Pierce® 660 nm Protein Assay Reagent (Thermo Fisher Scientific) following the manufacturer's instructions.

2.8. Statistical Analysis

Data were statistically analyzed by IBM Social Sciences Software Program (SPSS, version 26.0, Chicago, IL, USA) and Rstudio v4.04 (Rstudio, Inc., Boston, MA, USA) with R version 3.6.1 (R Core Team 2021, R Foundation for Statistical Computing, Vienna, Austria). The normality and homoscedasticity of the data were tested by Shapiro–Wilk's and Levene's test, respectively. In this case, we applied a two-way ANOVA test and, if significant differences were found, Tukey's post hoc test was carried out. Non-parametric Aligned Rank Transform for non-parametric factorial ANOVA (ART-ANOVA) followed by emmeans post hoc (Tukey-adjusted p value) were applied, using the ARTool [45,46] and emmeans [47] packages, respectively, for Rstudio.

When significant differences were detected ($p \leq 0.05$), the p values obtained in the two-way ANOVA or the ART-ANOVA for the variables exercise (E), hesperidin supplementation (H), and the interaction between them (E $\times$ H) are indicated in bold in the legend box. Changes due to the exercise are represented in the figure using different letters above the bars. When the E $\times$ H interaction was significant, changes between particular groups were represented with an asterisk above the respective bars.

3. Results

3.1. Changes in Morphometric Variables

The training program, both in non-supplemented animals and those supplemented with hesperidin, induced a slight but significant increase in the body mass index (BMI) and Lee index (Figure 1a,b, respectively). The final exhaustion test reduced both BMI and Lee index with respect to the values present in trained rats (TE and TE24 groups vs. T group). Such a decrease must be attributed to the weight lost (mainly water) due to exhaustion. BMI and Lee index did not change due to hesperidin supplementation.

3.2. Changes in Cecal Microbiota

The content of total bacteria in the cecum was established in sedentary and runner rats from non-supplemented groups and hesperidin-supplemented groups (Figure 2a). Neither exercise nor hesperidin supplementation significantly changed the total content of cecal bacteria. Likewise, we did not find significant variations in the abundance of *Lactobacillus*, *Bifidobacterium* and *Clostridium coccoides/Eubacterium rectale* groups in cecum due to hesperidin supplement or exercise (Figure 2b–d).

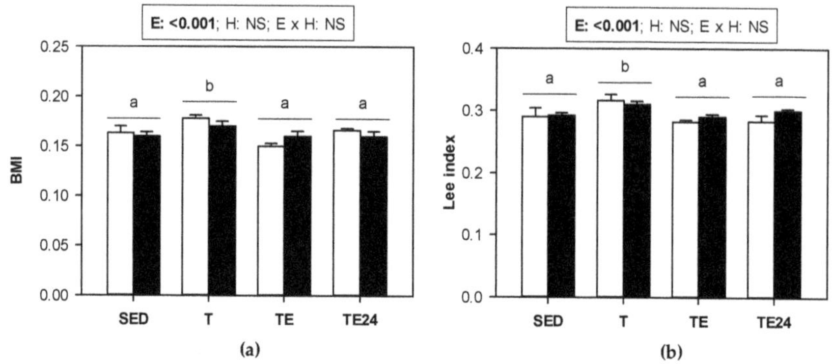

Figure 1. (**a**) Body mass index (BMI) and (**b**) Lee index in sedentary (SED) and runner rats (training for five weeks) that were classified into T group, with values obtained 24 h after a regular training session, TE group, with values obtained immediately after a final exhaustion test, and TE24 group, with values obtained 24 h after the final exhaustion test. The non-supplemented groups are represented by white bars and the hesperidin-supplemented groups by black bars. E = exercise; H: hesperidin supplementation; E × H = interaction between E and H; NS = no significant differences. Data are expressed as mean ± SEM (n = 8). Different letters between groups indicate significant differences ($p < 0.05$) due to exercise.

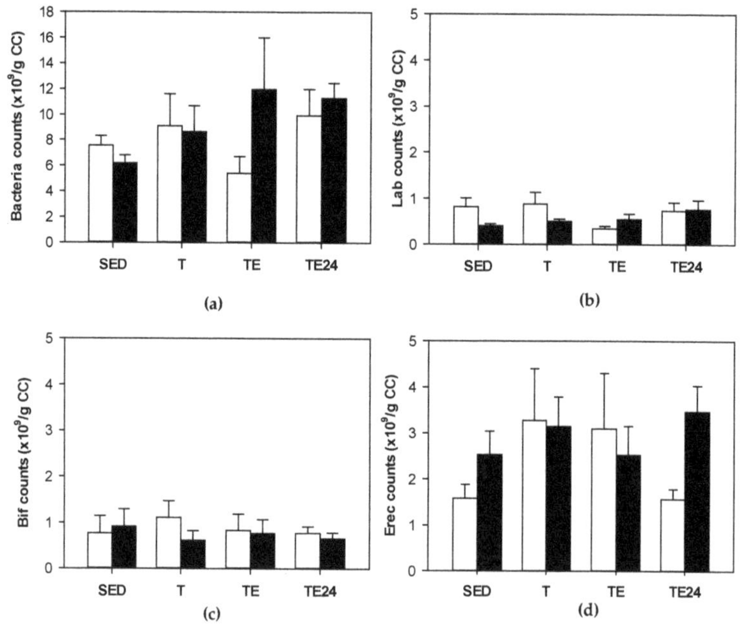

Figure 2. Total bacteria (counts $\times 10^9$/g cecal content) (**a**); *Lactobacillus* (counts $\times 10^9$/g cecal content) (**b**); *Bifidobacterium* (counts $\times 10^9$/g cecal content) (**c**); *Clostridium coccoides/Eubacterium rectale* (counts $\times 10^9$/g cecal content) (**d**). The non-supplemented groups are represented by white bars and the hesperidin-supplemented groups by black bars. CC = cecal content; SED = sedentary groups; T = trained groups; TE = T groups with an additional exhaustion test; TE24 = TE groups 24 h after the exhaustion test. Data are expressed as mean ± SEM (n = 8).

3.3. Changes in the Number of IgA-Coated Bacteria

Overall, the amount of cecal bacteria coated to IgA did not change exclusively due to exercise or hesperidin supplementation (Figure 3). Nevertheless, in those animals supplemented with hesperidin, there was an increase in the proportion of bacteria coated to IgA 24 h after carrying out the final exhaustion test. In these animals, the number of IgA-coated bacteria was almost three-fold higher than that in the non-supplemented group.

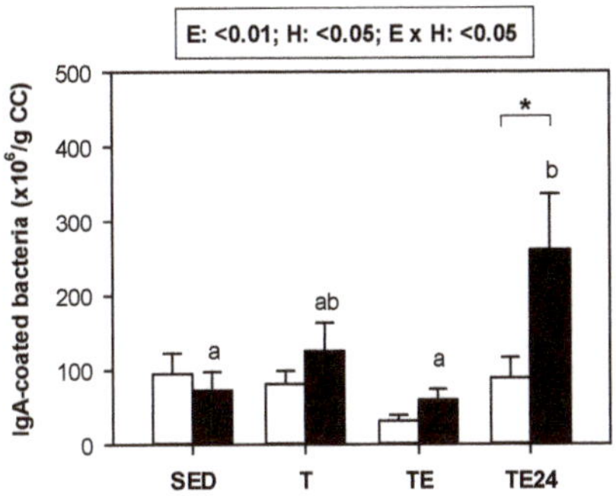

Figure 3. Number of cecal bacteria coated to IgA ($\times 10^6$/g cecal content). The non-supplemented groups are represented by white bars and the hesperidin-supplemented groups by black bars. CC = cecal content; SED = sedentary groups; T = trained groups; TE = T groups with an additional exhaustion test; TE24 = TE groups 24 h after the exhaustion test. E = exercise; H: hesperidin supplementation; E $\times$ H = interaction between E and H. Data are expressed as mean $\pm$ SEM (n = 8). As E $\times$ H interaction was significant, Tukey's post hoc test was carried out, and changes between particular groups are represented with different letters above those groups that showed different results, and changes due to hesperidin supplementation are represented with an asterisk.

3.4. Lymphocyte Composition of MLNs

Figure 4 summarizes the proportion of the main lymphocyte subsets found in MLNs of sedentary and runner animals that were supplemented or not with hesperidin. As we can see, in non-supplemented animals, exercise did not modify the proportion of T and B lymphocytes. However, in those groups that received hesperidin, T cell proportion increased by 14%, and B cell proportion diminished 24 h after exhaustion with respect to values obtained immediately after the exhaustion test (Figure 4a,b). When considering the ratio between Th and cytotoxic (Tc) cells in MLNs, an increase in the ratio was observed in trained animals, regardless of hesperidin supplementation. However, immediately after the final session of exhausting exercise, the ratio was lower and, 24 h later, it was recovered (Figure 4c).

The minor population of NKT cells, which represents about 1% of T cells, did undergo some changes due to exercise just in the hesperidin-supplemented animals. There was an increase in the percentage of NKT cells after exhaustion in hesperidin-supplemented animals, which showed a lower proportion of these cells than the non-supplemented group in the sedentary condition (about 60% of that of non-supplemented group) (Figure 4d).

The phenotype of activated and regulatory T cells was also considered in the Th lymphocytes of MLNs. No changes due to exercise or hesperidin supplement were observed in these two subsets (Table 1).

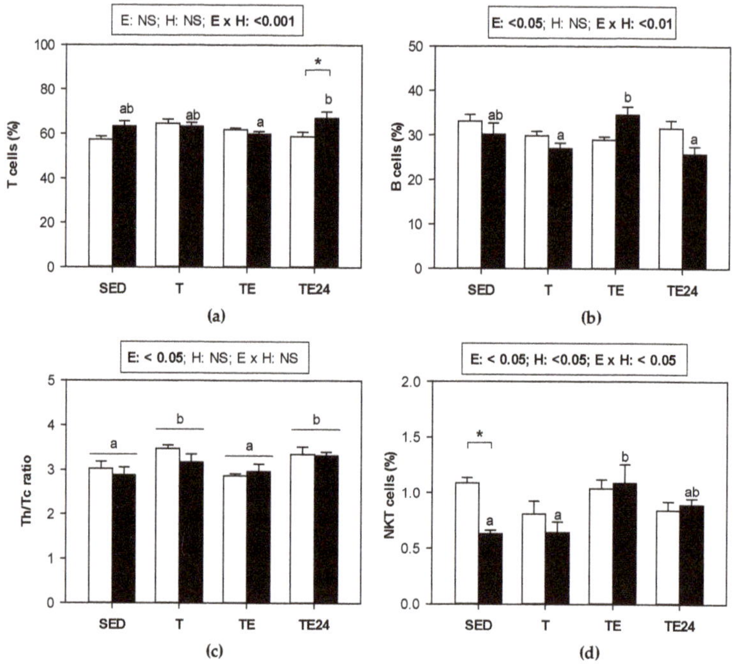

Figure 4. Proportion of the main lymphocyte populations in MLNs: (**a**) T cells, (**b**) B cells, (**c**) Th/Tc ratio, and (**d**) NKT cells. The non-supplemented groups are represented by white bars and the hesperidin-supplemented groups by black bars. SED = sedentary groups; T = trained groups; TE = T groups with an additional exhaustion test; TE24 = TE groups 24 h after the exhaustion test. E = exercise; H: hesperidin supplementation; E × H = interaction between E and H; NS = no significant differences. Data are expressed as mean ± SEM (n = 8). When E × H interaction was significant, Tukey's post hoc test was carried out, and changes between particular groups are represented with different letters above those groups that showed different results, and changes due to hesperidin supplementation are represented with an asterisk.

Table 1. Proportion of the activated and regulatory Th cells in MLNs.

Cells	Diet	SED (%)	T (%)	TE (%)	TE24 (%)
CD4+CD25+ cell proportion in Th cells	non-S [1]	2.19 ± 0.183	2.16 ± 0.167	2.39 ± 0.222	2.05 ± 0.149
	Hesp [2]	2.27 ± 0.117	2.06 ± 0.217	2.52 ± 0.282	2.27 ± 0.131
CD4+CD25+ Foxp3+ cell proportion in Th cells	non-S	1.79 ± 0.110	2.00 ± 0.1061	1.65 ± 0.260	1.75 ± 0.200
	Hesp	2.06 ± 0.114	1.93 ± 0.158	1.78 ± 0.310	1.98 ± 0.119

[1] non-S = non-supplemented group, [2] Hesp = hesperidin-supplemented group.

Figure 5 summarizes the proportion of the minor lymphocyte subsets found in MLNs of sedentary and runner animals that were supplemented or not with hesperidin. NK cell proportion underwent an increase just after exhaustion in both non-supplemented and hesperidin-supplemented animals (Figure 5a). On the other hand, although the proportion of γδT lymphocytes did not vary by either exercise or hesperidin supplementation in MLNs (Figure 5b), when considering the distribution between CD8αα+ and CD8αβ+ cell subsets, some changes associated with exercise appeared. In particular, we observed that 24 h after the final exhaustion test, the proportion of γδT CD8αα+ cells decreased whereas that of γδT CD8αβ+ cells increased (Figure 5c,d).

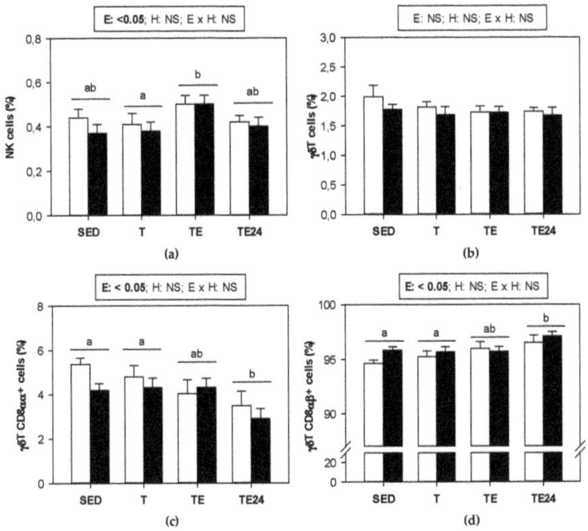

Figure 5. Proportion of the minor lymphocyte populations in mesenteric lymph nodes: (**a**) NK cells, (**b**) γδT cells, (**c**) γδT CD8αα+ cells, and (**d**) γδT CD8αβ+ cells. The non-supplemented groups are represented by white bars and the hesperidin-supplemented groups by black bars. SED = sedentary groups; T = trained groups; TE = T groups with an additional exhaustion test; TE24 = TE groups 24 h after the exhaustion test. E = exercise; H: hesperidin supplementation; E × H = interaction between E and H; NS = no significant differences. Data are expressed as mean ± SEM (n = 8). Statistical analysis (ART-ANOVA test): Different letters indicate significant differences due to exercise.

3.5. Lymphoproliferative Activity of MLNs

The proliferative activity of lymphocytes from MLNs was established (Figure 6). Training for 5 weeks tended to decrease their proliferative activity in both non-supplemented and hesperidin-supplemented groups. After the final exhaustion test, the lymphoproliferative activity increased.

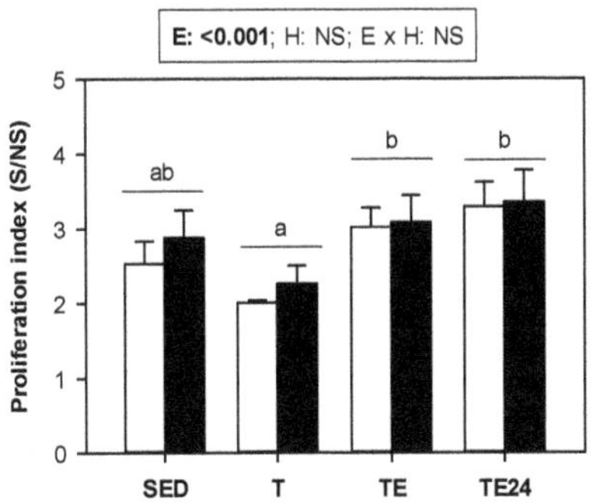

Figure 6. Proliferative activity of MLNs. The non-supplemented groups are represented by white bars and the hesperidin-supplemented groups by black bars. SED = sedentary groups; T = trained groups; TE = T groups with an additional exhaustion test; TE24 = TE groups 24 h after the exhaustion test. E = exercise; H: hesperidin supplementation; E × H = interaction between E and H; NS = no significant differences. Data are expressed as mean ± SEM (n = 8). Statistical analysis (ART-ANOVA test): Different letters indicate significant differences due to exercise.

3.6. Changes in Mucosal IgA

In the intestinal compartment, training or the final exhaustion test did not modify the IgA content in either the GW, cecal content, or feces (Figure 7a–c, respectively). In contrast, hesperidin supplementation did produce, overall, an increase in the IgA concentration found in small intestine and the cecum. In fact, in this last compartment, trained rats showed a three-fold increase in IgA content due to hesperidin supplementation. No significant changes were observed in feces (Figure 7c).

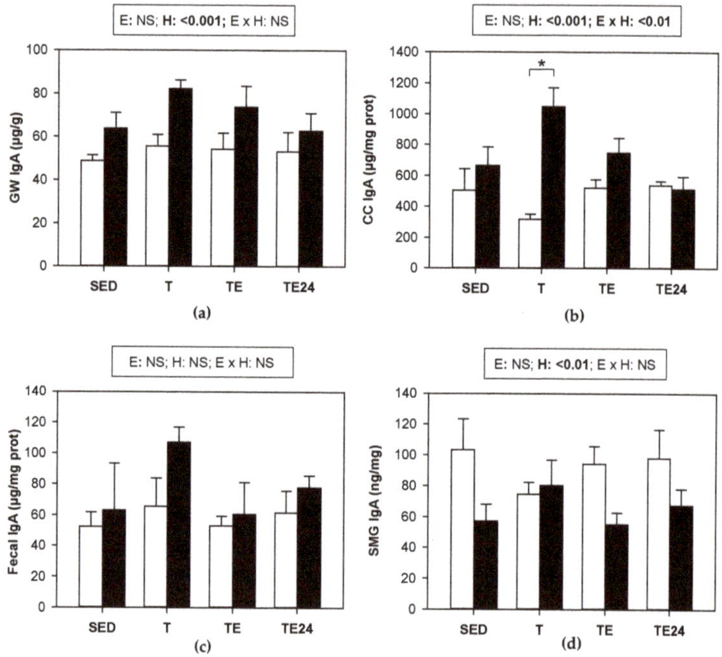

Figure 7. IgA concentration in: (**a**) gut wash; (**b**) cecal content; (**c**) fecal homogenate; (**d**) submaxillary glands. The non-supplemented groups are represented by white bars and the hesperidin-supplemented groups by black bars. GW = gut wash; CC = cecal content; SMG = submaxillary glands; SED = sedentary groups; T = trained groups; TE = T groups with an additional exhaustion test; TE24 = TE groups 24 h after the exhaustion test; E = exercise; H: hesperidin supplementation; E × H = interaction between E and H; NS = no significant differences. Data are expressed as mean ± SEM (n = 8). Statistical analysis (ART-ANOVA test): global effect of hesperidin supplementation is indicated in the corresponding box above. When E × H interaction was significant, an emmeans post hoc test (Tukey-adjusted p value for non-parametric data) was carried out, changes due to hesperidin supplementation are represented with an asterisk.

We also studied the IgA content in the SMGs. The IgA concentration was not significantly modified by training or final exhaustion test. Nevertheless, in general, supplementation with hesperidin decreased the IgA content in this mucosal compartment in most of the studied situations (Figure 7d).

4. Discussion

Although moderate physical exercise is good for the prevention of many chronic diseases, prolonged intense exercise may have negative health consequences, many of which may be mediated by physiological pathways activated by chronic stress. In consequence, among other effects, chronic exposure to stress exerts detrimental effects on immune function [4,50] and also the gastrointestinal ecosystem [13]. In these situations, nutritional supplements, such as polyphenol intake [51,52], could be useful for attenuating such conditions. Here we assessed the effect of hesperidin supplementation on the functionality of the intestinal immune system and the cecal microbiota of rats submitted to an intense training and a final exhaustion test. Previously, we have demonstrated that hesperidin supplementation during training partially improved exercise performance, enhanced the innate immunity by means of increasing the NK cell function as well as the proportion of phagocytic cells, and attenuated the release of some cytokines by macrophages. Moreover, hesperidin prevented the leukocytosis induced by exhaustion and promoted a higher proportion of Th cells in the thymus, blood, and spleen, 24 h after the exhaustion test [40].

Firstly, we focused on the gut microbiota which, in addition to its role in nutrient availability and vitamin synthesis, forms an intestinal barrier that helps to prevent pathogen colonization [53]. In addition, in the last decade, there has been an exponentially growing interest in the gut–brain axis, suggesting that microbiota has modulatory roles in neurodevelopment, brain function, and neurodegenerative diseases, and environmental factors such as diet and stress could modify this axis in a bidirectional pathway [54]. Although some researchers showed changes in the intestinal microbiota composition due to intense exercise [15,55,56], mostly studied by sequencing techniques, we could not detect alterations in the targeted bacteria considered in any of the three exercise conditions studied. The lack of significant changes in the gut microbiota could be due to the technique used (FISH-FCM) with low sensitivity and applied to only a few bacterial groups. In further studies, it would be necessary to apply a more sensitive assay, such as 16S rRNA sequencing methodology, that will provide more sensitive values and a wider range of information or even the Shotgun sequencing in order to obtain information regarding functionality in addition to the taxonomic data. On the other hand, we found that hesperidin supplementation did not significantly affect cecal microbiota in either sedentary or runner rats. These results agree with those reported by Unno et al. [57] and with an in vitro study showing that hesperidin had no effect on *Lactobacillus* spp. cultures [58]. However, in healthy humans, the consumption of orange juice for two months produced an increase in *Lactobacillus* spp. [44], and the intake of orange juice for 60 days increased the abundance of *Bifidobacteriaceae* [59]. In addition, a previous study in Lewis rats using a similar dosage of hesperidin increased the total bacteria counts and also the amount of *Bifidobacterium* and *Lactobacillus/Enterococcus* [42] but had no effect on the *C.coccoides/E.rectale* abundance in the cecum content of rats. The differences found for *Bifidobacterium* and *Lactobacillus/Enterococcus* bacteria could be partially due to the different rat strain (Lewis vs. Wistar) that will induce a different sensitivity to the flavanone intake.

Beyond gut microbiota, the intestinal immune system contains MLNs, where lymphocytes come into contact with dendritic cells charged with intestinal pathogens and will be easily mobilized to the blood. We found that training for 5 weeks did not significantly modify the proportion of total T cell population but increased the Th/Tc cell ratio in MLNs, which was normalized immediately after exhaustion, in agreement with previous results in this compartment [16] as well as in the spleen [10]. It is well known that exhaustion mobilizes T lymphocytes, and from the decrease in Th/Tc proportion in the TE group with respect to T group, it could be suggested that Th lymphocytes will be mobilized faster than Tc cells. The mobilization of Th cells from MLNs or spleen will increase blood Th cell proportion as previously reported in rats similarly trained and exhausted [10]. Later on, lymphopenia occurs due to movement back to the lymphoid organs, which could be responsible for the increase in the Th/Tc cell ratio in MLNs 24 h after exhaustion. In rats receiving hesperidin supplement, we did not find differences in the Th/Tc cell ratio with compared to non-supplemented animals but there was a higher T cell proportion and a lower B cell proportion 24 h after exhaustion. A higher T cell proportion by hesperidin has also been reported in blood from similarly trained and exhausted animals [40]. These increases in T cell proportions in MLNs and blood of rats that received hesperidin agree with the effect reported in the thymus, where it seems that the flavanone enhanced T cell maturation [40]. Likewise, it has been reported that, in non-trained rats, hesperidin supplement did increase T cell proportion (both Th and Tc lymphocytes) in MLNs [41,42].

On the other hand, the study of activated Th or regulatory T cells in MLNs revealed no changes due to exercise or hesperidin. However, the proportion of Tγδ cells with the phenotype CD8αα+ decreased 24 h after the final exhaustion, which was not prevented by hesperidin supplementation. Similarly, in both non-supplemented and hesperidin-supplemented animals, NK cell proportion increased immediately after exhaustion, which could be due to Th cell mobilization, decreasing the percentage of Th cells and reciprocally increasing other lymphocyte proportions. Lastly, the NKT cell proportion in sedentary rats was decreased by hesperidin supplementation, but those animals that received the flavanone, increased NKT cell proportion just after exhaustion, which could be a consequence of T cell mobilization in these animals.

The function of MLNs T lymphocytes, established by their proliferative ability, tended to decrease by chronic and intensive training. However, the final exhaustion, even 24 h later, increased their function, which is in agreement with results reported in similarly trained and exhausted animals [16], although other researchers reported a reduced T-cell proliferation [60]. Hesperidin supplementation did not change the results in terms of lymphocyte proliferation. In fact, previous results in non-trained animals treated with a similar dosage of hesperidin showed that it does not affect MLNs lymphocyte proliferative response or the release of cytokines from these cells [42].

In addition to the gut microbiota and the MLNs lymphocytes, one of the main players of the mucosal immune system is secretory IgA. This antibody plays an important role in the intestinal surface where it neutralizes toxins and viruses, blocks the excessive bacterial adherence or translocation, clears unwanted macromolecular structures at the epithelial surface, and directs sampling of luminal antigen [61]. Very prolonged bouts of exercise have been associated with a decreased secretory IgA production [62], which could explain, at least in part, the higher susceptibility to mucosal infections observed in athletes [63,64]. Here we found that, in the intestinal compartment, the applied intensive exercise conditions did not modify the IgA content, in agreement with a previous study [16], although in this last case, a decreased salivary IgA concentration was observed. Although exercise was not able to significantly modify mucosal IgA, hesperidin supplementation was able to enhance IgA concentration in the small intestine and cecal content, which was observed the most in chronically trained group (H-T group). In agreement, a similar hesperidin supplementation in nonrunner rats showed higher small IgA content in the intestine but not in the cecum [42], and the intake of a diet containing 0.5% hesperidin also increased fecal IgA concentration [41]. Likewise, it has been suggested that flavonoids such as hesperidin could increase mucosal IgA, among other immune-enhancing effects that could help in the treatment of COVID-19 [23].

The fact that 24 h after exhaustion, rats treated with hesperidin raised the amount of IgA-coated bacteria in the cecum is in agreement with the influence of hesperidin increasing IgA levels in the small intestine and the cecal content reported here and, previously, in cecum and feces [41,42]. This effect had already been observed in nonexercised animals [42], and could reflect the protective role of the flavanone in the mucosal immune system that appeared hours after exhaustion. Therefore, given this hesperidin intake consequence along with its ergogenic effect, it could be important to consider its supplementation in stressful situations, such as in some diseases, gastrointestinal complaints, and exhausting exercise.

5. Conclusions

Chronic intense exercise followed by an exhausting test in rats induces changes in the intestinal immune compartment as observed in the distribution and function of lymphocytes from mesenteric lymph nodes. Although the hesperidin supplementation applied here did not prevent the main alterations in lymphocyte subsets, it was able to enhance

IgA synthesis in the intestinal compartment, which could counteract the immune and gastrointestinal barrier alterations induced by intense and exhausting exercise.

Author Contributions: Conceptualization, F.J.P.-C. and M.C.; methodology, P.R.-I., S.E.-A., M.M.-C. and À.F.; formal analysis, P.R.-I.; investigation, P.R.-I., S.E.-A., M.M.-C. and À.F.; resources, M.C.; data curation, P.R.-I. and M.M.-C.; writing—original draft preparation, P.R.-I. and M.C.; writing—review and editing, all authors; supervision, F.J.P.-C. and M.C.; project administration, M.C.; funding acquisition, M.C. All authors have read and agreed to the published version of the manuscript.

Funding: This study was supported by the Spanish Ministry of Science and Innovation and AEI/FEDER, UE (AGL2016-76972-R). S.E.-A. was supported by a doctoral fellowship from the Generalitat de Catalunya (FI-DGR 2015). P.R.-I. holds a grant from the Spanish Ministry of Education, Culture, and Sport (FPU18-00807).

Institutional Review Board Statement: The animal study protocol was approved by the Ethical Committee for Animal Experimentation of the University of Barcelona and the Catalonia Government (CEEA/UB ref. 464/16 and DAAM 9257, respectively), in full compliance with national legislation following the EU-Directive 2010/63/EU for the protection of animals used for scientific purposes.

Informed Consent Statement: Not applicable.

Data Availability Statement: Data are contained within the article.

Acknowledgments: The authors are grateful to HealthTech BioActives for providing the hesperidin. The authors would like to thank Marta Périz, Ignasi Azagra, Mar Abril-Gil, and Blanca Grases-Pintó for their help with the laboratory work, and Karla Rio for her advice on statistical analysis.

Conflicts of Interest: The authors declare no conflict of interest.

References

1. Bermon, S.; Castell, L.M.; Calder, P.C.; Bishop, N.C.; Blomstrand, E.; Mooren, F.C.; Krüger, K.; Kavazis, A.N.; Quindry, J.C.; Senchina, D.S.; et al. Consensus statement immunutrition and exercise. *Exerc. Immunol. Rev.* **2017**, *23*, 8–50. [PubMed]
2. Walsh, N.P.; Gleeson, M.; Shephard, R.J.; Gleeson, M.; Woods, J.A.; Bishop, N.C.; Fleshner, M.; Green, C.; Pedersen, B.K.; Hoffman-Goetz, L.; et al. Position statement part one: Immune function and exercise. *Exerc. Immunol. Rev.* **2011**, *17*, 6–63. [PubMed]
3. Schwellnus, M.; Soligard, T.; Alonso, J.-M.; Bahr, R.; Clarsen, B.; Dijkstra, H.P.; Gabbett, T.J.; Gleeson, M.; Hägglund, M.; Hutchinson, M.R.; et al. How much is too much? (Part 2) International olympic committee consensus statement on load in sport and risk of illness. *Br. J. Sports Med.* **2016**, *50*, 1043–1052. [CrossRef] [PubMed]
4. Guimarães, T.T.; Terra, R.; Dutra, P.M.L.D. Chronic effects of exhausting exercise and overtraining on the immune response: Th1 and Th2 profile. *Motricidade* **2017**, *13*, 69–78. [CrossRef]
5. Kostrzewa-Nowak, D.; Buryta, R.; Nowak, R. Comparison of selected CD45+ cell subsets' response and cytokine levels on exhaustive effort among soccer players. *J. Med. Biochem.* **2019**, *38*, 256–267. [CrossRef]
6. Kostrzewa-Nowak, D.; Nowak, R. Differential Th cell-related immune responses in young physically active men after an endurance effort. *J. Clin. Med.* **2020**, *9*, 1795. [CrossRef]
7. Kaya, O. Effect of a four-week exercise program on the secretion of IFN-γ, TNF-α, IL-2 and IL-6 cytokines in elite Taekwondo athletes. *Biomed. Rep.* **2016**, *5*, 367–370. [CrossRef]
8. Docherty, S.; Harley, R.; McAuley, J.J.; Crowe, L.A.N.; Pedret, C.; Kirwan, P.D.; Siebert, S.; Millar, N.L. The effect of exercise on cytokines: Implications for musculoskeletal health: A narrative review. *BMC Sports Sci. Med. Rehabil.* **2022**, *14*, 1–14. [CrossRef]
9. Estruel-Amades, S.; Camps-Bossacoma, M.; Massot-Cladera, M.; Pérez-Cano, F.J.; Castell, M. Alterations in the innate immune system due to exhausting exercise in intensively trained rats. *Sci. Rep.* **2020**, *10*, 967. [CrossRef]
10. Estruel-Amades, S.; Ruiz-Iglesias, P.; Périz, M.; Franch, À.; Pérez-Cano, F.J.; Camps-Bossacoma, M.; Castell, M. Changes in lymphocyte composition and functionality after intensive training and exhausting exercise in rats. *Front. Physiol.* **2019**, *10*, 1491. [CrossRef]
11. Chow, L.S.; Gerszten, R.E.; Taylor, J.M.; Pedersen, B.K.; van Praag, H.; Trappe, S.; Febbraio, M.A.; Galis, Z.S.; Gao, Y.; Haus, J.M.; et al. Exerkines in health, resilience and disease. *Nat. Rev. Endocrinol.* **2022**, *18*, 273–289. [CrossRef] [PubMed]
12. De Oliveira, E.P.; Burini, R.C.; Jeukendrup, A. Gastrointestinal complaints during exercise: Prevalence, etiology, and nutritional recommendations. *Sport. Med.* **2014**, *44*, 79–85. [CrossRef] [PubMed]
13. Sponsiello, N.; Salamone, M.; Di Nardo, V.; Busa, F.; Andreani, N. Exercise and gastrointestinal complaints: An overview. *Prog. Nutr.* **2021**, *23*, e2021206. [CrossRef]

14. Morishima, S.; Aoi, W.; Kawamura, A.; Kawase, T.; Takagi, T.; Naito, Y.; Tsukahara, T.; Inoue, R. Intensive, prolonged exercise seemingly causes gut dysbiosis in female endurance runners. *J. Clin. Biochem. Nutr.* **2021**, *68*, 253–258. [CrossRef]
15. Aya, V.; Flórez, A.; Perez, L.; Ramírez, J.D. Association between physical activity and changes in intestinal microbiota composition: A systematic review. *PLoS ONE* **2021**, *16*, e0247039. [CrossRef]
16. Ruiz-Iglesias, P.; Estruel-Amades, S.; Camps-Bossacoma, M.; Massot-Cladera, M.; Castell, M.; Pérez-Cano, F.J. Alterations in the mucosal immune system by a chronic exhausting exercise in Wistar rats. *Sci. Rep.* **2020**, *10*, 17950. [CrossRef]
17. Zhang, M.; Zhu, S.; Yang, W.; Huang, Q.; Ho, C.T. The biological fate and bioefficacy of citrus flavonoids: Bioavailability, biotransformation, and delivery systems. *Food Funct.* **2021**, *12*, 3307–3323. [CrossRef] [PubMed]
18. Alam, F.; Mohammadin, K.; Shafique, Z.; Amjad, S.T.; bin Asad, M.H.H. Citrus flavonoids as potential therapeutic agents: A review. *Phyther. Res.* **2022**, *36*, 1417–1441. [CrossRef]
19. Tadros, F.J.; Andrade, J.M. Impact of hesperidin in 100% orange juice on chronic disease biomarkers: A narrative systematic review and gap analysis. *Crit. Rev. Food Sci. Nutr.* **2022**, *62*, 8335–8354. [CrossRef]
20. Pandey, P.; Khan, F. A mechanistic review of the anticancer potential of hesperidin, a natural flavonoid from citrus fruits. *Nutr. Res.* **2021**, *92*, 21–31. [CrossRef]
21. Farooqi, A.A.; Tahir, F.; Fakhar, M.; Butt, G.; Colombo Pimentel, T.; Wu, N.; Yulaevna, I.M.; Attar, R. Antimetastatic effects of Citrus-derived bioactive ingredients: Mechanistic insights. *Cell. Mol. Biol.* **2021**, *67*, 178–186. [CrossRef] [PubMed]
22. Yap, K.M.; Sekar, M.; Wu, Y.S.; Gan, S.H.; Rani, N.N.I.M.; Seow, L.J.; Subramaniyan, V.; Fuloria, N.K.; Fuloria, S.; Lum, P.T. Hesperidin and its aglycone hesperetin in breast cancer therapy: A review of recent developments and future prospects. *Saudi J. Biol. Sci.* **2021**, *28*, 6730–6747. [CrossRef] [PubMed]
23. Fazio, S.; Affuso, F.; Bellavite, P. A Review of the Potential Roles of Antioxidant and Anti-Inflammatory Pharmacological Approaches for the Management of Mild-to-Moderate Symptomatic COVID-19. *Med. Sci. Monit.* **2022**, *28*, e936292. [CrossRef] [PubMed]
24. Montenegro-Landívar, M.F.; Tapia-Quirós, P.; Vecino, X.; Reig, M.; Valderrama, C.; Granados, M.; Cortina, J.L.; Saurina, J. Polyphenols and their potential role to fight viral diseases: An overview. *Sci. Total Environ.* **2021**, *801*, 149719. [CrossRef]
25. Gour, A.; Manhas, D.; Bag, S.; Gorain, B.; Nandi, U. Flavonoids as potential phytotherapeutics to combat cytokine storm in SARS-CoV-2. *Phyther. Res.* **2021**, *35*, 4258–4283. [CrossRef]
26. Alam, S.; Sarker, M.M.R.; Afrin, S.; Richi, F.T.; Zhao, C.; Zhou, J.R.; Mohamed, I.N. Traditional Herbal Medicines, Bioactive Metabolites, and Plant Products Against COVID-19: Update on Clinical Trials and Mechanism of Actions. *Front. Pharmacol.* **2021**, *12*, 671498. [CrossRef]
27. Cao, Y.; Xie, L.; Liu, K.; Liang, Y.; Dai, X.; Wang, X.; Lu, J.; Zhang, X.; Li, X. The antihypertensive potential of flavonoids from Chinese Herbal Medicine: A review. *Pharmacol. Res.* **2021**, *174*, 105919. [CrossRef]
28. Adel Mehraban, M.S.; Tabatabaei-Malazy, O.; Rahimi, R.; Daniali, M.; Khashayar, P.; Larijani, B. Targeting dyslipidemia by herbal medicines: A systematic review of meta-analyses. *J. Ethnopharmacol.* **2021**, *280*, 114407. [CrossRef]
29. Yan, J.; Nie, Y.; Luo, M.; Chen, Z.; He, B. Natural Compounds: A Potential Treatment for Alcoholic Liver Disease? *Front. Pharmacol.* **2021**, *12*, 694475. [CrossRef]
30. Miles, E.A.; Calder, P.C. Effects of Citrus Fruit Juices and Their Bioactive Components on Inflammation and Immunity: A Narrative Review. *Front. Immunol.* **2021**, *12*, 712608. [CrossRef]
31. Jugran, A.K.; Rawat, S.; Devkota, H.P.; Bhatt, I.D.; Rawal, R.S. Diabetes and plant-derived natural products: From ethnopharmacological approaches to their potential for modern drug discovery and development. *Phyther. Res.* **2021**, *35*, 223–245. [CrossRef] [PubMed]
32. Overdevest, E.; Wouters, J.A.; Wolfs, K.H.M.; van Leeuwen, J.J.M.; Possemiers, S. Citrus flavonoid supplementation improves exercise performance in trained athletes. *J. Sports Sci. Med.* **2018**, *17*, 24–30. [PubMed]
33. Martínez-Noguera, F.J.; Marín-Pagán, C.; Carlos-Vivas, J.; Rubio-Arias, J.A.; Alcaraz, P.E. Acute effects of hesperidin in oxidant/antioxidant state markers and performance in amateur cyclists. *Nutrients* **2019**, *11*, 1898. [CrossRef] [PubMed]
34. Martínez-Noguera, F.J.; Marín-Pagán, C.; Carlos-Vivas, J.; Alcaraz, P.E. Effects of 8 weeks of 2S-hesperidin supplementation on performance in amateur cyclists. *Nutrients* **2020**, *12*, 3911. [CrossRef] [PubMed]
35. Martínez-Noguera, F.J.; Marín-Pagán, C.; Carlos-Vivas, J.; Alcaraz, P.E. 8-week supplementation of 2S-hesperidin modulates antioxidant and inflammatory status after exercise until exhaustion in amateur cyclists. *Antioxidants* **2021**, *10*, 432. [CrossRef]
36. Van Iersel, L.E.; Stevens, Y.R.; Conchillo, J.M.; Troost, F.J. The effect of citrus flavonoid extract supplementation on anaerobic capacity in moderately trained athletes: A randomized controlled trial. *J. Int. Soc. Sports Nutr.* **2021**, *18*, 2. [CrossRef] [PubMed]
37. De Oliveira, D.M.; Dourado, G.K.Z.S.; Cesar, T.B. Hesperidin associated with continuous and interval swimming improved biochemical and oxidative biomarkers in rats. *J. Int. Soc. Sports Nutr.* **2013**, *10*, 27. [CrossRef] [PubMed]
38. Tomazini Gonçalves, T.; Lazaro, C.M.; De Mateo, F.G.; Campos, M.C.B.; Mezencio, J.G.B.; Claudino, M.A.; Carvalho, P.d.O.; Webb, R.C.; Priviero, F.B.M. Effects of glucosyl-hesperidin and physical training on body weight, plasma lipids, oxidative status and vascular reactivity of rats fed with high-fat diet. *Diabetes Metab. Syndr. Obes. Targets Ther.* **2018**, *11*, 321–332. [CrossRef]
39. Estruel-Amades, S.; Massot-Cladera, M.; Garcia-Cerdà, P.; Pérez-Cano, F.J.; Franch, À.; Castell, M.; Camps-Bossacoma, M. Protective effect of hesperidin on the oxidative stress induced by an exhausting exercise in intensively trained rats. *Nutrients* **2019**, *11*, 783. [CrossRef]

40. Ruiz-Iglesias, P.; Estruel-Amades, S.; Camps-Bossacoma, M.; Massot-Cladera, M.; Franch, À.; Pérez-Cano, F.J.; Castell, M. Influence of Hesperidin on Systemic Immunity of Rats Following an Intensive Training and Exhausting Exercise. *Nutrients* **2020**, *12*, 1291. [CrossRef]
41. Camps-Bossacoma, M.; Franch, À.; Pérez-Cano, F.J.; Castell, M. Influence of hesperidin on the systemic and intestinal rat immune response. *Nutrients* **2017**, *9*, 580. [CrossRef]
42. Estruel-Amades, S.; Massot-Cladera, M.; Pérez-Cano, F.J.; Franch, À.; Castell, M.; Camps-Bossacoma, M. Hesperidin effects on gut microbiota and gut-associated lymphoid tissue in healthy rats. *Nutrients* **2019**, *11*, 324. [CrossRef]
43. Stevens, Y.; Van Rymenant, E.; Grootaert, C.; Van Camp, J.; Possemiers, S.; Masclee, A.; Jonkers, D. The intestinal fate of citrus flavanones and their effects on gastrointestinal health. *Nutrients* **2019**, *11*, 1464. [CrossRef] [PubMed]
44. Lima, A.C.D.; Cecatti, C.; Fidélix, M.P.; Adorno, M.A.T.; Sakamoto, I.K.; Cesar, T.B.; Sivieri, K. Effect of Daily Consumption of Orange Juice on the Levels of Blood Glucose, Lipids, and Gut Microbiota Metabolites: Controlled Clinical Trials. *J. Med. Food* **2019**, *22*, 202–210. [CrossRef] [PubMed]
45. Camps-Bossacoma, M.; Abril-Gil, M.; Saldaña-Ruiz, S.; Franch, À.; Pérez-Cano, F.J.; Castell, M. Cocoa diet prevents antibody synthesis and modifies lymph node composition and functionality in a rat oral sensitization model. *Nutrients* **2016**, *8*, 242. [CrossRef] [PubMed]
46. Ramiro-Puig, E.; Pérez-Cano, F.J.; Ramos-Romero, S.; Pérez-Berezo, T.; Castellote, C.; Permanyer, J.; Franch, À.; Izquierdo-Pulido, M.; Castell, M. Intestinal immune system of young rats influenced by cocoa-enriched diet. *J. Nutr. Biochem.* **2008**, *19*, 555–565. [CrossRef]
47. Massot-Cladera, M.; Pérez-Berezo, T.; Franch, A.; Castell, M.; Pérez-Cano, F.J. Cocoa modulatory effect on rat faecal microbiota and colonic crosstalk. *Arch. Biochem. Biophys.* **2012**, *527*, 105–112. [CrossRef]
48. Ruiz-Iglesias, P.; Massot-Cladera, M.; Rodríguez-Lagunas, M.J.; Franch, À.; Camps-Bossacoma, M.; Castell, M.; Pérez-Cano, F.J. A cocoa diet can partially attenuate the alterations in microbiota and mucosal immunity induced by a single session of intensive exercise in rats. *Front. Nutr.* **2022**, *9*, 861533. [CrossRef]
49. Massot-Cladera, M.; Franch, A.; Castellote, C.; Castell, M.; Pérez-Cano, F.J.; Franch, À.; Castellote, C.; Castell, M.; Pérez-Cano, F.J. Cocoa flavonoid-enriched diet modulates systemic and intestinal immunoglobulin synthesis in adult Lewis rats. *Nutrients* **2013**, *5*, 3272–3286. [CrossRef]
50. Walsh, N.P.; Gleeson, M.; Pyne, D.B.; Nieman, D.C.; Dhabhar, S.; Shephard, R.J.; Oliver, S.J.; Bermon, S.; Kajeniene, A. Position statement part two: Maintaining immune health. *Exerc. Immunol. Rev.* **2011**, *17*, 64–103.
51. Ruiz-Iglesias, P.; Gorgori-González, A.; Massot-Cladera, M.; Castell, M.; Pérez-Cano, F.J. Does flavonoid consumption improve exercise performance? Is it related to changes in the immune system and inflammatory biomarkers? A systematic review of clinical studies since 2005. *Nutrients* **2021**, *13*, 1132. [CrossRef] [PubMed]
52. Somerville, V.S.; Braakhuis, A.J.; Hopkins, W.G. Effect of Flavonoids on Upper Respiratory Tract Infections and Immune Function: A Systematic Review and Meta-Analysis. *Adv. Nutr.* **2016**, *7*, 488–497. [CrossRef]
53. Daniel, N.; Lécuyer, E.; Chassaing, B. Host/microbiota interactions in health and diseases—Time for mucosal microbiology! *Mucosal Immunol.* **2021**, *14*, 1006–1016. [CrossRef] [PubMed]
54. Gubert, C.; Kong, G.; Renoir, T.; Hannan, A.J. Exercise, diet and stress as modulators of gut microbiota: Implications for neurodegenerative diseases. *Neurobiol. Dis.* **2020**, *134*, 104621. [CrossRef]
55. Keohane, D.M.; Woods, T.; O'Connor, P.; Underwood, S.; Cronin, O.; Whiston, R.; O'sullivan, O.; Cotter, P.; Shanahan, F.; Molloy, M.G.M. Four men in a boat: Ultra-endurance exercise alters the gut microbiome. *J. Sci. Med. Sport* **2019**, *22*, P1059–P1064. [CrossRef] [PubMed]
56. Ruiz-Iglesias, P.; Massot-Cladera, M.; Estruel-Amades, S.; PérezCano, F.J.; Castell, M. Intensive Training and Sex Influence Intestinal Microbiota Composition: A Preclinical Approach. *Proceedings* **2020**, *61*, 11. [CrossRef]
57. Unno, T.; Hisada, T.; Takahashi, S. Hesperetin Modifies the Composition of Fecal Microbiota and Increases Cecal Levels of Short-Chain Fatty Acids in Rats. *J. Agric. Food Chem.* **2015**, *63*, 7952–7957. [CrossRef] [PubMed]
58. Duda-Chodak, A. The inhibitory effect of polyphenols on human gut microbiota. *J. Physiol. Pharmacol.* **2012**, *63*, 497–503.
59. Fidélix, M.; Milenkovic, D.; Sivieri, K.; Cesar, T. Microbiota modulation and effects on metabolic biomarkers by orange juice: A controlled clinical trial. *Food Funct.* **2020**, *11*, 1599–1610. [CrossRef]
60. Simpson, R.J.; Kunz, H.; Agha, N.; Graff, R. Exercise and the regulation of immune functions. In *Progress in Molecular Biology and Translational Science*; Elsevier Inc.: Amsterdam, The Netherlands, 2015; Volume 135, pp. 355–380.
61. Macpherson, A.J.; Yilmaz, B.; Limenitakis, J.P.; Ganal-Vonarburg, S.C. IgA function in relation to the intestinal microbiota. *Annu. Rev. Immunol.* **2018**, *36*, 359–381. [CrossRef]
62. Gleeson, M.; Williams, C. Intense exercise training and immune function. In Proceedings of the Nestle Nutrition Institute Workshop Series; S. Karger AG: Basel, Switzerland, 2013; Volume 76, pp. 39–50.

63. Krüger, K.; Mooren, F.-C.; Pilat, C. The immunomodulatory effects of physical activity. *Curr. Pharm. Des.* **2016**, *22*, 3730–3748. [CrossRef] [PubMed]
64. Alack, K.; Pilat, C.; Krüger, K. Current knowledge and new challenges in exercise immunology. *Dtsch. Z. Sportmed.* **2019**, *70*, 250–260. [CrossRef]

nutrients

MDPI

Article

Effects of Ingesting Both Catechins and Chlorogenic Acids on Glucose, Incretin, and Insulin Sensitivity in Healthy Men: A Randomized, Double-Blinded, Placebo-Controlled Crossover Trial

Aya Yanagimoto [1,*], Yuji Matsui [2], Tohru Yamaguchi [2], Masanobu Hibi [1], Shigeru Kobayashi [3] and Noriko Osaki [2]

1 Biological Science Laboratories, Kao Corporation, 2-1-3 Bunka, Sumida, Tokyo 131-8501, Japan
2 Health & Wellness Products Research Laboratories, Kao Corporation, 2-1-3 Bunka, Sumida, Tokyo 131-8501, Japan
3 Department of Surgery, Tokyo Rinkai Hospital, 1-4-2 Rinkai-cho, Edogawa, Tokyo 134-0086, Japan
* Correspondence: yanagimoto.aya@kao.com; Tel.: +81-3-5630-7476

Abstract: Epidemiologic studies have revealed that consuming green tea or coffee reduces diabetes risk. We evaluated the effects of the combined consumption of green tea catechins and coffee chlorogenic acids (GTC+CCA) on postprandial glucose, the insulin incretin response, and insulin sensitivity. Eleven healthy men were recruited for this randomized, double-blinded, placebo-controlled crossover trial. The participants consumed a GTC+CCA-enriched beverage (620 mg GTC, 373 mg CCA, and 119 mg caffeine/day) for three weeks; the placebo beverages (PLA) contained no GTC or CCA (PLA: 0 mg GTC, 0 mg CCA, and 119 mg caffeine/day). Postprandial glucose, insulin, glucagon-like peptide-1 (GLP-1), and glucose-dependent insulinotropic polypeptide (GIP) responses were measured at baseline and after treatments. GTC+CCA consumption for three weeks showed a significant treatment-by-time interaction on glucose changes after the ingestion of high-fat and high-carbohydrate meals, however, it did not affect fasting glucose levels. Insulin sensitivity was enhanced by GCT+CCA compared with PLA. GTC+CCA consumption resulted in a significant increase in postprandial GLP-1 and a decrease in GIP compared to PLA. Consuming a combination of GTC and CCA for three weeks significantly improved postprandial glycemic control, GLP-1 response, and postprandial insulin sensitivity in healthy individuals and may be effective in preventing diabetes.

Keywords: GIP; GLP-1; insulin sensitivity; type 2 diabetes; coffee chlorogenic acids; green tea catechins

Citation: Yanagimoto, A.; Matsui, Y.; Yamaguchi, T.; Hibi, M.; Kobayashi, S.; Osaki, N. Effects of Ingesting Both Catechins and Chlorogenic Acids on Glucose, Incretin, and Insulin Sensitivity in Healthy Men: A Randomized, Double-Blinded, Placebo-Controlled Crossover Trial. *Nutrients* **2022**, *14*, 5063. https://doi.org/10.3390/nu14235063

Academic Editors: Sonia de Pascual-Teresa and Luis Goya

Received: 10 November 2022
Accepted: 23 November 2022
Published: 28 November 2022

1. Introduction

Type 2 diabetes, a condition involving insulin resistance and impaired insulin secretion that progresses to a chronic hyperglycemic state [1], is a leading cause of death worldwide [1–3]. Ameliorating the sustained hyperglycemia resulting from whole-body insulin resistance is essential in preventing diabetes and its complications. Improving insulin resistance and controlling blood glucose by adopting healthy dietary habits and increasing physical activity in the early stages are essential factors for preventing diabetes onset; however, no effective non-pharmacologic methods for reducing insulin resistance have been established.

Blood glucose is strictly regulated by insulin, glucagon, and their feedback mechanisms. In addition, incretins, such as glucose-dependent insulinotropic polypeptide (GIP) and glucagon-like peptide-1 (GLP-1), are secreted from K cells in the upper part of the small intestine and from L cells in the lower part of the small intestine in response to dietary carbohydrates and fats, and they have important roles in regulating blood glucose levels via insulin secretion from pancreatic β-cells [4,5]. GLP-1 and GIP also have extrapancreatic effects and regulate physiologic functions related to glucose and energy homeostasis. The extrapancreatic effects of GLP-1 include increasing the glucose uptake capacity of the liver and muscle and improving whole-body insulin sensitivity [4,6], while extrahepatic GIP promotes fat accumulation via glucose and free fatty acid uptake in the adipocytes [7].

Green tea and coffee are popular beverages worldwide. Epidemiologic studies indicate that consuming green tea and coffee reduces type 2 diabetes risk. Individuals typically consuming over five cups of green tea [8] or three to six cups of coffee daily [9,10] are at a lower risk of type 2 diabetes. However, the effect of green tea or coffee on insulin resistance remains unclear [11–14], and the association of green tea and coffee with developing diabetes is also inconclusive. Green tea and coffee contain abundant polyphenols, such as green tea catechins (GTC) and coffee chlorogenic acids (CCA), respectively, which reduce the risk of developing type 2 diabetes and other metabolic diseases [15–19]. Venables et al. [20] reported that consuming 890 mg of green tea extract containing GTC increases insulin sensitivity by 13% in healthy individuals. Moreover, a single ingestion of CCA suppresses postprandial blood glucose levels [21] and promotes GLP-1 secretion [22]. Zuñiga et al. [23] reported that subjects with impaired glucose tolerance who consumed CCA (400 mg/d) for 12 weeks exhibited significantly reduced fasting blood glucose levels and a lower insulinogenic index. These reports suggest that due to their distinct mechanisms of action, the effects of GTC and CCA on glucose metabolism will be enhanced by their combined ingestion. To our knowledge, the effects of the combined consumption of GTC and CCA on hyperglycemia have not been evaluated.

Here, we hypothesized that the continuous combined consumption of GTC and CCA would improve the postprandial glycemic response in healthy men. We examined the effects of continuous combined consumption of GTC and CCA for three weeks on postprandial glucose, insulin and incretin responses, and insulin sensitivity in healthy men. The primary endpoint was the difference in the postprandial glucose response, secondary endpoints were changes in the insulin, GLP-1, and GIP responses; insulin sensitivity, evaluated using the Matsuda index; insulin resistance, evaluated using an index obtained by multiplying the glucose area under the curve (AUC) and the AUC of insulin over 2 h ($AUC_{ins \times glu}$); the homeostatic model assessment of insulin resistance index (HOMA-IR) and homeostasis model assessment β cell function (HOMA-β).

2. Materials and Methods

2.1. Subjects

Eleven healthy males (aged 20–60 years; fasting blood glucose $\leq$125 mg/dL) were recruited through emails on 26 September 2016, and our website described the study outline. Exclusion criteria were a history of diabetes or cardiovascular disease, hypertension, hypercholesterolemia, dyslipidemia, eating disorders, food allergies to the test food, excessive alcohol intake ($\geq$30 g/day), smoking habit, being a shift worker, or being determined to be unqualified by the physician in charge. The study was conducted in accordance with the ethical principles of the Declaration of Helsinki and was approved by the ethics committee of the Kao Corporation (No. 16–35, 8 September 2016; Tokyo, Japan). All the participants received verbal and written explanations and provided written informed consent prior to registration. The trial was registered with the University Hospital Medical Information Network (UMIN; http://www.umin.ac.jp/ (10 November 2022); Registration No. UMIN000039514).

2.2. Study Design

In this randomized, double-blinded, placebo-controlled crossover trial including two 3-week intervention periods with a 2-week wash-out period, participants consumed either a GTC- and CCA-enriched beverage (GTC+CCA) or a placebo beverage (PLA) containing no GTC or CCA between September 2016 and December 2016 in the Tokyo metropolitan area, Japan. The participants were enrolled and randomly allocated to the starting condition with a random number generator by an independent researcher using the Excel spreadsheet program (Microsoft Excel 2010, Microsoft Corp., Redmond, WA, USA). All persons involved in the trial, including the participants, researchers who performed the interventions, and the supervising investigator, were blinded to the randomization. Participants were asked to maintain their usual dietary habits and physical activities, and their intake of green tea and coffee was limited to once daily during the entire study period. During the wash-out period, food intake and exercise were not guided or restricted. After a 1-week run-in period, baseline measurements were conducted after overnight fasting for at least 12 h before the participants consumed a test meal [24]. During the 3-week intervention periods, the participants consumed either the GTC+CCA or PLA beverages once daily after breakfast. The daily diets three days before each examination period were assessed by registered dietitians. The participants recorded individual dietary details of all meals during the three days using a diet-recording form to demonstrate that their eating habits and exercise levels had not changed. The content of the diet according to the dietary form and pictures was analyzed by a dietician using Excel-Eiyou-kun Ver. 6 software (Kenpakusha, Tokyo, Japan), and Hymena & Co (Tokyo, Japan) calculated the total calorie, carbohydrate, fat, and protein intake. A day before the measurements, the participants were prohibited from doing strenuous exercises, were required to ingest pre-packaged meals provided by the study coordinator (total energy content: 2412 kcal/day, 14 E% from protein, 25 E% from fat, and 61 E% from carbohydrate), and were asked to go to bed before midnight without ingesting any food or beverage other than water after 9:00 pm. The energy of the pre-packaged meals was 25 E% for breakfast, 24 E% for lunch, 21 E% for a snack, and 30 E% for dinner. The subjects arrived at the laboratory before 8:00 am on measurement day. Height, weight, body fat (body fat scale, model TF-780, Tanita Corp., Tokyo, Japan), waist circumference, and blood pressure (digital blood pressure monitor, model HEM-1000, Omron, Japan) were measured. Blood was drawn from a brachial vein under fasting conditions (0 h) and at 0.5, 1, 1.5, 2, 3, and 4 h after the test meal consumption (ingested within 15 min; meal test S, Saraya Co., Ltd., Osaka, Japan) and subsequent intake of the test beverage. The test beverages and meal test S were distributed by the study coordinator immediately prior to ingestion. The test meal was a high-fat cookie made from flour, butter, maltose, chicken egg, and baking powder, with an energy content of 592 kcal (6 E% from protein, 43 E% from fat, and 50 E% from carbohydrates). The outcome measurements were glucose, insulin, glucagon, C-peptide, GLP-1, and GIP concentrations; body weight; body fat percentage; and waist circumference. Furthermore, the exploratory measurements were for free fatty acids, ketone bodies, triglycerides, total bile acid, cortisol, glycoalbumin, 1,5-anhydroglucitol, high-sensitivity C-reactive protein, low-density lipoprotein cholesterol (LDL-C), high-density lipoprotein cholesterol (HDL-C), and total cholesterol.

2.3. Test Beverages

The following test beverages were prepared at the Kao Corporation: GTC+CCA enriched beverage containing 620 mg GTC, 373 mg CCA, and 119 mg caffeine; and PLA beverage containing 0 mg GTC, 0 mg CCA, and 119 mg caffeine (Table 1). GTC and CCA in the active beverage and PLA beverages were determined by high-performance liquid chromatography and dispensed industrially into bottles. Table 1 shows the polyphenols composition of the GTC and CCA in active and PLA beverages.

Table 1. Composition of the test beverages.

		PLA	GTC+CCA
Catechin	mg	0	31
Epicatechin	mg	0	37
Gallocatechin	mg	0	134
Epigallocatechin	mg	0	120
Catechin gallate	mg	0	23
Epicatechin gallate	mg	0	41
Gallocatechin gallate	mg	0	108
Epigallocatechin gallate	mg	0	126
Total catechins	mg	0	620

Table 1. *Cont.*

		PLA	GTC+CCA
3-caffeoylquinic acid	mg	0	104
4-caffeoylquinic acid	mg	0	94
5-caffeoylquinic acid	mg	0	102
3-feruloylquinic acid	mg	0	25
4-feruloylquinic acid	mg	0	23
5-feruloylquinic acid	mg	0	25
Total chlorogenic acids	mg	0	373
Caffeine	mg	119	119

Abbreviations: GTC+CCA, green tea catechin + chlorogenic acid; PLA, placebo.

2.4. Analytical Procedure

Plasma and serum samples, collected from participants in the fasting state and after intake of the test meal and test beverages, were snap-frozen in liquid nitrogen and stored at −80 °C until the analyses. A plasma sample was used for glucose measurements, and a serum sample was used for insulin measurements. Samples for measurement of glucagon, active GLP-1, and total GIP were collected using BD P800 (Becton Dickenson, London, UK), and their levels were measured using ELISA kits (glucagon: Mercodia, Uppsala, Sweden; active GLP-1: Immune Biology Laboratories, Gunma, Japan; total GIP: Merck Millipore, Darmstadt, Germany). Coefficients of variation were intra-assay for glucagon: 2.3%, GLP-1: 5.3%, GIP: 9.6%; inter-assay for glucagon: 4.3%, GLP-1: 9.2%, GIP: 7.2%. All other indices were analyzed at LSI Medience, Co., Ltd. (Tokyo, Japan). Total AUCs for glucose, insulin, glucagon, C-peptide, GLP-1, and GIP were calculated using the trapezoidal rule. The Matsuda index of insulin sensitivity [25–27] was calculated from measurements obtained after the test meal consumption. The insulin and glucose measurements were used to calculate the HOMA-IR, HOMA-β, and $AUC_{ins \times glu}$ [28–30].

2.5. Statistical Analysis

Data were expressed as mean ± standard error unless otherwise indicated. The AUCs were calculated using the trapezoidal rule, maximum concentration (C_{max}), and effect size. The 95% confidence intervals (CI) for the primary and secondary outcome measures were calculated from values measured at each time point. The sample size was estimated based on the results of a preliminary study as 12, assuming a significance level of 5% and a statistical power of 80%. In the preliminary study, the effect size of the glucose AUC was −0.681 (mg/dL·2 h) with a standard deviation of 0.857 (mg/dL·2 h). The following statistical analyses were performed for the GTC+CCA and PLA conditions. Regarding AUC and C_{max}, differences between the first and second phases were determined for each subject, and comparisons were performed between the sequence group using an unpaired *t*-test to evaluate the effects of the interventions. Differences in the values at each time-point were determined for each participant and entered into a linear mixed-effect model with treatment, time, and treatment-by-time interaction as the fixed effects. An unstructured variance-covariance matrix was applied to compare time points, and empirical variances (EMPIRICAL option) were used to estimate the fixed effects. The significance of the fixed effects was evaluated, and comparisons between treatments at each time point were performed [31]. Pearson's correlation coefficients or Spearman's rank correlation coefficients analyzed the relationship between pairs of variables. To confirm the involvement of GLP-1 in altered insulin resistance due to GTC+CCA, we performed a post hoc analysis of correlations of the difference in insulin resistance with the difference in GLP-1 between GTC+CCA and PLA conditions. In all analyses, the significance level was 5%. Significant differences were tested using the following statistical analysis software: Microsoft Excel 14.0, SAS version 9.4 (SAS Institute Inc., Cary, NC, USA), and StatXact 8, CrossOver module (Cytel Software, Cambridge, MA, USA).

3. Results

3.1. Subjects

Eleven healthy men completed the trial, and all data collected were analyzed (Figure 1). The number of subjects (n = 11) assessed and enrolled in the study was one less than the estimated number of subjects needed (n = 12). The anthropometric measurements and blood glucose-related hormone levels at baseline are presented in Table 2. Based on three days of dietary records prior to the trial, the mean nutrient intake was: total energy intake for PLA: 2234 ± 85 kcal/d, GTC+CCA: 2219 ± 90 kcal/d; protein intake for PLA: 82.8 ± 3.0 g/d, GTC+CCA: 82.8 ± 4.4 g/d; fat intake for PLA: 74.5 ± 4.1 g/d, GTC+CCA: 71.6 ± 4.1 g/d; and CHO intake for PLA: 288.8 ± 10.5 g/d, GTC+CCA: 296.9 ± 11.4 g/d. Mean energy intake, protein, fat, and CHO during the last three days before the interventions did not differ significantly by treatment. Consumption of the GTC+CCA beverage for three weeks was associated with a slight but significant weight reduction compared to consumption of the PLA beverage ($p = 0.035$). No adverse events were observed.

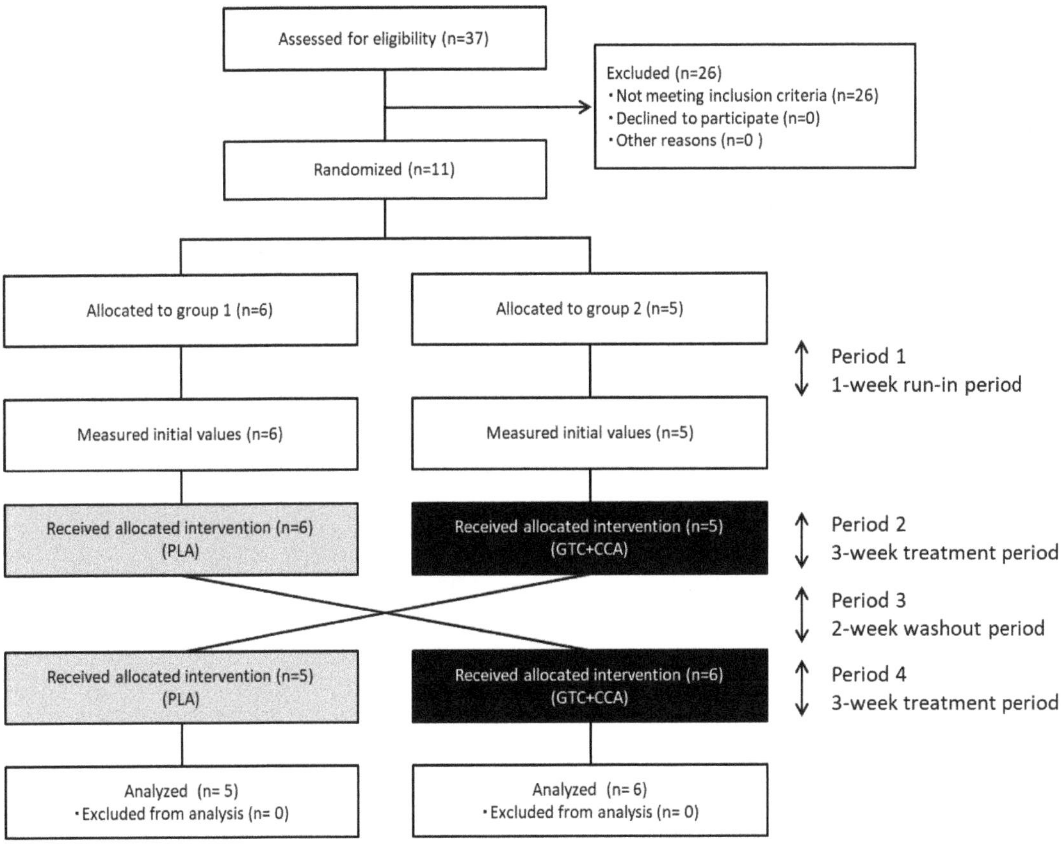

Figure 1. Diagram illustrating the flow of participants through each stage of the randomized crossover trial. Abbreviations: CCA, coffee chlorogenic acids; GTC, green tea catechins; PLA, placebo.

Table 2. Baseline physical characteristics, insulin-related indicators, and fasting blood constituents.

		Baseline
Age	y	41 ± 9
Height	cm	174.2 ± 4.0
Weight	kg	70.3 ± 9.4
Body mass index	kg/m^2	23.1 ± 2.7
Body fat	%	20.3 ± 4.6
Waist circumference	cm	84.5 ± 8.7
Systolic blood pressure	mmHg	129 ± 15
Diastolic blood pressure	mmHg	75 ± 12
Body temperature	°C	36.1 ± 0.4
Glucose	mg/dL	87 ± 5
Insulin	μU/mL	4.2 ± 1.8
HOMA-IR		0.90 ± 0.36
Glucose AUC_{2h} × Insulin AUC_{2h}		7366 ± 4021
Matsuda Index		12.3 ± 4.6
HOMA-β		67.3 ± 34.4
GA	%	13.3 ± 1.2
1,5-AG	μg/mL	24.3 ± 5.2
HbA_{1c}	%	5.4 ± 0.3
h-CRP	mg/dL	0.036 ± 0.052
LDL-C	mg/dL	113 ± 24
HDL-C	mg/dL	56 ± 13
TC	mg/dL	193 ± 22
AST	U/L	19 ± 5
ALT	U/L	20 ± 11
ALP	U/L	170 ± 33
γ-GTP	U/L	28 ± 14
T4	μg/dL	7.1 ± 1.0
T3	ng/dL	104 ± 20
TSH	μIU/mL	1.337 ± 0.629

Data are expressed as mean ± standard deviation (n = 11). Abbreviations: HOMA-IR, homeostatic model assessment of insulin resistance index; GA, glycoalbumin; 1,5-AG, 1,5-anhydroglucitol; HbA1c, hemoglobin A1C; h-CRP, high-sensitivity C-reactive protein; LDL-C, low-density lipoprotein cholesterol; HDL-C, high-density lipoprotein cholesterol; TC, total cholesterol; AST, aspartate aminotransferase; ALT, alanine aminotransferase; ALP, alkaline phosphatase; γ-GTP, γ-glutamyl transpeptidase; T4, thyroxine; T3, triiodothyronine; TSH, thyroid-stimulating hormone.

3.2. Blood Glucose and Insulin

The levels of fasting and postprandial glucose were within the normal range in all subjects. Comparisons of measurements between the PLA and GTC+CCA conditions using linear mixed-effect models revealed a treatment by time interaction (p = 0.002) for the change in blood glucose but no main treatment effect (p = 0.183) (Figure 2). The total AUC of glucose did not differ significantly between conditions (PLA: 367 ± 13 mg/dL·4h; GTC+CCA: 350 ± 10 mg/dL·4h; effect size [95% CI]: −15.55 [−50.78 to 19.68]; p = 0.3442), however, the C_{max} for postprandial blood glucose was significantly lower in the GTC+ PLA condition (PLA: 118.5 ± 5.2 mg/dL; GTC+CCA: 105.5 ± 4.6 mg/dL; effect size [95% CI]: −12.57 [−24.84 to −0.29]; p = 0.046). Analysis of insulin changes under the GTC+CCA condition compared to the PLA condition also revealed the main treatment effect (p = 0.003) and a treatment-by-time interaction (p = 0.028). Moreover, the total AUC of postprandial insulin (PLA: 75.6 ± 9.3 μU/mL·4h; GTC+CCA: 57.5 ± 6.5 μU/mL·4h; effect size [95% CI]: −17.02 (−27.21 to −6.82); p = 0.004] and C_{max} [PLA: 41.5 ± 5.6 μU/mL; GTC+CCA: 30.6 ± 4.1 μU/mL; effect size [95% CI]: −9.88 [−18.18 to −1.58]; p = 0.025) were significantly lower under the GTC+CCA condition. The change in C-peptide revealed a main effect of treatment (p = 0.012) and treatment-by-time interaction (p = 0.044). Additionally, the change in glucagon revealed the main treatment effect (p = 0.004) and a treatment-by-time interaction (p = 0.001) under the GTC+CCA condition compared with the PLA condition.

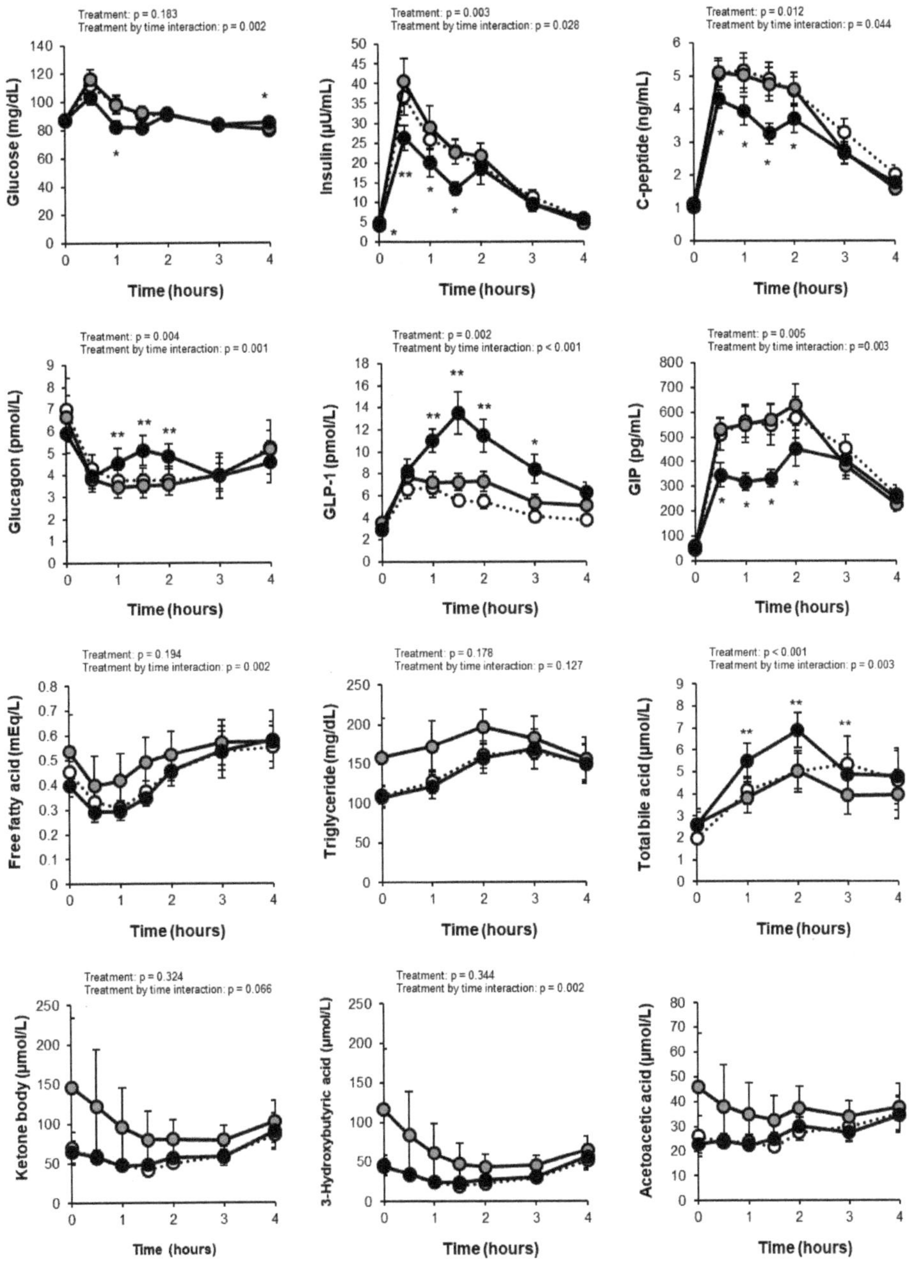

Figure 2. Changes in blood test results at fasting and after test meal consumption. White circles indicate the initial values. Gray and black circles indicate placebo and GTC+CCA, respectively. Treatment and treatment-by-time interactions were analyzed using linear mixed-effect models. Data are expressed as mean ± standard error. (n = 11) * $p < 0.05$, ** $p < 0.01$. The acetoacetic acid was not available for the statistical analysis due to a lack of convergence criteria. Abbreviations: AUC, area under the curve; CCA, coffee chlorogenic acids; GIP, glucose-dependent insulinotropic polypeptide; GLP-1, glucagon-like peptide-1; GTC, green tea catechins; PLA, placebo.

3.3. Incretins

The change in the postprandial GLP-1 revealed the main treatment effect ($p = 0.002$) and a treatment-by-time interaction ($p < 0.001$) in a mixed-effect model of the measurement data (Figure 2). Moreover, consuming GTC+CCA for three weeks significantly increased the postprandial GLP-1 AUC compared with the PLA (PLA: 25.3 ± 3.2 pmol/L·4h; GTC+CCA: 37.2 ± 4.2 pmol/L·4h; effect size [95% CI]: 11.50 [4.88 to 18.13]; $p = 0.003$). The change in the postprandial GIP also revealed a main treatment effect ($p = 0.005$) and a treatment-by-time interaction ($p = 0.003$). Furthermore, the postprandial GIP AUC_{4h} was significantly decreased under the GTC+CCA condition compared with the PLA condition (PLA: 1810.3 ± 159.2 pg/mL·4h; GTC+CCA: 1383.2 ± 105.4 pg/mL·4h; effect size [95% CI]: −413.5 [−664.6 to −162.4]; $p = 0.005$).

3.4. Insulin Sensitivity and Insulin Secretion Indices

Statistical analyses of the HOMA-IR (index of fasting insulin resistance), $AUC_{ins \times glu}$ (index of postprandial insulin resistance), Matsuda index (index of insulin sensitivity), and HOMA-β (index of insulin secretion) are presented in Table 3. Consuming the GTC+CCA beverage led to a significant decrease in the $AUC_{ins \times glu}$ of postprandial insulin resistance (PLA: 8367 ± 1686; GTC+CCA: 5630 ± 1032; effect size [95% CI]: −2558 [−4697 to −420]; $p = 0.024$), and a significant increase in the Matsuda index of insulin sensitivity (PLA: 11.3 ± 1.5; GTC+CCA: 14.6 ± 1.4; effect size [95% CI]: 2.96 [0.88 to 5.05]; $p = 0.011$). The fasting insulin resistance and secretion capacity indices were not significantly altered by consuming the GTC+CCA beverage.

Table 3. Differences in physical characteristics, insulin-related indicators, and fasting blood constituents between conditions after the intervention.

		PLA	GTC+CCA	Effect of Difference * Mean (95% CI)	*p*-Value
Weight	kg	70.7 ± 2.7	70.1 ± 2.7	−0.52 (−0.99, −0.05)	0.035
Body mass index	kg/m^2	23.2 ± 0.8	23.1 ± 0.8	−0.15 (−0.31, 0.01)	0.062
Body fat	%	20.4 ± 1.4	20.6 ± 1.6	0.28 (−0.47, 1.03)	0.422
Waist circumference	cm	85.0 ± 2.6	85.1 ± 2.6	0.15 (−0.66, 0.97)	0.681
Systolic blood pressure	mmHg	130.3 ± 4.2	124.8 ± 4.3	−5.32 (−15.91, 5.27)	0.285
Diastolic blood pressure	mmHg	80.3 ± 3.3	75.7 ± 3.7	−4.73 (−11.24, 1.77)	0.134
Body temperature	°C	36.3 ± 0.1	36.3 ± 0.1	0.02 (−0.15, 0.18)	0.823
HOMA-IR		1.07 ± 0.15	0.90 ± 0.08	−0.15 (−0.33, 0.03)	0.099
Glucose AUC_{2h} × Insulin AUC_{2h}		8367 ± 1686	5630 ± 1032	−2558 (−4697, −420)	0.024
Matsuda Index		11.3 ± 1.5	14.6 ± 1.4	2.96 (0.88, 5.05)	0.011
HOMA-β		76.0 ± 10.3	64.6 ± 7.4	−9.50 (−22.08, 3.09)	0.122
GA	%	13.2 ± 0.4	13.0 ± 0.4	−0.19 (−0.48, 0.10)	0.167
1,5-AG	μg/mL	25.7 ± 1.7	24.9 ± 1.8	−0.79 (−1.84, 0.25)	0.121
HbA_{1c}	%	5.3 ± 0.1	5.4 ± 0.1	0.04 (−0.03, 0.10)	0.245
h-CRP	mg/dL	0.04 ± 0.02	0.04 ± 0.01	−0.01 (−0.02, 0.01)	0.397
LDL-C	mg/dL	108.7 ± 7.0	110.9 ± 6.8	2.57 (−8.43, 13.56)	0.610
HDL-C	mg/dL	53.3 ± 2.7	52.1 ± 3.5	−1.30 (−4.41, 1.81)	0.369
TC	mg/dL	191.3 ± 7.5	186.9 ± 7.4	−3.48 (−20.55, 13.58)	0.655
AST	U/L	19.8 ± 1.6	18.9 ± 1.5	−0.97 (−3.39, 1.46)	0.390
ALT	U/L	17.3 ± 2.3	17.5 ± 2.2	0.28 (−3.48, 4.05)	0.869
ALP	U/L	176.2 ± 13.4	171.6 ± 10.5	−4.40 (−19.39, 10.59)	0.523
γ-GTP	U/L	26.8 ± 4.8	26.4 ± 4.3	−0.33 (−2.61, 1.94)	0.748
T4	μg/dL	7.1 ± 0.3	7.3 ± 0.3	0.18 (−0.46, 0.82)	0.543
T3	ng/dL	120.2 ± 7.1	121.2 ± 5.4	0.45 (−5.77, 6.67)	0.874
TSH	μIU/mL	1.3 ± 0.1	1.3 ± 0.1	0.01 (−0.22, 0.23)	0.942

Data are expressed as mean ± standard error. Differences between the PLA and GTC+CCA conditions were analyzed. * Difference means "CGA+CCA" treatment—"PLA" treatment (n = 11). Abbreviations: GTC-GCA, green tea catechin + chlorogenic acid; PLA, placebo; HOMA-IR, homeostatic model assessment of insulin resistance index; GA, glycoalbumin; 1,5-AG, 1,5-anhydroglucitol; HbA1c, hemoglobin A1C; h-CRP, high-sensitivity C-reactive protein; LDL-C, low-density lipoprotein cholesterol; HDL-C, high-density lipoprotein cholesterol; TC, total cholesterol; AST, aspartate aminotransferase; ALT, alanine aminotransferase; ALP, alkaline phosphatase; γ-GTP, γ-glutamyl transpeptidase; T4, thyroxine; T3, triiodothyronine; TSH, thyroid-stimulating hormone.

3.5. Blood Biochemical Values

The blood sample measurements are presented in Figure 2 and Table 3. Mixed-effect models of the measurement data revealed treatment by time interactions for postprandial free fatty acid ($p = 0.002$), 3-hydroxybutyric acid ($p = 0.002$), and cortisol ($p = 0.006$), and a main treatment effect ($p < 0.001$) and a treatment-by-time interaction ($p = 0.003$) for bile acid. None of the measured indices of fasting blood glucose control differed remarkably between both conditions, except that glycoalbumin decreased after GTC+CCA consumption in 64% of the subjects ($p = 0.167$).

3.6. Correlations between the Differences in Insulin Sensitivity with the Difference in GLP-1

Figure 3 illustrates the correlation analyses of the change in the GLP-1 AUC with the change in the $AUC_{ins \times glu}$ and the Matsuda index change between the GCT+CCA and PLA conditions. GLP-1 change positively correlated with a change in the Matsuda index of insulin sensitivity between the GTC+CCA and PLA conditions (r = 0.655, $p = 0.029$). Further, the differences in changes in the GLP-1 AUC between the GTC+CCA and PLA conditions tended to be negatively correlated with the differences in postprandial insulin resistance due to GTC+CCA consumption (r = −0.553, $p = 0.078$).

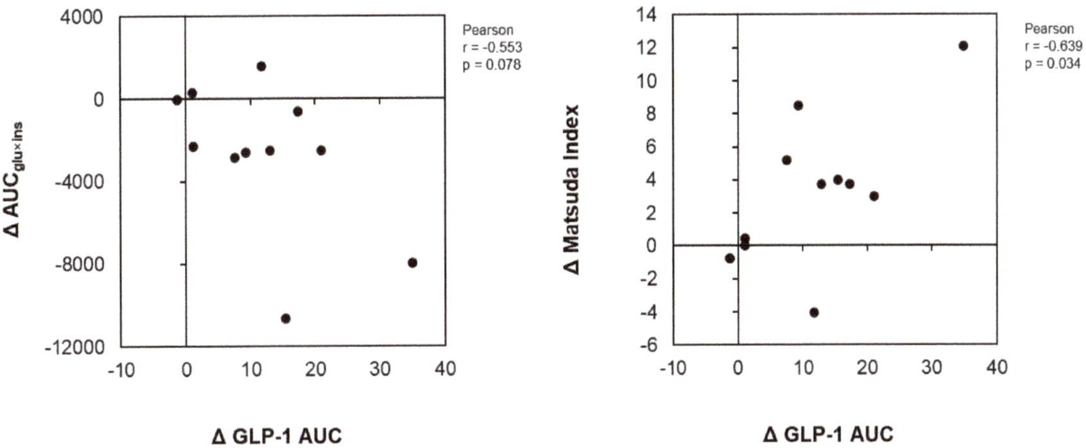

Figure 3. Correlations of differences in insulin resistance indices and AUC of GLP-1 after consuming GTC+CCA and PLA. ΔGLP-1, Δ $AUC_{glu \times ins}$, and ΔMatsuda index for each subject was calculated by subtracting the AUC of GLP-1, $AUC_{glu \times ins}$, and Matsuda index in the GTC+CCA treatment from those of the PLA treatment. The relationship between pairs of variables is presented using Pearson's correlation coefficient (n = 11). Abbreviations: AUC, area under the curve; CCA, coffee chlorogenic acids; GLP-1, glucagon-like peptide-1; GTC, green tea catechins; PLA, placebo.

4. Discussion

In this study, the effects of the continuous consumption of combined GTC and CCA on glucose metabolism, insulin sensitivity, and incretin secretion were evaluated. To the best of our knowledge, this study is the first to report the effects of the combined consumption of GTC and CCA, and our findings suggest that their combined consumption significantly improved postprandial whole-body insulin sensitivity, markedly increased GLP-1 secretion, decreased GIP secretion, and enhanced insulin sensitivity, possibly by increasing GLP-1 secretion.

The present study revealed that combined supplementation with GTC and CCA significantly decreased insulin resistance index ($AUC_{ins \times glu}$) values and improved insulin sensitivity index (Matsuda index) values after the high-fat and high-carbohydrate meal. Previous intervention studies have demonstrated that consuming green tea or GTC improves insulin resistance [15,17,20,32,33]. In animal studies, GTC consumption leads to a significant decrease in extracellular lipids in muscle, which is one of the mechanisms for improving insulin resistance [34]. A review on coffee and CCA also reported that short-term consumption of coffee, in contrast to habitual coffee consumption, might improve insulin sensitivity [35,36]. In animal studies, CCA increases sodium-glucose co-transporter-1 levels and promotes GLP-1 secretion [37]. Regarding the enhanced GLP-1 secretion following GTC and CCA consumption (additionally or synergistically), the increase observed in this study was remarkable compared to that in a previous study of single polyphenol consumption (GCT or CCA alone) that evaluated the association between improved postprandial hyperglycemia due to coffee polyphenols alone and increased GLP-1 secretion [22], or a study that examined the effects of CCA independently on postprandial GLP-1 secretion [38]. These findings indicate that combined supplementation with dietary polyphenols from green tea and coffee may effectively prevent diabetes onset due to their synergistic beneficial effects compared to single-polyphenol supplementation.

Among other food components, a single ingestion of green plant membranes and their continuous consumption for 90 days increases postprandial GLP-1 secretion [39]. The direct effect of green plant membrane ingestion on intestinal endocrine cells prolonged the process for digestion and absorption of dietary fat and carbohydrates in the small intestine and had an indirect effect via neural signals, such as cholecystokinin; these are suggested mechanisms underlying the increase in GLP-1 secretion [39]. The weight loss we observed after three weeks of combined GTC and CCA consumption may be partly due to appetite suppression resulting from the increased postprandial GLP-1 secretion or inhibitory effects of suppressed GIP secretion on fat accumulation. Furthermore, changes in daily food intake during the intervention period should be considered. Administration of GLP-1 analogs significantly reduces body weight, mainly via changes in food intake [40,41]. Therefore, it is critical to understand the neural systems involved in the effects of GLP-1 analogs to reduce food intake and body weight [42,43]. In this study, the increase in postprandial GLP-1 might be related to the weight loss observed, and further evaluation is needed.

GLP-1 secretion induced by the combined consumption of GTC and CCA may be promoted by the following two mechanisms. First, the glucose-sensing mechanism via sodium-glucose co-transporter-1 or a GLP-1 exocytosis mechanism due to glucose metabolism in L cells might be directly affected by GTC and CCA, or by nutritional components altered by GTC and CCA [34,44,45]. Second, the suppressive effect of GTC and CCA on glucose absorption in the small intestine might change the site of absorption to the lower part of the small intestine, thereby increasing the stimulation of L cells [37,46,47]. However, the GLP secretion mechanism associated with the combined consumption of GTC and CCA remains to be clarified.

This study had limitations. First, the sample size was small, with eleven men, and the participants were healthy, non-diabetic individuals. Thus, it is unclear whether GLP-1 secretion would be similarly promoted or if insulin resistance would be improved in people with severe insulin resistance or diabetes. In addition, since there were no women in the study group, the results of this study can only be related to men and cannot be inferred in general. The effects of the combined consumption of GTC and CCA should be confirmed by additional studies with larger sample sizes, including men and women with insulin resistance. Moreover, caffeine is one of the pharmacologically active components of green tea and coffee. While the effect of caffeine consumption to improve glucose tolerance has been suggested, its effect on glucose levels varies from study to study [48]. In contrast, several studies have reported that caffeine increases insulin secretion but does not necessarily improve glucose levels in oral glucose challenge tests [49–52], suggesting that caffeine could influence insulin clearance. In this study, the caffeine quantity was the same under the PLA and GTC+CCA conditions; nonetheless, we cannot rule out the possibility that caffeine, GTC, and CCA contributed to the observed effects.

5. Conclusions

Our findings demonstrated that the continuous, combined consumption of GTC and CCA for three weeks suppressed hyperglycemia and insulin after consuming a high-fat test meal containing 75 g of glucose and improved insulin sensitivity in healthy males. In addition, combined consumption of GTC and CCA promoted postprandial GLP-1 secretion and suppressed GIP secretion, suggesting that the increase in GLP-1 might at least partly account for improvements in insulin sensitivity.

Author Contributions: A.Y. and M.H. conceived and designed the study; A.Y. and Y.M. conducted the study; S.K. supervised the study; A.Y. and T.Y. analyzed the data and wrote the manuscript and had primary responsibility for the final content; M.H. and N.O. critically reviewed the manuscript. All authors have read and agreed to the published version of the manuscript.

Funding: This research was funded by the Kao Corporation (Tokyo, Japan).

Institutional Review Board Statement: The study was conducted in accordance with the Declaration of Helsinki and approved by the Ethics Committee of the Kao Corporation (No. 16-35, 8 September 2016).

Informed Consent Statement: Informed consent was obtained from all subjects involved in the study.

Data Availability Statement: The data used in this study are available from the corresponding author upon reasonable request.

Conflicts of Interest: A.Y., Y.M., T.Y., M.H. and N.O. are employees of the Kao Corporation. S.K. is an industrial physician of the Kao Corporation.

References

1. Akash, M.S.; Rehman, K.; Chen, S. Role of inflammatory mechanisms in pathogenesis of type 2 diabetes mellitus. *J. Cell. Biochem.* **2013**, *114*, 525–531. [CrossRef] [PubMed]
2. International Diabetes Federation. *IDF Diabetes Atlas*, 7th ed.; International Diabetes Federation: Brussels, Belgium, 2015.
3. Fujishima, M.; Kiyohara, Y.; Kato, I.; Ohmura, T.; Iwamoto, H.; Nakayama, K.; Ohmori, S.; Yoshitake, T. Diabetes and cardiovascular disease in a prospective population survey in Japan: The Hisayama Study. *Diabetes* **1996**, *45* (Suppl. S3), S14–S16. [CrossRef] [PubMed]
4. Baggio, L.L.; Drucker, D.J. Biology of incretins: GLP-1 and GIP. *Gastroenterology* **2007**, *132*, 2131–2157. [CrossRef] [PubMed]
5. Wölnerhanssen, B.K.; Meyer-Gerspach, A.C.; Schmidt, A.; Zimak, N.; Peterli, R.; Beglinger, C.; Borgwardt, S. Dissociable behavioral, physiological and neural effects of acute glucose and fructose ingestion: A pilot study. *PLoS ONE* **2015**, *10*, e0130280. [CrossRef] [PubMed]
6. Seino, Y.; Fukushima, M.; Yabe, D. GIP and GLP-1, the two incretin hormones: Similarities and differences. *J. Diabetes Investig.* **2010**, *1*, 8–23. [CrossRef] [PubMed]
7. Chen, S.; Okahara, F.; Osaki, N.; Shimotoyodome, A. Increased GIP signaling induces adipose inflammation via a HIF-1alpha-dependent pathway and impairs insulin sensitivity in mice. *Am. J. Physiol. Endocrinol. Metab.* **2015**, *308*, E414–E425. [CrossRef] [PubMed]
8. Iso, H.; Date, C.; Wakai, K.; Fukui, M.; Tamakoshi, A.; JACC Study Group. The relationship between green tea and total caffeine intake and risk for self-reported type 2 diabetes among Japanese adults. *Ann. Intern. Med.* **2006**, *144*, 554–562. [CrossRef] [PubMed]
9. Huxley, R.; Lee, C.M.; Barzi, F.; Timmermeister, L.; Czernichow, S.; Perkovic, V.; Grobbee, D.E.; Batty, D.; Woodward, M. Coffee, decaffeinated coffee, and tea consumption in relation to incident type 2 diabetes mellitus: A systematic review with meta-analysis. *Arch. Intern. Med.* **2009**, *169*, 2053–2063. [CrossRef] [PubMed]
10. van Dam, R.M.; Hu, F.B. Coffee consumption and risk of type 2 diabetes: A systematic review. *JAMA* **2005**, *294*, 97–104. [CrossRef] [PubMed]
11. Takahashi, M.; Miyashita, M.; Suzuki, K.; Bae, S.R.; Kim, H.K.; Wakisaka, T.; Matsui, Y.; Takeshita, M.; Yasunaga, K. Acute ingestion of catechin-rich green tea improves postprandial glucose status and increases serum thioredoxin concentrations in postmenopausal women. *Br. J. Nutr.* **2014**, *112*, 1542–1550. [CrossRef] [PubMed]
12. Ryu, O.H.; Lee, J.; Lee, K.W.; Kim, H.Y.; Seo, J.A.; Kim, S.G.; Kim, N.H.; Baik, S.H.; Choi, D.S.; Choi, K.M. Effects of green tea consumption on inflammation, insulin resistance and pulse wave velocity in type 2 diabetes patients. *Diabetes Res. Clin. Pract.* **2006**, *71*, 356–358. [CrossRef]
13. Brown, A.L.; Lane, J.; Coverly, J.; Stocks, J.; Jackson, S.; Stephen, A.; Bluck, L.; Coward, A.; Hendrickx, H. Effects of dietary supplementation with the green tea polyphenol epigallocatechin-3-gallate on insulin resistance and associated metabolic risk factors: Randomized controlled trial. *Br. J. Nutr.* **2009**, *101*, 886–894. [CrossRef] [PubMed]

14. Hsu, C.H.; Liao, Y.L.; Lin, S.C.; Tsai, T.H.; Huang, C.J.; Chou, P. Does supplementation with green tea extract improve insulin resistance in obese type 2 diabetics? A randomized, double-blind, and placebo-controlled clinical trial. *Altern. Med. Rev.* **2011**, *16*, 157–163. [CrossRef] [PubMed]
15. Legeay, S.; Rodier, M.; Fillon, L.; Faure, S.; Clere, N. Epigallocatechin gallate: A review of its beneficial properties to prevent metabolic syndrome. *Nutrients* **2015**, *7*, 5443–5468. [CrossRef] [PubMed]
16. Ferreira, M.A.; Silva, D.M.; de Morais, A.C., Jr.; Mota, J.F.; Botelho, P.B. Therapeutic potential of green tea on risk factors for type 2 diabetes in obese adults—A review. *Obes. Rev.* **2016**, *17*, 1316–1328. [CrossRef] [PubMed]
17. Thielecke, F.; Boschmann, M. The potential role of green tea catechins in the prevention of the metabolic syndrome—A review. *Phytochemistry* **2009**, *70*, 11–24. [CrossRef] [PubMed]
18. Tajik, N.; Tajik, M.; Mack, I.; Enck, P. The potential effects of chlorogenic acid, the main phenolic components in coffee, on health: A comprehensive review of the literature. *Eur. J. Nutr.* **2017**, *56*, 2215–2244. [CrossRef] [PubMed]
19. Meng, S.; Cao, J.; Feng, Q.; Peng, J.; Hu, Y. Roles of chlorogenic acid on regulating glucose and lipids metabolism: A review. *Evid. Based Complement. Altern. Med.* **2013**, *2013*, 801457. [CrossRef]
20. Venables, M.C.; Hulston, C.J.; Cox, H.R.; Jeukendrup, A.E. Green tea extract ingestion, fat oxidation, and glucose tolerance in healthy humans. *Am. J. Clin. Nutr.* **2008**, *87*, 778–784. [CrossRef]
21. Iwai, K.; Narita, Y.; Fukunaga, T.; Nakagiri, O.; Kamiya, T.; Ikeguchi, M.; Kikuchi, Y. Study on the postprandial glucose responses to a chlorogenic acid-rich extract of decaffeinated geen coffee beans in rats and healthy human subjects. *Food Sci. Technol. Res.* **2012**, *18*, 849–860. [CrossRef]
22. Jokura, H.; Watanabe, I.; Umeda, M.; Hase, T.; Shimotoyodome, A. Coffee polyphenol consumption improves postprandial hyperglycemia associated with impaired vascular endothelial function in healthy male adults. *Nutr. Res.* **2015**, *35*, 873–881. [CrossRef]
23. Zuñiga, L.Y.; Aceves-de la Mora, M.C.A.; González-Ortiz, M.; Ramos-Núñez, J.L.; Martínez-Abundis, E. Effect of chlorogenic acid administration on glycemic control, insulin secretion, and insulin sensitivity in patients with impaired glucose tolerance. *J. Med. Food* **2018**, *21*, 469–473. [CrossRef] [PubMed]
24. Harano, Y.; Miyawaki, T.; Nabiki, J.; Shibachi, M.; Adachi, T.; Ikeda, M.; Ueda, F.; Nakano, T. Development of cookie test for the simultaneous determination of glucose intolerance, hyperinsulinemia, insulin resistance and postprandial dyslipidemia. *Endocr. J.* **2006**, *53*, 173–180. [CrossRef]
25. Aloulou, I.; Brun, J.F.; Mercier, J. Evaluation of insulin sensitivity and glucose effectiveness during a standardized breakfast test: Comparison with the minimal model analysis of an intravenous glucose tolerance test. *Metabolism* **2006**, *55*, 676–690. [CrossRef]
26. Maki, K.C.; Kelley, K.M.; Lawless, A.L.; Hubacher, R.L.; Schild, A.L.; Dicklin, M.R.; Rains, T.M. Validation of insulin sensitivity and secretion indices derived from the liquid meal tolerance test. *Diabetes Technol. Ther.* **2011**, *13*, 661–666. [CrossRef]
27. Caumo, A.; Bergman, R.N.; Cobelli, C. Insulin sensitivity from meal tolerance tests in normal subjects: A minimal model index. *J. Clin. Endocrinol. Metab.* **2000**, *85*, 4396–4402. [CrossRef] [PubMed]
28. Matsuda, M.; DeFronzo, R.A. Insulin sensitivity indices obtained from oral glucose tolerance testing: Comparison with the euglycemic insulin clamp. *Diabetes Care* **1999**, *22*, 1462–1470. [CrossRef] [PubMed]
29. Matthews, D.R.; Hosker, J.P.; Rudenski, A.S.; Naylor, B.A.; Treacher, D.F.; Turner, R.C. Homeostasis model assessment: Insulin resistance and beta-cell function from fasting plasma glucose and insulin concentrations in man. *Diabetologia* **1985**, *28*, 412–419. [CrossRef]
30. Utzschneider, K.M.; Prigeon, R.L.; Faulenbach, M.V.; Tong, J.; Carr, D.B.; Boyko, E.J.; Leonetti, D.L.; McNeely, M.J.; Fujimoto, W.Y.; Kahn, S.E. Oral disposition index predicts the development of future diabetes above and beyond fasting and 2-h glucose levels. *Diabetes Care* **2009**, *32*, 335–341. [CrossRef] [PubMed]
31. Jones, B.; Kenward, M.G. *Design and Analysis of Cross-Over Trials*; Chapman & Hall/CRC: London, UK, 1989; Volume 5.
32. Shahwan, M.; Alhumaydhi, F.; Ashraf, G.M.; Hasan, P.M.Z.; Shamsi, A. Role of polyphenols in combating Type 2 Diabetes and insulin resistance. *Int. J. Biol. Macromol.* **2022**, *206*, 567–579. [CrossRef]
33. Li, X.; Zhang, Y.; Zhao, C.; Zhang, B.; Peng, B.; Zhang, Y.; Wang, J.; Wang, S. Positive effects of Epigallocatechin-3-gallate (EGCG) intervention on insulin resistance and gut microbial dysbiosis induced by bisphenol A. *J. Funct. Foods* **2022**, *93*, 105083. [CrossRef]
34. Murase, T.; Yokoi, Y.; Misawa, K.; Ominami, H.; Suzuki, Y.; Shibuya, Y.; Hase, T. Coffee polyphenols modulate whole-body substrate oxidation and suppress postprandial hyperglycaemia, hyperinsulinaemia and hyperlipidaemia. *Br. J. Nutr.* **2012**, *107*, 1757–1765. [CrossRef] [PubMed]
35. Akash, M.S.; Rehman, K.; Chen, S. Effects of coffee on type 2 diabetes mellitus. *Nutrition* **2014**, *30*, 755–763. [CrossRef]
36. Moon, S.M.; Joo, M.J.; Lee, Y.S.; Kim, M.G. Effects of Coffee Consumption on Insulin Resistance and Sensitivity: A Meta-Analysis. *Nutrients* **2021**, *13*, 3976. [CrossRef]
37. Peng, B.J.; Zhu, Q.; Zhong, Y.L.; Xu, S.H.; Wang, Z. Chlorogenic acid maintains glucose homeostasis through modulating the expression of SGLT-1, GLUT-2, and PLG in different intestinal segments of Sprague-Dawley rats fed a high-fat diet. *Biomed. Environ. Sci.* **2015**, *28*, 894–903. [CrossRef] [PubMed]
38. Johnston, K.L.; Clifford, M.N.; Morgan, L.M. Coffee acutely modifies gastrointestinal hormone secretion and glucose tolerance in humans: Glycemic effects of chlorogenic acid and caffeine. *Am. J. Clin. Nutr.* **2003**, *78*, 728–733. [CrossRef]

39. Montelius, C.; Erlandsson, D.; Vitija, E.; Stenblom, E.L.; Egecioglu, E.; Erlanson-Albertsson, C. Body weight loss, reduced urge for palatable food and increased release of GLP-1 through daily supplementation with green-plant membranes for three months in overweight women. *Appetite* **2014**, *81*, 295–304. [CrossRef] [PubMed]
40. de Boer, S.A.; Lefrandt, J.D.; Petersen, J.F.; Boersma, H.H.; Mulder, D.J.; Hoogenberg, K. The effects of GLP-1 analogues in obese, insulin-using type 2 diabetes in relation to eating behaviour. *Int. J. Clin. Pharm.* **2016**, *38*, 144–151. [CrossRef]
41. Spencer, N.J.; Hibberd, T.J. GLP-1 appetite control via intestinofugal neurons. *Cell Res.* **2022**, *32*, 711–712. [CrossRef] [PubMed]
42. Verdich, C.; Flint, A.; Gutzwiller, J.P.; Näslund, E.; Beglinger, C.; Hellström, P.M.; Long, S.J.; Morgan, L.M.; Holst, J.J.; Astrup, A. A meta-analysis of the effect of glucagon-like peptide-1 (7–36) amide on ad libitum energy intake in humans. *J. Clin. Endocrinol. Metab.* **2001**, *86*, 4382–4389. [CrossRef] [PubMed]
43. Flint, A.; Raben, A.; Astrup, A.; Holst, J.J. Glucagon-like peptide 1 promotes satiety and suppresses energy intake in humans. *J. Clin. Investig.* **1998**, *101*, 515–520. [CrossRef]
44. Matsumoto, N.; Ishigaki, F.; Ishigaki, A.; Iwashina, H.; Hara, Y. Reduction of blood glucose levels by tea catechin. *Biosci. Biotechnol. Biochem.* **1993**, *57*, 525–527. [CrossRef]
45. Hjorne, A.P.; Modvig, I.M.; Holst, J.J. The Sensory Mechanisms of Nutrient-Induced GLP-1 Secretion. *Metabolites* **2022**, *12*, 420. [CrossRef] [PubMed]
46. Shimizu, M. Interaction between food substances and the intestinal epithelium. *Biosci. Biotechnol. Biochem.* **2010**, *74*, 232–241. [CrossRef] [PubMed]
47. Shimizu, M.; Kobayashi, Y.; Suzuki, M.; Satsu, H.; Miyamoto, Y. Regulation of intestinal glucose transport by tea catechins. *BioFactors* **2000**, *13*, 61–65. [CrossRef]
48. da Silva, L.A.; Wouk, J.; Weber, V.M.R.; da Luz Eltchechem, C.; de Almeida, P.; Martins, J.C.L.; Malfatti, C.R.M.; Osiecki, R. Mechanisms and biological effects of caffeine on substrate metabolism homeostasis: A systematic review. *J. Appl. Pharm. Sci.* **2017**, *7*, 215–221. [CrossRef]
49. Graham, T.E.; Sathasivam, P.; Rowland, M.; Marko, N.; Greer, F.; Battram, D. Caffeine ingestion elevates plasma insulin response in humans during an oral glucose tolerance test. *Can. J. Physiol. Pharmacol.* **2001**, *79*, 559–565. [CrossRef] [PubMed]
50. Thong, F.S.; Graham, T.E. Caffeine-induced impairment of glucose tolerance is abolished by beta-adrenergic receptor blockade in humans. *J. Appl. Physiol.* **2002**, *92*, 2347–2352. [CrossRef]
51. Robinson, L.E.; Savani, S.; Battram, D.S.; McLaren, D.H.; Sathasivam, P.; Graham, T.E. Caffeine ingestion before an oral glucose tolerance test impairs blood glucose management in men with type 2 diabetes. *J. Nutr.* **2004**, *134*, 2528–2533. [CrossRef] [PubMed]
52. Colombo, R.; Papetti, A. Decaffeinated coffee and its benefits on health: Focus on systemic disorders. *Crit. Rev. Food Sci. Nutr.* **2021**, *61*, 2506–2522. [CrossRef] [PubMed]

nutrients

MDPI

Article

Modulatory Effect of Chlorogenic Acid and Coffee Extracts on Wnt/β-Catenin Pathway in Colorectal Cancer Cells

Hernán Villota [1], Gloria A. Santa-González [1], Diego Uribe [1], Isabel Cristina Henao [2], Johanna C. Arroyave-Ospina [3], Carlos J. Barrera-Causil [4] and Johanna Pedroza-Díaz [1,*]

1 Grupo de Investigación e Innovación Biomédica, Facultad de Ciencias Exactas y Aplicadas, Instituto Tecnológico Metropolitano, Medellín 050012, Colombia
2 Productos Naturales Marinos, Facultad de Ciencias Farmacéuticas y Alimentarias, Universidad de Antioquia UdeA, Medellín 050010, Colombia
3 Department of Gastroenterology and Hepatology, University Medical Center Groningen, University of Groningen, 9713 GZ Groningen, The Netherlands
4 Grupo de Investigación Davinci, Facultad de Ciencias Exactas y Aplicadas, Instituto Tecnológico Metropolitano, Medellín 050034, Colombia
* Correspondence: ninipedroza@itm.edu.co; Tel.: +57-604-440-5291

Abstract: The Wnt/β-Catenin pathway alterations present in colorectal cancer (CRC) are of special interest in the development of new therapeutic strategies to impact carcinogenesis and the progression of CRC. In this context, different polyphenols present in natural products have been reported to have modulatory effects against the Wnt pathway in CRC. In this study, we evaluate the effect of two polyphenol-rich coffee extracts and chlorogenic acid (CGA) against SW480 and HT-29 CRC cells. This involved the use of MTT and SRB techniques for cell viability; wound healing and invasion assay for the evaluation of the migration and invasion process; T cell factor (TCF) reporter plasmid for the evaluation of transciption factor (TCF) transcriptional activity; polymerase chain reaction (PCR) of target genes and confocal fluorescence microscopy for β-Catenin and E-Cadherin protein fluorescence levels; and subcellular localization. Our results showed a potential modulatory effect of the Wnt pathway on CRC cells, and we observed a reduction in the transcriptional activity of β-catenin. All the results were prominent in SW480 cells, where the Wnt pathway deregulation has more relevance and implies a constitutive activation of the signaling pathway. These results establish a starting point for the discovery of a mechanism of action associated with these effects and corroborate the anticancer potential of polyphenols present in coffee, which could be explored as chemopreventive molecules or as adjunctive therapy in CRC.

Keywords: chlorogenic acid; coffee extracts; colorectal cancer; Wnt/β-catenin pathway; therapeutic targets

Citation: Villota, H.; Santa-González, G.A.; Uribe, D.; Henao, I.C.; Arroyave-Ospina, J.C.; Barrera-Causil, C.J.; Pedroza-Díaz, J. Modulatory Effect of Chlorogenic Acid and Coffee Extracts on Wnt/β-Catenin Pathway in Colorectal Cancer Cells. *Nutrients* **2022**, *14*, 4880. https://doi.org/10.3390/nu14224880

Academic Editors: Sonia de Pascual-Teresa and Luis Goya

Received: 25 October 2022
Accepted: 14 November 2022
Published: 18 November 2022

1. Introduction

Polyphenols are natural organic compounds of particular interest in nutrition and functional food [1]. Dietary polyphenols play an important role in human health by regulating cellular metabolism, chronic disease, and cell proliferation [2,3]. To date, several polyphenols have been identified in different natural products, such as coffee which contains phenolic acids such as chlorogenic, caffeic, and feluric [4]; however, their impact on human health has not been fully characterized. Polyphenols from coffee regulate different biological processes, exhibiting chemopreventive, antioxidant, anti-inflammatory, and anticancer properties in in vitro and in vivo studies [5]. Some epidemiological studies also suggest that regular coffee consumption influences the prevention of cardiovascular diseases, obesity, diabetes, and some types of cancer [6,7].

Colorectal cancer (CRC) is one of the most frequent cancers globally, occupying third place after lung and prostate cancer in men and breast cancer in women. Additionally, 10% of all newly diagnosed cases and cancer-related deaths are associated with CRC [8]. Age, genetic and environmental factors play an important role in the development of CRC, and some studies have established a range of hereditary for CRC from 12% to 32%, suggesting that a large proportion of cases (68–82%) have modifiable risk factors, such as overweight and obesity, alcohol consumption, and processed meat intake, among others [9].

The development of CRC is strongly influenced by hereditary factors such as familial adenomatous polyposis (FAP) or Lynch syndrome; however, approximately 80% of cases are sporadic. The events leading to CRC develop slowly through a sequential progression and are staggered. Two pathways of CRC initiation have been described, the adenoma-carcinoma sequence, related to 60–85% of cases, and the alternative, or serrated, pathway, related to 15–40% of cases. These models include mutations in signaling pathways such as Wnt, mitogen-activated protein kinases (MAPK), transforming growth factor-beta (TGF-β), and p53. Tubular, tubulovillous, and villous adenomas are the most common lesions related to sporadic tumors in the adenoma-carcinoma sequence, where the most frequent molecular alterations are related to mutations in the adenomatous polyposis coli (APC) gene. On the other hand, the "serrated neoplastic pathway" describes the progression of serrated polyps, including sessile serrated adenomas and traditional serrated adenomas, with colorectal cancer. This pathway is associated with β-Catenin, BRAF mutation, and microsatellite instability, triggering CRC phenotypes such as CpG island methylator phenotype (CIMP) [10,11].

The Wnt pathway, widely studied for its importance in processes of embryogenesis, tissue homeostasis, and cell-cell adhesion, plays a crucial role in the initiation, progression, and metastasis of CRC. Wnt pathway activation depends on the alteration of its components and their functions. The transduction of the pathway includes processes such as the secretion of Wnt proteins, identification of Wnt co-receptors, silencing of the β-catenin destruction complex that includes proteins such as (APC and GSK3-β), translocation of β-catenin to the nucleus, recruitment of cofactors, and transcriptional activation of genes such as Cyclin D1 (*CCND1*), *MYC*, and *JUN* (Figure 1). Some aberrant functions in these processes can influence the development of cancer. In CRC, it has been shown that about 90% of cases are a consequence of damage to one of these Wnt pathway processes, especially APC loss-of-function mutations, β-catenin activation mutations leading to hyperactivation, and increased frizzled family receptors. However, APC and β-catenin mutations are generally mutually exclusive, with somatic APC mutations being found in more than 80% of sporadic colorectal tumors and β-catenin mutations in about 48% of tumors without APC mutation [12,13].

Recently, some studies have shown the anticancer activity of different polyphenols present in natural products such as grapes, apples, berries, herbs, spices, and drinks such as coffee. In colorectal cancer specifically, polyphenols show the capability of modulating the Wnt/β-catenin pathway using different regulatory mechanisms such as the decreased phosphorylation of GSK3-β by silibinin and epigalocatequin galato (EGCG) and destabilization of the β-catenin/TCF complex by resveratrol [14]. Coffee is the main natural source of chlorogenic acid (CGA), an hydroxycinnamic acid that is the most abundant polyphenol in natural products. Several in vitro and in vivo studies using CGA have shown its antioxidant activity, inhibition of mutagenic and carcinogenic *N*-nitroso compounds, DNA damage inhibition, and suppression of reactive oxygen species-mediated nuclear factor activation (NF-κB) [15]. Regarding the modulatory effect of CGA, some in vitro and in vivo studies have shown effects related to the regulation of the Wnt pathway, for example, inhibition of adipogenesis in murine fibroblast cells [16], inhibition of cell differentiation in human pulp stem cells [17], and antitumor activity in murine breast cancer models [18]. Additionally, we recently reported the biological effect of CGA and caffeic acid related to the inhibition of cell migration as a first approach to the study of the effect of polyphenols present in coffee in colorectal cancer models at the in vitro level [19]. However, the modulatory effect of coffee polyphenols on the Wnt pathway has not been studied in CRC models. For these reasons, in the present study, we explore the possible modulation of the Wnt/β-catenin pathway by CGA and two coffee extracts rich in polyphenols to evaluate the influence of treatments on cell proliferation, gene expression, Wnt transcriptional activity, and β-catenin and E-cadherin subcellular localization, using SW480 and HT-29 colorectal cancer cell lines.

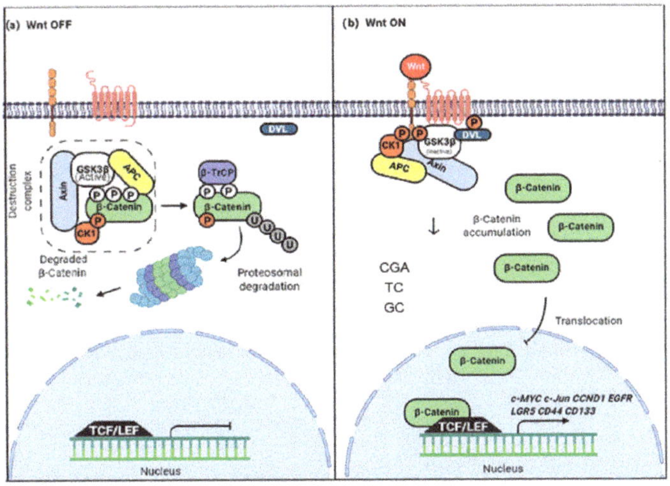

Figure 1. Mechanism of modulation of the Wnt/β-catenin pathway by coffee polyphenols. (**a**) Normal OFF state. The β-catenin degradation complex regulates the levels of β-catenin and controls its transcriptional activator activity (**b**) The normal ON state. The β-catenin degradation complex doesn't work, β-catenin levels increase, and translocation to the nucleus possibility its transcriptional activity. The possible mechanism of modulation associated with coffee polyphenols; CGA (Chrologenic acid), TC (Toasted coffee), and GC (Gree coffee) decrease the levels of β-catenin in the cytoplasm and nucleus, possibly through a β-catenin degradation mediation and nuclear translocation inhibition. Figure adapted from [14].

2. Materials and Methods

2.1. Chemical and Reagents

Neochlorogenic acid (99.3%), cryptochlorogenic acid (99.8%), and caffeic acid (98.5%) used in quantification analysis were purchased from Biopurify, Chengdu, China, and chlorogenic acid (99%) from was purchased from Extrasinthese, Genay, France. Chlorogenic acid ≥ 95% (titration) product reference (Ref.) C3878-1G, 3-(4,5-dimethylthiazol-2-yl)-2,5-diphenyltetrazolium bromide (MTT), Wnt CHIR 99021 inductor, and iCRT14 inhibitor were purchased from Sigma-Aldrich, Burlington, MA, USA. Acidified isopropyl alcohol, PBS, fetal bovine serum (FBS), Dulbecco's modified eagle's medium (DMEM), Roswell Park Memorial Institute (RPMI) 1640 medium, penicillin, and streptomycin were purchased from GIBCO, Grand Island, NY, USA. The TCF Reporter Plasmid Kit Ref. # 17-285 was purchased from Merck Millipore, Burlington, MA, USA.

2.2. Quantification of Chlorogenic Acids and Xanthines by HPLC-DAD

A total of 100 mg of each extract (GC and TC), obtained as previously described [19], were reconstituted with 70% ethanol centrifuged at 13,000 RMP for 10 min; the supernatant was transferred to a 20 mL volumetric flask, and the volume was adjusted. The supernatant was diluted using mobile phase, in proportions of 1 in 20, for the determination of neochlorogenic and cryptochlorogenic acids, and in proportions of 1 in 100 for the determination of chlorogenic acid. For the determination of xanthines, theobromine (99%), caffeine (199.8%), catechin (99%), and epicatechin (90%), the supernatant was diluted using mobile phase, in proportions of 1 in 20. An external standard method was used for quantification in an HPLC Agilent 1200 Series LC System CA, USA, using a Zorbax column SB-C18 from Merck, Burlington, MA, USA.

2.3. Cell Culture

Human cell lines SW480 (ATCC, Manassas, VA, USA; CCL-228) and HT-29 (ATCC, Manassas, VA, USA; HTB-38™) derived from colorectal adenocarcinoma, were used. SW480 and HT-29 cells were maintained in Dulbecco's Modified Eagle Medium (DMEM) and RPMI 1640 medium GIBCO, Grand Island, NY, USA, respectively, supplemented with 10% fetal bovine serum and antibiotics. The cellular passages were made at 70% of cell confluence. The cells were cultured under controlled conditions at 5% CO_2, 70% humidity, and 37 °C.

2.4. Cytotoxicity, Migration, and Invasion Studies

Two coffee extracts rich in green and toasted polyphenols (GC, TC) and CGA were prepared and used as previously reported [19]. In brief, all treatments were dissolved in a culture medium before treating the cells at different concentrations. The cytotoxicity of CGA and coffee extracts was assessed using MTT and SRB assays in 96 well plates at 2×10^4 cells/well. Cell viability was expressed as a percentage of the control as (absorbance of treated cells−absorbance of background controls)/(absorbance of nontreated cells−absorbance of background controls) ×100. For evaluation of cell migration, SW480 and HT-29 cells were seeded in 24-well plates at 4×10^5 cells/well, and a scratch was made with a 10 µL pipette tip, using subtoxic concentrations of 750 µg/mL (GC), 500 µg/mL (TC), and 187 µg/mL (CGA). The images were captured using an inverted microscope (Eclipse Ti-s) with a DsFi1c digital camera from Nikon, Tokyo, Japan (magnification: 10×) at 0, 24, 48, 72, 96, and 120 h. A total of 144 bright-field images were included for each biological replicate, after which, the images were analyzed for segmentation and quantification with Bio-EdiP from ITM, Medellin, Colombia [20]. A scratch in non-treated cells (NTC) was employed to support normal wound healing progression in vitro. For the invasion test, the commercial fluorometric Kit CytoSelect™ 96-well cell invasion assay from Cell Biolabs San Diego, CA, USA was used, treating the 2×10^4 cells/well for 24 h at concentrations of 750 µg/mL, 1500 µg/mL, and 2000 µg/mL for GC; 500 µg/mL, 750 µg/mL, and 1000 µg/mL for TC; and 150 µg/mL, 375 µg/mL, and 750 µg/mL for CGA. Migration and invasion tests were in serum-free media conditions, and bovine serum albumin (BSA) was used as a chemoattractant in the invasion test. The procedure was performed following the manufacturer's instructions. The suspension of HT-29 and SW480 cells in a culture medium without fetal bovine serum (FBS) was added to the upper chamber and in the lower chamber FBS, it was applied as a chemoattractant. Finally, the fluorescence was read with the fluorescence plate reader (GloMax-Multi Detection System by Promega, Madison, WI, USA) at 480 nm/520 nm.

2.5. Wnt Pathway Reporter Assay

For the Wnt pathway reporter assay, SW480 and HT-29 cells were seeded in 12 well plates at 3×10^5 cells/well. After 24 h, cells were transfected in serum and antibiotic-free media with 1 µg of TOP-Flash (plasmid that contains wild type TCF binding sites) or 1 µg of Fop-Flash (plasmid that contains mutated TCF binding sites), both from Millipore, using Lipofectamine® 2000 from Thermo Fisher, Waltham, MA, USA. Cells were treated 24 h after transfection with different concentrations of coffee extracts, CGA, or molecules to control the induction and inhibition of the Wnt pathway, such as CHIR 99021 and iCRT14, respectively. The luciferase activity was determined at 24 h of treatments using the Luciferase Assay System from Promega. The efficiency of transfection was determined by flow cytometry using GFP reporter plasmid. The promoter activity was expressed as the net of TOP-Flash relative light units after the substation of the associated FOP-flash relative light units. All assays were performed in triplicate.

2.6. RNA Isolation and RT-PCR

Total RNA was extracted with a Trizol solution (Ref. T9424), from Sigma, Burlington, MA, USA) according to the manufacturer's instructions. The concentration of RNA was determined by spectrophotometry (Nanodrop 2000/2000C, Thermofisher, Waltham, MA, USA). Two micrograms of total RNA were reverse-transcribed with the RevertAid Reverse Transcriptase kit (Ref. EP0442, Thermofisher, Waltham, MA, USA) with reaction conditions of 25 °C for 6 min, followed by 42 °C for 6 min and 70 °C for 6 min. Finally, cDNA from SW480 and HT-29 cells treated with GC, TC, and CGA for 24 h was stored at −20 °C.

2.7. Wnt Target Gene Analysis

For this purpose, quantitative PCR analysis was run in the Thermocycler CFX96 system (Bio-Rad, Hercules, CA, USA) with the Maxima SYBR Green qPCR Master Mix (2X) (Ref. K0252, Thermo Fisher) according to the manufacturer's protocol. Cycling conditions were 95 °C for 5 min, followed by 40 cycles of 95 °C for 15 s and 60 °C for 60 s. GAPDH was used as an endogenous reference. The $2^{-\Delta\Delta Ct}$ method was used to calculate the differences in the expression levels of CTNNB1, CDH1, and CCND1 during the different treatments. All experiments were run in triplicate to indicate intra-assay variation. The primer sequences were as follows: CTNNB1 forward (F): 5′-TCCGAATGTCTGAGGACAAGC-3′, reverse (R): 5′-CCAAGATCAGCAGTCTCATTCCA-3′; CDH1 (F): 5′-TCCTGGGCAGAGTGAATTTTG-3′, (R): 5′-CTGTAATCACACCATCTGTGC-3′; CCND1 (F): 5′-GAAGATCGTCGCCACCTG-3′, (R): 5′-TCGACATGGAGTCCCAGGA-3′; and GAPDH (F): 5′-AACGGGAAGCTTGTCATCAA-3′, (R): 5′-TGGACTCCACGACGTACTCA-3′.

2.8. Subcellular Localization of Wnt Proteins

To determine the subcellular localization of Wnt proteins, cell staining and confocal microscopy were carried out. Briefly, 15×10^4 cells were seeded in 24 well plates with circular glass covers overnight. The next day, the cells were treated with different concentrations of coffee extracts and CGA for 24 h. Afterward, the medium was removed, and cells were fixed in 4% formaldehyde in PBS for 15 min and permeabilized by 0.5% Triton X-100 in PBS for 5 min. After permeabilization, the samples were treated with a blocking solution (Invitrogen, Waltham, MA, USA) for 45 min and then incubated with primary antibodies (against β-catenin and E-cadherin) in TBS overnight. The following day, the cells were washed with PBS and incubated with secondary antibodies from Thermofisher, Waltham, MA, USA). (Alexa-Fluor™488-anti-rabbit and/or Alexa-Fluor™568-anti-mouse IgG) for 45 min. Each sample was washed, mounted, and monitored with an FV1000 Olympus laser scanner confocal microscope from Evident corporation, Tokyo, Japan.

2.9. Statistical Analysis

GraphPad 6 was used to perform statistical analysis (GraphPad Software (version number 6)) from GraphPad Software Inc, San Diego, CA. USA. The number of observations represents the categorical data. The variables with normal distributions were denoted by the mean and standard deviation. The variations in cell viability, open wound area, invasion, qPCR data, and immunofluorescence were analyzed using a two-way ANOVA. A p-value of ≤ 0.05 was considered statistically significant.

3. Results

3.1. Quantification of Chlorogenic Acids, Xanthines, and Catechins

The main metabolites in coffee extracts (Table 1) were determined by HPLC (high-performance liquid chromatography) equipped with a PDA (photodiode-array) detector. GC presented a concentration of chlorogenic acid 4.4 times greater than TC, while neochlorogenic and cryptochlorogenic acids, as well as caffeine and theobromine, showed a slightly higher concentration than TC. None of the studied extracts presented a quantifiable concentration of chatechin or epichatechin.

Table 1. The chlorogenic acids, xanthines, and catechins in green and toasted coffee extracts (GC, TC) by HPLC-DAD. All analyzes were performed on three independent samples. ND, not detectable. NA, not applicable. RSD, relative standard deviation.

Chlorogenic Acids	Neochlorogenic Acid		Chlorogenic Acid		Cryptochlorogenic Acid		Caffeic Acid	
	mg/100 g sample	RSD	mg/100 g sample	RSD	mg/100 g sample	RSD	mg/100 g sample	RSD
Green coffee	1114.50	1.205	17,715.79	4.451	2025.56	6.614	ND	NA
Toasted coffee	1485.62	2.023	3996.50	2.087	2095.93	1.929	ND	NA
Xanthines and catechins content	**Theobromine**		**Caffeine**		**Catechin**		**Epicatechin**	
	mg/100 g sample	RSD	mg/100 g sample	RSD	mg/100 g sample	RSD	mg/100 g sample	RSD
Green coffee	406.51	7.599	2878.03	6.225	ND	NA	ND	NA
Toasted coffee	563.84	2.796	3372.86	1.860	ND	NA	ND	NA

3.2. Cytotoxicity Activity

The cytotoxic activity of GC, TC, and chlorogenic acid (CGA) against SW480 and HT-29 cells was determined by the MTT and SRB methods for 24 and 48 h treatment times (Supplementary Figure S1), and 48 h treatment by MTT (Figure 2). All treatments showed similar cytotoxic activity with the two methods employed and were comparable with a dose/time-dependent tendency to decrease the cell viability. All treatments were more effective in SW480 cells than in HT-29 cells. A significant decrease in cell viability was observed with doses above 1500 μg/mL for GC and TC, and above 325 μg/mL for CGA. Even the highest dose (3000 μg/mL) at 48 h did not cause a significant decrease in cell viability in HT-29 cells. Figure 2A,B shows the difference between the cytotoxicity caused by GC and TC in SW-480 and HT-29 cells. Furthermore, for CGA (Figure 2C), only the highest concentration (3000 μg/mL) significantly decreased the viability of HT-29 cells. All treatments were evaluated by the SRB method (Supplementary Figure S1). Table 2 shows the IC50 values for all the evaluated treatments by MTT and SRB methods in HT-29 and SW480 cell lines. The data for the MTT analysis in SW480 cells were extracted from our previous report [19].

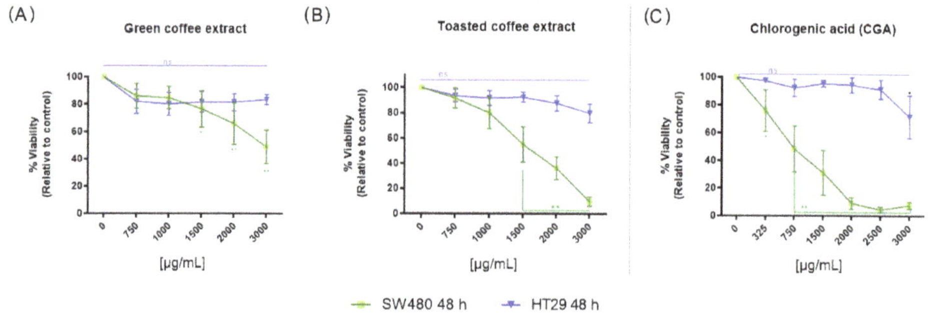

Figure 2. The cytotoxic activity at 48 h measured by MTT in the colorectal cancer cell lines SW480 and HT-29. Treatments with green coffee extract (GC) (**A**), toasted coffee extract (TC) (**B**), and CGA (**C**). Values are expressed as the mean ± SEM of at least three independent experiments. Two-way ANOVA, difference to non-treated cells, * $p \leq 0.05$, ** $p \leq 0.01$; ns: non-significant differences.

Table 2. Half maximal inhibitory concentration (IC50) values by MTT and SRB methods on SW480 and HT-29 cells treated with green and toasted coffee extracts (GC, TC) and chlorogenic acid (CGA).

IC $_{50}$ Value by MTT	SW480		HT-29	
	24 h	**48 h**	**24 h**	**48 h**
Green coffee	4325 µg/mL	2555 µg/mL	17,715 µg/mL	8416 µg/mL
Toasted coffee	3922 µg/mL	2226 µg/mL	9918 µg/mL	13,247 µg/mL
CGA	686.6 µg/mL	598.3 µg/mL	8114 µg/mL	6733 µg/mL
IC $_{50}$ Values by SRB	**24 h**	**48 h**	**24 h**	**48 h**
Green coffee	4676 µg/mL	2799 µg/mL	129,197 µg/mL	48,366 µg/mL
Toasted coffee	3656 µg/mL	1590 µg/mL	58,901 µg/mL	16,484 µg/mL
CGA	2844 µg/mL	1338 µg/mL	72,945 µg/mL	18,379 µg/mL

3.3. Wound Healing Assay

To determine the effect of treatments on cell migration inhibition, wound healing assays were performed using subtoxic doses of GC (750 µg/mL), TC (500 µg/mL), and CGA (187 µg/mL). Pictures to monitor the healing process were taken every 24 h. The cytotoxic activity results reported in the previous section are in accordance with the wound healing inhibition activity. All treatments have a higher impact on the decrease in the wound closure area in SW480 cells than in HT-29 cells (Figure 3). The wound closure rate was diminished in all treatments in both cell lines, even though the behavior is different. In SW480 (Figure 3A) the wounds healed at a slow rate with respect to the NTC (non-treated cells) during the first 48 h; after which, the wound closure rate decreased, reaching lower values than the initial condition. On the other hand, the results of HT-29 cells (Figure 3B) showed an effect comparable to the control condition, with a tendency to close the wound. CGA treatment significantly inhibited the rate of wound healing, followed by GC and CT, compared with the NTC. The data for the wound healing analysis in SW480 cells were extracted from our previous report [19].

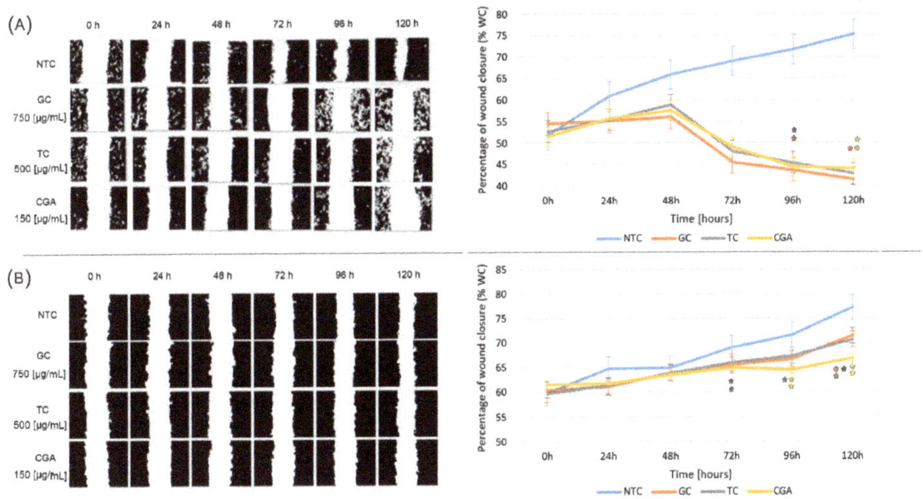

Figure 3. The inhibition of cellular migration on SW480 (**A**) and HT-29 (**B**) cell lines non-treated (blue) and treated with GC (orange), TC (grey), and CGA (yellow). Cell migration was observed with an inverted microscope (10× magnification) at intervals of 24 h over 120 h. Representative images are on the left panel. The right panel shows the quantitative analysis of cell migration by the percentage of wound closure. Values are expressed as the mean ± SEM of three independent experiments. Two-way ANOVA, difference to non-treated cells, * $p \leq 0.05$ and ** $p \leq 0.01$.

3.4. *Invasion Assay*

The CytoSelect™ 96-well cell invasion assay was used to evaluate the effect of treatments on cell invasion. Three subtoxic doses were selected for each treatment: 750 µg/mL, 1500 µg/mL, and 2000 µg/mL for GC; 500 µg/mL, 750 µg/mL, and 1000 µg/mL for TC; and 150 µg/mL, 375 µg/mL, and 750 µg/mL for CGA. To observe the basal level of invasion of both cell lines, two control conditions were used: with and without chemoattractant, to which no treatment was applied. Finally, the invasive cells that were able to cross the extracellular matrix protein membrane to the lower chamber were quantified with a fluorometric method, where fluorescence values are proportional to the number of invasive cells. The results of the controls showed a greater activation effect for the invasion in response to the chemoattractant in SW480 cells, which agrees with the migratory properties observed before for this cell line.

The results of the effect of the coffee extracts and CGA showed a powerful inhibition of the invasion of SW480 cells compared to the control with chemoattractant (Figure 4A), consistent with the cytotoxic and anti-migratory activities of the treatments on these cells. Additionally, TC 1000 µg/mL had an effect comparable to the control without chemoattractant while CGA 750 µg/mL exceeded the inhibition effect of invasion compared to this control. In contrast, the results for HT-29 cells (Figure 4B) showed a lower effect for the inhibition of cell invasion, considering that only the treatments with higher doses of TC and CGA had a significant impact compared to the control with a chemoattractant. However, these treatments had a higher effect on inhibition of invasion compared to the control without chemoattractant.

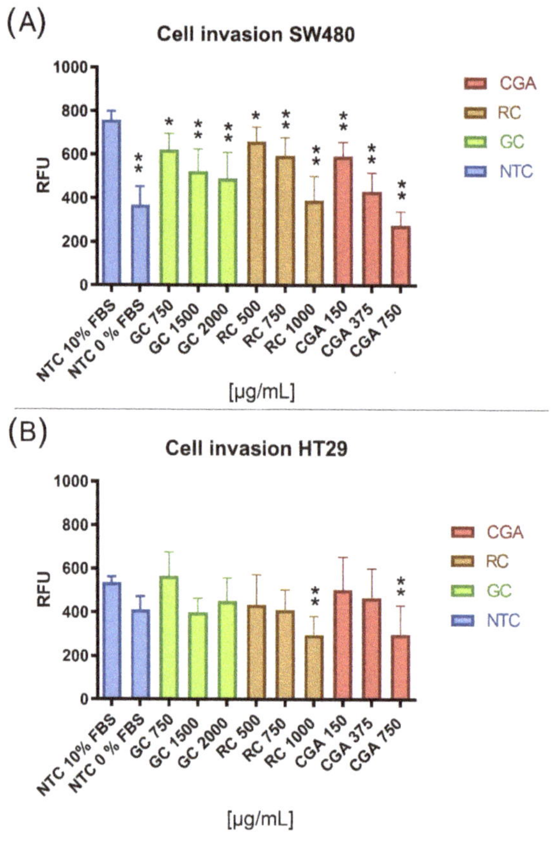

Figure 4. The inhibition of the invasion process of SW480 (**A**) and HT-29 (**B**) cell lines. The percentage of relative fluorescence units (RFU) is proportional to the number of invasive cells. FBS 10% was used as a

chemoattractant control and treatment exposure was 24 h. Values are expressed as the mean ± SEM of three independent experiments. Two-way ANOVA, the difference to non-treated cells, * $p \leq 0.05$ and ** $p \leq 0.01$.

3.5. *Top/Fop Flash Assay*

To determine if the treatments had an impact on the regulation of transcriptional activity of the Wnt/β-catenin pathway, HT-29, and SW480 cells were transfected with the TOP/FOP Flash reporter plasmid and treated with subtoxic doses of GC (1500 µg/mL), TC (750 µg/mL), and CGA (375 µg/mL) for 24 h. As controls of the regulation of the Wnt pathway, doses lower than the IC50 were employed. For positive and negative regulation of the pathway, the selective inducer CHIR 99021 (10 µM) and the selective inhibitor iCRT14 (40 µM) were included in the experiments. As we expected, the results in SW480 cells (Figure 5A) show a substantial effect on the induction of pathway activation with CHIR, while the inhibition effect of the transcriptional activity of the pathway by iCRT14 is significant compared to the NTC condition. Treatments with GC and TC did not have significant differences in the regulation of the pathway, while CGA had a comparable effect with the selective inhibitor. In the case of HT-29 cells (Figure 5B), the effect of the inducer was lower than in SW480, and treatments with GC, TC, and CGA had a trend of negative regulation of the pathway. However, only TC and CGA had significant values, and CGA had a comparable effect to the selective inhibitor.

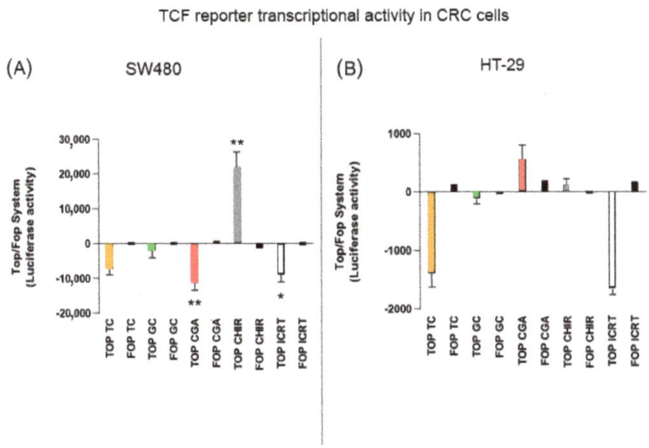

Figure 5. The inhibition of TCF promotor activation of the Wnt pathway plasmid reporter system in SW480 (**A**) and HT-29 (**B**) cell lines. CHIR 99021 was used as a pathway inductor and iCRT14 as an inhibitor. Treatment exposure was 24 h after transfection. Values are expressed as the mean ± SEM of three independent experiments. Two-way ANOVA, difference to non-treated cells, * $p \leq 0.05$ and ** $p \leq 0.01$.

3.6. *Changes in mRNA Expression Levels of CDH1, CTNNB1, and CCND1*

qPCR analysis was used to find possible differences in the mRNA expression levels of genes related to the Wnt/β-catenin pathway, such as *CDH1* (which encodes for E-cadherin), *CTNNB1* (which encodes for β-catenin), and *CCND1* (which encodes for cyclin D1). *GAPDH* was used to normalize expression levels. The treatments included two subtoxic doses of each treatment for 24 h: GC (1500 µg/mL and 2000 µg/mL), TC (750 µg/mL and 1000 µg/mL), and CGA (375 µg/mL and 750 µg/mL). The results are shown in Figure 6.

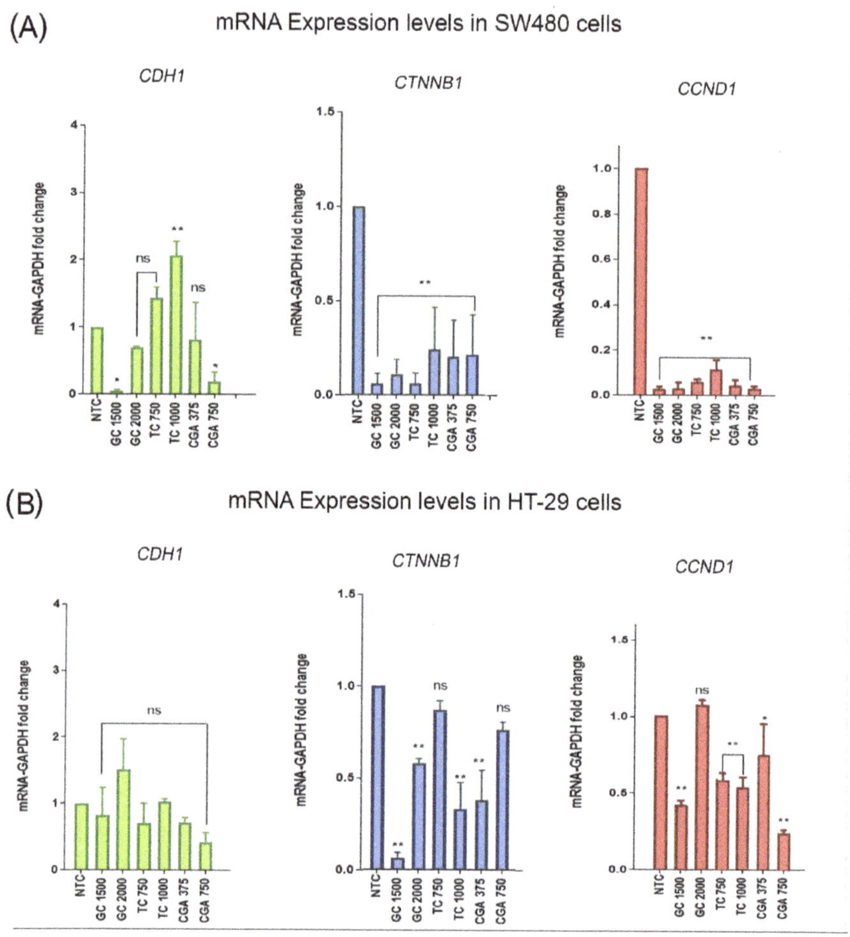

Figure 6. (CGA) Chlorogenic acid, (TC) toasted coffee extract, and (GC) green coffee extract modulation of the mRNA expression levels of *CDH1*, in green bars (which encodes for E-cadherin); *CTNNB1*, in blue bars (which encodes for β-catenin); and *CCND1*, in red bars (which encodes for Cyclin D1) in SW480 (**A**) and HT-29 (**B**) cell lines. Values are expressed as the mean ± SEM of three independent experiments. Two-way ANOVA, difference to non-treated cells, * $p \leq 0.05$ and ** $p \leq 0.01$. ns: non-significant differences.

In the case of E-cadherin in SW480 cells, we found a significant increase in mRNA levels with TC at the highest doses, while GC and CGA mainly had a negative regulatory effect. For HT-29 cells, we observed a tendency for mRNA levels to increase only in GC with 2000 μg/mL. For β-catenin mRNA levels in both SW480 and HT-29 cells, all treatment conditions showed a negative expression regulation effect, being stronger in SW480. Likewise, cyclin D1 expression levels decreased significantly in both cell lines, as expected in accordance with the result found regarding the negative regulation of β-catenin mRNA levels.

3.7. Subcellular Localization of β-Catenin and E-Cadherin Proteins

Confocal microscopy was performed to determine the effect of the treatments on the subcellular localization of β-catenin and E-cadherin proteins. SW480 and HT-29 cells were exposed for 24 h to GC 1500 μg/mL, TC 750 μg/mL, and CGA 375 μg/mL. The results of the previously described assays showed the increased sensitivity of SW480 cells for all treatments. Consistently with this, we observed that under control conditions without any treatment, the distribution of nuclear β-catenin was significantly different between SW480 and HT-29 (Figure 7). Specifically, the results showed high levels of nuclear β-catenin in SW480 cells, whereas we observed a specific distribution of cytoplasmic and cell membrane β-catenin in HT-29 cells, which correlates with a high percentage of β-catenin/E-cadherin co-localization.

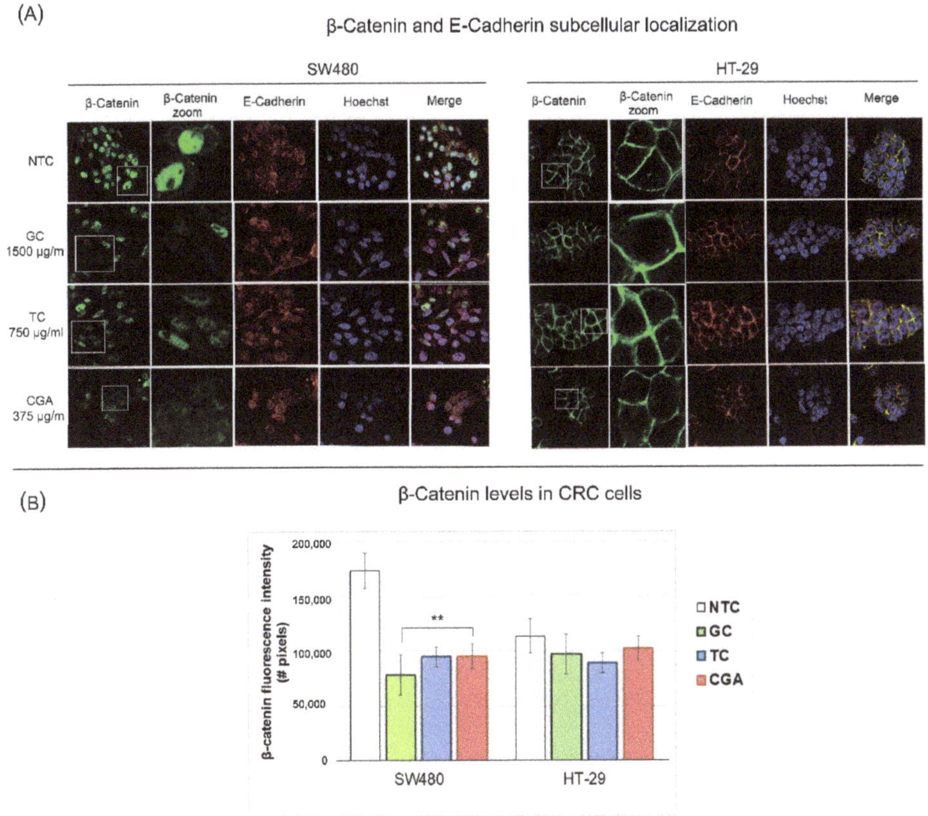

Figure 7. Subcellular localization of β-catenin (green) and E-cadherin (red) proteins in SW480 and HT29 cells treated with chlorogenic acid (CGA), roasted coffee (TC), and green coffee (GC) at 24 h. The panel (**A**) includes representative images, and the (**B**) panel is the β-catenin protein fluorescence intensity in both treated cell lines. The images were captured through an Olympus FV1000 confocal laser scanner microscope with a 60x objective and image scale of 150 μm. Values are expressed as the mean ± SEM of three independent experiments. Two-way ANOVA, difference to non-treated cells, ** $p \leq 0.01$; # = number.

The response to the treatments in SW480 cells (Figure 7) showed a decreased β-catenin fluorescence intensity compared to the control condition, which was most marked in GC, followed by TC and CGA, while the fluorescence intensity levels of E-cadherin remained similar. On the other hand, for HT-29 cells, no nuclear β-catenin staining was observed, and the levels of co-localization between β-catenin and E-cadherin proteins were found to be high in the control condition and maintained in treatments. Thus, during all the treatments, we did not find significant differences in the cellular distribution of β-catenin/E-cadherin, and the β-catenin fluorescence intensity was similar to the control condition.

4. Discussion

Phytochemicals and their derivatives are promising options for preventing and treating diseases such as cancer. In vitro and in vivo studies and clinical trials have shown their regulation of various physiological processes, such as inflammation reduction, cell differentiation regulation, and protection against oxidative damage of organelles and DNA [21–23]. In recent decades, polyphenols have been highlighted in the context of phytochemicals, and several recent studies have reported on different types of cancer, such as liver, breast, prostate, and CRC [24–26].

Diet is the primary source of polyphenols in humans. Coffee, tea, apple juice, and red wine are the most consumed beverages worldwide and represent the main sources of polyphenols in the diet [27,28]. Epidemiological studies on the relationship between colorectal cancer and coffee consumption are inconclusive because results vary by study design, cancer site, gender, and ethnicity. However, current evidence suggests an inverse relationship in case-control studies where the category of high coffee consumption is associated with a 15% to 21% decreased risk of colon cancer [29]. Among the bioactive components of coffee, which include caffeine, melanoidins, and diterpenes, the most abundant polyphenols are chlorogenic, caffeic, and ferulic acids, the levels of which may vary depending on the coffee species, degree of roasting, and preparation technique [30].

Exposure to coffee polyphenols could promote colorectal cancer chemoprevention through diverse mechanisms, such as the activation of anti-mutagenic pathways, modification of the microbiome, or modulation of signaling pathways critical for cancer development [31]. The Wnt pathway plays a critical role in the carcinogenesis and progression of colorectal cancer. Recent studies show the ability of some polyphenols to regulate the Wnt pathway, suggesting the potential of these molecules as a chemopreventive treatment or chemotherapeutic alternative against CRC [32]. However, the involvement of coffee polyphenols in the modulation of the Wnt pathway has not been explored in CRC models.

The present study evaluated the effect of two coffee extracts rich in polyphenols as well as the independent effect of CGA, the most abundant polyphenol in coffee. SW480 and HT-29 colorectal cancer cells were treated to analyze the effect of these using cellular and molecular approximations, such as cytotoxic potential, migration, invasion modulation, and Wnt pathway transcriptional activity.

In the first instance, MTT and SRB analyses were carried out to determine the cytotoxic effect of green coffee extract (GC), toasted coffee extract (TC), and CGA. The SRB method was used because the mechanism of action of the MTT method is related to the reduction capacity of mitochondrial enzymes, and there are reports of interference caused by some polyphenols due to their antioxidant activity [33]. On the other hand, the metabolic activity of MTT reduction can differ between cell lines and be related to the induction of injury and cell death. At the same time, the SRB method, based on binding to slightly acidic total proteins, is considered a complementary test for determining the cytotoxic potential [34,35]. In this way, we obtained a complete perspective of the effects of the evaluated treatments. Both methods show a similar trend of decreased cell viability which is more evident in SW480 cells. This trend is most marked in the treatments with CGA, followed by TC and GC (Table 2). However, the MTT (Figure 2) method showed minor IC50 values in the evaluated cell lines that could be related to the mechanism of action, perhaps mainly to metabolic pathways.

TC treatment showed a higher cytotoxic effect in this study than in GC (Figure 2). During the roasting process, the number of polyphenols such as CGA in coffee diminished, and flavonoids increased [30]. The quantification results demonstrated this effect of the roasting process, showing that CGA levels decreased in TC, and xanthine, theobromine, and caffeine increased (Table 1). These conditions, together with the generation of new bioactive molecules such as melanoidins during the roasting process [36,37], could be related to the biological effect of TC.

The IC50 values for CGA in the CRC cell lines found in this study were higher than in previous reports. However, the highest sensitivity continues to be found in SW480 cells, as in the report of Aires et al. where the exposure of HT-29 and SW480 cells to resveratrol for 72 h resulted in IC50 values of 70 μM and 40 μM, respectively [38]. On the other hand, quercetin showed similar behavior, with IC50 values of 75 μM for HT-29 cells and a 70% inhibition of viability with a dose of 80 μM in SW480 cells [39,40].

This effect was evident in the wound healing cell migration assay, where a more significant effect on migration inhibition was observed in SW480 cells with treatments at subtoxic doses (Figure 3A). The effect was comparable in all treatments, where we found a migration inhibition behavior up to 120 h of follow-up. In HT-29 cells (Figure 3B), the effect observed on migration inhibition was less pronounced than in SW480, and CGA had a higher inhibition. Reports on the migration inhibition activity of curcumin on these cell lines showed that HT-29 cells are less sensitive than SW480 cells [41]. Another related report showed the same effect on SW480 cells compared to DLD1 cells, where resveratrol decreased the migration capability, mainly in SW480 [42]. In cell invasion experiments, the behavior was similar, considering that we found the treatments to have a more powerful effect in SW480 cells, where the inhibition of cell invasion was dose-dependent, and the effect of treatments was notable and comparable to the control without chemoattractant (0% SFB) (Figure 4A). The experiments showed that the invasive capabilities of HT-29 cells are less significant than those of SW480, even with chemoattractant (10% SFB); in addition, the treatments with significant differences in decreasing the invasion were CGA and TC (Figure 4B).

After observing the biological effect of treatments on the evaluated in vitro models, differences were found at the cytotoxic level and in the modulation of cell migration and invasion capabilities. Therefore, we performed some experiments to determine whether these effects could be related to the modulation of the Wnt pathway. For this, we initially used the reporter assay system Top/Fop flash, which uses a plasmid that includes the TCF transcription factor sequence. The increase and translocation of β-catenin to the nucleus implies the formation of the β-catenin/TCF complex, inducing the expression of target genes of the Wnt pathway [43]. The results obtained (Figure 5) show that decreasing the transcriptional activation of the plasmid has an effect that is more pronounced with the CGA treatment compared with the specific inhibitor iCRT14 in both cell lines, except in the case of HT-29 cells where the effect of CGA was comparable with TC. Previous reports evaluated the levels of the basal intrinsic activation of the Wnt pathway using the Top/Fop flash system and showed a change up to 19-fold higher in Wnt pathway activation in SW480 vs. HT-29 cells [44]. Meanwhile, another report described a difference of 1.7 for HT-29 to 52.1 for SW480 [45]. This difference in the level of basal activity of the Wnt pathway between cell lines could explain the different effects of the Wnt inductor CHIR and the response to treatments where the induction of transcriptional activity was much higher in SW480 cells only. In addition, the effect on the decrease in transcriptional activity was evident and comparable with other studies that report the effect of some polyphenol extracts and compounds on CRC cells, i.e., the HS7 fraction of the TC. These include camphoratus extract, which showed an inhibition effect on the Wnt pathway on SW480, HCT116, and HT-29 cells, decreasing the transcriptional activity by between 50 and 60% [46]. Another report showed that silibinin significantly inhibits the activity of the pathway at 24 h of treatment with 100 μM in SW480 cells [47].

The regulation of the Wnt pathway was explored with a reporter assay. We performed qPCR experiments to determine the influence of the treatments on the expression of genes related to the pathway, such as CTNNB1 (which encodes for β-catenin), CDH1 (which encodes for E-cadherin), and the Wnt pathway target gene CCND1 (which encodes for cyclin-D1). The results showed that CDH1 expression levels increase in SW480 cells with TC at 1000 μg/mL and decrease with GC and CGA (Figure 6). In HT-29 cells, all treatments had significant differences. For CTNNB1 and CCND1, a significant decrease in the expression levels was observed, particularly in SW480 cells, while for CTNNB1 mRNA levels in HT-29 cells, some doses of treatments with TC and CGA had no significant differences in comparison with the control. A previous report showed the basal expression level of the *CTNNB1* gene on the CCD18-CO normal colon fibroblast and SW480, Caco-2, and HT-29 colorectal cancerous cells. *CTNNB1* was downregulated in HT-29 and Caco-2 cells, and the levels in SW480 and CCD18-CO were higher and similar. Previous studies showed the effect of natural compounds on regulating the expression of these genes related to the pathway. For example, a study reported the effect of two metabolites of ellagitannins, MPhA and MPhb in Caco-2 and CCD18-CO cells, where the treatments did not alter the expression level of *CTTNB1* in either cell line [48]. Another study reported the effect of lupeol on *CTNNB1* expression levels in SW480 and HCT116 after 24 h of treatment, where the expression was significantly downregulated only in HCT116 cells at the higher doses of 80 μM; however, the protein levels in SW480 also decreased. For cyclin D1 in SW480 cells, the gene expression was downregulated from doses of 40 μM and 20 μM in HCT116 cells [49]. In addition, the downregulation of *CTNNB1* and *CDH1* gene expression was significant at 24 h in HT-29 cells, using doses of 120 μM and 50 μM of phenethyl isothiocyanate and sulforaphane, the major isothiocyanates of broccoli [50].

On the other hand, a series of studies report the effect of some natural compounds on the expression of cyclin D1 in models of colorectal cancer, such as diospyros kaki thunb (DKC), which decreased protein and mRNA levels at concentrations of 50 μg/mL, mainly in SW480 cells, after 24 h [51]. In addition, safflower seed (*Carthamus tinctorius* L.) was recently shown to decrease the expression of cyclin D1 mRNA at concentrations of 100 μg/mL at 24 h, with LoVo and HT-29 cells being more sensitive to treatment compared to SW480 and HCT116 [52].

Finally, we performed immunostaining experiments to determine the location at the subcellular level of E-cadherin and β-catenin. For HT-29 cells (Figure 7), we did not observe changes in the subcellular location or co-localization of these proteins with the treatments, with β-catenin mainly in the membrane. On the other hand, in SW480 cells, it was possible to show that the treatments had the effect of reducing the amount of β-catenin, in addition to increasing the amount of β-catenin/E-cadherin co-localization, confirming a potential modulatory effect on the Wnt pathway. Very few studies on polyphenols have addressed the regulation of the Wnt pathway through fluorescence microscopy techniques. One of these few studies showed the effect of some synthetic derivatives of sibylline, where concentrations of 4 μM for 24 h increased the levels of E-cadherin and decreased β-catenin in HCT116 cells [53].

In summary, in the present work, we observed a biological effect of chlorogenic acid and polyphenol-rich coffee extracts on cell migration and invasion, possibly related to transcriptional regulation of the Wnt pathway, which was most pronounced in SW480 cells. The difference in the sensitivity of this cell line compared to HT-29 cells, which were less sensitive, could be related to aspects such as the genetic background of each cell line, which influences the activity levels of the Wnt pathway, and the different protein levels, which are related to transport and detoxification processes.

First, the genetic and epigenetic background of SW480 promotes a higher activation of the Wnt pathway than HT-29. The effect of the treatments on the modulation of the Wnt pathway would be more evident in the cell line with the highest activation of the pathway. The consensus molecular subtypes (CMS) of CRC establish four molecular subtypes based on gene expression profiles that have specific implications in the clinical context, independent of the stage of the disease [54]. This classification is influenced mainly by the tumoral microenvironment; however, in vitro models could show diverse CMS. Colorectal cell lines have been used to identify multiple and specific molecular aberrations for all CMS of CRC. In this context, a multi-omics study established that the HT-29 cell line in the CMS3 metabolic type is related to tubulovillous adenomas with serrated features and more prevalent KRAS mutations. The SW480 cell line was classified in the CMS4 mesenchymal class, a relatively aggressive phenotype related to serrated adenomas, high levels of TGF-β, and very high somatic copy number alteration (SCNA), and are known to be pro-inflammatory and pro-angiogenesis. In the clinical context, these classifications are determined at advanced stages [55]. Mutations of the APC protein indicate the difference in the Wnt pathway activity levels in both evaluated cell lines. APC is truncated at the carboxyl-terminal end at residue 1338 in SW480 and residue 1555 in HT-29 [56]. APC truncated mutations in SW480 compromise some crucial domains in APC, such as the β-catenin inhibitory domain (CID). This has high relevance for β-catenin targeting for ubiquitination and also plays a role in the interaction between the SAMP binding site in APC with axin for the degradation complex, conformation, and stabilization [57]. The reduction in APC truncated protein through RNA interference technology decreases the proliferation of six types of colorectal cancer cells and decreases tumor growth in vivo [57]. On the other hand, another study suggests mechanisms that explain how the truncated APC with CID loss promotes β-catenin deubiquitination by the reverse binding of β-TrCP and USP7 [58].

The implications of APC for the regulation of phosphorylation, ubiquitination, degradation, nuclear transport of β-catenin, and the affectation levels of the APC protein are reflected in the Wnt pathway activation, where these mutations imply inhibition of the degradation of β-catenin in SW480 but not in HT-29, DLD-1, and the wild type APC HCT116 cells.

Secondly, protein levels related to the transport of substances, mainly export, as well as cell detoxification processes between the cell lines evaluated, were reported, including high levels of *MDR1* (multidrug resistance gene) in SW480 and an undetectable protein level of WB in HT-29 cells [38]. The MDR1 coding by the *ABCB1* gene is a member of the superfamily of ATP-binding cassette (ABC) transporters, and the inhibition effect of polyphenols has been reported [59]. Moreover, the BCRP (breast cancer resistance protein) detected in HT-29 cells works as an efflux pump with broad substrate recognition [60] and facilitates the detoxification or expulsion of the treatments. This acts to reduce its effects, while a dynamic of decreased levels of MDR1 in SW480 cells due to polyphenol treatments favors its effect by increasing the intracellular amounts.

Furthermore, some studies have reported different levels of UDP-glucuronosyltransferases (UGTs) in colorectal cancer cells. These UGTs are involved in the glucuronidation process that facilitates the elimination of substances in cells, including phenolic compounds. Low levels of UGT proteins are associated with high drug-induced toxicity, and, in contrast, a high level of UGT is related to the loss of the bio-availability of treatments, premature glucuronidation, and lack of efficiency [61]. In this context, different isoforms of UGT1A are expressed in HT-29 cells and not in HCT116, influencing the intracellular accumulation of tanshinone IIA (TSA), a phytochemical from the Chinese medical herb Salvia miltiorrhiza bunge (danshen), and reducing its antitumoral effect in HT29 cells [62]. On the other hand, HT-29 cells are resistant to treatment with ganetespib, a specific inhibitor of HSP90, while SW480 and HCT116 cells are susceptible to this treatment. Another study evidenced a high correlation between IC50 levels and UGT1A expression, and this effect was reverted with UGT1A knockdown siRNA-mediated in HT29 cells [63].

In this study, we were able to detect the activation levels of the Wnt pathway in both cell lines by analyzing the control conditions, such as cell migration and invasion. The Wnt pathway has been reported to be related to tumor progression, promoting migration and invasion processes. The behavior of these phenomena is higher in untreated SW480 cells [64]. The invasion capability of multiple CRC cells was reported, and the difference in the number of invasive cells was 49.7 for HT-29 compared to 169.5 for SW480. In addition, the level of expression of proteins related to the invasion process, such as vimentin, *N*-cadherin, and ZEB1, was very high in SW480 and low in HT-29 cells [65]. Similarly, in the reporter assay for the evaluation of transcriptional activity, we observed that the specific inhibitor of GSK-3β, the Wnt inducer CHIR, was much more effective in SW480 cells. In contrast, in HT-29 cells, the activation level of the pathway was lower with the inductor treatment, suggesting a possible mechanism of phosphorylation-ubiquitination degradation different from HT-29. This process could be related to the previously-mentioned truncated APC protein, where APC mutations present in SW480 compromise the β-catenin inhibitory domain (CID) [58,66]. In addition, in the controls of the immunostaining tests, the difference in the quantity and subcellular location of β-catenin between both cell lines was evident. We observed a more significant presence of nuclear β-catenin exclusively in SW480 cells. Furthermore, our results showed that coffee polyphenol extracts and CGA reduce the fluorescence intensity of β-catenin in SW480 cells and downregulate the cyclin D1 expressions, suggesting a potential Wnt/b-catenin pathway modulation. This pathway has been established as an important therapeutic target in CRC, and natural and synthetic molecules have been established as modulators and drug candidates for prevention and treatment strategies [67].

5. Conclusions

The study of the effect of bioactive molecules on natural products has been of great interest, particularly regarding the polyphenols present in the diet and their effect on CRC and the Wnt pathway. In the present study, the effect of two coffee extracts rich in polyphenols (green coffee and toasted coffee) and CGA on the inhibition of cell viability and modulation of the migratory and invasive properties of SW480 and HT-29 CRC cells was demonstrated. This effect was also accompanied by the potential modulation activity of the Wnt/β-catenin pathway, evidenced through a decrease in the expression of related genes, *CTNNB1* (which encodes for β-catenin), CDH1 (which encodes for E-cadherin), and the Wnt pathway target gene *CCND1* (which encodes for cyclin-D1). In addition, decreased reporter activation for the TCF4 promoter and changes in β-catenin protein fluorescence levels were observed. For all treatments, SW480 cells had a higher sensitivity than HT-29 cells, and the activity of CGA treatments was more evident, followed by TC and GC. These differences in sensitivity between cell lines could be related to the level of mutations in the *APC* gene, which are in turn directly related to the regulation of β-catenin and Wnt pathway modulation. In addition, the differences in the expression profiles of proteins involved in the cellular detoxification of polyphenols contribute to the higher impact of treatment on SW480 cells. Our results provide an exciting starting point on the effect of polyphenols in coffee in the context of CRC and the Wnt pathway. More studies are necessary to determine the specific regulatory mechanism of these molecules on the pathway and the possible implications of the metabolism and the interaction with the microbiota in the in vivo context.

Supplementary Materials: The following supporting information can be downloaded at: https://www.mdpi.com/article/10.3390/nu14224880/s1, Figure S1: Cytotoxicity activity measured by MTT (A) and SRB (B) in the colorectal cancer cell lines SW480 and HT-29 at 24–48 h.

Author Contributions: Conceptualization, G.A.S.-G., J.P.-D. and D.U.; formal analysis and investigation, J.P.-D., D.U., H.V., G.A.S.-G., I.C.H.; writing—original draft preparation, H.V., G.A.S.-G., J.P.-D. and D.U.; writing—review and editing, I.C.H.; supervision, G.A.S.-G., J.P.-D., D.U., J.C.A.-O. and C.J.B.-C.; project administration, J.P.-D. All authors have read and agreed to the published version of the manuscript.

Funding: This research was funded by a Minciencias grant (project code: 115080763215 CT 811-2018) and was supported by the Instituto Tecnológico Metropolitano and Universidad de Antioquia.

Institutional Review Board Statement: Not applicable.

Informed Consent Statement: Not applicable.

Data Availability Statement: Not applicable.

Acknowledgments: The authors thank Natucafé for providing the coffee beans to obtain the extracts evaluated in this work.

Conflicts of Interest: The authors declare no conflict of interest.

References

1. Williamson, G. The role of polyphenols in modern nutrition. *Nutr. Bull.* **2017**, *42*, 226. [CrossRef] [PubMed]
2. Fraga, C.G.; Croft, K.D.; Kennedy, D.O.; Tomás-Barberán, F.A. The effects of polyphenols and other bioactives on human health. *Food Funct.* **2019**, *10*, 514–528. [CrossRef] [PubMed]

3. Cory, H.; Passarelli, S.; Szeto, J.; Tamez, M.; Mattei, J. The Role of Polyphenols in Human Health and Food Systems: A Mini-Review. *Front. Nutr.* **2018**, *5*, 87. [CrossRef] [PubMed]
4. Nuhu, A.A. Bioactive Micronutrients in Coffee: Recent Analytical Approaches for Characterization and Quantification. *ISRN Nutr.* **2014**, *2014*, 384230. [CrossRef]
5. Iczbiński, P.L.; Bukowska, B. Tea and coffee polyphenols and their biological properties based on the latest in vitro investigations. *Ind. Crops Prod.* **2022**, *175*, 114265. [CrossRef]
6. Burdan, F. *Coffee in Health and Disease Prevention*; Elsevier: Amsterdam, The Netherlands, 2015.
7. Poole, R.; Kennedy, O.J.; Roderick, P.; Fallowfield, J.A.; Hayes, P.C.; Parkes, J. Coffee consumption and health: Umbrella review of meta-analyses of multiple health outcomes. *BMJ* **2018**, *360*, k194. [CrossRef]
8. Sung, H.; Ferlay, J.; Siegel, R.L.; Laversanne, M.; Soerjomataram, I.; Jemal, A.; Bray, F. Global Cancer Statistics 2020: GLOBOCAN Estimates of Incidence and Mortality Worldwide for 36 Cancers in 185 Countries. *CA Cancer J. Clin.* **2021**, *71*, 209–249. [CrossRef]
9. Dekker, E.; Tanis, P.J.; Vleugels, J.L.A.; Kasi, P.M.; Wallace, M.B. Colorectal cancer. *Lancet* **2019**, *394*, 1467–1480. [CrossRef]
10. Nguyen, L.H.; Goel, A.; Chung, D.C. Pathways of Colorectal Carcinogenesis. *Gastroenterology* **2020**, *158*, 291–302. [CrossRef]
11. Tariq, K.; Ghias, K. Colorectal cancer carcinogenesis: A review of mechanisms. *Cancer Biol. Med.* **2016**, *13*, 120. [CrossRef]
12. Yu, F.; Yu, C.; Li, F.; Zuo, Y.; Wang, Y.; Yao, L.; Wu, C.; Wang, C.; Ye, L. Wnt/β-catenin signaling in cancers and targeted therapies. *Signal Transduct. Target. Ther.* **2021**, *6*, 307. [CrossRef] [PubMed]
13. Bian, J.; Dannappel, M.; Wan, C.; Firestein, R. Transcriptional Regulation of Wnt/β-Catenin Pathway in Colorectal Cancer. *Cells* **2020**, *9*, 2125. [CrossRef] [PubMed]
14. Villota, H.; Röthlisberger, S.; Pedroza-Díaz, J. Modulation of the Canonical Wnt Signaling Pathway by Dietary Polyphenols, an Opportunity for Colorectal Cancer Chemoprevention and Treatment. *Nutr. Cancer* **2021**, *74*, 384–404. [CrossRef]
15. Lu, H.; Tian, Z.; Cui, Y.; Liu, Z.; Ma, X. Chlorogenic acid: A comprehensive review of the dietary sources, processing effects, bioavailability, beneficial properties, mechanisms of action, and future directions. *Compr. Rev. Food Sci. Food Saf.* **2020**, *19*, 3130–3158. [CrossRef] [PubMed]
16. Liu, M.; Qin, J.; Cong, J.; Yang, Y. Chlorogenic Acids Inhibit Adipogenesis: Implications of Wnt/ β-Catenin Signaling Pathway. *Int. J. Endocrinol.* **2021**, *2021*, 2215274. [CrossRef] [PubMed]
17. Hu, X.; Wang, L.; He, Y.; Wei, M.; Yan, H.; Zhu, H. Chlorogenic Acid Promotes Osteogenic Differentiation of Human Dental Pulp Stem Cells Through Wnt Signaling. *Stem Cells Dev.* **2021**, *30*, 641–650. [CrossRef] [PubMed]
18. Xu, R.; Kang, Q.; Ren, J.; Li, Z.; Xu, X. Antitumor Molecular Mechanism of Chlorogenic Acid on Inducting Genes GSK-3 β and APC and Inhibiting Gene β-Catenin. *J. Anal. Methods Chem.* **2013**, *2013*, 951319. [CrossRef] [PubMed]
19. Villota, H.; Moreno-Ceballos, M.; Santa-González, G.; Uribe, D.; Castañeda, I.; Preciado, L.; Pedroza-Díaz, J. Biological Impact of Phenolic Compounds from Coffee on Colorectal Cancer. *Pharmaceuticals* **2021**, *14*, 761. [CrossRef]
20. Cardona, A.; Ariza-Jiménez, L.; Uribe, D.; Arroyave, J.C.; Galeano, J.; Cortés-Mancera, F.M. Bio-EdIP: An automatic approach for in vitro cell confluence images quantification. *Comput. Methods Programs Biomed.* **2017**, *145*, 23–33. [CrossRef]
21. Choudhari, A.S.; Mandave, P.C.; Deshpande, M.; Ranjekar, P.; Prakash, O. Phytochemicals in cancer treatment: From preclinical studies to clinical practice. *Front. Pharmacol.* **2020**, *10*, 1614. [CrossRef]
22. Rahman, M.A.; Hannan, M.A.; Dash, R.; Rahman, M.H.; Islam, R.; Uddin, M.J.; Sohag, A.A.; Rahman, M.H.; Rhim, H. Phytochemicals as a Complement to Cancer Chemotherapy: Pharmacological Modulation of the Autophagy-Apoptosis Pathway. *Front. Pharmacol.* **2021**, *12*, 718. [CrossRef] [PubMed]
23. Forni, C.; Facchiano, F.; Bartoli, M.; Pieretti, S.; Facchiano, A.; D'Arcangelo, D.; Norelli, S.; Valle, G.; Nisini, R.; Beninati, S.; et al. Beneficial Role of Phytochemicals on Oxidative Stress and Age-Related Diseases. *Biomed. Res. Int.* **2019**, *2019*, 8748253. [CrossRef] [PubMed]
24. Niedzwiecki, A.; Roomi, M.W.; Kalinovsky, T.; Rath, M. Anticancer Efficacy of Polyphenols and Their Combinations. *Nutrients* **2016**, *8*, 552. [CrossRef] [PubMed]
25. Zhou, Y.; Zheng, J.; Li, Y.; Xu, D.-P.; Li, S.; Chen, Y.-M.; Li, H.-B. Natural polyphenols for prevention and treatment of cancer. *Nutrients* **2016**, *8*, 515. [CrossRef]
26. Pashirzad, M.; Johnston, T.P.; Sahebkar, A. Therapeutic Effects of Polyphenols on the Treatment of Colorectal Cancer by Regulating Wnt β-Catenin Signaling Pathway. *J. Oncol.* **2021**, *2021*, 3619510. [CrossRef]
27. Fukushima, Y.; Tashiro, T.; Kumagai, A.; Ohyanagi, H.; Horiuchi, T.; Takizawa, K.; Sugihara, N.; Kishimoto, Y.; Taguchi, C.; Tani, M.; et al. Coffee and beverages are the major contributors to polyphenol consumption from food and beverages in Japanese middle-aged women. *J. Nutr. Sci.* **2014**, *3*, 10. [CrossRef]
28. González, S.; Fernández, M.; Cuervo, A.; Lasheras, C. Dietary intake of polyphenols and major food sources in an institutionalised elderly population. *J. Hum. Nutr. Diet.* **2014**, *27*, 176–183. [CrossRef]
29. Schmit, S.L.; Rennert, H.S.; Rennert, G.; Gruber, S.B. Coffee Consumption and the Risk of Colorectal Cancer. *Cancer Epidemiol. Biomark. Prev.* **2016**, *25*, 634–639. [CrossRef]
30. Król, K.; Gantner, M.; Tatarak, A.; Hallmann, E. The content of polyphenols in coffee beans as roasting, origin and storage effect. *Eur. Food Res. Technol.* **2020**, *246*, 33–39. [CrossRef]
31. Bułdak, R.J.; Hejmo, T.; Osowski, M.; Bułdak, Ł.; Kukla, M.; Polaniak, R.; Birkner, E. The Impact of Coffee and Its Selected Bioactive Compounds on the Development and Progression of Colorectal Cancer In Vivo and In Vitro. *Molecules* **2018**, *23*, 3309. [CrossRef]

32. Bruggisser, R.; von Daeniken, K.; Jundt, G.; Schaffner, W.; Tullberg-Reinert, H. Interference of plant extracts, phytoestrogens and antioxidants with the MTT tetrazolium assay. *Planta Med.* **2002**, *68*, 445–448. [CrossRef] [PubMed]
33. Riss, T.L.; Moravec, R.A.; Niles, A.L.; Duellman, S.; Benink, H.A.; Worzella, T.J.; Minor, L. Cell Viability Assays. In *Assay Guidance Manual*; Eli Lilly & Company and the National Center for Advancing Translational Sciences: Bethesda, MD, USA, 2016. Available online: https://europepmc.org/article/NBK/nbk144065 (accessed on 13 November 2022).
34. Keepers, Y.P.; Pizao, P.E.; Peters, G.J.; van Ark-Otte, J.; Winograd, B.; Pinedo, H.M. Comparison of the sulforhodamine B protein and tetrazolium (MTT) assays for in vitro chemosensitivity testing. *Eur. J. Cancer* **1991**, *27*, 897–900. [CrossRef]
35. Iriondo-DeHond, A.; Casas, A.R.; del Castillo, M.D. Interest of Coffee Melanoidins as Sustainable Healthier Food Ingredients. *Front. Nutr.* **2021**, *8*, 733. [CrossRef] [PubMed]
36. Langner, E.; Rzeski, W. Biological Properties of Melanoidins: A Review. *Int. J. Food Prop.* **2014**, *17*, 344–353. [CrossRef]
37. Aires, V.; Colin, D.J.; Doreau, A.; Di Pietro, A.; Heydel, J.-M.; Artur, Y.; Latruffe, N.; Delmas, D. P-Glycoprotein 1 Affects Chemoactivities of Resveratrol against Human Colorectal Cancer Cells. *Nutrients* **2019**, *11*, 2098. [CrossRef]
38. Shan, B.-E.; Wang, M.-X.; Li, R. Quercetin Inhibit Human SW480 Colon Cancer Growth in Association with Inhibition of Cyclin D_1 and Survivin Expression through Wnt/β-Catenin Signaling Pathway. *Cancer Investig.* **2009**, *27*, 604–612. [CrossRef]
39. Atashpour, S.; Fouladdel, S.; Movahhed, T.K.; Barzegar, E.; Ghahremani, M.H.; Ostad, S.N.; Azizi, E. Quercetin induces cell cycle arrest and apoptosis in CD133+ cancer stem cells of human colorectal HT29 cancer cell line and enhances anticancer effects of doxorubicin. *Iran. J. Basic Med. Sci.* **2015**, *18*, 635. [CrossRef]
40. Radhakrishnan, V.M.; Kojs, P.; Young, G.; Ramalingam, R.; Jagadish, B.; Mash, E.A.; Martinez, J.D.; Ghishan, F.K.; Kiela, P.R. pTyr421 Cortactin Is Overexpressed in Colon Cancer and Is Dephosphorylated by Curcumin: Involvement of Non-Receptor Type 1 Protein Tyrosine Phosphatase (PTPN1). *PLoS ONE* **2014**, *9*, e85796. [CrossRef]
41. Chung, S.S.; Dutta, P.; Austin, D.; Wang, P.; Awad, A.; Vadgama, J.V. Combination of resveratrol and 5-flurouracil enhanced anti-telomerase activity and apoptosis by inhibiting STAT3 and Akt signaling pathways in human colorectal cancer cells. *Oncotarget* **2018**, *9*, 32943–32957. [CrossRef]
42. Zhang, Y.; Wang, X. Targeting the Wnt/β-catenin signaling pathway in cancer. *J. Hematol. Oncol.* **2020**, *13*, 165. [CrossRef]
43. Yu, J.; Yang, K.; Zheng, J.; Zhao, W.; Sun, X. Synergistic tumor inhibition of colon cancer cells by nitazoxanide and obeticholic acid, a farnesoid X receptor ligand. *Cancer Gene Ther.* **2020**, *28*, 590–601. [CrossRef] [PubMed]
44. Kuroda, T.; Rabkin, S.D.; Martuza, R.L. Effective Treatment of Tumors with Strong β-Catenin/T-Cell Factor Activity by Transcriptionally Targeted Oncolytic Herpes Simplex Virus Vector. *Cancer Res.* **2006**, *66*, 10127–10135. [CrossRef] [PubMed]
45. Yeh, C.-T.; Yao, C.-J.; Yan, J.-L.; Chuang, S.-E.; Lee, L.-M.; Chen, C.-M.; Yeh, C.-F.; Li, C.-H.; Lai, G.-M. Apoptotic cell death and inhibition of Wnt/ β-catenin signaling pathway in human colon cancer cells by an active fraction (hs7) from taiwanofungus camphoratus. *Evid.-Based Complement. Altern. Med.* **2011**, *2011*, 750230. [CrossRef] [PubMed]
46. Kaur, M.; Velmurugan, B.; Tyagi, A.; Agarwal, C.; Singh, R.P.; Agarwal, R. Silibinin suppresses growth of human colorectal carcinoma SW480 cells in culture and xenograft through down-regulation of beta-catenin-dependent signaling. *Neoplasia* **2010**, *12*, 415–424. [CrossRef]
47. González-Sarrías, A.; Núñez-Sánchez, M.Á.; Tomé-Carneiro, J.; Tomás-Barberán, F.A.; García-Conesa, M.T.; Espín, J.C. Comprehensive characterization of the effects of ellagic acid and urolithins on colorectal cancer and key-associated molecular hallmarks: MicroRNA cell specific induction of CDKN1A (p21) as a common mechanism involved. *Mol. Nutr. Food Res.* **2016**, *60*, 701–716. [CrossRef]
48. Wang, Y.; Hong, D.; Qian, Y.; Tu, X.; Wang, K.; Yang, X.; Shao, S.; Kong, X.; Lou, Z.; Jin, L. Lupeol inhibits growth and migration in two human colorectal cancer cell lines by suppression of Wnt-β-catenin pathway. *Onco Targets Ther.* **2018**, *11*, 7987–7999. [CrossRef]
49. Pereira, L.P.; Silva, P.; Duarte, M.; Rodrigues, L.; Duarte, C.M.M.; Albuquerque, C.; Serra, A.T. Targeting Colorectal Cancer Proliferation, Stemness and Metastatic Potential Using Brassicaceae Extracts Enriched in Isothiocyanates: A 3D Cell Model-Based Study. *Nutrients* **2017**, *9*, 368. [CrossRef]
50. Bin Park, S.; Park, G.H.; Song, H.M.; Son, H.-J.; Um, Y.; Kim, H.-S.; Jeong, J.B. Anticancer activity of calyx of Diospyros kaki Thunb. through downregulation of cyclin D1 via inducing proteasomal degradation and transcriptional inhibition in human colorectal cancer cells. *BMC Complement. Altern. Med.* **2017**, *17*, 445. [CrossRef]
51. Park, G.H.; Hong, S.C.; Jeong, J.B. Anticancer Activity of the Safflower Seeds (*Carthamus tinctorius* L.) through Inducing Cyclin D1 Proteasomal Degradation in Human Colorectal Cancer Cells. *Korean J. Plant Resour.* **2016**, *29*, 297–304. [CrossRef]
52. Amawi, H.; Hussein, N.A.; Ashby, C.R.J.; Alnafisah, R.; Sanglard, L.M.; Manivannan, E.; Karthikeyan, C.; Trivedi, P.; Eisenmann, K.M.; Robey, R.W.; et al. Bax/Tubulin/Epithelial-mesenchymal pathways determine the efficacy of silybin analog HM015k in colorectal cancer cell growth and metastasis. *Front. Pharmacol.* **2018**, *9*, 520. [CrossRef]
53. Guinney, J.; Dienstmann, R.; Wang, X.; De Reyniès, A.; Schlicker, A.; Soneson, C.; Marisa, L.; Roepman, P.; Nyamundanda, G.; Angelino, P.; et al. The consensus molecular subtypes of colorectal cancer. *Nat. Med.* **2015**, *21*, 1350–1356. [CrossRef] [PubMed]
54. Berg, K.C.G.; Eide, P.W.; Eilertsen, I.A.; Johannessen, B.; Bruun, J.; Danielsen, S.A.; Bjørnslett, M.; Meza-Zepeda, L.A.; Eknæs, M.; Lind, G.E.; et al. Multi-omics of 34 colorectal cancer cell lines—A resource for biomedical studies. *Mol. Cancer* **2017**, *16*, 116. [CrossRef] [PubMed]
55. Yang, J.; Zhang, W.; Evans, P.M.; Chen, X.; He, X.; Liu, C. Adenomatous Polyposis Coli (APC) Differentially Regulates β-Catenin Phosphorylation and Ubiquitination in Colon Cancer Cells. *J. Biol. Chem.* **2006**, *281*, 17751–17757. [CrossRef] [PubMed]

56. Chandra, S.H.V.; Wacker, I.; Appelt, U.K.; Behrens, J.; Schneikert, J. A common role for various human truncated adenomatous polyposis coli isoforms in the control of beta-catenin activity and cell proliferation. *PLoS ONE* **2012**, *7*, e34479. [CrossRef] [PubMed]
57. Novellasdemunt, L.; Foglizzo, V.; Cuadrado, L.; Antas, P.; Kucharska, A.; Encheva, V.; Snijders, A.P.; Li, V.S. USP7 Is a Tumor-Specific WNT Activator for APC-Mutated Colorectal Cancer by Mediating β-Catenin Deubiquitination. *Cell Rep.* **2017**, *21*, 612–627. [CrossRef] [PubMed]
58. Nabekura, T.; Kawasaki, T.; Furuta, M.; Kaneko, T.; Uwai, Y. Effects of Natural Polyphenols on the Expression ofDrug Efflux Transporter P-Glycoprotein in Human IntestinalCells. *ACS Omega* **2018**, *3*, 1621. [CrossRef]
59. Kim, J.H.; Park, J.M.; Roh, Y.J.; Kim, I.W.; Hasan, T.; Choi, M.G. Enhanced efficacy of photodynamic therapy by inhibiting ABCG2 in colon cancers. *BMC Cancer* **2015**, *15*, 504. [CrossRef]
60. Wu, B.; Kulkarni, K.; Basu, S.; Zhang, S.; Hu, M. First-pass metabolism via UDP-glucuronosyltransferase: A barrier to oral bioavailability of phenolics. *J. Pharm. Sci.* **2011**, *100*, 3655–3681. [CrossRef]
61. Liu, M.; Wang, Q.; Liu, F.; Cheng, X.; Wu, X.; Wang, H.; Wu, M.; Ma, Y.; Wang, G.; Hao, H. UDP-Glucuronosyltransferase 1A Compromises Intracellular Accumulation and Anti-Cancer Effect of Tanshinone IIA in Human Colon Cancer Cells. *PLoS ONE* **2013**, *8*, e79172. [CrossRef]
62. Landmann, H.; Proia, D.A.; He, S.; Ogawa, L.S.; Kramer, F.; Beißbarth, T.; Grade, M.; Gaedcke, J.; Ghadimi, M.; Moll, U.; et al. UDP glucuronosyltransferase 1A expression levels determine the response of colorectal cancer cells to the heat shock protein 90 inhibitor ganetespib. *Cell Death Dis.* **2014**, *5*, e1411. [CrossRef]
63. Zhou, C.; Li, Y.; Wang, G.; Niu, W.; Zhang, J.; Wang, G.; Zhao, Q.; Fan, L. Enhanced SLP-2 promotes invasion and metastasis by regulating Wnt/β-catenin signal pathway in colorectal cancer and predicts poor prognosis. *Pathol. Res. Pract.* **2019**, *215*, 57–67. [CrossRef] [PubMed]
64. Feng, B.; Dong, T.T.; Wang, L.L.; Zhou, H.M.; Zhao, H.C.; Dong, F.; Zheng, M.H. Colorectal Cancer Migration and Invasion Initiated by microRNA-106a. *PLoS ONE* **2012**, *7*, e43452. [CrossRef]
65. Zhang, L.; Shay, J.W. Multiple Roles of APC and its Therapeutic Implications in Colorectal Cancer. *JNCI J. Natl. Cancer Inst.* **2017**, *109*, djw332. [CrossRef] [PubMed]
66. Sferrazza, G.; Corti, M.; Brusotti, G.; Pierimarchi, P.; Temporini, C.; Serafino, A.; Calleri, E. Nature-derived compounds modulating Wnt/β-catenin pathway: A preventive and therapeutic opportunity in neoplastic diseases. *Acta Pharm. Sin. B* **2020**, *10*, 1814. [CrossRef]
67. Yu, W.-K.; Xu, Z.-Y.; Yuan, L.; Mo, S.; Xu, B.; Cheng, X.-D.; Qin, J.-J. Targeting β-Catenin Signaling by Natural Products for Cancer Prevention and Therapy. *Front. Pharmacol.* **2020**, *11*, 984. [CrossRef]

nutrients

Article

Urinary Metabolomics Study on the Protective Role of Cocoa in Zucker Diabetic Rats via ^{1}H-NMR-Based Approach

Elisa Fernández-Millán [1,2,*], Sonia Ramos [3], David Álvarez-Cilleros [3], Sara Samino [2,4], Nuria Amigó [2,4,5], Xavier Correig [2,4,6], Mónica Chagoyen [7], Carmen Álvarez [1,2] and María Ángeles Martín [2,3,*]

1 Department of Biochemistry and Molecular Biology, Faculty of Pharmacy (UCM), 28040 Madrid, Spain
2 CIBER of Diabetes and Associated Metabolic Disease (CIBERDEM), ISCIII, 28029 Madrid, Spain
3 Institute of Food Science, Technology and Nutrition (ICTAN-CSIC), 28040 Madrid, Spain
4 Pere Virgili Institute for Health Research (IISPV), 43007 Tarragona, Spain
5 Biosfer Teslab, 43201 Reus, Spain
6 Department of Electronic Engineering, Rovira i Virgili University (URV), 43003 Tarragona, Spain
7 National Centre for Biotechnology (CNB-CSIC), 28049 Madrid, Spain
* Correspondence: elfernan@ucm.es (E.F.-M.); amartina@ictan.csic.es (M.Á.M.)

Abstract: Cocoa constitutes one of the richest sources of dietary flavonoids with demonstrated anti-diabetic potential. However, the metabolic impact of cocoa intake in a diabetic context remains unexplored. In this study, metabolomics tools have been used to investigate the potential metabolic changes induced by cocoa in type 2 diabetes (T2D). To this end, male Zucker diabetic fatty rats were fed on standard (ZDF) or 10% cocoa-rich diet (ZDF-C) from week 10 to 20 of life. Cocoa supplementation clearly decreased serum glucose levels, improved glucose metabolism and produced significant changes in the urine metabolome of ZDF animals. Fourteen differential urinary metabolites were identified, with eight of them significantly modified by cocoa. An analysis of pathways revealed that butanoate metabolism and the synthesis and degradation of branched-chain amino acids and ketone bodies are involved in the beneficial impact of cocoa on diabetes. Moreover, correlation analysis indicated major associations between some of these urine metabolites (mainly valine, leucine, and isoleucine) and body weight, glycemia, insulin sensitivity, and glycated hemoglobin levels. Overall, this untargeted metabolomics approach provides a clear metabolic fingerprint associated to chronic cocoa intake that can be used as a marker for the improvement of glucose homeostasis in a diabetic context.

Keywords: polyphenols; urine metabolites; type 2 diabetes; branched-chain aminoacids; untargeted metabolomics

Citation: Fernández-Millán, E.; Ramos, S.; Álvarez-Cilleros, D.; Samino, S.; Amigó, N.; Correig, X.; Chagoyen, M.; Álvarez, C.; Martín, M.Á. Urinary Metabolomics Study on the Protective Role of Cocoa in Zucker Diabetic Rats via ^{1}H-NMR-Based Approach. *Nutrients* **2022**, *14*, 4127. https://doi.org/10.3390/nu14194127

Academic Editor: Andrea Fabbri

Received: 15 September 2022
Accepted: 30 September 2022
Published: 4 October 2022

1. Introduction

Type 2 Diabetes (T2D) is the most prevalent metabolic disease in the world, with a large socio-sanitary impact due to its chronic macro- and micro-vascular complications [1]. T2D is characterized by the elevation of blood glucose resulting from defects in insulin secretion, insulin action, or both. [2]. There are several glucose-lowering drugs available for the treatment of diabetes; however, most of them have potential adverse effects and are not completely effective in preventing the progression of the disease [3]. Therefore, more research in this area is crucial to develop alternative therapeutic agents or strategies that might reduce the harmful effect of diabetes and its associated complications.

Natural products rich in phenolic compounds represent promising candidates for the treatment of diabetes, due to their good effectiveness in glucose metabolism, and their low cost and toxicity [4]. Particularly, flavonoids, the most abundant phenolic compounds in fruits and vegetables, have been described as multitarget agents with great therapeutic potential for diabetes management [5]. In this regard, cocoa constitutes one of the richest sources of dietary flavonoids, mainly flavanols which is gaining importance in research since it has been proven to reduce the pathogenesis of diabetes and its complications [6]. Notably, our previous studies have demonstrated that cocoa supplementation improves the glucose metabolism in diabetic animals [7] and has beneficial effects on associated diabetic complications, mainly arterial stiffness [8] and nephropathy [9]. Several mechanisms have been described as being involved in these positive outcomes [10], including a possible prebiotic effect of cocoa [11]. However, the precise underlying mechanisms of these antidiabetic actions are not fully understood. In this sense, knowing how cocoa modulates metabolic pathways significantly affected in diabetes could be essential to further clarify its final biological actions.

Over the last years, metabolomics approaches have been successfully applied to evaluate metabolite alterations related to metabolic disease, such as diabetes [12], as well as to investigate the metabolic response to a nutritional intervention by characterizing the disturbed metabolites between control and treated groups [13]. Untargeted metabolomic approaches are especially useful, as they offer information on the levels of endogenous small molecule metabolites, presented in a biological sample without a priori information [14]. This comprehensive analysis unravels metabolite profile modifications and detects alterations in biological pathways and biochemical processes, providing further insight into the molecular mechanisms involved in the nutritional intervention. Particularly, nuclear magnetic resonance (NMR) is the most common analytical technique used to reveal profiles of metabolites involved in the tricarboxylic acid (TCA) cycle, amino acid, and carbohydrates metabolism, all of them suggested as significant pathways disturbed in diabetic models [15]. Indeed, the NMR study of biofluids has been proven to be useful to investigate the anti-diabetic activity of several plants' bioactive compounds [15–18] as well as dietary flavonoids [19,20]. However, at present, there are no studies evaluating the metabolic impact of cocoa consumption in a diabetic setting, which could certainly help to better understand the mechanisms underlying its anti-diabetic activity.

Therefore, this study was aimed to identify potential metabolic changes that could be involved in the beneficial effect of cocoa on diabetes. To this end, we have performed an untargeted NMR-based metabolomics analysis along with a multivariate analysis in an in vivo model of T2D, the Zucker diabetic fatty (ZDF) rats. Our strategic approach has revealed significant differences in the urine metabolic profiles between control (ZDF) and cocoa supplemented (ZDF-C) diabetic animals. In addition, main differential metabolites and metabolic pathways affected by the cocoa intake in diabetic animals have been identified. Overall, these results provide new insights on the mechanism underlying the impact of cocoa on T2D, improving the functional understanding of its anti-diabetic effects.

2. Materials and Methods

2.1. Diets, Animals, and Experimental Design

Diets were prepared from an AIN-93G formulation (Panlab S.L., Barcelona, Spain). A cocoa rich-diet (10%) was produced by adding 100 g/kg of natural Forastero cocoa powder (a kind gift from Idilia Foods, Barcelona, Spain) to AIN-93G diet. The total polyphenolic content of the cocoa powder, as determined with the Folin-Ciocalteu method, was 2.4 g/100 g on a dry matter basis. The main cocoa flavonoids were determined by LC-MS, as previously described [8]. Monomeric epicatechin (382.0 mg/100 g) and catechin (115.2 mg/100 g) were the major flavanols in the extract, together with appreciable amounts of procyanidins B1 (35.2 mg/100 g) and B2 (132.2 mg/100 g). The resulting cocoa diet was isoenergetic and its composition is given in Supplementary Table S1.

Male ZDF rats and their Zucker lean controls (ZL) were purchased from Charles River Laboratories (L'arbresle, France) at 9 weeks of age. Animals were acclimated for one week under standard controlled conditions (21 °C ± 1 °C; 12 h day/night cycle). After that, the ZDF rats were randomly sorted into two groups (eight animals per group) that received the standard AIN-93G diet (ZDF), or the same control diet supplemented with 10% of cocoa (ZDF-C), for 10 weeks. The lean Zucker rats (ZL) (n = 6) received the standard AIN-93G diet. During the experiment, food and water were available *ad libitum*. Body weight and blood glucose in overnight fasted animals were followed weekly during the entire study and food intake was monitored three times per week. The animals were treated according to the European (2010/63/EU) and Spanish (RD 53/2013) legislation on Care and Use of Experimental Animals and the experiments were approved by the Ethics Committee from Comunidad de Madrid (PROEX 304/15).

2.2. Biochemical Determinations

At 20 weeks of age, the animals were fasted overnight, and blood samples were collected for biochemical analysis. Blood glucose was determined using an Accounted Glucose Analyser (LifeScan España, Madrid, Spain). Serum insulin and Hb1Ac were analyzed with ELISA kits (Rat Insulin, Mercodia, Uppsala, Sweden; HbA1c Kit Spinreact, BioAnalitica, Madrid, Spain). The fasting plasma concentrations of both glucose and insulin were used to calculate the indices of homeostatic model assessment of the insulin resistance (HOMA-IR) and insulin secretion (HOMA-B) according to the following formulas: HOMA-IR = fasting insulin (mU/mL) X fasting glucose (mM)/22.5 and HOMA-B = 20 X fasting insulin (mU/mL)/[fasting glucose (mM) − 3.5], respectively. Triacylglycerols (TG), HDL-Cholesterol and LDL-Cholesterol were determined in serum by kits (BioSystems, Madrid, Spain) as described elsewhere [8].

One week before the end of the study, a glucose tolerance test was performed. Briefly, after overnight fasting, 35% glucose solution (2 g/kg of body weight) (Sigma Chemical, Madrid, Spain) was administrated to the rats by intraperitoneal injection. Blood samples were obtained from the tail vein before the glucose load (t = 0) and at, 30, 60, 90, and 120 min, after the glucose administration and glucose levels were measured using an Accounted Glucose Analyzer (LifeScan España, Madrid, Spain). The integrated glucose response (area under the curve, AUC) over a period of 120 min after glucose overload was also calculated.

2.3. Preparation of Urine Samples and NMR Analysis

The 23 h urine samples were collected by means of metabolic cages at the end of the experiment, and were prepared and analyzed using standardized and optimized protocols, as previously described [21]. Briefly, the urine samples were thawed at room temperature before 540 µL of urine was mixed with 60 µL of buffer (KH_2PO_4, 1.5 M, pH 7.4 made up in ^2H_2O) containing 5.8 mM mM trimethylsilyl propionate (TSP) and 2 mM NaN_3. Following centrifugation (12,000× *g*, 4 °C, 5 min) to remove solids, 550 µL of sample were transferred into 5 mm SampleJet NMR tubes and immediately loaded onto a refrigerated SampleJet robot (Bruker Biospin, Germany) and maintained at 4 °C until NMR analysis

The NMR area associated with the concentration of each metabolite was obtained after the spectral analysis by using an in-house lineshape fitting based on an algorithm developed to deconvolute the pre-processed NMR spectra by using Lorentzian and Gaussian functions to minimize the fitting error. The NMR areas were transformed into concentration units by using specific conversion factors depending on the proton numbers of the molecular structure generating the signal and a TSP internal standard as previously described [22]. Finally, the metabolite concentrations were normalized by using PQN normalization [23] to avoid urine dilution effects. Resonances assignment and metabolite identification was verified using the Chenomx and the Human Metabolome Database (HMDB).

2.4. Statistical and Multivariate Analysis

The data from biochemical parameters were tested for homogeneity of variances by the test of Levene; for multiple comparisons, one-way ANOVA was followed by a Tukey test when variances were homogeneous, or by the Tamhane test when variances were not homogeneous. All the data were presented as mean ± standard deviation (SD). For the metabolomics study, the median and interquartile range concentration of each metabolite was compared by using non-parametric Wilcoxon rank-sum text. A GraphPad Prism version 8.00 (GraphPad software, Inc., La Jolla, CA, USA) was used. The level of significance was $p < 0.05$.

Multivariate analysis was used to analyze the metabolomics differences of urine of the different groups. Principal components analysis (PCA) was performed to describe the metabolic profiles among groups. This unsupervised method allows us to identify the maximum number of uncorrelated principal components that together explain the maximum amount of variance in the NMR metabolomic data set. Prior to running the PCA, we auto scaled the input data and then transformed to create a composite score for each principal component. The loading analysis of the principal components was used to identify the most relevant molecular components. Orthogonal partial least squares discriminant analysis (OPLS-DA) was applied as the supervised regression modelling for discriminating groups. The validity of the models against overfitting was estimated by the parameter R2Y, and the predictive ability was described by Q2 values. Furthermore, the differential metabolites of variable importance in the projection (VIP) from the OPLS-DA were utilized to screen the biomarkers. Relationship strength between differential metabolites and physiological parameters was assessed using the two tailed Pearson's correlation test. The correlation was considered significant when the absolute value of Pearson's correlation coefficient r was > 0.5. Finally, identified metabolites were submitted to the metabolite set enrichment analysis (MSEA) module in MetaboAnalyst 5.0 using HMDB identifiers. The Kyoto Encyclopedia of Genes and Genomes (KEGG) was used as input.

3. Results

3.1. Effect of Cocoa Supplementation on Physiological Parameters

The metabolic characteristics of the ZL and ZDF animals are shown in Table 1. At 20 weeks of life, the body weight of ZDF and ZDF-C rats were significantly increased as compared to non-diabetic ZL animals. Likewise, both ZDF groups showed hyperglycemia, hyperinsulinemia, increased HbA1c levels, and glucose intolerance (AUC), which confirmed their hyperphagic and diabetic state. However, all these parameters were significantly improved in diabetic animals that were supplemented with cocoa. Moreover, cocoa intake was also effective in reducing insulin resistance (HOMA-IR) and glucose intolerance (AUC) in ZDF rats, as well as increasing beta-cell function (HOMA-B). On the contrary, the serum levels of HDL-Cholesterol, LDL-Cholesterol, and TG, were significantly elevated in both ZDF groups in comparison to the ZL group, and there were no differences between those animals fed with standard (ZDF) or with cocoa diet (ZDF-C). These results indicate that a cocoa rich diet significantly improved glucose homeostasis but not lipid profile, in 20-weeks old ZDF rats.

Table 1. Biological parameters of Zucker lean rats (ZL), Zucker diabetic fatty rats fed with control diet for 10 weeks (ZDF), and Zucker diabetic fatty rats fed with cocoa diet for 10 weeks (ZDF-C). Data represent the means ± SD of 6–8 animals. Different letters denote statistically significant differences, $p < 0.05$.

	ZL	ZDF	ZDF-C
Body weight (g)	329 ± 41 ^{a}	444 ± 82 ^{b}	400 ± 12 ^{c}
Glucose levels (mg/dL)	86 ± 90 ^{a}	238 ± 71 ^{b}	118 ± 11 ^{c}
Insulin levels (ng/mL)	0.40 ± 0.02 ^{a}	4.36 ± 0.50 ^{b}	1.14 ± 0.22 ^{c}
HbA1c (%)	4.38 ± 0.19 ^{a}	10.40 ± 1.58 ^{b}	6.09 ± 0.79 ^{c}
AUC (mmol/L/min)	1803 ± 96 ^{a}	4044 ± 60 ^{b}	3049 ± 33 ^{c}
HOMA IR	2.60 ± 0.20 ^{a}	90.96 ± 14.26 ^{b}	12.22 ± 1.36 ^{c}
HOMA B	149.81 ± 22.21 ^{a}	169.52 ± 46.20 ^{a}	227.83 ± 14.32 ^{b}
TG (mmol/L)	0.39 ± 0.09 ^{a}	2.74 ± 0.24 ^{b}	2.87 ± 0.31 ^{b}
HDL-Cholesterol (mmol/L)	2.27 ± 0.22 ^{a}	2.85 ± 0.37 ^{b}	3.08 ± 0.32 ^{b}
LDL-Cholesterol (mmol/L)	0.92 ± 0.09 ^{a}	2.28 ± 0.33 ^{b}	2.82 ± 0.31 ^{c}

3.2. Metabolic Analysis of Urine Samples

An untargeted approach using ^{1}H-NMR techniques was used to explore the metabolomics changes in the urine of ZL, ZDF and ZDF-C animals. A total of 37 metabolites were identified and quantified in the urine of lean and diabetic rats (Supplementary Table S2). A representative

^{1}H-NMR spectrum of one of the urine samples with the metabolite assignment is shown in Supplementary Figure S1.

To obtain a preliminary understanding of the overall differences in metabolites between groups, and the degree of variability among the samples within the group, we achieved an exploratory PCA on the entire dataset. As shown in Figure 1A, the first two components (PC1 and PC2) were enough to identify specific group molecular characteristics, explaining 55% of the variance in the data (37% and 18%, respectively). PC1 showed a clear separation among non-diabetic ZL and the two diabetic groups, driven by higher concentration of creatinine, 2-oxoisocaproate and nicotinamide-N-oxide in the ZL group, together with a lower concentration of glucose and 4-hydroxy phenylacetate (Figure 1B). On the other side, PC2 was not enough to separate ZDF-C and ZDF groups, but it can be achieved using PC1 and PC2 together. Altogether, our results strongly indicated that cocoa supplementation modify the urinary metabolic profile of ZDF diabetic animals.

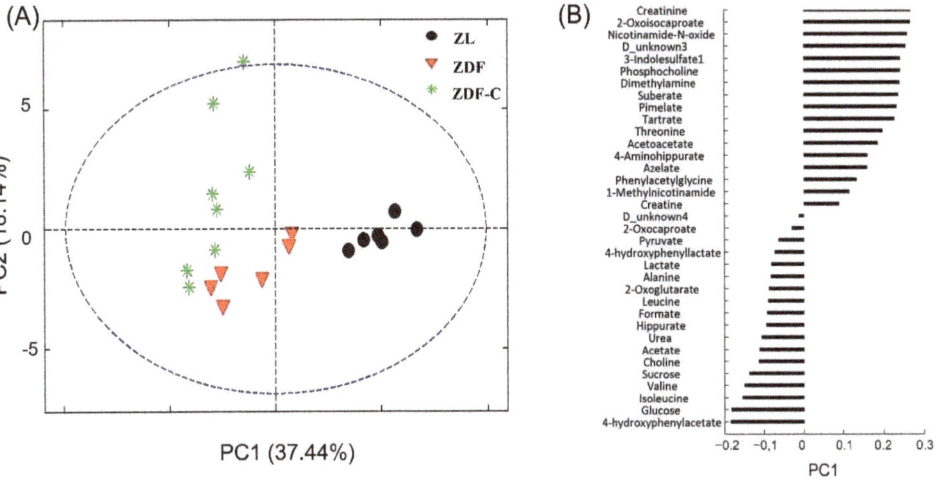

Figure 1. (**A**) Principal components analysis (PCA) score plots from urine samples from normal lean (ZL), diabetic (ZDF), and cocoa-supplemented diabetic (ZDF-C) rats, at 20 weeks of life. (**B**) Loading plot of PC1.

3.3. *Identification of Potential Biomarkers Associated with the Cocoa Ingestion in ZDF Rats*

Based on this initial observation, the OPLS-DA supervised method was employed to explain the differences caused by cocoa intake in diabetic Zucker rats. As shown in Figure 2A, a clear separation between the two groups (ZDF and ZDF-C) was observed as demonstrated by the OPLS-DA scores scatter plot. Moreover, R2Y and Q2 parameters were relatively high with good predictive ability and reliability (R2Y = 0.812 and Q2 = 0.729, respectively), which was able to manifest the changed trend in metabolites between groups. In order to discover the potential metabolites contributing to the group classification, the variable importance in the projection (VIP) values of the obtained multivariate analysis OPLS-DA model was used. Establishing a cut-off VIP values above 1.0 and the significance level $p < 0.05$, we found 14 endogenous metabolites that could be identified as potential biomarkers (Figure 2B). For the cocoa intake, the most common metabolites enrichment were valine, isoleucine, leucine, 2-oxoglutarate, alanine, and hippurate. On the other hand, 3-indolelsulfate, D_unknow4 metabolite, acetoacetate, urea, 4-hidroxyphenillactate, suberate, formate, and glucose, were enriched in the diabetic control rats.

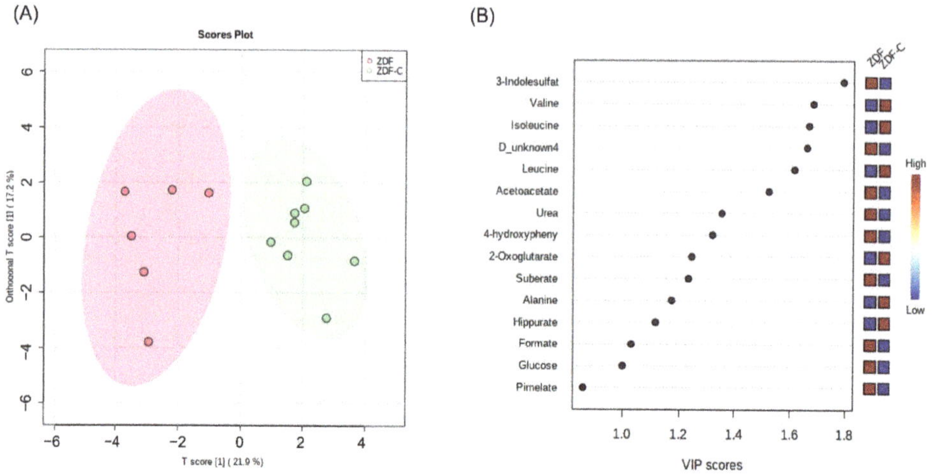

Figure 2. (**A**) Orthogonal partial least squares discriminant analysis (OPLS-DA) score plots from urine samples of diabetic (ZDF) and cocoa-supplemented diabetic (ZDF-C) rats at 20 weeks of life. (**B**) VIP values derived from OPLS-DA. The blue and red boxes on the right indicate whether the mean metabolite abundance is increased (red) or decreased (blue) in ZDF-C vs. ZDF.

For these 14 metabolites, we further assessed their differential urine concentration between groups, yielding eight metabolites (valine, leucine, isoleucine, acetoacetate, urea, hippurate, 3- indolelsulfate, and D_unknow4) that were significantly modified by the cocoa intake (Figure 3).

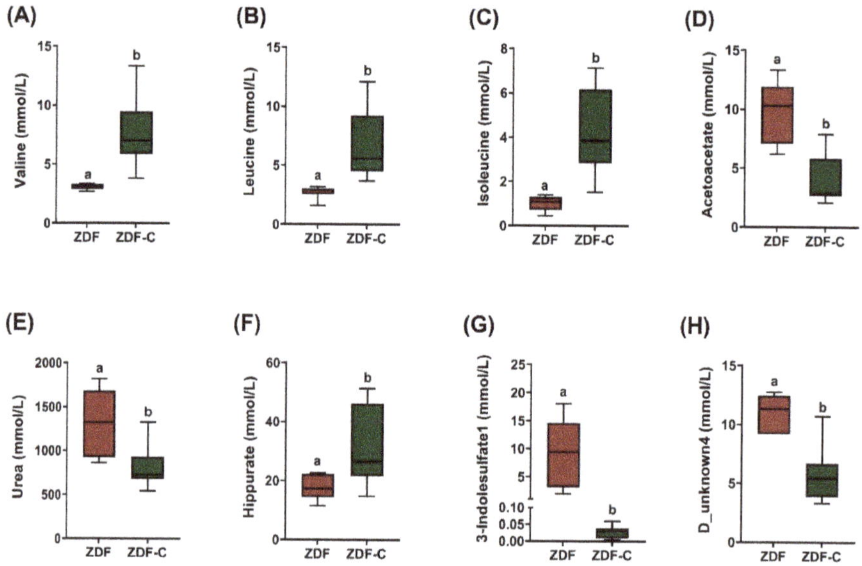

Figure 3. Altered urine metabolites between the diabetic (ZDF) and cocoa-supplemented diabetic (ZDF-C) rats at 20 weeks of life: (**A**) valine, (**B**) leucine, (**C**) isoleucine, (**D**) acetoacetate, (**E**) urea, (**F**) hippurate, (**G**) 3- indolelsulfate, and (**H**) D_unknow4 metabolite levels. Data represent the means ± SD of 6–8 animals. Different letters denote statistically significant differences, $p < 0.05$.

3.4. Correlation Analysis between Urine Metabolites and Biomarkers Associated with Cocoa Intake

Next, we investigate the relationship between the significant metabolites identified in the urine of diabetic animals and the clinical parameters related to glucose and lipid metabolisms. To this end, a Pearson's correlation analysis was conducted. The results were presented as a correlation heatmap, where green color indicates a positive correlation and red color indicates a negative one (Figure 4). A significant positive association was found between the levels of acetoacetate, 3-indolelsulfate, and the D_unknown4 metabolite and increased body weight, glycaemia, and HbA1c, as well as with insulin resistance (HOMA-IR) and glucose intolerance (AUC). However, valine and isoleucine were negatively correlated with all these parameters. Likewise, the levels of leucine showed significantly negative correlations with glucose, HbA1c, and AUC. The results also showed that urea exhibited a positive and significant association with the increased levels of glucose and HbA1c. Interestingly, hippurate levels, despite of all the other metabolites, were positively linked with an increase in the beta cell function (HOMA-B). On the other hand, increased LDL levels were positively associated with valine, isoleucine, and leucine.

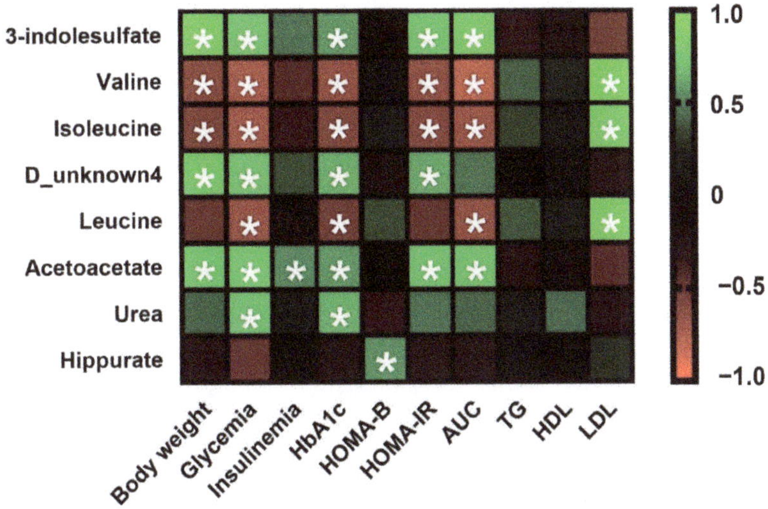

Figure 4. Heatmap of correlation between the main significantly altered urine metabolites and biochemical biomarkers related to diabetes. Pearson correlation values were used for the matrix. Green color indicates a positive correlation, whereas red color indicates a negative correlation; the intensity of the color represents the degree of association. * Represents adjusted $p < 0.05$.

3.5. Analysis of Biomarker Networks and Reconstruction of Metabolic Pathways

To further explore the metabolic pathways associated to the cocoa intake in diabetic animals, we performed a metabolite set enrichment analysis (MSEA) using the KEGG database as a framework (MetaboAnalyst 5.0). The enriched analysis showed the main metabolic routes modulated by the cocoa intake under diabetic conditions (Figure 5). Among them, butanoate metabolism, valine, leucine, and isoleucine degradation and biosynthesis, pantothenate and CoA biosynthesis, and synthesis and degradation of ketone bodies, were significantly enriched ($p < 0.05$) (Table 2). Therefore, these pathways represent the potential targeted pathways of cocoa intake in the diabetic condition. Based on the above results, a correlation network graph responding to the cocoa intake was constructed (Figure 6).

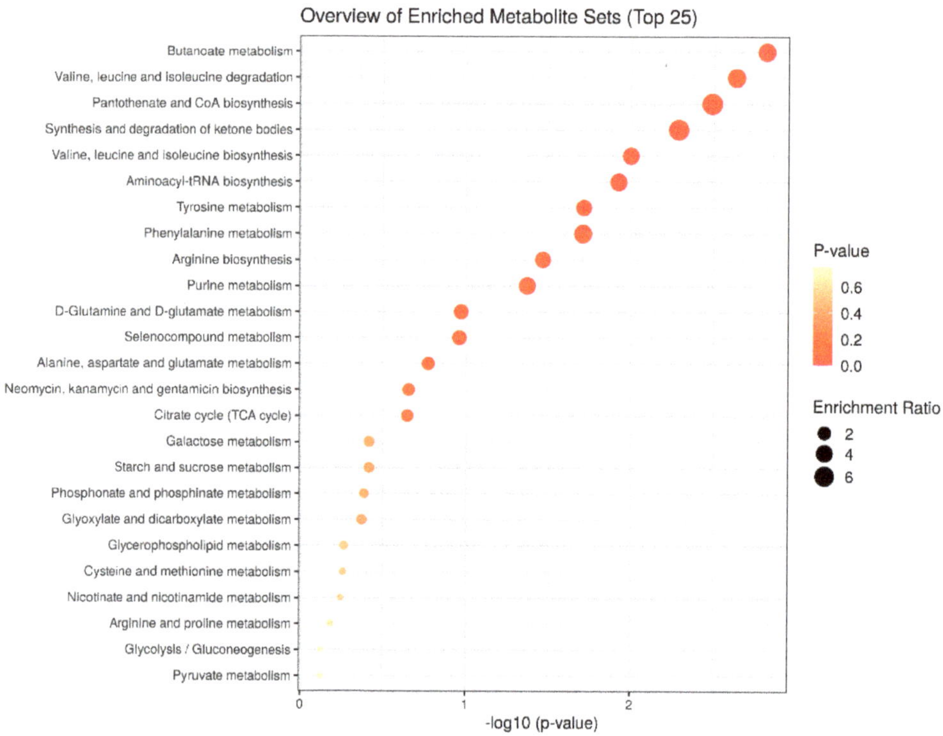

Figure 5. Metabolite set enrichment analysis of significantly altered urine metabolites according to the KEGG database.

Table 2. Top pathways enriched with metabolites having significantly altered abundance in Zucker diabetic fatty rats fed with cocoa diet (ZDF-C), as identified by the pathway analysis using MetaboAnalyst.

Top Pathways	Total Compounds	Hits	*p* Value	FDR *	Metabolites Identified
Butanoate metabolism	15	2	0.0015	0.026	Acetoacetate, 2-oxoglutarate
Valine, leucine and isoleucine degradation	40	5	0.0023	0.026	Isoleucine, Leucine, Valine, Acetoacetate, 4-Methyl-2-oxopentanoate
Pantothenate and CoA biosynthesis	19	1	0.0032	0.026	Valine
Synthesis and degradation of ketone bodies	5	1	0.0051	0.031	Acetoacetate
Valine, leucine and isoleucine biosynthesis	8	5	0.01	0.047	Isoleucine, Leucine, Valine, Threonine, Acetoacetate,
Aminoacyl-tRNA biosynthesis	48	5	0.019	0.047	Isoleucine, Leucine, Valine, Alanine, Lysine

* FDR is the *p* value adjusted using False Discovery Rate.

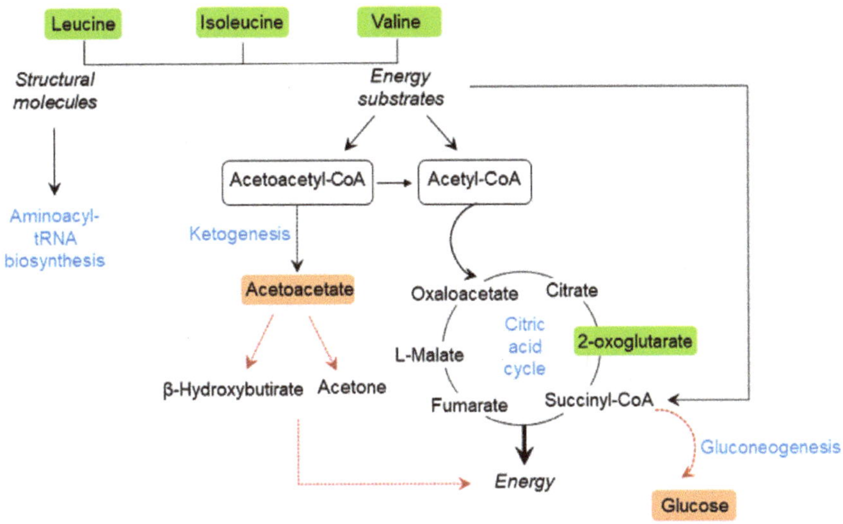

Figure 6. Schematic representation depicting the interrelationships of the disturbed metabolic pathways identified by ^{1}H NMR urine analysis. Green (significantly increased), and red (not significantly increased), as compared to ZDF ($p < 0.05$ is significant).

4. Discussion

The anti-diabetic potential of cocoa has previously been described both in pre-diabetic [7] and in diabetic ZDF animals [8]. Cocoa exerts these beneficial effects via multiple mechanisms, including antioxidant and anti-inflammatory effects, as well as by increasing both insulin secretion and insulin action [10]. Likewise, cocoa intake can modify the composition of the gut microbiota in ZDF rats, and these changes have been closely associated with the improved glucose homeostasis [11]. Herein, we show for the first time that chronic cocoa supplementation for 10 weeks significantly modified the urinary profile of diabetic fatty rats. These findings are in concordance with those of Massot-Cladera et al. [24] who exposed that the consumption of a cocoa-rich diet for 3 weeks resulted in a different urinary metabolic pattern in normal Wistar rats. In contrast, in a recent human intervention study, chronic consumption ofprocyanidin-rich cocoa (10 weeks) was found to have a marginal impact on serum metabolome in male endurance athletes [25]. Likewise, the biological response of a free-living population to the daily consumption of dark chocolate for two weeks only had a significant metabolic impact in subjects with essential high anxiety [26]. Together, these data seem to indicate that the effects of cocoa on the human metabolome are highly dependent on the stress situations of individuals, highlighting the beneficial implications of cocoa on the metabolic response to stress. According to this, our results further support the positive effects of the cocoa consumption on stress-associated metabolic abnormalities, like those that appear in animals with diabetes.

In the present study, we applied NMR-based metabolomics and multivariate analysis, including PCA and OPLS-DA, to study the metabolic profile of ZDF rats and discriminate the protective effect of a cocoa-rich diet. A total of 14 potential biomarkers associated with diabetes and obesity were identified, which were related to disturbed metabolic pathways involving valine, leucine and isoleucine biosynthesis, synthesis and degradation of ketone bodies, butanoate metabolism, pantothenate and CoA biosynthesis and aminoacyl-tRNA biosynthesis, among others. The chronic supplementation of diet with 10% of cocoa powder interfered in these metabolic pathways by restoring potential biomarkers back to normality or even improving them, which might be part of the protective mechanism of cocoa on the disrupted diabetic metabolism.

Overt diabetes is characterized by an increased conversion of proteins into glucose via hepatic gluconeogenesis, contributing to worsen the hyperglycemia. In agreement, many reports have shown a positive correlation between insulin resistance and elevated plasma levels of branched chain amino acids (BCAAs, such as valine, leucine, isoleucine) [27–29]. The mechanistic explanation for this correlation seems to include an enhanced turnover of BCAAs, rather than an increase in their biosynthesis [28]. Moreover, blood accumulation of BCAAs leads by itself to the development of further insulin resistance [27,30], whereas the decreased dietary consumption of BCAAs without calorie restriction enhances energy expenditure and improves insulin sensitivity in mice [31].

However, not many studies have examined the level of amino acids in the urine of diabetic patients. Salek et al. [32] showed in db/db diabetic mice a decrease in the excretion of several amino acids, including valine and leucine. Interestingly, an increase of the amino acid concentration in the urine has been reported in diabetic patients following rosiglitazone treatment [33], suggesting that this change could be associated to the pharmacological treatment that, by reducing insulin resistance, also decreases the demand for amino acids as substrates for gluconeogenesis. Likewise, urinary metabolomic data indicate that physical activity improves insulin sensitivity in diabetic patients and raises BCAA excretion [34]. In the present study, we have observed an inverse correlation between the urine concentration of BCAAs, mainly valine and isoleucine, with body weight, HOMA-IR and glucose tolerance test AUC. Therefore, the increased amino acid excretion observed in cocoa-fed diabetic rats is suggestive of a less active hepatic gluconeogenesis and, accordingly, improvedperipheral sensitivity to insulin [35].

Consistent with the common increased protein breakdown in diabetic patients [32,36], the high levels of urea, the final product of amino acid catabolism, were also present in ZDF urinary samples in comparison with non-diabetic rats, but cocoa administration entirely normalized this parameter. The mechanism involved in this phenomenon might include the amelioration of insulin resistance and the improvement of insulin production and secretion, as we have previously described [7].

In addition, leucine and isoleucine are also considered ketogenic amino acids since they are degraded entirely or in part into acetoacetyl-CoA and/or acetyl-CoA. In the liver, acetoacetyl-CoA is converted to acetoacetate and then to acetone and β-hydroxybutyrate. Their ability to form ketone bodies is particularly evident in inadequately controlled diabetes, in which the liver produces large amounts of ketone bodies from both fatty acids and the ketogenic amino acids. In contrast, by observing the reduction of acetoacetate in the urine of ZDF-C rats, we can speculate that the cocoa rich-diet is able to prevent the enhancement of ketone body synthesis associated to insulin resistant states. Moreover, the low levels of urinary acetoacetate in the present study are correlated with the low levels of T2D biomarkers, such as, body weight, glycemia, insulinemia, HbA1c, HOMA-IR values, or glucose intolerance.

Interestingly, metabolite enrichment analyses and KEGG pathway map also revealed significant differing metabolites belonging to the butanoate metabolism, including not only acetoacetate but also 2-oxoglutarate. Butanoate metabolism, also known as butyrate metabolism, outlines the metabolic destiny of short-chain fatty acids or short-chain alcohols produced by the intestinal bacterial fermentation of fiber. Many of these molecules are ultimately used in the production of ketone bodies, as the aforementioned acetoacetate, the synthesis of lipids, or as precursors of the tricarboxylic acid (TCA) cycle or glutamate synthesis. The supplementation of diet for 10-weeks with cocoa significantly increased urine 2-oxoglutarate levels in ZDF rats compared with the values observed in ZL animals ($p = 0.001$), but not in ZDF rats receiving the standard diet. Thus, the increased presence of 2-oxoglutarate, one of the most important TCA cycle intermediates, may be due to an enhancement in overall mitochondrial biogenesis induced by the cocoa, or increased TCA substrate availability related to the previously mentioned amelioration of ketogenesis and gluconeogenesis. The effects of cocoa and its constituents on energy metabolism have been extensively studied [24,26] with dissimilar results. Massot-Cladera et al. [24]

observed that the administration of a diet containing 10% cocoa to healthy Wistar rats decreased urinary excretion of energy intermediates, such as citrate, 2-oxoglutarate, and N-methylnicotinamide, which is opposite to our current data obtained in diabetic rats. These differential results highlight that the effects of cocoa intake on health are highly determined by the (patho)physiological state of individuals.

Moreover, diabetic rats receiving the cocoa-rich diet also excreted greater amounts of the glycine conjugate of benzoic acid, hippurate, compared with those receiving the standard diet. Although hippurate is a normal constituent of the endogenous urinary metabolite profile, enhanced excretion has been associated with gut microbial metabolism of polyphenol-rich components of the diet, such as vegetables, tea, coffee [37,38], and cocoa [24]. It has been proposed that, in addition to the availability of precursors in the diet, the absence or presence of urinary hippurate is also influenced by variations of the intestinal microbiota. In agreement, we and others have previously described a marked gut dysbiosis in ZDF rats [11,39], whereas chronic cocoa supplementation modified the gut microbiota to a healthier profile in diabetic rats [11]. Interestingly, decreased hippurate excretion was found in a ^{1}H-NMR-based metabonomic analysis of urine samples from obese individuals [40] or type 2 diabetic patients [32], as well as in Zucker (fa/fa) obese rats compared to Zucker (fa/-) lean controls [32]. On the contrary, weight loss has been linked to higher hippurate excretion in both humans [40] and animals [41]. Consistent with this, we have described that a cocoa-rich diet is able to reduce the body weight of ZDF rats without modifying the daily food intake, probably by variations in the activity of gut microbiota and basal energy expenditure [11]. The beneficial effects of increased hippurate production on glucose metabolism needs further studies, but it has been reported that its precursor, hippuric acid, potentiates glucose stimulated insulin secretion from beta cells [42], which is in agreement with the positive correlation observed herein between hippurate and HOMA-B values.

Conversely, cocoa-fed diabetic rats excreted lower amounts of other metabolites arising from gut microbial-host metabolism, such as 3-indolesulfate, a derived product of tryptophan [43]. The study of Dou et al. [44] showed that the uremic solute 3-indolesulfate was able to cause oxidative stress in endothelial cells by increasing nicotinamide adenine dinucleotide phosphate (NADPH)-oxidase activity and ROS production. Obesity and diabetes are considered chronic inflammatory diseases closely related with oxidative stress [45], and cocoa consumption has long been demonstrated to have antioxidant effects and to improve the metabolic profile in diabetes [6,8]. Mechanistically, we previously reported that cocoa intake prevented ROS production through downregulating the protein expression of NADPH oxidase in ZDF rat aorta, which may be beneficial in decreasing the risk of the cardiovascular disease associated to diabetes [8]. Thus, our present results, in agreement with those of other authors [24], indicate that the reduced urinary concentration of 3-indolesulfate observed in ZDF-C rats reflects the antioxidant capacity of cocoa.

5. Conclusions

This study shows, for the first time, that a controlled and realistic supplementation of diet with cocoa powder induced changes in the urinary metabolome of diabetic animals (Figure 6). The use of a holistic approach, such as untargeted metabolomics, allowed us to disclose how cocoa can modulate the endogenous metabolism, thus providing further insights into the positive effects that chronic cocoa intake has in a diabetic context.

Specifically, the novelty of this study is that the treatment of diabetic animals with cocoa markedly increased urinary BCAAs levels, while reducing acetoacetate values, possibly indicating a decrease in gluconeogenesis as well as in ketogenesis and, therefore, an improvement of insulin sensitivity and glycemic control (Figure 6). It is conceivable that improved insulin action would also lead to the stimulation of protein synthesis and the further reduction of circulating BCAAs. Since BCAA catabolic intermediates feed into other metabolic pathways, such as the TCA cycle, when in excess, BCAAs may disturb the mitochondrial function. Consequently, it is reasonable to infer that cocoa's beneficial effects could be due in part to its impact on energy metabolism.

Finally, it is worth noting that while the ZDF rats have an obesity phenotype, not all changes observed in this model may be generalizable to the entire human forms of T2D. Additional studies are needed to determine which of these pathways and changes observed in this model, in addition to an elevation of BCAA excretion, are also observed in humans.

Supplementary Materials: The following supporting information can be downloaded at: https://www.mdpi.com/article/10.3390/nu14194127/s1, Table S1: composition of the experimental control and cocoa-rich diets. Table S2: metabolites identified in the urine of lean and diabetic rats. Figure S1: representative 600-MHz ^{1}H-NMR spectrum of a urine sample including metabolite assignment.

Author Contributions: Conceptualization, M.Á.M. and E.F.-M.; data curation, S.S., N.A. and X.C.; formal analysis, S.S., N.A., D.Á.-C. and M.C.; funding acquisition, M.Á.M., E.F.-M., S.R. and C.Á.; investigation, M.Á.M., D.Á.-C. and S.R.; methodology, M.Á.M., D.Á.-C. and S.R.; project administration, M.Á.M., E.F.-M., S.R. and C.Á.; writing original draft M.Á.M. and E.F.-M.; writing—review and editing M.Á.M., E.F.-M. and S.R. All authors have read and agreed to the published version of the manuscript.

Funding: This work was supported by the grant RTI2018-095059-B-I00 and PID2020-116134RB-I00 and funded by MCIN/AEI/10.13039/501100011033/ and by "ERDF A way of making Europe".

Institutional Review Board Statement: All the experiments were conducted according to the European Union (2010/63/EU) and Spanish (RD 53/2013) legislation and with the approval of the Animal Care and Use Committee of the Community of Madrid (PROEX 304/15).

Informed Consent Statement: Not applicable.

Data Availability Statement: Data are available upon request to the authors.

Conflicts of Interest: The authors declare no conflict of interest.

References

1. International Diabetes Federation. *IDF Diabetes Atlas*, 10th ed.; International Diabetes Federation: Brussels, Belgium, 2021; Available online: https://www.diabetesatlas.org (accessed on 21 March 2020).
2. American Diabetes Association; Classification and Diagnosis of Diabetes: Standards of Medical Care in Diabetes. *Diabetes Care* **2021**, *44*, S15–S33.
3. Palanisamy, S.; Yien, E.L.H.; Shi, L.W.; Si, L.Y.; Qi, S.H.; Ling, L.S.C.; Lun, T.W.; Chen, Y.N. Systematic review of efficacy and safety of newer antidiabetic drugs approved from 2013 to 2017 in controlling HbA1c in diabetes patients. *Pharmacy* **2018**, *6*, 57. [CrossRef] [PubMed]
4. Márquez Campos, E.; Jakobs, L.; Simon, M.-C. Antidiabetic Effects of Flavan-3-ols and Their Microbial Metabolites. *Nutrients* **2020**, *12*, 1592. [CrossRef]
5. Dinda, B.; Dinda, M.; Roy, A.; Dinda, S. Dietary plant flavonoids in prevention of obesity and diabetes. *Adv. Protein Chem. Struct. Biol.* **2020**, *120*, 159–235.
6. Ramos, S.; Martín, M.A.; Goya, L. Effects of Cocoa Antioxidants in Type 2 Diabetes Mellitus. *Antioxidants* **2017**, *6*, 84. [CrossRef] [PubMed]
7. Fernández-Millán, E.; Cordero-Herrera, I.; Ramos, S.; Escrivá, F.; Alvarez, C.; Goya, L.; Martín, M.A. Cocoa-rich diet attenuates beta cell mass loss and function in young Zucker diabetic fatty rats by preventing oxidative stress and beta cell apoptosis. *Mol. Nutr. Food Res.* **2015**, *59*, 820–824. [CrossRef] [PubMed]
8. Álvarez-Cilleros, D.; López-Oliva, M.E.; Morales-Cano, D.; Barreira, B.; Pérez-Vizcaíno, F.; Goya, L.; Ramos, S.; Martín, M.A. Dietary cocoa prevents aortic remodeling and vascular oxidative stress in diabetic rats. *Mol. Nutr. Food Res.* **2019**, *30*, e1900044. [CrossRef] [PubMed]
9. Álvarez-Cilleros, D.; López-Oliva, E.; Goya, L.; Martín, M.A.; Ramos, S. Cocoa intake attenuates renal injury in Zucker Diabetic fatty rats by improving glucose homeostasis. *Food Chem. Toxicol.* **2019**, *127*, 101–109. [CrossRef]
10. Martín, M.A.; Goya, L.; Ramos, S. Antidiabetic actions of cocoa flavanols. *Mol. Nutr. Food Res.* **2016**, *60*, 1756–1769. [CrossRef]
11. Álvarez-Cilleros, D.; Ramos, S.; López-Oliva, M.E.; Escrivá, F.; Álvarez, C.; Fernández-Millán, E.; Martín, M.A. Cocoa diet modulates gut microbiota composition and improves intestinal health in Zucker diabetic rats. *Food Res. Int.* **2020**, *132*, 109058. [CrossRef]
12. Arneth, B.; Arneth, R.; Shams, M. Metabolomics of Type 1 and Type 2 Diabetes. *Int. J. Mol. Sci.* **2019**, *20*, 2467. [CrossRef] [PubMed]
13. Zhang, A.; Sun, H.; Wang, P.; Han, Y.; Wang, X. Recent and potential developments of biofluid analyses in metabolomics. *J. Proteome* **2012**, *75*, 1079–1088. [CrossRef] [PubMed]
14. Klassen, A.; Faccío, A.T.; Canuto, G.A.; da Cruz, P.L.; Ribeiro, H.C.; Tavares, M.F.; Sussulini, A. Metabolomics: Definitions and significance in systems biology. *Adv. Exp. Med. Biol.* **2017**, *965*, 3–17. [PubMed]

15. Hasanpour, M.; Iranshahy, M.; Iranshahi, M. The application of metabolomics in investigating anti-diabetic activity of medicinal plants. *Biomed. Pharmacother.* **2020**, *128*, 110263. [CrossRef] [PubMed]
16. Sajak, A.A.B.; Mediani, A.; Dom, N.S.M.; Machap, C.; Hamid, M.; Ismail, A.; Khatib, A.; Abas, F. Effect of Ipomoea aquatica ethanolic extract in streptozotocin (STZ) induced diabetic rats via ^{1}H NMR-based metabolomics approach. *Phytomedicine* **2017**, *36*, 201–209. [CrossRef]
17. Azam, A.A.; Pariyani, R.; Ismail, I.S.; Ismail, A.; Khatib, A.; Abas, F.; Shaari, K. Urinary metabolomics study on the protective role of Orthosiphon stamineus in Streptozotocin induced diabetes mellitus in rats via ^{1}H NMR spectroscopy. *BMC Complement. Altern. Med.* **2017**, *17*, 278. [CrossRef]
18. Mediani, A.; Abas, F.; Maulidiani, M.; Khatib, A.; Tan, C.P.; Ismail, I.S.; Shaari, K.; Ismail, A.; Lajis, N. Metabolic and biochemical changes in streptozotocin induced obese-diabetic rats treated with Phyllanthus niruri extract. *J. Pharm. Biomed. Anal.* **2016**, *128*, 302–312. [CrossRef]
19. Chen, K.; Wei, X.; Zhang, J.; Pariyani, R.; Jokioja, J.; Kortesniemi, M.; Linderborg, K.M.; Heinonen, J.; Sainio, T.; Zhang, Y.; et al. Effects of Anthocyanin Extracts from Bilberry (*Vaccinium myrtillus* L.) and Purple Potato (*Solanum tuberosum* L. Var. 'Synkeä Sakari') on the Plasma Metabolomic Profile of Zucker Diabetic Fatty Rats. *J. Agric. Food Chem.* **2020**, *68*, 9436–9450. [CrossRef]
20. Singh, A.K.; Raj, V.; Keshari, A.K.; Rai, A.; Kumar, P.; Rawat, A.; Maity, B.; Kumar, D.; Prakash, A.; De, A.; et al. Isolated mangiferin and naringenin exert antidiabetic effect via PPARγ/GLUT4 dual agonistic action with strong metabolic regulation. *Chem. Biol. Interact.* **2018**, *280*, 33–44. [CrossRef]
21. Dona, A.C.; Jiménez, B.; Schäfer, H.; Humpfer, E.; Spraul, M.; Lewis, M.R.; Pearce, J.T.M.; Holmes, E.; Lindon, J.C.; Nicholson, J.K. Precision high-throughput proton NMR spectroscopy of human urine, serum, and plasma for large-scale metabolic phenotyping. *Anal. Chem.* **2014**, *86*, 9887–9894. [CrossRef]
22. Serkova, N.; Fuller, T.F.; Klawitter, J.; Freise, C.E.; Niemann, C.U. ^{1}H-NMR–based metabolic signatures of mild and severe ischemia/reperfusion injury in rat kidney transplants. *Kidney Int.* **2005**, *67*, 1142–1151. [CrossRef] [PubMed]
23. Dieterle, F.; Ross, A.; Schlotterbeck, G.; Senn, H. Probabilistic quotient normalization as robust method to account for dilution of complex biological mixtures. Application in ^{1}H NMR metabonomics. *Anal Chem.* **2006**, *78*, 4281–4290. [CrossRef] [PubMed]
24. Massot-Cladera, M.; Mayneris-Perxachs, J.; Costabile, A.; Swann, J.R.; Franch, À.; Pérez-Cano, F.J.; Castell, M. Association between urinary metabolic profile and the intestinal effects of cocoa in rats. *Br. J. Nutr.* **2017**, *117*, 623–634. [CrossRef]
25. Tabone, M.; García-Merino, J.A.; Bressa, C.; Guzman, N.E.R.; Rocha, K.H.; Chu Van, E.; Castelli, F.A.; Fenaille, F.; Larrosa, M. Chronic Consumption of Cocoa Rich in Procyanidins Has a Marginal Impact on Gut Microbiota and on Serum and Fecal Metabolomes in Male Endurance Athletes. *J. Agric. Food Chem.* **2022**, *70*, 1878–1889. [CrossRef] [PubMed]
26. Martin, F.P.; Rezzi, S.; Peré-Trepat, E.; Kamlage, B.; Collino, S.; Leibold, E.; Kastler, J.; Rein, D.; Fay, L.B.; Kochhar, S. Metabolic effects of dark chocolate consumption on energy, gut microbiota, and stress-related metabolism in free-living subjects. *J. Proteome Res.* **2009**, *8*, 5568–5579. [CrossRef] [PubMed]
27. Jang, C.; Oh, S.F.; Wada, S.; Rowe, G.C.; Liu, L.; Chan, M.C.; Rhee, J.; Hoshino, A.; Kim, B.; Ibrahim, A.; et al. A branched-chain amino acid metabolite drives vascular fatty acid transport and causes insulin resistance. *Nat. Med.* **2016**, *22*, 421–426. [CrossRef]
28. Menni, C.; Fauman, E.; Erte, I.; Perry, J.R.; Kastenmuller, G.; Shin, S.Y.; Petersen, A.K.; Hyde, C.; Psatha, M.; Ward, K.J.; et al. Biomarkers for type 2 diabetes and impaired fasting glucose using a nontargeted metabolomics approach. *Diabetes* **2013**, *62*, 4270–4276. [CrossRef]
29. Wang, T.J.; Larson, M.G.; Vasan, R.S.; Cheng, S.; Rhee, E.P.; McCabe, E.; Lewis, G.D.; Fox, C.S.; Jacques, P.F.; Fernandez, C.; et al. Metabolite profiles and the risk of developing diabetes. *Nat. Med.* **2011**, *17*, 448–453. [CrossRef]
30. Bloomgarden, Z.J. Diabetes and branched-chain amino acids: What is the link? *Diabetes* **2018**, *10*, 350–352. [CrossRef]
31. Cummings, N.E.; Williams, E.M.; Kasza, I.; Konon, E.N.; Schaid, M.D.; Schmidt, B.A.; Poudel, C.; Sherman, D.S.; Yu, D.; Arriola Apelo, S.I.; et al. Restoration of metabolic health by decreased consumption of branched-chain amino acids. *J. Physiol.* **2018**, *596*, 623–645. [CrossRef]
32. Salek, R.M.; Maguire, M.L.; Bentley, E.; Rubtsov, D.V.; Hough, T.; Cheeseman, M.; Nunez, D.; Sweatman, B.C.; Haselden, J.N.; Cox, R.D.; et al. A metabolomic comparison of urinary changes in type 2 diabetes in mouse, rat, and human. *Physiol. Genom.* **2007**, *29*, 99–108. [CrossRef] [PubMed]
33. Van Doorn, M.; Kemme, M.; Ouwens, M.; van Hoogdalem, E.J.; Jones, R.; Romijn, H.; de Kam, M.; Schoemaker, R.; Burggraaf, K.; Cohen, A. Evaluation of proinflammatory cytokines and inflammation markers as biomarkers for the action of thiazolidinediones in Type 2 diabetes mellitus patients and healthy volunteers. *Br. J. Clin. Pharmacol.* **2006**, *62*, 391–402. [CrossRef] [PubMed]
34. Benetti, E.; Liberto, E.; Bressanello, D.; Bordano, V.; Rosa, A.C.; Miglio, G.; Haxhi, J.; Pugliese, G.; Balducci, S.; Cordero, C. Sedentariness and Urinary Metabolite Profile in Type 2 Diabetic Patients, a Cross-Sectional Study. *Metabolites* **2020**, *10*, 205. [CrossRef] [PubMed]
35. Cordero-Herrera, I.; Martín, M.A.; Escrivá, F.; Álvarez, C.; Goya, L.; Ramos, S. Cocoa-rich diet ameliorates hepatic insulin resistance by modulating insulin signaling and glucose homeostasis in Zucker diabetic fatty rats. *J. Nutr. Biochem.* **2015**, *26*, 704–712. [CrossRef] [PubMed]
36. Wohl, P.; Krušinová, E.; Klementová, M.; Wohl, P.; Kratochvílová, S.; Pelikánová, T. Urinary urea nitrogen excretion during the hyperinsulinemic euglycemic clamp in type 1 diabetic patients and healthy subjects. *Physiol. Res.* **2008**, *57*, 247–252. [CrossRef]
37. Clarke, E.D.; Collins, C.E.; Rollo, M.E.; Kroon, P.A.; Philo, M.; Haslam, R.L. The relationship between urinary polyphenol metabolites and dietary polyphenol intakes in young adults. *Br. J. Nutr.* **2022**, *127*, 589–598. [CrossRef]

38. Clifford, M.N.; Copeland, E.L.; Bloxsidge, J.P.; Mitchell, L.A. Hippuric acid as a major excretion product associated with black tea consumption. *Xenobiotica* **2000**, *30*, 317–326. [CrossRef] [PubMed]
39. Cabello-Olmo, M.; Oneca, M.; Torre, P.; Sainz, N.; Moreno-Aliaga, M.J.; Guruceaga, E.; Díaz, J.V.; Encio, I.J.; Barajas, M.; Araña, M. A Fermented Food Product Containing Lactic Acid Bacteria Protects ZDF Rats from the Development of Type 2 Diabetes. *Nutrients* **2019**, *11*, 2530. [CrossRef]
40. Calvani, R.; Miccheli, A.; Capuani, G.; Tomassini Miccheli, A.; Puccetti, C.; Delfini, M.; Iaconelli, A.; Nanni, G.; Mingrone, G. Gut microbiome-derived metabolites characterize a peculiar obese urinary metabotype. *Int. J. Obes.* **2010**, *34*, 1095–1098. [CrossRef]
41. Wang, Y.; Lawler, D.; Larson, B.; Ramadan, Z.; Kochhar, S.; Holmes, E.; Nicholson, J.K. Metabonomic investigations of aging and caloric restriction in a life-long dog study. *J. Proteome Res.* **2007**, *6*, 1846–1854. [CrossRef]
42. Bitner, B.F.; Ray, J.D.; Kener, K.B.; Herring, J.A.; Tueller, J.A.; Johnson, D.K.; Tellez Freitas, C.M.; Fausnacht, D.W.; Allen, M.E.; Thomson, A.H.; et al. Common gut microbial metabolites of dietary flavonoids exert potent protective activities in β-cells and skeletal muscle cells. *J. Nutr. Biochem.* **2018**, *62*, 95–107. [CrossRef] [PubMed]
43. Meyer, T.W.; Hostetter, T.H. Uremic solutes from colon microbes. *Kidney Int.* **2012**, *81*, 949–954. [CrossRef] [PubMed]
44. Dou, L.; Jourde-Chiche, N.; Faure, V.; Cerini, C.; Berland, Y.; Dignat-George, F.; Brunet, P. The uremic solute indoxyl sulfate induces oxidative stress in endothelial cells. *J. Thromb. Haemost.* **2007**, *5*, 1302–1308. [CrossRef] [PubMed]
45. Rani, V.; Deep, G.; Singh, R.K.; Palle, K.; Yadav, U.C. Oxidative stress and metabolic disorders: Pathogenesis and therapeutic strategies. *Life Sci.* **2016**, *148*, 183–193. [CrossRef]

nutrients

MDPI

Article

Effect of Olive Pomace Oil on Cardiovascular Health and Associated Pathologies

Susana González-Rámila, Beatriz Sarriá, Miguel Ángel Seguido, Joaquín García-Cordero, Laura Bravo-Clemente and Raquel Mateos *

Department of Metabolism and Nutrition, Institute of Food Science, Technology and Nutrition (ICTAN-CSIC), Spanish National Research Council (CSIC), José Antonio Nováis 10, 28040 Madrid, Spain
* Correspondence: raquel.mateos@ictan.csic.es

Abstract: Background: olive pomace oil (OPO) is a nutritionally relevant fat due to its high oleic acid content (C18:1) and the presence of a wide range of minor bioactive components. Although numerous in vitro and preclinical studies have been developed to study some of its characteristic components, the health effect of prolonged OPO consumption is unknown. Methods: a randomised, blinded, cross-over, controlled clinical trial was carried out in 31 normocholesterolemic and 37 hypercholesterolemic subjects. Participants consumed 45 g/day of OPO or sunflower oil (SO) for 4 weeks, each preceded by a 3-week run-in/wash-out phase with corn oil (CO). Results: regular consumption of OPO and SO had no statistically significant effect on any of the markers related to lipid profile, blood pressure, and endothelial function in both groups, except for eNOS levels, which were close to statistical significance due to the effect of oil (OPO and SO) ($p = 0.083$). A decrease in visceral fat ($p = 0.028$) in both groups was observed after OPO intake, accompanied by an increment of leptin ($p = 0.017$) in the hypercholesterolemic group. Conclusion: reducing visceral fat after prolonged OPO intake might contribute to improve cardiometabolic status, with a potentially positive effect on the vascular tone. Further clinical trials are needed to confirm the present results.

Keywords: cardiovascular health; clinical trial; endothelial function; olive pomace oil; sunflower oil; visceral fat

Citation: González-Rámila, S.; Sarriá, B.; Seguido, M.Á.; García-Cordero, J.; Bravo-Clemente, L.; Mateos, R. Effect of Olive Pomace Oil on Cardiovascular Health and Associated Pathologies. *Nutrients* **2022**, *14*, 3927. https://doi.org/10.3390/nu14193927

Academic Editors: Egeria Scoditti and Alice J. Owen

Received: 8 August 2022
Accepted: 19 September 2022
Published: 22 September 2022

1. Introduction

Cardiovascular disease is the prevalent cause of morbidity and mortality worldwide, affecting millions of individuals every year. One of the most relevant environmental factors contributing to the development of this pathology is the diet [1]. In this sense, the Mediterranean Diet (MD) is considered one of the best models of healthy eating due to its beneficial effects on chronic non-communicable diseases, such as cardiovascular disorders, diabetes, obesity, and inflammation [2]. The main source of fat in this dietary pattern is olive oil [2]. This oil is obtained exclusively from olives and, depending on the technological process to which it is subjected, different commercial categories are obtained: extra virgin olive oil (EVOO), virgin olive oil (VOO), olive oil (OO), mild and intense), and olive pomace oil (OPO) [3]. The health benefits attributed to olive oil consumption, particularly reducing the risk of cardiovascular diseases, have been related to EVOO and VOO consumption due to its preferably monounsaturated fat and its high content of phenolic compounds [4,5]. However, in countries such as Spain, where the olive oil consumption is deeply rooted, the main competitors in the olive market are seed oils (high oleic sunflower oil (HOSO), sunflower oil (SO), etc.) [6]. These oils are mainly used for frying, and their low economic costs are the main reason for their elevated consumption [7,8]. This situation represents an opportunity for OPO because its composition gives the ideal characteristics for frying and, in addition, its market value is quite competitive [9,10].

OPO is obtained from alperujo, a solid by-product consisting of pieces of olive skin, pulp, pits, and stones [3,11]. This oil is characterised by its high oleic acid content, and its specific composition in minor components, mainly triterpenic acids and dialcohols, squalene, tocopherols, sterols, aliphatic fatty alcohols, and phenolic compounds [3,9]. The refining process carried out on this oil causes the loss of some of the triterpenic acids and phenolic compounds contained in alperujo [3]. However, concentrations of the minor components present in the refined pomace oil are found in amounts that are likely to induce beneficial effects on health, particularly on cardiovascular health [3].

In fact, numerous in vitro and preclinical studies have evaluated one or more components of OPO's minor fraction with promising results [12–15]. In addition, the long-term effect of OPO consumption at nutritional doses has been evaluated in two clinical trials developed by our research group. In the first clinical trial, OPO was compared to HOSO [16], whereas in the second, OPO was compared to SO, a polyunsaturated fat. This present study is focused on the second study. Recently, the comparison of the two clinical trials has been published [17].

Thus, the aim of this study was to determine the possible beneficial role of OPO on biomarkers of cardiovascular health and associated pathologies (hypertension, inflammation, diabetes, and obesity) in healthy volunteers and subjects at cardiovascular risk (hypercholesterolemic volunteers), in order to comparatively test the dietary treatment in both groups. The effect of OPO was compared to SO (control fat), given the high consumption of this seed oil. Corn oil (CO), a seed oil other than SO, was used as run in/wash out oil.

2. Materials and Methods

2.1. Study Design

The study was a randomized, blind, crossover, and controlled clinical trial with a duration of 14 weeks. After a 3-week run-in stage, in which all volunteers consumed CO to wash out the effect of oil consumed in their habitual diet, half of the participants were randomly assigned to the OPO and the other half to the SO oil, both in the healthy and in the at-risk groups. After the first 4-week intervention, a 3-week wash-out period with CO followed, and subsequently, participants consumed the other oil (OPO or SO) during the same period of 4 weeks (Figure 1). Randomization of participants to OPO or SO (control) interventions was performed using Microsoft® Excel 2016 software in a 1:1 ratio. Assignment of codes to participants, randomization and allocation to each oil were carried out by different members of the research team. To blind participants, oils were presented in identical plastic bottles with different taps for each type of oil (OPO, SO, and CO).

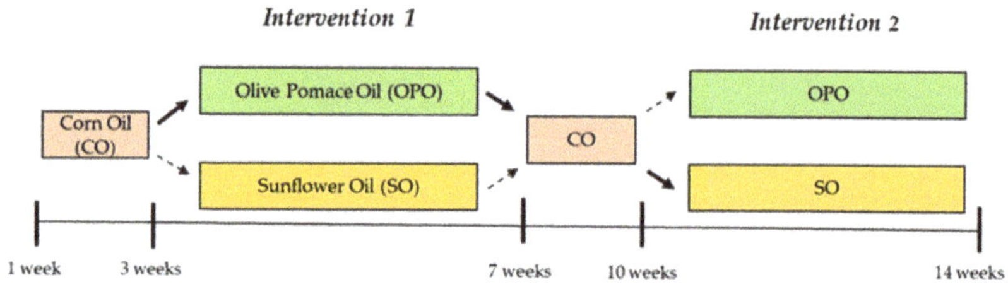

Figure 1. Design of the intervention study with olive pomace oil (OPO) and sunflower oil (SO). Corn oil (CO) was used during run-in and wash-out.

2.2. Participants and Setting

Of the 109 subjects interviewed, the volunteers who met the inclusion criteria (age between 18–55 years, body mass index (BMI) between 18–25 kg/m^2; specific inclusion criteria for healthy and at-risk volunteers were based on serum lipid levels as stated below), were 43 women and 29 men, aged 19 to 59 years old, with a BMI between 19.0 and 32.0 kg/m^2. The study groups were formed according to the serum total and LDL-cholesterol (LDL-C) concentrations determined in a pre-screen consented blood analysis. Healthy participants were assigned to the normocholesterolemic group when total cholesterol (TC) and LDL-

C levels were below 200 mg/dL and 135 mg/dL, respectively. At-risk volunteers were assigned to the hypercholesterolemic group when TC was between 200–300 mg/dL and LDL-C between 135–175 mg/dL (Figure 2). The exclusion criteria established were: suffering from acute or chronic pathologies, except hypercholesterolemia for the risk group, having digestive disorders/pathologies (gastric ulcer, Crohn's disease, inflammatory bowel syndrome, etc.), smoking, pregnant women, vegetarians, on antibiotic treatment three months before starting the study, or taking medication, hormones, or vitamins or dietary supplements.

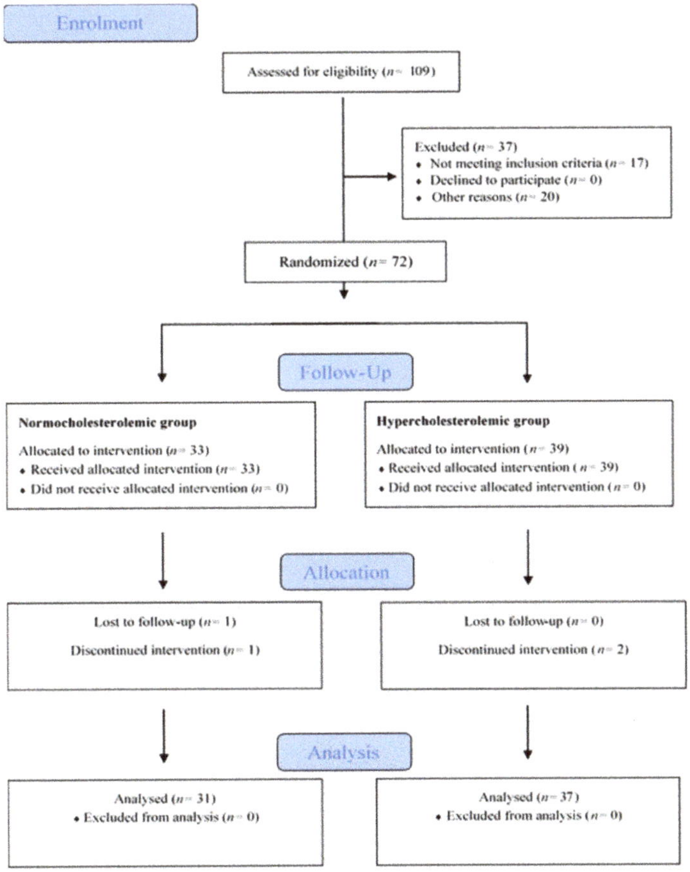

Figure 2. Study flow diagram (Consolidated Standards of Reporting Trials, CONSORT 2010).

The recruitment was conducted between June and September 2019 mainly through our database of participants in previous studies, social networks and by placing flyers at the Institute of Food Science, Technology and Nutrition (ICTAN) and at the Complutense University of Madrid campus (UCM). Volunteers interested in the study received additional information by e-mail and telephone. Finally, those who met the inclusion criteria were invited to a meeting to explain the study in detail, and answer questions raised by the attendees. The study was approved by the Clinical Research Ethics Committee of the Hospital Universitario Puerta de Hierro, at Majadahonda in Madrid (Spain), and by the Bioethics Committee of the Consejo Superior de Investigaciones Científicas (CSIC). It also followed the guidelines laid down in the Declaration of Helsinki for experiments in humans.

Before starting the study, we ensured that we obtained signed informed consent from all participants. The study was registered in Clinical Trials (NCT04998695).

2.3. Intervention

The intervention study was conducted during October-December 2019 at the Human Nutrition Unit (HNU) of the ICTAN. All volunteers consumed 45 g/day of oil according to the intervention phase (OPO, SO, and CO) to cover 20% of the daily energy intake of monounsaturated fats (equivalent to 44–67 g/day for 2000–3000 Kcal/d, respectively) following the Spanish Society of Community Nutrition (SENC) recommendations. To this end, one litre of oil per week was provided for family consumption to avoid using any other culinary oil. Participants were required to maintain their dietary and lifestyle habits unchanged, except for some foods rich in mono- and polyunsaturated fat (olives, sunflower seeds, nuts, avocado, margarine, butter, and mayonnaise, except if prepared with the study oils), which were restricted during the study.

2.4. Chemical Characterization of the Study Oils

The oils used in the study were analysed according to the following standardized methods:
ISO 12228-2:2014 method for the determination of sterols.
Regulation (EEC) N°. 2568/91 Annex V for determining triterpenic alcohols.
Regulation (EEC) N°. 2568/91 Annex XIX for determining aliphatic alcohols.
Regulation (EEC) N°. 2568/91 Annex X for determining fatty acid composition.
ISO 9936:2016 for determining tocopherols and tocotrienols.

Triterpenic acids were analysed following the method of Pérez-Camino & Cert [18], and squalene was determined by gas chromatography [19]. Phenols were analysed by high-performance liquid chromatography with on-line diode array detection (HPLC-DAD) according to the procedure developed by Mateos et al. [20].

2.5. Dietary Assessment and Compliance

As indicated in the intervention section, participants were asked to maintain their dietary habits and lifestyle unchanged during the study. To assess if they did, at each visit, participants completed a 24-h questionnaire for the day prior to the visit, and a detailed record of food intake for 72 h, in which they recorded the ingredients, units, or quantity of food, as well as the time and place where they ate. A Manual of Nutrition and Dietetics [21] was provided to the participants to facilitate the interpretation of home food measurements and usual portions. Compliance with food restrictions and the correct intake of oil was controlled by weekly calling and/or emailing participants.

Macronutrients (proteins, carbohydrates, and lipids), dietary fibre and vitamin E intake were calculated using the program DIAL (Department of Nutrition and Bromatology, Faculty of Pharmacy, Complutense University of Madrid, Madrid, Spain), as well as to estimate the total caloric intake.

2.6. Blood Sample Collection

Fasting blood samples were collected after an overnight fast using BD Vaccuette® tubes (Greiner Bio-One GmbH, Kirchdorf an der Krems, Austria) with EDTA or without anticoagulant to separate plasma and serum, respectively. Subsequently, samples were centrifuged, aliquoted, and stored at −80 °C until analysis.

2.7. Primary Outcomes and Other Outcomes Measures

Primary outcomes were markers related to cardiovascular health: lipid profile, blood pressure, and endothelial function. Secondary outcomes were biomarkers associated with diabetes, obesity, and inflammation.

2.7.1. Biochemical Analysis

Biochemical analysis of metabolic markers in serum was performed following procedures of reference as established by the Spanish Society of Clinical Biochemistry and Molecular Pathology (Sociedad Española de Bioquímica Clínica y Patología Molecular (SEQC). TC, triglycerides (TG), high-density lipoprotein (HDL) cholesterol, apolipoprotein A1 (Apo A1), and apolipoprotein B (Apo B) were determined using a Roche Cobas Integra 400 plus analyser (Roche Diagnostics, Mannheim, Germany). In addition, the Friedewald formula was used to estimate LDL-C and VLDL (very low-density lipoprotein) content, and Apo B/Apo A1, LDL/HDL, and TC/HDL ratios were calculated. Alanine (ALAT) and aspartate (ASAT) aminotransferases were analysed spectrophotometrically following standard procedures.

2.7.2. Blood Pressure

Systolic blood pressure (SBP) and diastolic blood pressure (DBP) were measured with an OMRON® M2 HEM-7121-E sphygmomanometer (OMRON HEALTHCARE Co., Ltd., Kyoto, Japan). Participants rested in a sitting position for 15 min before measurement, which was taken in triplicate in the non-prevailing arm, waiting 5 min between each.

2.7.3. Endothelial Function

Circulating levels of E-selectin, P-selectin, and the enzyme endothelial nitric oxide synthase (eNOS) were determined in plasma by ELISA following the protocols of Cloud-Clone Corp. (Katy, TX, USA). Spectrophotometric reading was performed using a Bio-Tek® Synergy™ HT Multi-Detection microplate reader controlled by BioTek®Gen5 software version 2.01.14 (BioTek Instruments, Winooski, VT, USA). Intercellular (ICAM-1) and vascular (VCAM-1) adhesion molecules were determined in serum samples with Bio-Plex® Pro Human Cytokine ICAM-1 and VCAM-1 kits (Bio-Rad, Hercules, CA, USA) using a MAGPIX™ Multiplex fluorescence reader operating with the Bio-Plex Pro Wash Station and the Bio-Plex Manager™ MP software for data processing (Luminex Corporation, Austin, TX, USA).

2.7.4. Diabetes and Obesity Biomarkers Analysis

Glucose, insulin, and glycosylated haemoglobin (HbA1c) were determined in serum samples following recommendations of the Spanish Society of Clinical Biochemistry and Molecular Pathology (SEQC). From the glucose and insulin data, insulin resistance (HOMA-IR) and pancreatic beta-cell function (HOMA-β) were calculated according to the mathematical model known as HOMA (Homeostasis Model Assessment) proposed by Matthews et al. [22]: HOMA-IR = [Glucose (mg/dL) × Insulin (mU/L)]/405; HOMA-β = [(360 × Insulin (mU/L)/(Glucose (mg/dL) − 63)]. In addition, insulin sensitivity was determined using the QUICKI ("Quantitative Insulin Sensitivity Check Index"), based on a logarithmic model calculated from fasting glucose and insulin concentrations using the following equation QUICKI = 1/[log Insulin (mU/L) + log Glucose (mg/dL)].

The hormones insulin, glucagon, incretin gastric inhibitory polypeptide (GIP), as well as glucagon-like peptide type 1 (GLP-1), C-peptide, and ghrelin; along with the adipokines leptin, resistin, plasminogen activator inhibitor-1 (PAI-1), visfatin, adiponectin, and adipsin were analysed in serum samples using the Bio-Plex® Pro™ Human Diabetes Panel 10-Plex kit (Bio-Rad, Hercules, CA, USA) in the MAGPIX™ Multiplex fluorescence reader and Bio-Plex Manager™ MP software (Luminex Corporation, Austin, TX, USA).

2.7.5. Anthropometry and Body Composition

Height and body perimeters (waist, abdomen, and brachial) were measured in triplicate using a wall-mounted height measuring rod (Soehnle Professional, GmBH, Backnang, Germany) and a tape measure (Fisaude ADE, Madrid, Spain), respectively. Weight, visceral, and body fat percentages were estimated by single-frequency tetrapolar electrical bioimpedance using a Tanita® BC 601 segmental body composition analyser with a digital scale included (Tanita Europe BV, Amsterdam, The Netherlands).

2.7.6. Inflammatory Biomarker Analysis

Pro-inflammatory (IL-1β, IL-2, IL-6, IL-7, IL-8, IL-12(p70), IL-17) and anti-inflammatory (IL-4, IL-10 and IL-13) interleukins (IL), interferon-gamma (IFN-γ), and tumour necrosis factor alpha (TNF-α), as well as monocyte chemoattractant protein 1 (MCP-1), macrophage inflammatory protein 1 beta (MIP-1β), granulocyte colony-stimulating factor (G-CSF), and granulocyte monocyte colony-stimulating factor (GM-CSF) were analysed in serum samples using Bio-Plex Pro™ Human Cytokine Grp I Panel 17-Plex kits in the MAGPIX™ Multiplex fluorescence reader and Bio-Plex Manager™ MP software (Luminex Corporation, Austin, TX, USA). In addition, high-sensitivity C-reactive protein (CRP) was also determined in serum samples with an automated ultra-sensitive turbidimetric method (AU2700 Chemistry Analyzer, Olympus Corp., Japan).

2.7.7. Antioxidant Capacity and Oxidation Biomarkers

Antioxidant activity was measured in serum samples by the ABTS radical cation [23] and the oxygen radical absorbance capacity (ORAC) methods [24], and the reducing capacity was determined by the ferric reducing/antioxidant power (FRAP) assay [25]. Trolox was used as standard, and results were expressed as μM of Trolox equivalent (TE). Low-density lipoprotein oxidation (LDLox) levels were determined in serum samples by ELISA assay according to the protocols of the Cloud-Clone Corp. kit (Katy, TX, USA). These parameters were analysed using a Bio-Tek® Synergy™ HT Multi-Detection plate reader (Highland Park, Winooski, Vermont 05404-0998 USA) controlled by BioTek®Gen5 software version 2.01.14.

Malondialdehyde (MDA) levels, a biomarker of lipid oxidation, was determined in serum samples by high-performance liquid chromatography (HPLC) following the methodology proposed by Mateos et al. [26]. For this purpose, a 1200 series HPLC equipment (Agilent Technologies, Santa Clara, CA, USA) and a Nucleosil 120 C18 column (25 mm × 0.46 mm, particle size 5 μm, TeknoKroma, Barcelona, Spain) were used.

2.8. Sample Size Calculation and Statistical Analysis

To estimate the sample size, the G*Power 3.1.9.7 program was used, considering TC concentration as the main variable and the study design [randomized, blind, crossover, and controlled clinical trial, in which all subjects consumed both test (OPO) and control oils (SO)]. Other premises considered, following previous studies with a similar design, were: a statistical power of 80%, a level of significance of 0.05, two tail, a standard deviation of 25, mean of pre-post differences of 13.4 units, and an effect size of 0.54 [27]. A sample size of 30 volunteers was established. Finally, considering their TC and LDL-C concentration, 72 subjects were recruited and allocated in the normocholesterolemic or hypercholesterolemic groups, although only n = 31 healthy and n = 37 at risk volunteers completed the study. These numbers were higher than the sample size of 30 initially calculated and allowed for adequate statistical analysis.

For the statistical design, the following factors were considered as two fixed effects: group (normocholesterolemic/hypercholesterolemic) and treatment (OPO/SO, repeated measures), and the order of oil intake (starting with OPO or SO within each group) was considered as a random effect.

The statistical models applied to analyse the results of this study were:

1. A general linear repeated measures model to study energy, macronutrient, and micronutrient intake throughout the study, considering that the order of intake of the test and control oils would not affect the overall dietary pattern of the volunteers. In each group (normocholesterolemic and hypercholesterolemic), baseline, initial (pre-treatment), and final (post-treatment) results with OPO and SO were compared. Results are shown as mean ± standard error of the mean (SEM).

2. A linear mixed model was applied to study the rate of change [(final value—initial value)/initial value] of each variable. This statistical model considers the order of intake of the oils, presenting the data in a correlated and non-constant variability form. This statistical model was also applied to the initial and final mean values. The statistical model was full factorial, considering that group (normocholesterolemic hypercholesterolemic), treatment oil (OPO and SO) and interaction group*treatment. Pre- and post-treatment data are shown as mean ± standard error of the mean (SEM), and the rate of change is expressed as a percentage ± SEM.

Normality of data distribution was verified by the Kolmogorov-Smirnov test, and a box-plot analysis was performed for all variables before statistical analysis. In addition, the Bonferroni test (within each group) was applied to compare pairwise the effect of the intake of each oil (OPO and SO). The significance level was set at $p < 0.05$. Data were analysed using SPSS software (version 27.0; SPSS, Inc., IBM Company, Armonk, NY, USA).

3. Results

3.1. Chemical Composition of the Study Oils

The chemical characterisation of OPO, SO, and CO is shown in Table 1. OPO was a monounsaturated fat with an oleic acid (C18:1) content of 74.32%, followed by palmitic acid (C16:0), and linoleic acid (C18:2) with a content of 10.78% and 9.02%, respectively. Linoleic acid (C18:2) was the main fatty acid in SO (58.46%) and in corn oil (CO) (50.52%), followed by oleic acid (C18:1), with a content of 29.76% in SO and 35.36% in CO. As for minor components, squalene was present in the three oils, with a higher concentration in OPO (675 ppm) compared to SO (314 ppm) and CO (548 ppm). In the case of tocopherols, while OPO (195 mg/kg) and SO (217 mg/kg) showed a high content of α-tocopherol (vitamin E), CO stood out for its γ-tocopherol content (205 mg/kg). Tocotrienols were only detected in CO (19 mg/kg). Regarding sterols composition, the content was highest in CO (8962 ppm), followed by OPO (3344.2 ppm) and SO (2820.5 ppm). However, it should be noted that triterpenic alcohols (745 mg/kg) and triterpenic acids (191 mg/kg), as well as aliphatic alcohols (1681 mg/kg), were mainly detected in OPO, compared to their low or no content in SO and CO, respectively. Finally, the phenol content was below 2 mg/kg in the three oils (Table 1).

Table 1. Chemical composition of olive pomace oil (OPO), sunflower oil (SO), and corn oil (CO).

	OPO	SO	CO
	Fatty acids (%)		
C 12:0 (Lauric acid)	-	<0.01	0.01
C 14:0 (Myristic acid)	0.02	0.08	0.04
C 16:0 (Palmitic acid)	10.78	6.43	10.60
C 16:1 (Palmitoleic acid)	0.84	0.14	0.15
C 17:0 (Margaric acid)	0.06	0.03	0.07
C 17:1 (Margaroleic acid)	0.10	0.03	0.05
C 18:0 (Stearic acid)	3.10	3.63	2.03
C 18:1 (Oleic acid)	74.32	29.76	35.36
C 18:2 (Linoleic acid)	9.02	58.46	50.52
C 20:0 (Arachidic acid)	0.48	0.26	0.48
C 18:3 (Linolenic acid)	0.68	0.09	0.04
C 20:1 (Eicosenoic acid)	0.33	0.16	0.31
C 22:0 (Behenic acid)	0.19	0.69	0.17
C 22:1 (Erucic acid)	-	<0.01	<0.01
C 24:0 (Lignoceric acid)	0.08	0.24	0.17
Trans Oleic (*t*-C18:1)	0.28	0.03	0.02
Trans Linoleic + *Trans* Linolenic (*t*-C18:2 + *t*-C18:3)	0.15	0.19	1.15

Table 1. *Cont.*

	OPO	SO	CO
	Squalene (ppm)		
Squalene (ppm)	675	314	548
	Tocopherols (mg/kg)		
α-Tocopherol (Vit. E)	195	217	33
β-Tocopherol	<2	9	<2
γ-Tocopherol	<2	<2	205
δ -Tocopherol	<2	<2	6
	Tocotrienols (mg/kg)		
A-Tocotrienol	-	-	10
β-Tocotrienol	-	-	<2
Υ-Tocotrienol	-	-	5
Δ-Tocotrienol	-	-	<2
	Sterols (%)		
Cholesterol	0.19	0.12	0.21
Brassicasterol	<0.10	<0.1	0.62
24-Methylcholesterol	<0.10	0.12	0.92
Campesterol	3.06	8.40	20.75
Campestanol	0.16	0.07	0.92
Stigmasterol	1.23	6.90	6.85
Δ7-Campesterol	<0.10	2.70	<0.10
Δ5,23-Stigmastadienol	0.32	0.34	<0.10
Clerosterol	1.08	0.66	0.70
β-Sitosterol	88.64	53.67	61.09
Sitostanol	1.44	0.64	2.11
Δ5-Avenasterol	1.82	3.08	3.90
Δ5,24-Stigmastadienol	1.35	1.08	0.51
Δ7-Stigmastenol	0.50	16.29	0.63
Δ7-Avenasterol	0.19	5.93	0.78
Δ-Sitosterol apparent	94.65	59.47	68.31
Total Sterols (ppm)	3344.2	2820.5	8962.0
	Triterpenic alcohols (mg/kg)		
Erythrodiol + Uvaol	745	<1.0	<1.0
	Phenols (mg/kg)		
Total phenols	<1.0	<1.0	<1.0
	Triterpenic acids (mg/kg)		
Oleanolic acid	187	<2.0	<2.0
Ursolic acid	<2.0	<2.0	<2.0
Maslinic acid	<2.0	<2.0	<2.0
	Aliphatic alcohols (mg/kg)		
C22 + C24 + C26 + C28	1681	38	29

3.2. Baseline Characteristics of Participants and Dietary Control

Of the 109 screened for assessment, only 72 participants were recruited. Two participants dropped out due to incompatibility with their work, one due to medical prescription, and one was lost to follow-up (the study flow diagram is shown in Figure 2). Thus, 68 participants (31 normocholesterolemic and 37 hypercholesterolemic) successfully completed the study. Although an effort was made to ensure equal representation of both sexes in the two groups, at the end there were more normocholesterolemic women (23) than men (8), although hypercholesterolemic men (19) and women (18) were balanced. The baseline characteristics of participants are presented in Table 2. Two hypercholesterolemic volunteers presented unusually high inflammatory biomarkers (outliers) and thus were excluded in the statistical analysis of inflammation results.

Table 2. Baseline characteristics of participants.

	Normocholesterolemic (n = 31)	Hypercholesterolemic (n = 37)
Men, n	8	19
Women, n	23	18
Age (years)	30 ± 2	41 ± 2
BMI (kg/m^2)	23 ± 2	26 ± 1
Waist circumference (cm)	78 ± 2	85 ± 2
Total cholesterol (mg/dL)	172 ± 3	239 ± 5
LDL-cholesterol (mg/dL)	93 ± 3	148 ± 5
Systolic blood pressure (mmHg)	110 ± 2	121 ± 2
Diastolic blood pressure (mmHg)	74 ± 1	81 ± 2

Values represent mean ± SEM. BMI: body mass index; LDL: low-density lipoprotein.

Table 3 shows energy, macronutrient, micronutrient, and dietary fibre intake during the study. Energy, protein, carbohydrate, and lipid values did not change significantly in both population groups after OPO and SO intervention ($p < 0.05$) (Table 3). These data confirm that the volunteers followed the instructions of not changing their dietary habits during the study. The total caloric intake showed mean values between 1906 and 2081 Kcal/day, slightly below normal limits [28]. In relation to macronutrient intake, the mean values for proteins (84 g/day), carbohydrates (188 g/day), and lipids (92 g/day) represented 16.7%, 37.5%, and 41.3%, respectively. In accordance with the recommended daily intake and nutritional targets for the Spanish population, the data for proteins and lipids were found to be above the recommendations and, in the case of carbohydrates, their values were under the recommendation (50–60% of the total diet) (Table 3) [28].

Saturated fatty acids (SFA) intake values ranged between 28 and 31 g/day (12.6 and 14% of the total diet), above 7–8% of the recommended total daily energy [28]. However, OPO and SO consumption did not show significant variations in SFA values ($p < 0.05$). As for monounsaturated fatty acids (MUFA), significant changes ($p < 0.001$) were observed in both groups after OPO and SO intake. According to the multiple comparisons test, the results showed that, after regular consumption of OPO, there was a significantly increased consumption of this nutrient (with a mean value of 44.5 g/day), reaching 20% of the total energy of the diet [28]. In contrast, MUFA values remained constant after the SO intervention, with an average intake of 29.5 g/day, which represented 13.3% of the total energy [28] (Table 3). This result was expected considering that OPO is a rich source of MUFA compared to SO, whose fat is mostly polyunsaturated (Table 1). Polyunsaturated fatty acid (PUFA) intake showed significant changes ($p < 0.001$) in the normocholesterolemic and hypercholesterolemic groups after the intervention with OPO and SO. The Bonferroni test revealed that, in line with the dietary intervention, after OPO consumption, there was a significant decrease in PUFA intake (11–13 g/day, equivalent to 5–6% of total dietary energy), and an increase after regular intake of SO (29–31 g/day, equivalent to 13–14% of total dietary energy). It is worth mentioning that in the run-in and wash-out stages (before the start with OPO and SO intervention), the mean PUFA values (22–27 g/day, equivalent to 10–12%, respectively) were higher than those observed with basal diet consumption (13.5 g/day, equivalent to 6% of total energy). The reason was that the CO used during these stages (run-in and wash-out) is a rich source of PUFA. The recommendation of 5% PUFA of total dietary energy was close to those achieved after the OPO intervention (5–6% of total dietary energy) and the basal stage (6% of total energy) (Table 3).

Table 3. Energy intake and dietary components during the intervention trial with the two oils, olive pomace oil (OPO) and sunflower oil (SO) *.

	Normocholesterolemic n = 31					p Value **	Hypercholesterolemic n = 37					p Value **
		OPO		SO				OPO		SO		
	Baseline	Initial	Final	Initial	Final	Oil	Baseline	Initial	Final	Initial	Final	Oil
Energy (kcal/day)	1976 ± 70	1906 ± 77	1949 ± 78	1996 ± 85	1982 ± 82	0.314	2081 ± 79	2061 ± 78	2042 ± 72	2025 ± 72	2043 ± 74	0.455
Proteins (g/day)	84 ± 4	78 ± 3	80 ± 3	84 ± 5	81 ± 4	0.621	92 ± 4	88 ± 4	83 ± 4	88 ± 4	81 ± 3	0.103
Carbohyd-rates (g/day)	187 ± 10	182 ± 12	181 ± 11	186 ± 11	182 ± 10	0.942	195 ± 12	199 ± 10	195 ± 9	188 ± 10	187 ± 10	0.427
Lipids (g/day)	89 ± 4	88 ± 4	90 ± 4	93 ± 4	93 ± 5	0.400	91 ± 4	90 ± 4	93 ± 4	94 ± 4	98 ± 4	0.124
SFA (g/day)	28 ± 2	28 ± 2	28 ± 2	31 ± 2	28 ± 2	0.287	29 ± 2	28 ± 2	28 ± 2	28 ± 2	28 ± 2	0.773
MUFA (g/day)	38 ± 2 [a]	30 ± 2 [b]	44 ± 3 [a]	31 ± 1 [b]	28 ± 2 [b]	0.000	41 ± 2 [a]	30 ± 1 [b]	45 ± 2 [a]	31 ± 1 [b]	31 ± 2 [b]	0.000
PUFA (g/day)	15 ± 1 [a]	22 ± 2 [b]	11 ± 1 [c]	23 ± 2 [bd]	29 ± 2 [d]	0.000	12 ± 1 [a]	23 ± 1 [b]	13 ± 1 [a]	27 ± 1 [bc]	31 ± 2 [d]	0.000
Cholest-erol (mg/day)	325 ± 23	308 ± 23	337 ± 23	319 ± 19	340 ± 29	0.668	372 ± 21	311 ± 22	317 ± 24	312 ± 21	351 ± 23	0.067
Dietary fibre (g/day)	20 ± 2	17 ± 1	17.5 ± 0.1	18 ± 1	20 ± 1	0.137	22 ± 2	22 ± 1	22 ± 1	22 ± 1	22 ± 1	0.964
Vitamin E (mg/day)	9.7 ± 0.8 [a]	14 ± 1 [b]	17 ± 1 [b]	14 ± 1 [b]	22 ± 2 [c]	0.000	120 ± 14 [a]	121 ± 13 [b]	137 ± 18 [c]	146 ± 13 [c]	114 ± 11 [d]	0.000

* Values represent mean ± SEM. Data were analysed using a general linear repeated measures model. According to the Bonferroni test, values with different superscript letters correspond to significant differences within the normocholesterolemic N or hypercholesterolemic H groups. p values correspond to the effect of taking OPO or SO. Significance level was $p < 0.05$. SFA: saturated fatty acids. MUFA: monounsaturated fatty acids. PUFA: polyunsaturated fatty acids. ** Oil (Baseline vs. OPO vs. SO).

3.3. Blood Pressure

No significant differences were found in SBP and DBP after the dietary intervention with OPO and SO. However, there were significant differences in SBP and DBP between the normocholesterolemic and hypercholesterolemic groups, when initial and final values were analysed ($p < 0.05$) (Supplementary Table S1). According to the European Society of Cardiology (ESC), all participants maintained blood pressure values within the normal range of <120 mmHg for SBP and <80 mmHg for DBP

3.4. Blood Biochemistry: Lipid Profile and Liver Function

As shown in Table 4, the volunteers' lipid profiles did not change ($p < 0.05$) after prolonged consumption of OPO and SO in either group (normocholesterolemic and hypercholesterolemic). However, the trend towards a decrease in circulating levels of TC, TG, LDL-C, and VLDL-C after OPO intervention in the healthy group is noteworthy. On the other hand, when the linear mixed model was applied to the initial and final values, TC, TG, LDL-C, VLDL-C, LDL-C/HDL-C, TC/HDL-C, Apo A1, Apo B, and Apo B/Apo A1 ratio showed differences between the normocholesterolemic and hypercholesterolemic groups ($p < 0.05$). In relation to the ALAT and ASAT enzymes, there were some differences between the groups at the initial and end of the intervention, with higher values ($p < 0.05$) in the hypercholesterolemic subjects. When the rate of change was analysed, ASAT enzyme also showed significantly higher values in the hypercholesterolemic group compared to the healthy group ($p = 0.045$). However, none of the studied oils had a significant effect on these enzymes.

Table 4. Effect of olive pomace oil (OPO) and sunflower oil (SO) consumption on lipid profile and liver function *.

	Normocholesterolemic n = 31		**Hypercholesterolemic** n = 37		***p* Value**		
	OPO	**SO**	**OPO**	**SO**	**Oil**	**N/H**	**N/H × Oil**
			Total-cholesterol (mg/dL)				
Initial	168 ± 5	168 ± 5	221 ± 5	221 ± 6	0.923	0.000	0.868
Final	164 ± 5	169 ± 5	220 ± 5	220 ± 6	0.642	0.000	0.679
Rate of change	−1.4 ± 2.0	1.1 ± 1.8	−0.3 ± 1.6	−0.1 ± 1.6	0.425	0.907	0.813
			Triglycerides (mg/dL)				
Initial	73 ± 5	76 ± 6	111 ± 9	111 ± 11	0.778	0.000	0.884
Final	70 ± 5	72 ± 6	109 ± 9	114 ± 11	0.694	0.000	0.975
Rate of change	−2.6 ± 0.04	−1.8 ± 5.0	1.2 ± 4.0	6.9 ± 5.7	0.486	0.215	0.784
			HDL-cholesterol (mg/dL)				
Initial	60 ± 2	60 ± 2	65 ± 3	64 ± 3	0.857	0.146	0.814
Final	59 ± 2	60 ± 2	64 ± 2	63 ± 2	0.676	0.179	0.405
Rate of change	−1.2 ± 1.9	0.8± 2.2	0.9 ± 2.0	−2.0 ± 1.9	0.838	0.754	0.342
			LDL-cholesterol (mg/dL)				
Initial	93 ± 4	92 ± 4	133 ± 5	134 ± 5	0.919	0.000	0.859
Final	91 ± 4	94 ± 4	134 ± 4	134 ± 5	0.622	0.000	0.801
Rate of change	−0.3 ± 2.8	3.3 ± 2.9	1.1 ± 1.7	1.0 ± 2.2	0.547	0.733	0.657
			VLDL-cholesterol (mg/dL)				
Initial	15 ± 1	15 ± 1	24 ± 3	23 ± 3	0.942	0.000	0.713
Final	14 ± 1	14 ± 1	22 ± 2	23 ± 3	0.766	0.000	0.902
Rate of change	−3.0± 4.3	−2.8 ± 4.7	0.2 ± 4.5	5.7 ± 5.8	0.267	0.267	0.783
			LDL-cholesterol/HDL-cholesterol				
Initial	1.6 ± 0.1	1.6 ± 0.1	2.2 ± 0.1	2.2 ± 0.1	0.852	0.000	0.823
Final	1.6 ± 0.1	1.6 ± 0.1	2.3 ± 0.1	2.2 ± 0.1	0.788	0.000	0.874
Rate of change	1.7 ± 3.2	4.3 ± 3.7	1.4 ± 2.3	4.4 ± 3.0	0.315	0.945	0.968
			Total Cholesterol/HDL-cholesterol				
Initial	2.9 ± 0.1	2.9 ± 0.1	3.6 ± 0.1	3.7 ± 0.2	0.992	0.000	0.990
Final	2.9 ± 0.1	2.9 ± 0.1	3.7 ± 0.2	3.6 ± 0.2	0.778	0.000	0.866
Rate of change	0.26 ± 1.8	1.32 ± 2.2	−0.3 ± 1.8	2.8 ± 2.0	0.942	0.262	0.615
			Apolipoprotein (Apo A1) (mg/dL)				
Initial	161 ± 3	163 ± 3	175 ± 4	177 ± 4	0.635	0.000	0.963
Final	165 ± 4	167 ± 3	183 ± 4	179 ± 5	0.397	0.001	0.333
Rate of change	2.98 ± 1.9	3.0 ± 1.7	4.6 ± 1.1	1.0 ± 1.3	0.195	0.849	0.298
			Apo B (mg/dL)				
Initial	72 ± 3	73 ± 3	101 ± 3	100 ± 3	0.974	0.000	0.772
Final	77 ± 3	79 ± 3	108 ± 3	108 ± 4	0.833	0.000	0.580
Rate of change	8.8 ± 2.4	9.5 ± 2.3	8.3 ± 1.9	9.0 ± 1.9	0.768	0.846	0.953
			Apo B/Apo A1				
Initial	0 46 ± 0.02	0.45 ± 0.02	0.59 ± 0.03	0.56 ± 0.02	0.458	0.000	0.757
Final	0.47 ± 0.02	0.48 ± 0.02	0.61 ± 0.03	0.63 ± 0.03	0.875	0.000	0.970
Rate of change	4.3 ± 3.2	6.3 ± 2.7	5.6 ± 2.4	11.3 ± 2.6	0.126	0.420	0.578
			ALAT (UI/L)				
Initial	21 ± 2	21 ± 2	24 ± 3	27 ± 3	0.504	0.052	0.608
Final	19 ± 2	20 ± 2	26 ± 3	28 ± 3	0.498	0.005	0.868
Rate of change	107.7 ± 5.0	109.41 ± 7.5	119.8 ± 5.7	117.0 ± 7.1	0.921	0.119	0.782

Table 4. *Cont.*

	Normocholesterolemic $n = 31$		Hypercholesterolemic $n = 37$		p Value		
	OPO	SO	OPO	SO	Oil	N/H	N/H × Oil
			ASAT (UI/L)				
Initial	20.6 ± 0.9	21.0 ± 1.2	22.7 ± 1.3	24.1 ± 1.3	0.536	0.033	0.322
Final	18.5 ± 0.8	19.2 ± 0.7	22.5 ± 1.2	24.2 ± 1.3	0.275	0.000	0.635
Rate of change	101.7 ± 3.1	106.9 ± 4.7	113.5 ± 4.6	112.4 ± 4.7	0.559	0.045	0.418

* Values represent mean ± SEM. The table shows the initial (pre-treatment) and final (post-treatment) mean values. The rate of change was calculated from initial and final values as [(final value-initial value)/initial value] and expressed as percentage. Data were analyzed using a linear mixed model. p values in the first column correspond to the effect of taking the oil (OPO or SO), those of the penultimate column to the effect of the group [normocholesterolemic (N) or hypercholesterolemic (H)], and the last column to the interaction of oil and group. Significance level was set at $p < 0.05$. Apo: Apolipoprotein. ALAT: Alanine aminotransferase. ASAT: Aspartate aminotransferase.

3.5. Inflammatory Biomarkers

After OPO and SO intervention, statistically significant changes were observed in IL-4 ($p = 0.021$) and IL-13 ($p = 0.023$). According to Table 5, IL-4 showed important changes by the interaction of the normo- and hypercholesterolemic groups with the oils (OPO or SO), with an increase and a decrease after OPO and SO intake, respectively. With respect to IL-13 ($p = 0.023$), there were significant changes after prolonged consumption of both vegetable oils, with an increase after the OPO intake, and a decrease after the SO consumption (Table 5).

Table 5. Effect of olive pomace oil (OPO) and sunflower oil (SO) consumption on inflammatory biomarkers *.

	Normocholesterolemic $n = 31$		Hypercholesterolemic $n = 35$ **		p Value		
	OPO	SO	OPO	SO	Oil	N/H	N/H × Oil
			CRP (mg/dL)				
Initial	0.23 ± 0.07	0.25 ± 0.10	0.14 ± 0.03	0.34 ± 0.17	0.222	0.823	0.860
Final	0.17 ± 0.05	0.11 ± 0.03	0.17 ± 0.04	0.16 ± 0.04	0.311	0.579	0.438
Rate of change *	39.1 ± 3.1	32.1 ± 0.3	15.1 ± 0.7	17.5 ± 1.1	0.476	0.410	0.815
			IL-1β (pg/mL)				
Initial	0.97 ± 0.06	0.95 ± 0.03	0.89 ± 0.02	0.91 ± 0.03	0.785	0.097	0.691
Final	0.92 ± 0.03	0.91 ± 0.03	0.89 ± 0.02	1.05 ± 0.15	0.311	0.980	0.335
Rate of change	−2.4 ± 2.8	−3.2 ± 2.8	4.6 ± 2.3	0.8 ± 14.1	0.449	0.202	0.259
			IL-2 (pg/mL)				
Initial	16.4 ± 0.7	16.6 ± 0.5	15.4 ±0.3	15.9 ± 0.5	0.308	0.147	0.766
Final	16.0 ± 0.5	15.6 ± 0.5	15.3 ± 0.4	15.6 ± 0.5	0.573	0.488	0.433
Rate of change	0.1 ± 2.3	−5.2 ± 2.3	−0.4 ± 1.8	−1.4 ± 2.3	0.193	0.544	0.370
			IL-4 (pg/mL)				
Initial	2.68 ± 0.06	2.64 ± 0.06	2.57 ± 0.05	2.61 ± 0.05	0.907	0.217	0.430
Final	2.64 ± 0.06	2.59 ± 0.07	2.55 ± 0.04	2.62 ± 0.07	0.515	0.762	0.137
Rate of change	0.7 ± 2.3	−3.0 ± 2.3	2.5 ± 1.6	−1.6 ± 2.1	0.778	0.229	0.021
			IL-6 (pg/mL)				
Initial	5.2 ± 0.2	5.4 ± 0.3	5.1 ± 0.2	5.0 ± 0.2	0.947	0.206	0.258
Final	5.3 ± 0.3	5.1 ± 0.3	5.0 ± 0.2	5.2 ± 0.2	0.986	0.571	0.668
Rate of change	3.5 ± 4.4	−2.1 ± 4.4	−3.0 ± 4.3	5.0 ± 3.3	0.816	0.885	0.169

Table 5. *Cont.*

	Normocholesterolemic $n = 31$		Hypercholesterolemic $n = 35$ **		p Value		
	OPO	SO	OPO	SO	Oil	N/H	N/H × Oil
			IL-7 (pg/mL)				
Initial	33 ± 1	34 ± 1	33 ± 1	34 ± 1	0.388	0.613	0.782
Final	34 ± 1	34 ± 1	33 ± 1	33 ± 1	0.692	0.610	0.856
Rate of change	2.3 ± 2.5	−1.2 ± 2.5	0.6 ± 1.7	−1.0 ± 2.3	0.252	0.777	0.641
			IL-8 (pg/mL)				
Initial	22 ± 2	24 ± 2	20 ± 1	21 ± 1	0.634	0.057	0.861
Final	21 ± 1	21 ± 1	20 ± 1	20 ± 1	0.871	0.633	0.874
Rate of change	−5.6 ± 3.8	−3.2 ± 3.8	3.0 ± 2.5	−3.0 ± 2.2	0.204	0.113	0.568
			IL-10 (pg/mL)				
Initial	10.9 ± 0.2	11.1 ± 0.3	10.8 ± 0.3	10.7 ± 0.2	0.319	0.260	0.677
Final	11.0 ± 0.2	10.8 ± 0.2	10.9 ± 0.3	10.6 ± 0.2	0.883	0.552	0.233
Rate of change	1.9 ± 1.6	−2.6 ± 1.6	−0.3 ± 1.1	0.9 ± 1.6	0.310	0.686	0.080
			IL-12 (p70) (pg/mL)				
Initial	12.1 ± 0.3	12.4 ± 0.3	11.7 ± 0.3	11.9 ± 0.2	0.287	0.091	0.838
Final	12.2 ± 0.3	11.9 ± 0.3	11.7 ± 0.2	12.0 ± 0.3	0.957	0.565	0.253
Rate of change	1.5 ± 2.5	−4.1 ± 2.5	0.9 ± 1.7	1.0 ± 1.7	0.200	0.293	0.182
			IL-13 (pg/mL)				
Initial	3.0 ± 0.1	3.2 ± 0.2	2.8 ± 0.1	2.9 ± 0.1	0.345	0.045	0.992
Final	3.1 ± 0.1	3.0 ± 0.1	2.8 ± 0.1	2.9 ± 0.1	0.826	0.142	0.437
Rate of change	2.8 ± 2.8	−5.2 ± 2.8	1.0 ± 1.9	−0.7 ± 3.0	0.023	0.527	0.159
			IL-17 (pg/mL)				
Initial	13.9 ± 0.4	14.0 ± 0.3	13.4 ± 0.3	13.5 ± 0.2	0.475	0.199	0.875
Final	14.0 ± 0.4	13.8 ± 0.4	13.3 ± 0.2	13.5 ± 0.3	0.936	0.143	0.489
Rate of change	0.8 ± 2.1	−0.7 ± 2.1	−0.2 ± 1.6	0.3 ± 1.6	0.773	0.997	0.570
			G-CSF (pg/mL)				
Initial	137 ± 3	140 ± 2	134 ± 2	134 ± 2	0.456	0.041	0.453
Final	137 ± 2	137 ± 3	133 ± 2	135 ± 2	0.875	0.150	0.547
Rate of change	0.9 ± 1.7	−1.7 ± 1.7	−0.4 ± 1.0	0.8 ± 1.2	0.618	0.711	0.185
			GM-CSF (pg/mL)				
Initial	4.7 ± 0.2	4.7 ± 0.2	4.7 ± 0.3	4.6 ± 0.2	0.726	0.929	0.911
Final	4.7 ± 0.2	4.5 ± 0.2	4.5 ± 0.2	4.8 ± 0.3	0.746	0.772	0.329
Rate of change	0.9 ± 2.6	−2.4 ± 2.6	−1.6 ± 2.3	4.4 ± 3.3	0.651	0.472	0.114
			MCP-1 (pg/mL)				
Initial	39 ± 3	37 ± 4	33 ± 2	33 ± 2	0.909	0.088	0.949
Final	35 ± 6	34 ± 3	31 ± 1	32 ± 2	0.574	0.243	0.416
Rate of change	−2.9 ± 3.8	−6.5 ± 3.8	−1.7 ± 3.2	0.3 ± 3.2	0.816	0.273	0.497
			MIP-1β (ng/mL)				
Initial	69 ± 6	73 ± 5	71 ± 8	73 ± 8	0.572	0.790	0.793
Final	67 ± 4	66 ± 4	71 ± 8	71 ± 9	0.834	0.453	0.845
Rate of change	4.9 ± 3.9	−8.3 ± 3.9	−1.3 ± 4.1	1.4 ± 6.3	0.361	0.774	0.133

Table 5. *Cont.*

	Normocholesterolemic $n = 31$		Hypercholesterolemic $n = 35$ **		p Value		
	OPO	SO	OPO	SO	Oil	N/H	N/H × Oil
			TNF-α (pg/mL)				
Initial	23.1 ± 0.9	23.7 ± 0.8	23.3 ± 1.4	23.9 ± 1.4	0.633	0.897	0.978
Final	23.0 ± 0.8	22.4 ± 0.8	23.3 ± 1.5	23.4 ± 1.5	0.888	0.584	0.644
Rate of change	0.5 ± 1.8	−5.0 ± 1.8	0.0 ± 1.4	−1.5 ± 2.0	0.083	0.567	0.382

* Values represent mean ± SEM. The table shows the initial (pre-treatment) and final (post-treatment) mean values. The rate of change was calculated from initial and final values as [(final value-initial value)/initial value] and expressed as percentage. Data were analyzed using a linear mixed model. p values in the first column correspond to the effect of taking the oil (OPO or SO), those of the penultimate column to the effect of the group [normocholesterolemic (N) or hypercholesterolemic (H)], and the last column to the interaction of oil and group. Significance level was set at $p < 0.05$. CRP: C reactive protein. IL: Interleukin. G-CSF: Granulocyte colony-stimulating factor. GM-CSF: Granulocyte–macrophage colony-stimulating factor. IFN-γ: Interferon gamma. MCP-1: Monocyte chemoattractant protein-1. MIP-1β: Macrophage inflammatory protein 1 beta. TNF-α: Tumor necrosis factor alpha. ** Two hypercholesterolemic volunteers were excluded for statistical analysis due to outliers.

3.6. *Biomarkers of Endothelial Function*

Endothelial function biomarkers (Table 6) showed no significant changes throughout the study according to the linear mixed model applied on the rates of change ($p < 0.05$). However, it is noteworthy that circulating eNOS concentrations were close to the level of significance (set at $p < 0.05$) in both groups, with an increasing and decreasing trend after OPO and SO intake, respectively.

Table 6. Effect of olive pomace oil (OPO) and sunflower oil (SO) consumption on endothelial function biomarkers *.

	Normocholesterolemic $n = 31$		Hypercholesterolemic $n = 37$		p Value		
	OPO	SO	OPO	SO	Oil	N/H	N/H × Oil
			eNOS (ng/mL)				
Initial	0.21 ± 0.03	0.25 ± 0.05	0.22 ± 0.03	0.21 ± 0.03	0.926	0.647	0.597
Final	0.22 ± 0.04	0.20 ± 0.03	0.19 ± 0.03	0.20 ± 0.03	0.944	0.708	0.644
Rate of change	16.03 ± 0.14	−6.67 ± 0.09	15.98 ± 0.14	−4.76 ± 0.12	0.083	0.903	0.912
			E-selectin (ng/mL)				
Initial	11 ± 2	15 ± 2	12 ± 2	13 ± 2	0.260	0.939	0.540
Final	11 ± 2	14 ± 3	11 ± 2	10 ± 1	0.588	0.193	0.045
Rate of change	18.0 ± 0.1	14.4 ± 0.1	5.4 ± 0.1	16.5 ± 0.2	0.552	0.342	0.960
			P-selectin (ng/mL)				
Initial	214 ± 18	190 ± 12	195 ± 11	195 ± 11	0.412	0.537	0.469
Final	206 ± 15	202 ± 16	179 ± 12	196 ± 11	0.635	0.226	0.448
Rate of change	0.60 ± 0.05	6.86 ± 0.05	−1.72 ± 0.06	2.85 ± 0.04	0.289	0.519	0.871
			ICAM-1 (pg/mL)				
Initial	3093 ± 495	3332 ± 502	2954 ± 439	3291 ± 579	0.546	0.854	0.553
Final	2900 ± 452	3302 ± 566	3270 ± 557	3311 ± 643	0.705	0.770	0.601
Rate of change	−0.09 ± 0.05	−1.84 ± 0.04	9.30 ± 0.06	−1.88 ± 0.05	0.120	0.742	0.473

Table 6. *Cont.*

	Normocholesterolemic $n = 31$		Hypercholesterolemic $n = 37$		p Value		
	OPO	SO	OPO	SO	Oil	N/H	N/H × Oil
	VCAM-1 (pg/mL)						
Initial	11,049 ± 2375	11,961 ± 2422	10,716 ± 2081	12,911 ± 2894	0.426	0.909	0.822
Final	11,280 ± 2523	13,101 ± 3152	12,271 ± 2480	13,917 ± 3574	0.522	0.822	0.621
Rate of change	5.44 ± 0.07	10.51 ± 0.08	19.90 ± 0.08	6.11 ± 0.09	0.373	0.856	0.073

* Values represent mean ± SEM. The table shows the initial (pre-treatment) and final (post-treatment) mean values. The rate of change was calculated from initial and final values as [(final value-initial value)/initial value] and expressed as percentage. Data were analyzed using a linear mixed model. p values in the first column correspond to the effect of taking the oil (OPO or SO), those of the penultimate column to the effect of the group [normocholesterolemic (N) or hypercholesterolemic (H)], and the last column to the interaction of oil and group. Significance level was set at $p < 0.05$. eNOS: Endothelial nitric oxide synthase. E-selectin: Endothelial selectin. P-selectin: Platelet selectin. ICAM-1: Intercellular adhesion molecule 1. VCAM-1: Vascular cell adhesion molecule 1.

3.7. Diabetes Markers: Glycaemia, Insulin Levels, Glycosylated Haemoglobin Concentration, Insulin Resistance/Sensitivity Indices (HOMA-IR/QUICKI), and Pancreatic Beta-Cell Function (HOMA-β)

To assess insulin resistance and glycaemic homeostasis, mean glucose, insulin, and HbA1c values were analysed. As shown in Table 7, glucose ($p = 0.007$) showed significant differences after OPO and SO dietary intervention due to the interaction of the group (normo- and hypercholesterolemic) with the studied oils (OPO and SO). According to the Bonferroni test, there were differences between OPO and SO in subjects with normal cholesterol levels. Thus, glucose values increased by 1.3% after OPO intake and decreased by 2.6% after SO intake. In the case of insulin, there were no statistically significant variations throughout the study, although it was close to being significant due to the effect of consuming OPO and SO ($p = 0.077$). Regarding Hb1Ac ($p = 0.021$), the differences occurred between groups, with slightly higher values in the hypercholesterolemic group. Importantly, despite the observed changes in glucose and Hb1Ac, the values of both parameters were within normal levels for the adult population (<6% and <120 mg/dL, respectively) established by the Spanish Diabetes Federation (Federación Española de Diabetes, FEDE). The mathematical models proposed by Matthews et al. [22] to assess insulin resistance (HOMA-IR) and pancreatic beta-cell function (HOMA-β), as well as the QUICKI index calculated for insulin sensitivity, did not show statistically significant differences after consumption of OPO and SO ($p < 0.05$). However, differences were observed between healthy and at-risk subjects in HOMA-IR ($p = 0.020$) and QUICKI ($p = 0.011$) values. Finally, after linear mixed model analysis of initial and final values, HOMA-IR ($p = 0.006$) and QUICKI index ($p = 0.005$) revealed differences between normocholesterolemic and hypercholesterolemic subjects.

Table 7. Effect of olive pomace oil (OPO) and sunflower oil (SO) consumption on diabetes markers *.

	Normocholesterolemic $n = 31$		Hypercholesterolemic $n = 37$		p Value		
	OPO	SO	OPO	SO	Oil	N/H	N/H × Oil
	Glucose (mg/dL)						
Initial	80 ± 1	82 ± 1	83 ± 2	81 ± 1	0.938	0.259	0.130
Final	81 ± 1	79 ± 2	83 ± 1	83 ± 1	0.433	0.403	0.944
Rate of change	1.3 ± 1.2 [a]	−2.6 ± 1.3 [b]	−0.3 ± 1.2	3.1 ± 1.3	0.825	0.073	0.007

Table 7. *Cont.*

	Normocholesterolemic $n = 31$		Hypercholesterolemic $n = 37$		*p* Value		
	OPO	**SO**	**OPO**	**SO**	**Oil**	**N/H**	**N/H × Oil**
			Insulin (μUI/mL)				
Initial	7.3 ± 0.8	8.3 ± 0.8	10.2 ± 1.5	8.5 ± 0.9	0.977	0.243	0.579
Final	6.2 ± 0.6	6.8 ± 0.7	9.1 ± 0.9	9.1 ± 1.2	0.918	0.007	0.449
Rate of change	−5.6 ± 6.5	−12.5 ± 5.6	4.7 ± 7.6	8.8 ± 5.9	0.077	0.802	0.825
			HbA1c (%)				
Initial	5.30 ± 0.05	5.33 ± 0.04	5.33 ± 0.04	5.36 ± 0.04	0.433	0.403	0.944
Final	5.30 ± 0.04	5.29 ± 0.04	5.33 ± 0.03	5.32 ± 0.04	0.837	0.488	0.949
Rate of change	0.1 ± 0. 5	−0.6 ± 0. 6	0.0 ± 0. 4	−0.8 ± 0. 4	0.994	0.021	0.258
			HOMA-IR				
Initial	1.4 ± 0.2	1.7 ± 0.2	2.2 ± 0.4	1.7 ± 0.2	0.886	0.186	0.674
Final	1.3 ± 0.1	1.3 ± 0.1	1.9 ± 0.2	1.9 ± 0.3	0.978	0.006	0.560
Rate of change	−3.1 ± 7.4	−14.3 ± 5.8	6.0 ± 8.6	12.4 ± 6.2	0.952	0.020	0.102
			HOMA-β				
Initial	174 ± 21	191 ± 27	188 ± 20	140 ± 36	0.503	0.619	0.767
Final	143 ± 14	185 ± 43	186 ± 19	166 ± 15	0.669	0.607	0.185
Rate of change	−8.5 ± 5.3	4.9 ± 5.2	7.1 ± 7.2	−1.7 ± 6.9	0.665	0.633	0.218
			QUICKI				
Initial	0.371 ± 0.005	0.363 ± 0.005	0.358 ± 0.006	0.365 ± 0.006	0.931	0.348	0.200
Final	0.379 ± 0.005	0.376 ± 0.006	0.359 ± 0.005	0.363 ± 0.006	0.943	0.005	0.658
Rate of change	2.2 ± 1.3	3.9 ± 1.2	0.7 ± 0.9	−0.7 ± 1.1	0.891	0.011	0.096

* Values represent mean ± SEM. The table shows the initial (pre-treatment) and final (post-treatment) mean values. The rate of change was calculated from initial and final values as [(final value-initial value)/initial value] and expressed as percentage. Data were analyzed using a linear mixed model. *p* values in the first column correspond to the effect of taking the oil (OPO or SO), those of the penultimate column correspond to the effect of the group [normocholesterolemic (N) or hypercholesterolemic (H)], and the last column to the interaction of oil and group. Significance level was set at $p < 0.05$. HbA1c: haemoglobin A1c. HOMA-IR: Homeostatic model to assessment insulin resistance. HOMA-β: Homeostatic model to assess β-cell functionality. QUICKI: Quantitative insulin sensitivity check index.

3.8. Obesity Biomarkers

To determine the possible beneficial role of OPO and SO consumption on diabetes and obesity, several biomarkers closely related to these two pathologies were analysed (Table 8). According to the linear mixed model applied to rates of change, the dietary intervention with OPO and SO showed significant changes in the hormones ghrelin ($p = 0.031$) and leptin ($p = 0.017$). Specifically, ghrelin values increased in both groups after OPO and SO intervention, being more marked in cardiovascular risk subjects after SO intake. Regarding leptin, values in healthy subjects decreased by −8.0% after OPO intervention and slightly increased by 1.7% after SO intervention, while there was an increase of 8.6% and 18.6% after OPO and SO intake, respectively, in at-risk subjects. All other variables analysed showed no changes. In addition, belonging to the normocholesterolemic or hypercholesterolemic group had significant effects on C-peptide (initial and final stages), insulin (final stages), and visfatin (final stages) when initial and final values were analysed ($p < 0.05$). On the other hand, adiponectin showed changes due to the effect of the oils (OPO and SO) when final values were compared ($p = 0.029$).

Table 8. Effect of olive pomace oil (OPO) and sunflower oil (SO) consumption on diabetes and obesity biomarkers *.

	Normocholesterolemic $n = 31$		**Hypercholesterolemic** $n = 37$		***p* Value**		
	OPO	**SO**	**OPO**	**SO**	**Oil**	**N/H**	**N/H × Oil**
			C-peptide (pg/mL)				
Initial	548 ± 45	546 ± 43	687 ± 63	634 ± 43	0.582	0.030	0.788
Final	489 ± 33	504 ± 38	639 ± 46	630 ± 43	0.937	0.001	0.873
Rate of change	−6.2 ± 4.7	−3.3 ± 4.7	1.9 ± 5.8	1.8 ± 3.7	0.697	0.259	0.723
			Ghrelin (pg/mL)				
Initial	319 ± 42	297 ± 26	271 ± 26	252 ± 23	0.642	0.159	0.971
Final	299 ± 38	309 ± 30	250 ± 23	296 ± 26	0.161	0.330	0.543
Rate of change	3.8 ± 7.9	6.0 ± 5.8	2.3 ± 7.4	41.2 ± 20.7	0.031	0.348	0.166
			GIP (pg/mL)				
Initial	807 ± 168	840 ± 188	667 ± 29	635 ± 36	0.639	0.281	0.549
Final	770 ± 162	837 ± 188	657 ± 24	647 ± 38	0.847	0.581	0.571
Rate of change	−1.9 ± 5.7	0.2 ± 2.9	0.4 ± 3.0	3.1 ± 2.9	0.554	0.513	0.975
			GLP-1 (pg/mL)				
Initial	255 ± 12	252 ± 13	259 ± 9	251 ± 10	0.624	0.852	0.656
Final	261 ± 11	257 ± 10	256 ± 9	256 ± 10	0.842	0.775	0.896
Rate of change	−3.1 ± 7.7	−4.4 ± 7.8	0.1 ± 2.8	4.0 ± 3.4	0.992	0.902	0.274
			Glucagon (pg/mL)				
Initial	865 ± 71	807 ± 64	833 ± 54	858 ± 62	0.806	0.871	0.512
Final	906 ±74	824 ± 71	903 ± 55	869 ± 53	0.351	0.751	0.679
Rate of change	16.8 ± 14.7	5.5 ± 7.1	14.54 ± 6.1	7.97 ± 5.6	0.956	0.480	0.627
			Insulin (pg/mL)				
Initial	176 ± 25	200 ± 23	246 ± 33	211 ± 24	0.739	0.114	0.198
Final	150 ± 16	162 ± 20	235 ± 32	234 ± 32	0.841	0.005	0.818
Rate of change	−0.6 ± 9.1	−9.3 ± 9.2	12.0 ± 10.3	15.0 ± 8.1	0.886	0.065	0.807
			Leptin (ng/mL)				
Initial	3.3 ± 0.4	3.2 ± 0.4	3.6 ± 0.5	3.0 ± 0.4	0.535	0.940	0.645
Final	2.9 ± 0.4	3.2 ± 0.4	3.2 ± 0.4	3.4 ± 0.4	0.593	0.519	0.911
Rate of change	−8.0± 5.4	1.7 ± 6.3	8.6 ± 6.8	18.6 ± 98.3	0.017	0.180	0.947
			PAI-1 (ng/mL)				
Initial	6.9 ± 0.3	7.0 ± 0.3	7.2 ± 0.5	6.9 ± 0.4	0.815	0.853	0.608
Final	7.1 ± 0.3	6.9 ± 0.3	7.1 ± 0.4	7.2 ± 0.4	0.847	0.565	0.590
Rate of change	4.5 ± 3.3	−1.5 ± 2.8	5.7 ± 6.3	9.3 ± 8.1	0.805	0.459	0.444
			Resistin (ng/mL)				
Initial	6.0 ± 0.6	6.6 ± 0.7	6.5 ± 0.6	7.2 ± 1.0	0.179	0.641	0.377
Final	6.1 ± 0.4	6.4 ± 0.6	6.1 ± 0.6	6.3 ± 0.5	0.744	0.969	0.858
Rate of change	22.1 ± 0.217.5	8.1 ± 10.4	5.5 ± 11.7	11.7 ± 24.7	0.826	0.221	0.719
			Visfatin (ng/mL)				
Initial	1.53 ± 0.09	1.36 ± 0.08	1.63 ± 0.06	1.56 ± 0.07	0.100	0.097	0.596
Final	1.44 ± 0.07	1.48 ± 0.16	1.62 ± 0.07	1.62 ± 0.07	0.764	0.050	0.971
Rate of change	−5.3 ± 2.2	3.9 ± 20.6	0.5 ± 3.0	5.7 ± 4.2	0.724	0.091	0.470
			Adiponectin (ng/mL)				
Initial	39 ± 8	56 ± 12	71 ± 17	78 ± 21	0.157	0.264	0.461
Final	49 ± 9	61 ± 20	91 ± 29	63 ± 15	0.029	0.793	0.916
Rate of change	20.4 ± 33.1	63.1 ± 28.2	54.4 ± 28.8	−69.8 ± 36.9	0.205	0.081	0.098

Table 8. *Cont.*

	Normocholesterolemic $n = 31$		Hypercholesterolemic $n = 37$		p Value		
	OPO	SO	OPO	SO	Oil	N/H	N/H × Oil
			Adipsin (pg/mL)				
Initial	660 ± 68	677 ± 107	853 ± 92	815 ± 91	0.866	0.072	0.739
Final	619 ± 73	600 ± 65	716 ± 64	664 ± 61	0.523	0.190	0.619
Rate of change	4.6 ± 13.2	9.6 ± 11.1	−6.0 ± 1.0	−6.5 ± 9.1	0.969	0.260	0.621

* Values represent mean ± SEM. The table shows the initial (pre-treatment) and final (post-treatment) mean values. The rate of change was calculated from initial and final values as [(final value-initial value)/initial value] and expressed as percentage. Data were analyzed using a linear mixed model. p values in the first column correspond to the effect of taking the oil (OPO or SO), those of the penultimate column correspond to the effect of the group [normocholesterolemic (N) or hypercholesterolemic (H)], and the last column to the interaction of oil and group. Significance level was set at $p < 0.05$. GIP: Gastric inhibitory polypeptide. GLP-1: Glucagon like peptide 1. PAI-1: Plasminogen activator inhibitor-1.

3.9. Anthropometric Measurements and Body Composition

Table 9 shows how 4-week consumption of OPO and SO significantly affected visceral fat percentage ($p = 0.028$). In detail, there was a decrease of −4.5% (normocholesterolemic group) and −1.4% (hypercholesterolemic group) after OPO intake, and an increase of 2.2% (normocholesterolemic group) and 6.6% (hypercholesterolemic group) after SO intake. The other parameters assessed were not affected after the intake of both vegetable oils (OPO and SO). However, when healthy subjects were compared with those at-risk, differences in BMI ($p = 0.050$) and body fat ($p = 0.058$) values were observed. These differences between groups were also observed in weight (initial and final stages), BMI (initial and final stages), visceral fat (initial and final stages), waist circumference (initial and final stages), hip circumference (late stages), and arm circumference (initial and final stages) ($p < 0.05$), when the statistical analysis was performed on initial and final values (Table 9).

Table 9. Effect of olive pomace oil (OPO) and sunflower oil (SO) consumption on anthropometric measurements and body composition *.

	Normocholesterolemic $n = 31$		Hypercholesterolemic $n = 37$		p Value		
	OPO	SO	OPO	SO	Oil	N/H	N/H × Oil
			Weight (kg)				
Initial	64.1 ± 2.1	64.0 ± 2.1	74.1 ± 2.6	74.1 ± 2.7	0.990	0.000	0.967
Final	63.9 ± 2.1	64.0 ± 2.1	74.2 ± 2.7	74.2 ± 2.7	0.945	0.000	0.967
Rate of change	−0.3± 0.3	0.0 ± 0.2	0.1 ± 0.2	0.1 ± 0.2	0.660	0.272	0.545
			BMI (kg/m^2)				
Initial	23.9 ± 0.7	23.7 ± 0.7	26.0 ± 0.7	26.0 ± 0.7	0.893	0.002	0.976
Final	23.8 ± 0.7	23.8 ± 0.7	26.1 ± 0.7	26.1 ± 0.7	0.930	0.001	0.980
Rate of change	−0.4 ± 0. 3	0.3 ± 0.5	0.2 ± 0.2	0.4 ± 0.3	0.365	0.050	0.805
			Body fat (%)				
Initial	25.5 ± 1.5	24.8 ± 1.5	26.2 ± 0.7	26.0 ± 0.7	0.803	0.512	0.845
Final	25.9 ± 1.3	25.2 ± 1.5	26.1 ± 0.7	25.8 ± 0.7	0.844	0.805	0.818
Rate of change	1.1 ± 2.3	−1.0 ± 1.1	3.3 ± 2.4	3.8 ± 2.2	0.848	0.058	0.714
			Visceral fat (%)				
Initial	3.5 ± 0.5	3.3 ± 0.5	7.3 ± 0.7	7.0 ± 0.7	0.794	0.000	0.935
Final	3.4 ± 0.5	3.4 ± 0.5	7.2 ± 0.7	7.1 ± 0.7	0.822	0.000	0.867
Rate of change	−4.5 ± 2.7	2.2 ± 4.0	−1.4 ± 1.5	6.6 ± 6.1	0.028	0.490	0.636

Table 9. *Cont.*

	Normocholesterolemic $n = 31$		Hypercholesterolemic $n = 37$		*p* Value		
	OPO	SO	OPO	SO	Oil	N/H	N/H × Oil
			Waist circumference (cm)				
Initial	75.8 ± 1.8	76.6 ± 2.2	84.5 ± 2.3	85.6 ± 2.5	0.545	0.000	0.863
Final	74.7 ± 1.8	76.0 ± 2.0	83.7 ± 2.3	85.1 ± 2.3	0.548	0.000	0.997
Rate of change	−1.2 ± 0. 9	−0.7 ± 0. 9	−1.0± 0. 4	−0.5 ± 0. 8	0.467	0.935	0.884
			Hip circumference (cm)				
Initial	96.1 ± 1.5	96.0 ± 1.7	99.7 ± 1.2	98.8 ± 1.6	0.973	0.053	0.886
Final	97.6 ± 1.4	96.7 ± 1.7	100.2 ± 1.3	99.2 ± 1.6	0.848	0.001	0.948
Rate of change	0.6 ± 0.5	0.6 ± 0. 9	0.6 ± 0. 4	1.1 ± 0.9	0.534	0.616	0.538
			Arm circumference (cm)				
Initial	28.5 ± 0.6	28.3 ± 0.6	30.5 ± 0.6	30.4 ± 0.6	0.848	0.001	0.948
Final	28.3 ± 0.6	28.6 ± 0.7	30.3 ± 0.6	30.5 ± 0.6	0.945	0.002	0.920
Rate of change	−0.5 ± 0. 5	0.1 ± 0. 4	−0.7 ± 0. 5	0.2 ± 0. 4	0.130	0.962	0.559

* Values represent mean ± SEM. The table shows the initial (pre-treatment) and final (post-treatment) mean values. The rate of change was calculated from initial and final values as [(final value-initial value)/initial value] and expressed as percentage. Data were analyzed using a linear mixed model. *p* values in the first column correspond to the effect of taking the oil (OPO or SO), those of the penultimate column correspond to the effect of the group [normocholesterolemic (N) or hypercholesterolemic (H)], and the last column to the interaction of oil and group. Significance level was set at $p < 0.05$. BMI: Body mass index.

3.10. Antioxidant Capacity and Oxidation Biomarkers

The consumption of OPO and SO had no effect on any of the parameters used to assess the oxidative status of the participants ($p < 0.05$). However, belonging to the normocholesterolemic or hypercholesterolemic group influenced FRAP values ($p = 0.013$). In addition, initial and final FRAP and LDLox values showed variations between the normocholesterolemic and hypercholesterolemic groups, with slightly higher values in the at-risk group ($p < 0.05$) (Supplementary Table S2).

4. Discussion

The MD is one of the best-studied diets because of its beneficial, protective effect against chronic and inflammatory diseases. One of the most well-known and important characteristics of this diet is the use of olive oil as the principal source of energy from fat [2]. All varieties of olive oil (EVOO, VOO, OO, and OPO) are characterised by their high oleic acid content and their wide variety of bioactive compounds, while chemical composition depends on the extraction method [3]. OPO is obtained from the residue remaining after mechanical extraction of the VOO [9]. Although it is refined to make it suitable for consumption, it remains a rich source of triterpenic acids and dialcohols, squalene, tocopherols, sterols, and aliphatic fatty alcohols [3]. Numerous in vitro and preclinical studies have been conducted on some of the minor components of OPO showing promising results in the prevention of cardiovascular diseases and their risk factors, modulating lipid profile, improving endothelial function, inducing hypotensive effects, lowering inflammation levels, and improving biomarkers related to the prevention of diabetes and obesity [3]. However, the possible effects on human health following prolonged OPO consumption remain unknown. Therefore, a randomized, blinded, crossover, controlled clinical trial with OPO, analysing numerous biomarkers related to heart disease and associated pathologies (obesity, diabetes, and inflammation) has been carried out.

4.1. Chemical Composition of the Study Oils

OPO is mainly a monounsaturated fat thanks to its high oleic acid (C18:1) content in comparison with SO and CO, and it is richer in linoleic acid (C18:2). These results were in line with those recently reported by Holgado et al. [29], who characterized three different brands of OPO and SO. The oleic acid (72.02–73.8% for OPO; 28.38–32.16% for SO) and linoleic acid (9.5–11.1% for OPO; 56.6–60.0% for SO) content was similar to those quantified in our study oils (Table 1). These authors also analysed other minor components, with some differences between the OPO used in this study and those characterized by Holgado et al. [29]. The three different OPO analysed by these authors showed a higher content of squalene (742–1538 mg/kg), tocopherols (301–446 mg/kg), and phenolic compounds (8–15 mg/kg), and a lower amount of aliphatic fatty alcohols (1677–2269 mg/kg) than the OPO used in this intervention study (squalene: 675 mg/kg; tocopherols: 201 mg/kg; phenolic compounds < 2 mg/kg; aliphatic fatty alcohols: 1681 mg/kg, Table 1). Regarding CO composition, oleic acid, linoleic acid, and total sterol content shower a similar level to that reported by Güneşer et al. [30].

4.2. Effect of Nutritional Intervention on Lipid Profile, Blood Pressure, and Markers of Liver Function and Endothelial Function

Prolonged consumption of OPO and SO did not induce changes ($p < 0.05$) in any of the biomarkers analysed related to lipid profile and liver function (Table 4). Although the hypolipidemic effect of OPO had not been tested before, numerous in vivo and in vitro studies performed with OPO components (mainly triterpenic acids and aliphatic fatty alcohols) showed a favourable influence on various lipid profile parameters [3]. In addition, oleic acid, as the main fatty acid in all categories of olive oil (EVOO, VOO, OO, and OPO), allows these oils to be classified as monounsaturated fats with the potential to reduce coronary heart disease risk, as had already been recognized by the FDA (Food and Drug Administration 2004) [31]. In this regard, a systematic review and meta-analysis developed by Pastor et al. [32] to assess the effect of hydroxytyrosol, oleic acid, or a combination of both (olive oil) on metabolic syndrome revealed that oleic acid consumption had a beneficial effect on lipid profile. Therefore, the slight downward trend observed in serum levels of TC (in both groups), and TG, LDL-C, and VLDL-C in healthy subjects after OPO intake (Table 4), yet not reaching significant differences, could be a consequence of consuming this monounsaturated fat (OPO) rich in bioactive compounds. In relation to SO, its effect on the lipid profile was investigated in 200 patients with heart disease, where consumption of SO versus coconut oil (a source of saturated fat) showed no significant differences in lipid profile markers (TC, LDL-C, HDL-C, VLDL, Apo B/Apo A ratio) [33], in agreement with the results obtained in the present study. As shown in Table 4, the intervention with SO also resulted in slight, non-significant decreases in TG and VLDL levels in healthy subjects and a reduction in TC in the risk group. This downward trend may be related to the high linoleic acid content of SO, since consumption of this PUFA has been associated with reduced CVD mortality [34].

On the other hand, the mean values of ASAT and ALAT enzymes did not show variations in both groups (normocholesterolemic and hypercholesterolemic) after OPO and SO intake (Table 4). This result was to be expected considering that an increase in the concentration and/or activity of these enzymes is related to adverse events [35], which were not observed in any volunteers. Moreover, blood pressure not only remained within the normal range in all volunteers (with values between 130 and 80 mmHg for SBP and DBP, respectively) [36], but also throughout the study after the OPO and SO dietary intervention (Supplementary Table S1). Several animal studies conducted to evaluate the effect of triterpenic acids (specifically oleanolic acid) on endothelial function demonstrated a protective action against blood pressure dysregulation and disorders associated with hypertension [14,37]. However, considering that participants in the present study were not hypertensive, the lack of effect on blood pressure could be due to being normotensive.

Similarly, markers of endothelial function did not show significant changes after the dietary intervention with OPO and SO. Only eNOS was close to reaching the significance level for the effect of oil ($p = 0.083$), with a downward and upward trend after OPO and SO consumption, respectively (Table 6). This enzyme is responsible for the synthesis of nitric oxide (NO), a key element in vasodilation and the maintenance of vascular tone [38]. Therefore, this result reinforces the possible beneficial effect of OPO on endothelial function previously observed in hypertensive animals fed OPO enriched in triterpenic acids (oleanolic acid and maslinic acid) [13,38]. The slight non-significant decrease detected after SO consumption could be related to its high content in linoleic acid, based on a study in endothelial progenitor cells (EPCs) in which incubation with linoleic acid (the main fatty acid in SO) caused a reduction in eNOS expression [39].

4.3. Effect of Nutritional Intervention on Inflammation Markers

Given the relationship between cardiovascular risk and inflammatory processes, different biomarkers of inflammation were analysed. According to the linear mixed model applied to the rates of change, OPO resulted in a slight increase in serum levels of the anti-inflammatory cytokines IL-4 ($p = 0.021$) and IL-13 ($p = 0.023$) concentrations, showing a decrease after SO consumption in both normo- and hypercholesterolemic humans (Table 5). IL-4 acts as an anti-inflammatory cytokine by blocking the synthesis of IL-1, TNF-α, IL-6, and macrophage inflammatory protein [40]. It also promotes Th2 lymphocyte differentiation, B-lymphocyte proliferation and differentiation, and is a potent inhibitor of apoptosis [40,41]. The anti-inflammatory cytokine IL-13 modulates the production of IL-1, TNF-alpha, IL-8, and macrophage inflammatory protein. In addition, IL-13 stimulates B cell growth and differentiation, inhibits Th1 cells and the production of inflammatory cytokines [42]. IL-4 and IL-13 have been studied for their involvement in the pathogenesis of allergic disorders such as asthma [40,41] and atopic dermatitis [42] through the activation of eosinophils and the production of immunoglobulin E (IgE) by B cells [43]. Although the assessed inflammation parameters showed significant differences in IL-4 and IL-13, their association with allergic disorders such as asthma suggests that these changes could be driven by factors external to the dietary intervention.

The behaviour of some minor components of OPO, such as squalene [43] or triterpenic acids (ursolic and oleanolic acids) [44,45] on the inflammatory response has already been evaluated in vitro and in animal models. The slight upward trend of IL-4 and IL-13 after OPO intake agrees with that observed in a study by Sánchez-Quesada et al. [43] after treatment with 1 μM squalene of immune cells differentiated into pro-inflammatory M1 macrophages. In contrast, studies carried out in murine models with ursolic acid [45,46] and oleanolic acid [44] have shown a reduction in IL-4 [46] and IL-13 [44–46] levels. The relationship between consumption of OPO enriched with triterpenic acids (maslinic and oleanolic acids) and the production of inflammatory mediators has been evaluated in several studies in murine animals [13,14]. In these studies, a lower expression of TNF-α and MCP-1 was found when the animals were fed with OPO enriched with oleanolic and maslinic acids. In the study (Table 5), TNF-α value showed minimal variation after OPO ingestion. However, these results cannot be compared with those obtained in the present trial because they were performed in cultured M1 macrophages [43] or in murine models with asthma [44–46], in contrast to our study, where participants did not have a baseline inflammatory state.

The slight decrease observed in IL-4 after SO intake was also observed in a clinical trial conducted in 78 patients after administration of 2.5 g/day of conjugated linoleic acid (CLA), 400 mg/day of vitamin E, or their mixture in adults with rheumatoid arthritis [47]. All other parameters analysed did not reach the level of significance (Table 5). However, in line with Claro-Cala et al. [14], MCP-1 values showed a slight decreasing trend in both groups of volunteers after OPO ingestion without reaching the level of significance. As mentioned above, more nutritional intervention studies with OPO are required to determine if the findings of this study are replicated in other trials.

4.4. *Effect of Nutritional Intervention on Diabetes and Obesity Markers*

Regarding the effect of OPO and SO consumption on parameters associated with insulin resistance and glycaemic homeostasis, Table 7 shows that mean values of glucose, insulin and HbA1c remained within the normal range for the adult population [48]. This outcome was expected as participants did not suffer from type 2 diabetes mellitus and suggests the lack of insulin resistance in the study population. However, significant changes were observed in HbA1c ($p = 0.021$) by group effect (normocholesterolemic and hypercholesterolemic group), and in glucose values by dietary intervention effect (OPO and SO) ($p = 0.007$) (Table 7). According to the Bonferroni test, changes in glucose values ($p = 0.007$) were observed in normocholesterolemic subjects, with a slight increase of 1.3% after OPO intake, and a decrease of 2.6% after SO intake (Table 7). This tendency was contrary to that in hypercholesterolemic subjects, with a slight decrease of 0.3% after OPO intake and an increase of 3.1% after SO intake. However, considering that glucose levels in volunteers of both groups were within the normal range for the adult population (70 and 130 mg/dL) [48], these small changes would likely have minor physiological relevance. The effect of SO on blood glucose levels has been assessed in a recent clinical trial where consumption of 25 mL/day of SO for seven weeks showed no changes in this marker of diabetes [49]. Similarly, a clinical trial conducted with conjugated linoleic acid (3.9 g/day) versus high oleic sunflower oil (3.9 g/day) in 62 healthy subjects for 12 weeks showed no changes in glucose concentrations [50].

The effect of OPO intake on glucose levels has also been evaluated in obese rats after administration of OPO enriched in oleanolic and maslinic acids. The results demonstrated improved oral glucose tolerance compared to obese control mice [14]. Oleanolic acid, which is substantially more abundant in OPO (187 mg/kg) than in SO (<2 mg/kg), has been shown to exert a beneficial effect on glucose metabolism after administration of 55 mL/day of oleanolic acid-enriched olive oil in a clinical trial in 176 pre-diabetic subjects [51]. In addition, animal studies have evaluated the effect of consuming oleanolic acid, either 10 mg/kg daily [52] or 80 mg/kg every three days [53]. The results supported the hypoglycaemic role of this compound by demonstrating a decrease in blood glucose levels [52,53]. Another characteristic component of the triterpene fraction of OPO, ursolic acid, has also shown antidiabetic activity by reducing blood glucose levels and improving glucose tolerance [54]. These results show that OPO rich in triterpenic compounds could positively contribute to glucose metabolism, and more clinical trials in hyperglycaemic volunteers consuming OPO daily are needed to understand the effect of this type of oil.

Finally, the calculation of mathematical equations to measure insulin resistance (HOMA-IR) ($p = 0.020$) and insulin sensitivity (QUICKI index) ($p = 0.011$) showed statistically significant differences between the normocholesterolemic and hypercholesterolemic groups in agreement with the mentioned results of glucose and insulin (Table 7).

Among the obesity-related markers, ghrelin ($p = 0.031$) and leptin ($p = 0.017$), known for their role in appetite regulation, showed significant changes due to the study oils effect (Table 8). The orexigenic hormone ghrelin increased in both groups after dietary intervention with OPO and SO, the increase being more pronounced after SO intake (18.6%) in the risk group. As for the anorexigenic hormone leptin, except for an 8.0% decrease in the normocholesterolemic group after OPO intake, this hormone increased subsequently after intervention with OPO (in the risk group) and SO in both groups of volunteers (Table 8). Studies in obese mice after administration of oleanolic acid (10 mg/kg) and ursolic acid (50 mg/L, in the drinking water) revealed a decrease in ghrelin levels [52,55], contrary to that observed in the present trial after OPO intervention. In relation to leptin, results reported in the literature are contradictory. De Melo et al. [52] and Rao et al. [55] observed an increase in plasma levels of this hormone in mice treated with oleanolic and ursolic acids. In contrast, a study developed by Jia et al. [54] showed decreased levels of leptin after oral administration of ursolic acid (50 and 200 mg/kg), in line with what we observed in healthy volunteers after OPO administration. However, the amount of triterpenic compounds provided by OPO in the present study (equivalent to 8.4 mg/day of oleanolic acid and trace amounts of ursolic acid) is significantly lower than those administered in the aforementioned studies, which would justify the results of the present clinical study.

4.5. *Effect of Nutritional Intervention on Anthropometric Measurements*

One of the most remarkable results associated with OPO and SO consumption was observed in visceral fat ($p = 0.028$), a marker of cardiovascular and metabolic risk. According to Table 9, OPO consumption produced a slight decrease of 4.5% and 1.4% in both groups of volunteers (healthy and at-risk groups), and an increase after SO intake of 2.2% and 6.6% in the normocholesterolemic and hypercholesterolemic groups, respectively. The beneficial effect on anthropometric parameters following consumption of a monounsaturated fat was also observed in an intervention study, where consumption of up to 20 g/day of olive oil for 6 months led to a significant decrease in body weight and BMI [56]. Similarly, administration of oleanolic acid [52] or ursolic acid [55] to obese mice, as well as with OPO enriched in triterpenic acids (oleanolic and maslinic acids) [14], showed not only a decrease in body weight but also a decrease in visceral adiposity. Given that visceral obesity is associated with increased adipocytokine production, pro-inflammatory activity, increased risk of developing diabetes, dyslipidaemia, hypertension, and atherosclerosis [57], this finding supports the idea that OPO consumption could be a nutritional tool to prevent the onset of cardiometabolic diseases. The upward trend observed in the present study following the intake of SO is consistent with the study in which SO and its effect on visceral fat accumulation was evaluated in humans (39 subjects). The study involved overfeeding with PUFA-rich muffins, which showed a lower increase in visceral fat compared with subjects fed saturated fat-rich muffins [58]. Finally, BMI was statistically significantly different between the normocholesterolemic and hypercholesterolemic groups, with higher mean levels in participants with elevated cholesterol levels, supporting the association between altered lipid profile levels (such as high TC levels) and elevated indicators of overweight/obesity. However, no differences were observed due to the consumption of both oils studied (Table 9).

4.6. *Effect of Nutritional Intervention on Oxidative Status*

Three complementary methods such as FRAP, ABTS, and ORAC were applied to determine the antioxidant capacity after nutritional intervention with OPO and SO. According to the linear repeated measures model, consumption of OPO and SO had no effect on any of the parameters used to evaluate the oxidative status of the volunteers (Supplementary Table S2). Significant differences in FRAP values were only observed for the effect of belonging to the normocholesterolemic or hypercholesterolemic group ($p = 0.013$). Another marker assessed for oxidative status after OPO and SO consumption was LDLox. Oxidation of this lipoprotein, which plays an important role in the initiation and progression of atherosclerosis, also showed no significant variations throughout the trial (Supplementary Table S2). This parameter has been evaluated in different in vitro assays, carried out with different compounds present in OPO, such as erythrodiol, uvaol, oleanolic acid, and maslinic acid, with positive results in the reduction of lipid peroxidation [3]. Furthermore, a diet supplemented with a by-product rich in squalene resulted in lower serum LDLox levels in postmenopausal women [59]. In the present study, a slight decrease of −10.0% (Supplementary Table S2) in LDLox concentrations was observed after OPO intervention in at-risk subjects, although without reaching significant differences. These results are in line with those described by Conterno et al. [60] in hypercholesterolemic subjects, where the 8-week consumption of olive pomace powder-enriched cookies tended to reduce LDLox levels, again without reaching the level of significance. These results evidence the interest in conducting further clinical trials with OPO in subjects with high cholesterol levels to confirm the potential effect on LDLox.

On the other hand, the consumption of both vegetable oils (OPO and SO) had no significant effect on serum MDA levels, a widely used marker of lipid peroxidation, in both study groups (Supplementary Table S2). There are previous studies demonstrating how triterpenic acids (oleanolic, ursolic, and maslinic acids) [61–64] and squalene (components of OPO) [65] decrease MDA levels in various animal models of hypertension, inflammation, or diabetes. However, the high, pharmacological dose of pure triterpenic acids and squalene administered to the experimental animals, as well as the pathologies of the murine models used, limit potential comparisons with the present nutritional human intervention study.

Finally, it is important to note that both oils evaluated in this study (OPO and SO) are foods with a high vitamin E content, a nutrient that has been shown to contribute to the protection of cells against oxidative damage (Commission Regulation (EC) No. 432/2012) [66]. Due to the similarity in the vitamin E content in OPO and SO (Table 1), it is possible that no differences were observed in the oxidative status parameters evaluated.

4.7. *Strengths and Limitations*

One of the strengths of this novel clinical trial carried out with OPO was the identification of numerous biomarkers related to cardiovascular disease and its associated pathologies. This approach provides a broad understanding of the role of OPO on health in humans. However, the present study also had limitations on the blinding of volunteers, considering that the colour, smell, and taste of the two oils are different, it was not possible to prevent volunteers (familiar with both oils) from knowing/guessing which oil they were consuming in each stage. Another limitation was that the results reported in the present trial do not allow the generation of clinical evidence because the type I error of 0.05 and statistical power of 80% (1-beta) were taken as a reference in an incidental sample of volunteers controlled only by cholesterol levels. To generate clinical evidence, it would have been necessary to adjust (reduce) alpha and to perform the present study in a population with many other controllable effects (in addition to cholesterol levels).

5. Conclusions

Regular consumption of OPO and SO had no statistically significant effect on any of the markers related to lipid profile, blood pressure, and endothelial function in normocholesterolemic and hypercholesterolemic participants. Only eNOS level change was close to being significant due to the effect of oil (OPO and SO) ($p = 0.083$); further studies would be necessary to assess the potential effect of OPO on this biomarker, of great interest considering the involvement of eNOS on the synthesis of nitric oxide, the main factor responsible for vasodilation and maintenance of vascular tone. This result was reinforced by a significant decrease in visceral fat ($p = 0.028$) after OPO intake in both groups, accompanied by the increment of leptin level in the hypercholesterolemic group ($p = 0.017$). These results suggest a potential beneficial effect of sustained consumption of OPO on biomarkers that may have a positive effect on cardiometabolic health. However, more clinical trials on at-risk populations are needed to confirm these health effects.

Supplementary Materials: The following supporting information can be downloaded at: https://www.mdpi.com/article/10.3390/nu14193927/s1, Table S1: Effect of olive pomace oil (OPO) and sunflower oil (SO) consumption on blood; Table S2. Effect of olive pomace oil (OPO) and sunflower oil (SO) consumption on antioxidant capacity and lipid peroxidation.

Author Contributions: Conceptualization, L.B.-C., R.M. and B.S.; methodology, B.S., R.M., S.G.-R., M.Á.S. and J.G.-C.; software, S.G.-R., M.Á.S. and J.G.-C.; validation, R.M.; analysis, S.G.-R., M.Á.S. and J.G.-C.; investigation, R.M., B.S. and L.B.-C.; resources, R.M., B.S. and L.B.-C.; data curation, S.G.-R. and M.Á.S.; writing—original draft preparation, S.G.-R. and R.M.; writing-review and editing, R.M., B.S. and L.B.-C.; supervision, R.M. and B.S.; project administration, L.B.-C., R.M. and B.S.; funding acquisition R.M. and L.B.-C. All authors have read and agreed to the published version of the manuscript.

Funding: Interprofesional del Aceite de Orujo de Oliva (ORIVA) financed the study (Ref. 20193239) and the predoctoral contract of S.G.R. (Ref. 20184930). The Community of Madrid partly financed the predoctoral contracts of M.A.S. (PEJD-2018-PRE/SAL-9104) and J.G.-C. (PEJD-2017-PRE/BIO-4225). No other conflicts of interests are declared. ORIVA had no part on the design, conductance, analysis, or interpretation of results.

Institutional Review Board Statement: The study was conducted in accordance with the Declaration of Helsinki, and approved by the Ethics Committee of Hospital Universitario Puerta de Hierro de Majadahonda (protocol code PI 110/19 date of approval 22 July 2019) and the Bioethics Committee of CSIC.

Informed Consent Statement: The study was approved by the Bioethics Committee of Consejo Superior de Investigaciones Científicas (CSIC) and the Clinical Research Ethics Committee of Hospital Universitario Puerta del Hierro (protocol code PI 110/19, date of approval 22 July 2019). It was retrospectively registered in ClinicalTrials.gov (NCT04998695). Informed consent was obtained from all subjects involved in the study.

Data Availability Statement: The data presented in this study are available on request from the corresponding author. The data are not publicly available due to privacy concerns.

Acknowledgments: We are grateful to volunteers for their participation in the study and to Laura Barrios for her statistical advice.

Conflicts of Interest: The authors declare no conflict of interest.

Abbreviations

BMI: body mass index; CLA: conjugated linoleic acid; CSIC: Consejo Superior de Investigaciones Científicas; CO: corn oil; DBP: diastolic blood pressure; eNOS: endothelial nitric oxide synthase; EPCs: endothelial progenitor cells; (ESC) European society of cardiology; EVOO: extra virgin olive oil; FRAP: ferric reducing/antioxidant power; GLP-1: glucagon-like peptide type 1; HbA1c: glycosylated haemoglobin; G-CSF: granulocyte colony-stimulating factor; GM-CSF: granulocyte monocyte colony-stimulating factor; HPLC: high-performance liquid chromatography; HOMA-IR: homeostatic model to assess β-cell functionality; HOMA-IR: homeostatic model to assessment insulin resistance; IgE: immunoglobulin E; GIP: incretin gastric inhibitory polypeptide; ICTAN: institute of food science, technology and nutrition; ICAM-!: intercellular adhesion molecules; IFN-γ: interferon-gamma; IL: interleukins; LDL-C: low density lipoprotein; LDL-ox: low density lipoprotein oxidation; MIP-1β: macrophage inflammatory protein 1 beta; MDA: malondialdehyde; MD: Mediterranean diet; MCP-1: monocyte chemoattractant protein 1; MUFA: monounsaturated fatty acids; NO: nitric oxide; OO: olive oil; OPO: olive pomace oil; ORAC: oxygen radical absorbance capacity; PAI-1: plasminogen activator inhibitor-1; PUFA: polyunsaturated fatty acid; SO: sunflower oil; SBP: systolic blood pressure; TC: total cholesterol; TRL: triglyceride-rich lipoproteins (TRL); TG: triglycerides; TNF-α: tumour necrosis factor alpha; VCAM-1: vascular adhesion molecules; VOO: virgin olive oil; VLDL: very low-density lipoprotein.

References

1. D'Innocenzo, S.; Biagi, C.; Lanari, M. Obesity and the Mediterranean Diet: A review of evidence of the role and sustainability of the Mediterranean Diet. *Nutrients* **2019**, *11*, 1306. [CrossRef]
2. Martínez-González, M.A.; Salas-Salvadó, J.; Estruch, R.; Corella, D.; Fitó, M.; Ros, E.; PREDIMED investigators. Benefits of the Mediterranean Diet: Insights from the PREDIMED study. *Prog. Cardiovasc. Dis.* **2015**, *58*, 50–60. [CrossRef] [PubMed]
3. Mateos, R.; Sarriá, B.; Bravo, L. Nutritional and other health properties of olive pomace oil. *Crit. Rev. Food Sci. Nutr.* **2020**, *60*, 3506–3521. [CrossRef]
4. Marcelino, G.; Hiane, P.A.; Freitas, K.C.; Santana, L.F.; Pott, A.; Donadon, J.R.; Guimarães, R. Effects of olive oil and its minor components on cardiovascular diseases, inflammation, and gut microbiota. *Nutrients* **2019**, *11*, 1826. [CrossRef]
5. Mazzocchi, A.; Leone, L.; Agostoni, C.; Pali-Schöll, I. The secrets of the Mediterranean Diet. Does [only] olive oil matter? *Nutrients* **2019**, *11*, 2941. [CrossRef] [PubMed]
6. ANIERAC: Asociación Nacional de Industriales Envasadores y Refinadores de Aceites Comestibles. Available online: https://anierac.org/consumo-en-espana/?lang=es (accessed on 1 January 2022).

7. García-Aloy, M.; Hulshof, P.J.M.; Estruel-Amades, S.; Osté, M.C.J.; Lankinen, M.; Geleijnse, J.M.; de Goede, J.; Ulaszewska, M.; Mattivi, F.; Bakker, S.J.L.; et al. Biomarkers of food intake for nuts and vegetable oils: An extensive literature search. *Genes Nutr.* **2019**, *14*, 7. [CrossRef] [PubMed]
8. Aceites de Semillas: Palma, Colza, Soja y Girasol Lideran la Producción y el Consumo Mundial. Available online: https://www.mercasa.es/media/publicaciones/58/pag_065-070_Murcia.pdf (accessed on 2 February 2022).
9. Ruiz-Méndez, M.V.; Márquez-Ruiz, G.; Holgado, F.; Velasco, J. Stability of bioactive compounds in olive-pomace oil at frying temperature and incorporation into fried foods. *Foods* **2021**, *10*, 2906. [CrossRef]
10. Olimerca. Available online: https://www.olimerca.com/precios/tipoInforme/1 (accessed on 2 February 2022).
11. Sánchez-Moral, P.; Ruíz-Méndez, M.V. Production of pomace olive oil. *Grasas y Aceites* **2006**, *57*, 47–55. [CrossRef]
12. Lee, W.; Yang, E.J.; Ku, S.K.; Song, K.S.; Bae, J.S. Anti-inflammatory effects of oleanolic acid on LPS-induced inflammation in vitro and in vivo. *Inflammation* **2013**, *36*, 94–102. [CrossRef]
13. Valero-Muñoz, M.; Martín-Fernández, B.; Ballesteros, S.; de la Fuente, E.; Quintela, J.C.; Lahera, V.; de las Heras, N. Protective effect of a pomace olive oil concentrated in triterpenic acids in alterations related to hypertension in rats: Mechanisms involved. *Mol. Nutr. Food Res.* **2014**, *58*, 376–383. [CrossRef]
14. Claro-Cala, C.M.; Quintela, J.C.; Pérez-Montero, M.; Miñano, J.; de Sotomayor, M.A.; Herrera, M.D.; Rodríguez-Rodríguez, A.R. Pomace olive oil concentrated in triterpenic acids restores vascular function, glucose tolerance and obesity progression in mice. *Nutrients* **2020**, *12*, 323. [CrossRef] [PubMed]
15. Cabello-Moruno, R.; Martinez-Force, E.; Montero, E.; Perona, J.S. Minor components of olive oil facilitate the triglyceride clearance from postprandial lipoproteins in a polarity-dependent manner in healthy men. *Nutr. Res.* **2014**, *34*, 40–47. [CrossRef] [PubMed]
16. González-Rámila, S.; Sarriá, B.; Seguido, M.A.; García-Cordero, J.; Mateos, R.; Bravo, L. Olive pomace oil can improve blood lipid profile: A randomized, blind, crossover, controlled clinical trial in healthy and at-risk volunteers. *Eur. J. Nutr.* 2022, *in press*.
17. González-Rámila, S.; Mateos, R.; García-Cordero, J.; Seguido, M.A.; Bravo-Clemente, L.; Sarriá, B. Olive pomace oil versus high oleic sunflower oil and sunflower oil: A comparative study in healthy and cardiovascular risk humans. *Foods* **2022**, *11*, 2186. [CrossRef] [PubMed]
18. Pérez-Camino, M.C.; Cert, A. Quantitative determination of hydroxy pentacyclic triterpene acids in vegetable oils. *J. Agric. Food Chem.* **1999**, *47*, 1558–1562. [CrossRef]
19. Giacometti, J. Determination of aliphatic alcohols, squalene, alpha-tocopherol and sterols in olive oils: Direct method involving gas chromatography of the unsaponifiable fraction following silylation. *Analyst* **2001**, *126*, 472–475. [CrossRef]
20. Mateos, R.; Espartero, J.L.; Trujillo, M.; Ríos, J.J.; León-Camacho, M.; Alcudia, F.; Cert, A. Determination of phenols, flavones, and lignans in virgin olive oils by solid-phase extraction and high-performance liquid chromatography with diode array ultraviolet detection. *J. Agric. Food Chem.* **2001**, *49*, 2185–2192. [CrossRef]
21. Manual de Nutrición y Dietética. Available online: https://eprints.ucm.es/id/eprint/22755/1/Manual-nutricion-dietetica-CARBAJAL.pdf (accessed on 20 January 2022).
22. Matthews, D.R.; Hosker, J.P.; Rudenski, A.S.; Naylor, B.A.; Treacher, D.F.; Turner, R.C. Homeostasis model assessment: Insulin resistance and β-cell function from fasting plasma glucose and insulin concentrations in man. *Diabetologia* **1985**, *28*, 412–419. [CrossRef]
23. Re, R.; Pellegrini, N.; Proteggente, A.; Pannala, A.; Yang, M.; Rice-Evans, C. Antioxidant activity applying an improved ABTS radical cation decolorization assay. *Free Radic. Biol. Med.* **1999**, *26*, 1231–1237. [CrossRef]
24. Huang, D.; Ou, B.; Hampsch-Woodill, M.; Flanagan, J.A.; Prior, R.L. High-throughput assay of oxygen radical absorbance capacity (ORAC) using a multichannel liquid handling system coupled with a microplate fluorescence reader in 96-well format. *J. Agric. Food Chem.* **2012**, *50*, 4437–4444. [CrossRef]
25. Benzie, I.F.; Strain, J.J. The ferric reducing ability of plasma (FRAP) as a measure of "antioxidant power": The FRAP assay. *Anal. Biochem.* **1996**, *239*, 70–76. [CrossRef]
26. Mateos, R.; Lecumberri, E.; Ramos, S.; Goya, L.; Bravo, L. Determination of malondialdehyde (mda) by high-performance liquid chromatography in serum and liver as a biomarker for oxidative stress: Application to a rat model for hypercholesterolemia and evaluation of the effect of diets rich in phenolic antioxidants from fruits. *J. Chromatogr. B* **2005**, *827*, 76–82.
27. Martínez-López, S.; Sarriá, B.; Mateos, R.; Bravo-Clemente, L. Moderate consumption of a soluble green/roasted coffee rich in caffeoylquinic acids reduces cardiovascular risk markers: Results from a randomized, cross-over, controlled trial in healthy and hypercholesterolemic subjects. *Eur. J. Nutr.* **2019**, *58*, 865–878. [CrossRef] [PubMed]
28. Moreiras, O.; Carbajal, A.; Cabrera, L.; Cuadrado, C. Ingestas diarias recomendadas de energía y nutrientes para la población española. In *Tabla de Composición de Alimentos*, 18th ed.; Moreiras, O., Carbajal, A., Cabrera, L., Cuadrado, C., Eds.; Ediciones Pirámides (Grupo Anaya, SA): Madrid, Spain, 2016; pp. 127–131.
29. Holgado, F.; Ruiz-Méndez, M.V.; Velasco, J.; Márquez-Ruiz, G. Performance of olive-pomace oils in discontinuous and continuous frying. Comparative behavior with sunflower oils and high-oleic sunflower oils. *Foods* **2021**, *10*, 3081. [CrossRef]
30. Güneşer, B.A.; Yılmaz, E.; Ok, S. Cold pressed versus refined winterized corn oils: Quality, composition and aroma. *Grasas y Aceites* **2017**, *68*, e194. [CrossRef]
31. Food and Drugs Administration (FDA). Available online: https://www.fda.gov/ (accessed on 9 June 2022).
32. Pastor, R.; Bouzas, C.; Tur, J.A. Beneficial effects of dietary supplementation with olive oil, oleic acid, or hydroxytyrosol in metabolic syndrome: Systematic review and meta-analysis. *Free Radic. Biol. Med.* **2021**, *172*, 372–385. [CrossRef]

33. Vijayakumar, M.; Vasudevan, D.M.; Sundaram, K.R.; Krishnan, S.; Vaidyanathan, K.; Nandakumar, S.; Chandrasekhar, R.; Mathew, N. A randomized study of coconut oil versus sunflower oil on cardiovascular risk factors in patients with stable coronary heart disease. *Indian Heart J.* **2016**, *68*, 498–506. [CrossRef]
34. Jandacek, R.J. Linoleic acid: A nutritional quandary. *Healthcare* **2017**, *5*, 25. [CrossRef]
35. Weiß, J.; Rau, M.; Geier, A. Non-alcoholic fatty liver disease: Epidemiology, clinical course, investigation, and treatment. *Dtsch. Arztebl. Int.* **2014**, *111*, 447–452.
36. de la Sierra, A. New American and European hypertension guidelines, reconciling the differences. *Cardiol. Ther.* **2019**, *8*, 157–166. [CrossRef]
37. Rodríguez-Rodríguez, R. Oleanolic acid and related triterpenoids from olives on vascular function: Molecular mechanisms and therapeutic perspectives. *Curr. Med. Chem.* **2015**, *22*, 1414–1425. [CrossRef] [PubMed]
38. Rodríguez-Rodríguez, R.; Herrera, M.D.; de Sotomayor, M.A.; Ruiz-Gutiérrez, V. Pomace olive oil improves endothelial function in spontaneously hypertensive rats by increasing endothelial nitric oxide synthase expression. *Am. J. Hypertens.* **2007**, *20*, 728–734. [CrossRef] [PubMed]
39. Guo, W.X.; Yang, Q.D.; Liu, Y.H.; Xie, X.Y.; Wang, M.; Niu, R.C. Palmitic and linoleic acids impair endothelial progenitor cells by inhibition of Akt/eNOS pathway. *Arch. Med. Res.* **2008**, *39*, 434–442. [CrossRef]
40. Junttila, I.S. Tuning the Cytokine Responses: An Update on Interleukin (IL)-4 and IL-13 receptor complexes. *Front. Immunol.* **2018**, *9*, 888. [CrossRef] [PubMed]
41. Gour, N.; Wills-Karp, M. IL-4 and IL-13 signalling in allergic airway disease. *Cytokine* **2015**, *75*, 68–78. [CrossRef]
42. Gonçalves, F.; Freitas, E.; Torres, T. Selective IL-13 inhibitors for the treatment of atopic dermatitis. *Drugs Context* **2021**, *10*, 2021-1-7. [CrossRef]
43. Sánchez-Quesada, C.; López-Biedma, A.; Toledo, E.; Gaforio, J.J. Squalene stimulates a key innate immune cell to foster wound healing and tissue repair. *Evid. Based Complement. Alternat. Med.* **2018**, *2018*, 9473094. [CrossRef]
44. Kim, S.H.; Hong, J.H.; Lee, Y.C. Oleanolic acid suppresses ovalbumin-induced airway inflammation and Th2-mediated allergic asthma by modulating the transcription factors T-bet, GATA-3, RORγt and Foxp3 in asthmatic mice. *Int. Immunopharmacol.* **2014**, *18*, 311–324. [CrossRef]
45. Kim, S.H.; Hong, J.H.; Lee, Y.C. Ursolic acid, a potential PPARγ agonist, suppresses ovalbumin-induced airway inflammation and Penh by down-regulating IL-5, IL-13, and IL-17 in a mouse model of allergic asthma. *Eur. J. Pharmacol.* **2013**, *701*, 131–143. [CrossRef]
46. Sun, N.; Han, Z.; Wang, H.; Guo, Z.; Deng, C.; Dong, W.; Zhuang, G.; Zhang, R. Effects of ursolic acid on the expression of Th1-Th2-related cytokines in a rat model of allergic rhinitis after PM2.5 exposure. *Am. J. Rhinol. Allergy* **2020**, *34*, 587–596. [CrossRef]
47. Aryaeian, N.; Djalali, M.; Shahram, F.; Djazayery, A.; Eshragian, M.R. Effect of conjugated linoleic acid, vitamin E, alone or combined on immunity and inflammatory parameters in adults with active rheumatoid arthritis: A randomized controlled trial. *Int. J. Prev. Med.* **2014**, *5*, 1567–1577. [PubMed]
48. Nathan, D.M.; Buse, J.B.; Davidson, M.B.; Ferrannini, E.; Holman, R.R.; Sherwin, R.; Zinman, B.; American Diabetes Association; European Association for the Study of Diabetes. Medical management of hyperglycaemia in type 2 diabetes mellitus: A consensus algorithm for the initiation and adjustment of therapy: A consensus statement from the American Diabetes Association and the European Association for the Study of Diabetes. *Diabetologia* **2009**, *52*, 17–30. [CrossRef]
49. Akrami, A.; Nikaein, F.; Babajafari, S.; Faghih, S.; Yarmohammadi, H. Comparison of the effects of flaxseed oil and sunflower seed oil consumption on serum glucose, lipid profile, blood pressure, and lipid peroxidation in patients with metabolic syndrome. *J. Clin. Lipidol.* **2018**, *12*, 70–77. [CrossRef]
50. Lambert, E.V.; Goedecke, J.H.; Bluett, K.; Heggie, K.; Claassen, A.; Rae, D.E.; West, S.; Dugas, J.; Dugas, L.; Meltzeri, S.; et al. Conjugated linoleic acid versus high-oleic acid sunflower oil: Effects on energy metabolism, glucose tolerance, blood lipids, appetite and body composition in regularly exercising individuals. *Br. J. Nutr.* **2007**, *97*, 1001–1011. [CrossRef]
51. Santos-Lozano, J.M.; Rada, M.; Lapetra, J.; Guinda, Á.; Jiménez-Rodríguez, M.C.; Cayuela, J.A.; Ángel-Lugo, A.; Vilches-Arenas, Á.; Gómez-Martín, A.M.; Ortega-Calvo, M.; et al. Prevention of type 2 diabetes in prediabetic patients by using functional olive oil enriched in oleanolic acid: The PREDIABOLE study, a randomized controlled trial. *Diabetes Obes. Metab.* **2019**, *21*, 2526–2534. [CrossRef]
52. de Melo, C.L.; Queiroz, M.G.; Fonseca, S.G.; Bizerra, A.M.; Lemos, T.L.; Melo, T.S.; Santos, F.A.; Rao, V.S. Oleanolic acid, a natural triterpenoid improves blood glucose tolerance in normal mice and ameliorates visceral obesity in mice fed a high-fat diet. *Chem. Biol. Interact.* **2010**, *185*, 59–65. [CrossRef]
53. Gamede, M.; Mabuza, L.; Ngubane, P.; Khathi, A. The effects of plant-derived oleanolic acid on selected parameters of glucose homeostasis in a diet-induced pre-diabetic rat model. *Molecules* **2018**, *23*, 794. [CrossRef]
54. Jia, Y.; Kim, S.; Kim, J.; Kim, B.; Wu, C.; Lee, J.H.; Jun, H.J.; Kim, N.; Lee, D.; Lee, S.J. Ursolic acid improves lipid and glucose metabolism in high-fat-fed C57BL/6J mice by activating peroxisome proliferator-activated receptor alpha and hepatic autophagy. *Mol. Nutr. Food Res.* **2015**, *59*, 344–354. [CrossRef]
55. Rao, V.S.; de Melo, C.L.; Queiroz, M.G.; Lemos, T.L.; Menezes, D.B.; Melo, T.S.; Santos, F.A. Ursolic acid, a pentacyclic triterpene from *Sambucus australis*, prevents abdominal adiposity in mice fed a high-fat diet. *J. Med. Food.* **2011**, *14*, 1375–1382. [CrossRef]

56. Nigam, P.; Bhatt, S.; Misra, A.; Chadha, D.S.; Vaidya, M.; Dasgupta, J.; Pasha, Q.M. Effect of a 6-month intervention with cooking oils containing a high concentration of monounsaturated fatty acids (olive and canola oils) compared with control oil in male Asian Indians with nonalcoholic fatty liver disease. *Diabetes Technol. Ther.* **2014**, *16*, 255–261. [CrossRef]
57. Amato, M.C.; Giordano, C.; Galia, M.; Criscimanna, A.; Vitabile, S.; Midiri, M.; Galluzzo, A.; AlkaMeSy Study Group. Visceral adiposity index: A reliable indicator of visceral fat function associated with cardiometabolic risk. *Diabetes Care* **2010**, *33*, 920–922. [CrossRef] [PubMed]
58. Rosqvist, F.; Iggman, D.; Kullberg, J.; Cedernaes, J.; Johansson, H.E.; Larsson, A.; Johansson, L.; Ahlström, H.; Arner, P.; Dahlman, I.; et al. Overfeeding polyunsaturated and saturated fat causes distinct effects on liver and visceral fat accumulation in humans. *Diabetes* **2014**, *63*, 2356–2368. [CrossRef]
59. Casado-Díaz, A.; Túnez-Fiñana, I.; Mata-Granados, J.M.; Ruiz-Méndez, M.V.; Dorado, G.; Romero-Sánchez, M.C.; Navarro-Valverde, C.; Quesada-Gómez, J.M. Serum from postmenopausal women treated with a by-product of olive-oil extraction process stimulates osteoblastogenesis and inhibits adipogenesis in human mesenchymal stem-cells (MSC). *Exp. Gerontol.* **2017**, *90*, 71–78. [CrossRef]
60. Conterno, L.; Martinelli, F.; Tamburini, M.; Fava, F.; Mancini, A.; Sordo, M.; Pindo, M.; Martens, S.; Masuero, D.; Vrhovsek, U.; et al. Measuring the impact of olive pomace enriched biscuits on the gut microbiota and its metabolic activity in mildly hypercholesterolaemic subjects. *Eur. J. Nutr.* **2019**, *58*, 63–81. [CrossRef]
61. Iskender, H.; Dokumacioglu, E.; Terim Kapakin, K.A.; Yenice, G.; Mohtare, B.; Bolat, I.; Hayirli, A. Effects of oleanolic acid on inflammation and metabolism in diabetic rats. *Biotech. Histochem.* **2021**, *97*, 269–276. [CrossRef]
62. Madlala, H.P.; Van Heerden, F.R.; Mubagwa, K.; Musabayane, C.T. Changes in renal function and oxidative status associated with the hypotensive effects of oleanolic acid and related synthetic derivatives in experimental animals. *PLoS ONE* **2015**, *10*, e0128192. [CrossRef]
63. Liu, B.; Piao, X.; Guo, L.; Liu, S.; Chai, F.; Gao, L. Ursolic acid protects against ulcerative colitis via anti-inflammatory and antioxidant effects in mice. *Mol. Med. Rep.* **2016**, *13*, 4779–4785. [CrossRef]
64. Mkhwanazi, B.N.; Serumula, M.R.; Myburg, R.B.; Van Heerden, F.R.; Musabayane, C.T. Antioxidant effects of maslinic acid in livers, hearts and kidneys of streptozotocin-induced diabetic rats: Effects on kidney function. *Ren. Fail.* **2014**, *36*, 419–431. [CrossRef]
65. Chen, Y.; Gu, Y.; Zhao, H.; Zhang, H.; Zhou, Y. Effects of graded levels of dietary squalene supplementation on the growth performance, plasma biochemical parameters, antioxidant capacity, and meat quality in broiler chickens. *Poult. Sci.* **2020**, *99*, 5915–5924. [CrossRef]
66. European Commission. Commission Regulation (EC) No 432/2012. *Off. J. Eur. Union* **2012**, *L136*, 1–40.

nutrients

MDPI

Review

Why Should Pistachio Be a Regular Food in Our Diet?

Raquel Mateos [1,*], María Desamparados Salvador [2], Giuseppe Fregapane [2] and Luis Goya [1,*]

1 Institute of Food Science, Technology and Nutrition (ICTAN-CSIC), Spanish National Research Council (CSIC), José Antonio Nováis 10, 28040 Madrid, Spain
2 Facultad de Ciencias y Tecnologías Químicas, Universidad de Castilla-La Mancha, Camilo José Cela n° 10, 13071 Ciudad Real, Spain
* Correspondence: raquel.mateos@ictan.csic.es (R.M.); luisgoya@ictan.csic.es (L.G.)

Abstract: The pistachio is regarded as a relevant source of biologically active components that, compared to other nuts, possess a healthier nutritional profile with low-fat content composed mainly of monounsaturated fatty acids, a high source of vegetable protein and dietary fibre, remarkable content of minerals, especially potassium, and an excellent source of vitamins, such as vitamins C and E. A rich composition in terms of phytochemicals, such as tocopherols, carotenoids, and, importantly, phenolic compounds, makes pistachio a powerful food to explore its involvement in the prevention of prevalent pathologies. Although pistachio has been less explored than other nuts (walnut, almonds, hazelnut, etc.), many studies provide evidence of its beneficial effects on CVD risk factors beyond the lipid-lowering effect. The present review gathers recent data regarding the most beneficial effects of pistachio on lipid and glucose homeostasis, endothelial function, oxidative stress, and inflammation that essentially convey a protective/preventive effect on the onset of pathological conditions, such as obesity, type 2 diabetes, CVD, and cancer. Likewise, the influence of pistachio consumption on gut microbiota is reviewed with promising results. However, population nut consumption does not meet current intake recommendations due to the extended belief that they are fattening products, their high cost, or teething problems, among the most critical barriers, which would be solved with more research and information.

Keywords: pistachio; prevalent chronic diseases; nuts; nutritional value; health benefits; barriers; facilitator

Citation: Mateos, R.; Salvador, M.D.; Fregapane, G.; Goya, L. Why Should Pistachio Be a Regular Food in Our Diet? *Nutrients* **2022**, *14*, 3207. https://doi.org/10.3390/nu14153207

Academic Editor: Justyna Godos

Received: 14 July 2022
Accepted: 2 August 2022
Published: 5 August 2022

Publisher's Note: MDPI stays neutral with regard to jurisdictional claims in published maps and institutional affiliations.

1. Introduction

Nuts are regarded as a relevant source of biologically active components, mainly thanks to their great proportion of unsaturated and essential fatty acids as well as phenolic components [1–3]. In fact, in July 2003, the health claim: "scientific evidence suggests but does not prove that eating 1.5 oz (42.5 g) per day of most nuts, such as pistachios, as part of a diet low in saturated fat and cholesterol may reduce the risk of heart disease", was ratified by the Food and Drug Administration of United States [4].

Since then, more studies have been carried out to determine the impact of frequent consumption of nuts on our health. Among them, the PREDIMED study [5] is one of the most valued interventions, which evaluated the effect of a Mediterranean diet supplemented with either 30 g of nuts or olive oil in 7447 participants at high risk of cardiovascular disease (CVD) in comparison to the control diet. Results showed that the Mediterranean diet that included the daily consumption of nuts reduced the risk of suffering a myocardial infarction, stroke, or death from a cardiovascular cause. Thus, improving glycaemic control, lipid levels, blood pressure, obesity biomarkers, and endothelial function, among others, could minimise the risk of suffering CVD [3,6]. Indeed, in a recent overview, the health benefits of nuts regarding the management of dysmetabolic conditions, such as obesity and type 2 diabetes mellitus (T2DM), closely associated with CVD, have been broadly proven [7].

Regarding the most consumed types of nuts, almonds and walnuts accounted for half of the total tree nut estimated consumption worldwide in 2019 (30% and 20%, respectively), followed by cashews, pistachios and hazelnuts, which accounted for 18%, 15% and 11%, respectively, according to International Nut & Dried Fruits [8]. Following Europe as the leading consumer, Asia and North America were the second and third largest consuming regions, with similar market shares. In the period 2010–2019, nuts were mainly consumed in high- and middle-income economies (56% and 39% of the world share for tree nuts in 2019), while peanuts were mostly consumed in middle-income countries, which represented 91% of the global share [8].

It is worth drawing attention to the pistachio among the nuts. *Pistacia vera* L. is a plant of the Anacardiaceae family which yields a distinctive green fruit of great gastronomic reputation called pistachio. Its cultivation began in western Asia but later spread to the Mediterranean basin through Iran [9]. Season 2020/2021 was largely an "on year" and, as a result, global production totalled over 1 million metric tons (in-shell basis), the highest amount of the last decade, representing a 54% increase from the previous year and a 68% increase over the previous 10-year average. The United States was the top supplier, accounting for 47% of the global share. It was the second "on year" in a row for Iran, and Turkey showed an exceptionally high bumper crop. The US, Turkey, and Iran produced 97% of the world's pistachio. Syria (2%), Greece (1%), and others, such as Italy, Spain, Greece, Tunisia, Afghanistan, China, and Australia, account for the remaining 1% [8]. Although pistachio has been the least explored nut, there is broad evidence confirming its beneficial health effects and, particularly, its positive contribution to minimizing the risk for CVD [10,11], among other pathologies.

Due to the health benefits associated with the intake of nuts, their consumption is promoted as a regular food in most of the food-based dietary guidelines [12] in the context of a healthy diet. However, the population does not always consume them. For example, the Global Burden of Disease Study 2017 estimated 21 g per day as the optimal intake of nuts and seeds, while their actual consumption barely reached 12% of the recommended level [13].

This review aims to revalue pistachio consumption, the least explored nut but with the same or even more healthy potential than the rest due to its interesting nutritional composition. A review of the studies focused on evaluating the biological properties of pistachio, world consumption, and other uses will be developed. In addition, barriers and facilitators of pistachio consumption will be analysed to understand why the dietary requirements for nut consumption are not being met, attempting to shed light and help make it happen.

2. Nutritional Value of Pistachios

The pistachio is a low-water (3–6%) and nutritionally rich nut mainly because of its high fat (48–63%) and protein concentration (18–22%), together with the dietary fibre (8–12%; Table 1) [14,15]. In fact, the daily intake of nuts recommended (1.5 oz equivalent to 42.5 g; 4) in the form of pistachios is approximately 15% of the Dietary Reference Intake (DRI) for proteins, 11–18% of DRI for male and female respectively for dietary fibre, and 24% of DRI for fat. Lipids, although present in great amounts, have an equilibrated content of mono- (56–77%) and polyunsaturated (14–33%; Table 1) fatty acids, that may help to reduce LDL-cholesterol and hence the coronary heart disease risk [2,16,17].

Dry roasted pistachios have a lower fat content (45.82 g/100 g) than other nuts (Table 1), mainly monounsaturated fatty acid (24.53 g) followed by polyunsaturated fatty acid (13.35 g) and saturated fatty acid (5.64 g). Of the fatty acids, oleic is the main monounsaturated fatty acid (MUFA) in pistachios followed by linoleic acid (C18:2), which represents more than half of the total fatty acid content (<60 g) (Table 1). Pistachio shows a similar lipid profile to almonds and hazelnuts while walnut fat is headed by linoleic acid (54.0–65.0 g/100 g) followed by oleic (C18:2) and linolenic (C18:3) acids with similar contents. Pistachios are also a worthy source of vegetable protein (about 21%) as almonds

(21%), and higher than other nuts, such as hazelnuts and/or walnuts. The amount of total carbohydrates is low to moderate (28%), but pistachios are rich in dietary fibre, mainly insoluble fibre (about 10% versus less than 1% of soluble fibre).

Table 1. Composition (g/100 g) in nutrients of dry roasted nuts. Source: USDA National Nutrient Database for Standard Reference (2020). MUFA: monousaturated fatty acids; SFA: saturated fatty acids; PUFA: polyunsaturated fatty acids.

g/100 g	Pistachio	Walnut	Almond	Hazelnut
Water	1.85	4.39	2.41	2.52
Energy (Kcal)	572	643	598	646
Lipids	45.82	60.71	52.54	62.40
SFA	5.64	5.36	4.10	4.51
PUFA	13.35	44.18	12.96	8.46
MUFA	24.53	8.37	33.08	46.61
C16:0	8.0–13.0	6.0–8.0	4.0–13.0	4.0–9.0
C18:0	0.5–2.0	1.0–3.0	2.0–10.0	1.0–4.0
C16:1	0.5–1.0	0.1–0.2	0.2–0.6	0.1–0.3
C18:1	45.0–70.0	13.0–21.0	48.0–80.0	66.0–85.0
C18:2	16.0–37.0	54.0–65.0	15.0–34.0	5.7–25.0
C18:3	0.1–0.4	13.0–14.0	N.D.	0.0–0.2
Proteins	21.05	14.29	20.96	15.03
Carbohydrates	28.28	17.86	21.01	17.60
Fiber	10.30	7.10	10.90	9.40
Sugars	7.74	3.57	4.86	4.89

Pistachio also contains a remarkable content of minerals, such as magnesium, calcium, potassium, and phosphorus, and is documented as an important dietary source of potassium (1025 mg/100 g versus ~700 mg of almond and hazelnut or 450 mg of walnut per 100 g dry roasted nuts). Pistachios are rich in vitamins, especially vitamins C (5.60 mg/100 g) and E (mainly as γ-tocopherol) (2.17 g/100 g). Others, such as vitamin A, vitamin B (except B12), vitamin K, and folate, are existent in pistachios, which could contribute to their respective Recommended Dietary Allowance (RDA).

Furthermore, the analysis of phytochemicals in this kind of nut has shown the content of a diversity of bioactive polar and non-polar components, like tocopherols, phytosterols, and phenolics [3,15], reaching the top fifty foods possessing higher antioxidant activity [18]. Thus, pistachios are the nuts with the highest content of phytosterols compared with the other widely consumed nuts (2790 mg/Kg versus 1990–1130 mg/kg present in walnut, almond, and hazelnut), including β-sitosterol, Δ5-avenasterol, campesterol, and stigmasterol [19]. Pistachio is characterized by its carotenoid content, particularly luteolin and zeaxanthin with an amount of about 2760 µg/100 g and for β-carotene of about 200 µg/100 g. Compared with hazelnuts, pistachio exhibited 16-fold and eight-fold higher levels of lutein/zeaxanthin and β-carotene, respectively [20]. Finally, pistachios are also a rich source of phenolic compounds, which will be addressed in a separate section given their nutritional and biological relevance.

3. Antioxidant Phenolic Composition of Pistachios

Nutrient databases, like the Phenol-Explorer and the flavonoid and proanthocyanidin databases (USDA) [21,22] report the content of polyphenol of many foods. Furthermore, the concentration of various families of phenolic components in nuts, including pistachio, as well as their antioxidant capacity and evidence for healthy effects have been recently reviewed [1,23].

Walnuts, pecans, and pistachios are the kinds of nuts with the higher phenolic amounts, on the contrary almonds, peanuts, and hazelnut present lower levels (e.g., [1,23]. A variety-dependent content of total polar phenolics (TPP) is found in pistachios, ranging from 1600 mg/kg for the Kastel cultivars to more than three times more, 4900 mg/kg for Larnaka. As reported, polar phenolic components can be grouped into different families. Flavanols being the most abundant phenolics found (about 90% of total, from 1500 mg/kg to 4500 mg/kg), whereas the Kastel and Larnaka cultivars showed the lowest and highest concentrations, respectively [24]. Other families, e.g., anthocyanins (from 54 to 218 mg/kg), flavonols (from 76 to 130 mg/kg), flavanones (from 12 to 71 mg/kg), and gallotannins (from 4 to 46 mg/kg), are also measured. These findings are similar to those reported by Alasalvar & Bolling [1] or Chang et al. [23] (Table 2).

Nuts are rather consumed roasted because this operation enhance their color, flavor, and crunchy characteristics. However, roasting conditions may lead to a relevant decrease of the antioxidant capacity in certain nuts (walnut and hazelnut), but in pistachio and almond its activity apparently keeps stable or is modestly improved [23,25]. This effect may be justified by the decrease of phenolics due to effect of heating which is countervailed by the generation of antioxidant-active components formed by the Maillard reactions. Therefore, nuts must be roasted under appropriate settings to keep their great polyphenols content, antioxidant capacity, as well as sensory properties.

It is important to highlight that the antioxidant capacity shown by different nuts depends on the type of assay employed. Therefore, it is recommended to use more than one type of antioxidant assay to take into account the different antioxidant mechanism and the restrains of each method [26]. In fact, various in vitro chemical assays (e.g., ABTS, FRAP, DPPH, ORAC, etc.) and biological methods (generally based on the oxidation of lipoprotein) are employed to measure the antioxidant capacity of nuts [23].

Table 2. Total phenolic content and antioxidant activities of different varieties of pistachios.

Type of Pistachio	Status of Pistacho	TPP	Antioxidant Assay	Value of Antioxidant Capacity	References
Larnaka (Spanish cultivar)	Natural	4900 mg/Kg fw	ORAC	330 mmol of TE/Kg	[24]
			DPPH	35 mmol of TE/Kg	
Kastel (Spanish cultivar)	Natural	1600 mg/Kg fw	ORAC	89 mmol of TE/Kg	[24]
			DPPH	13 mmol of TE/Kg	
Kerman (Spanish cultivar)	Natural	1900 mg/Kg fw	ORAC	61 mmol of TE/Kg	[24]
			DPPH	10 mmol of TE/Kg	
Uzun (Turkish cultivar)	Natural	26.2 mg/100 g fw	DPPH	8.05 µmol of TE/g	[27]
	Roasted	32.4–42.4 mg/100 g fw	DPPH	9.76–11.5 µmol of TE/g	[27]
Ohadi (Turkish cultivar)	Natural	9.23–10.55 mg/Kg GAE	DPPH	4.34–5.56 mmol TE/Kg	[28]
			ABTS	4.11–5.95 mmol TE/Kg	[28]
	Roasted	10.35–11.23 mg/Kg GAE	DPPH	3.53–6.32 mmol TE/Kg	[28]
			ABTS	5.80–7.35 mmol TE/Kg	[28]

Table 2. *Cont.*

Type of Pistachio	Status of Pistacho	TPP	Antioxidant Assay	Value of Antioxidant Capacity	References
Uzum (Turkish cultivar)	Natural	9.19–11.46 mg/Kg GAE	DPPH	7.16–13.58 mmol Trolox/Kg	[28]
			ABTS	15.69–28.28 mmol Trolox/Kg	[28]
	Roasted	10.46–12.73 mg/Kg GAE	DPPH	13.5–18.00 mmol Trolox/Kg	[28]
			ABTS	26.81–35.86 mmol Trolox/Kg	[28]

DPPH: 2,2′-diphenyl-1-picrylhydrazyl; FRAP: ferric reducing antioxidant power; ABTS: 2,2′-azino-bis(3-ethylbenzthiazoline-6-sulphonic acid); ORAC: Oxygen radical absorbance capacity; TPP: total polar phenolics; fw: fresh weight; GAE: gallic acid equivalent; TE: trolox equivalent.

Regarding the TPP antioxidant activity of pistachios, a wide cultivar-dependent interval was obtained; showing Larnaka the greatest Trolox Equivalent Antioxidant Capacity (TEAC) for both ORAC (330 mmol/kg) and DPPH (35 mmol/kg) methods. On the contrary, Kastel and Kerman are the cultivars with the lowest TEAC data, 89 and 61 mmol/kg (ORAC), and 13 and 10 mmol/kg (DPPH) respectively [24]. These findings are in agreement with those observed by evaluating the DPPH antioxidant capacity of the Uzun Turkish pistachio cultivar [27,28]. Furthermore, Wu and Prior [29] created a database concerning ORAC data in foods, giving 76 mmol/kg for pistachio nuts, similar to the value obtained for Kastel and Kerman varieties (Table 2).

4. Health Benefits of Pistachio Consumption

Pistachios have an exciting nutritional profile compared with the rest of the nuts due to their lower energy content and highest levels of γ-tocopherol, phytoesterols, carotenoids, minerals, such magnesium and potassium, and vitamins K and B. The nutrients mentioned above undoubtedly contribute to the evidence that the regular intake of pistachios improves health [15]. Cardiometabolic disease involves dyslipidemia, insulin resistance, hypertension, and excessive visceral fat, which are behind T2DM and CVD [30]. Epidemiological and/or clinical trials have shown that nut consumption has a positive influence on health, by reducing the risk of suffering CVD [15,31,32], hypertension [33], T2DM [34] and obesity [35], among others. These beneficial effects are a consequence of their unique composition since nuts are nutrient-dense foods with healthy MUFA and PUFA fatty acid composition, dietary fibre, high-quality vegetable protein, vitamins, and minerals, along with carotenoids, phytosterols, and phenolic compounds previously described, with recognized benefits to human health [3,10,11,36]. Regarding pistachio, there is wide evidence confirming its beneficial health effects and, particularly, its positive contribution to minimizing the risk for CVD [10,11].

4.1. *Effects of Pistachio Consumption on Blood Lipids*

Hyperlipemia is an established risk for CVD. Pistachio intake has been related with the improvement of lipid profile, decreasing total cholesterol (TC) concentration [37–40], TC/high-density lipoprotein (HDL) ratio and low-density lipoprotein (LDL)/HDL ratio [37–39,41] in the pistachio-supplemented cluster compared with the control one, in both healthy [38–40] and patients with moderate hypercholesterolemia [37,41]. LDL concentrations also decreased significantly in the pistachio-supplemented group in some studies [39,40,42], whereas others observed a non-significant reduction [37,38,43]. Additionally, Sheridan et al. [41] observed a significant increase in circulating HDL concentration in those subjects who consumed pistachios. Recently, a systematic review of epidemiological evidence developed by Lippi et al. [44] showed the beneficial effects of pistachio intake for improving the blood lipid profile (Figure 1).

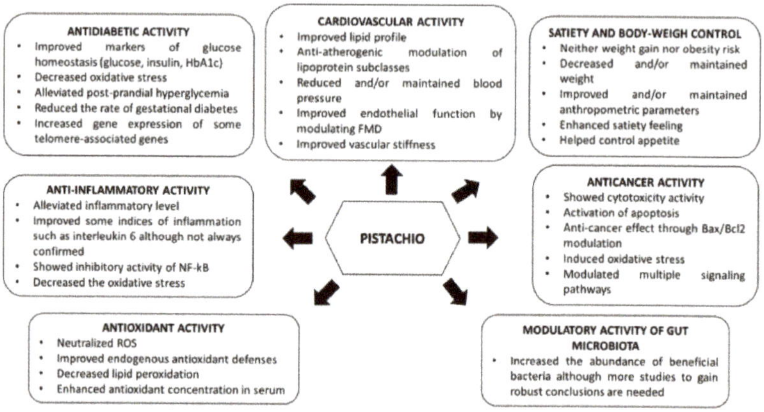

Figure 1. Health benefits of pistachio consumption.

Clinical trials were developed to study the effect of pistachio consumption on the concentration and size of lipoprotein subclasses (small, medium and large), other blood lipid markers of atherosclerosis different from the classical lipid profile [43,45], found a significant antiatherogenic variation of lipoprotein subclasses. Thus, much evidence suggests that pistachios may improve the blood lipid profile, contributing to decreased cardiovascular risk. Most interestingly, in a network meta-analysis of clinical trials comparing the effects of different types of tree nut consumption on blood lipids published this year, Liu et al. [46] concluded that diets enriched with pistachio and walnut could be better alternatives for lowering tryglycerides (TGs), LDL cholesterol, and TC compared to other nut-enriched diets. Another conclusion from this study is that more clinical trials are needed to confirm those claims and suggest changes in dietary and nutritional patterns for the ordinary population. Last year, a systematic review and meta-analysis of randomized controlled trials strongly proposed that pistachios may improve lipid profiles (TC, LDL, TG) and protect against cardiometabolic diseases [47]. Early this year, a comprehensive meta-analysis of randomized controlled trials (RCT) studying the effects of nut consumption on blood lipid profile concluded that, although there was no overall effect of nut consumption on lipid profile, pistachio consumption may reduce TC levels [48] (Figure 1).

4.2. *Effects of Pistachio on Blood Pressure and Endothelial Function*

Several prospective studies have revealed an inverse relationship between nut consumption and blood pressure (BP) or hypertension. In the particular case of pistachio, a beneficial effect on BP has also been observed in a clinical trial conducted on 28 dyslipidaemic individuals who followed for four weeks either a low-fat control diet, a diet having 10% of the total energy from pistachios, or a diet with 20% of the total energy from pistachios. A significant systolic blood pressure (SBP) reduction was observed, particularly after following the diet supplemented with 10% of the total energy as pistachios, thus, no dose-dependent effect was observed. In addition, no difference in diastolic blood pressure (DBP) was recorded [49]. Another more recent study conducted on T2DM subjects by Sauder et al. [50] showed a decrease in SBP after consuming a diet with 20% energy from pistachios for four weeks. However, three additional controlled feeding trials evaluated the BP lowering effects of pistachios as a secondary outcome and non-significant differences in both SBP and DBP between those subjects supplemented or un-supplemented with pistachios were observed [40,41,51]. Finally, a recent review and meta-analysis of more than 20 RCTs found that despite the intake of mixed nuts may reduce DBP, pistachios seemed to have the most potent effect on reducing both DBP and SBP [33]. Therefore, although there is some evidence that suggests that pistachios may reduce BP the non-consistent results encourage continuing to investigate this aspect (Figure 1).

Nuts consumption also improves the endothelial function. The antiatherogenic effect of hazelnut before and after consumption in 21 hypercholesterolemic subjects was investigated. The consumption of a hazelnut-enriched diet (contributing 18–20% of the total daily energy intake) for four weeks significantly improved flow-mediated dilation (FMD), TC, TG, LDL, and HDL as well oxidized-LDL (oxLDL), C-reactive protein (CRP), and soluble vascular cell adhesion molecule-1 (sVCAM-1) compared with the control diet, confirming the antiatherogenic effect of hazelnut-enriched diets by improving endothelial function, preventing LDL oxidation and inflammatory markers, in addition to their lipid and lipoprotein-lowering effects [52]. Another study conducted by Katz et al. [53] evaluated the effects of daily walnut consumption on endothelial function and other cardiac risk biomarkers in 46 overweight adults with visceral obesity. Results revealed that a daily intake of 56 g of walnuts for eight weeks improved endothelial function, without weight gain, by improving FMD compared with the control diet, as well as beneficial trends in SBP and maintenance of anthropometric values. Recently, a review summarized the effects of tree nut and peanut intake on vascular function, excluding FMD. A total of 16 studies were evaluated, although very heterogeneous in terms of dose, length of supplementation, study designs, etc. Conclusions were discrepant. Ten studies provided no significant changes, and six studies (one acute and five chronic studies) informed improvements in at least one measure of vascular function. In summary, nuts have the potential to improve vascular function, but the development of future studies is necessary to expand existing knowledge [54] (Figure 1).

Regarding pistachio, Kasliwal et al. [55] evaluated its effect on vascular health in 60 adults with mild dyslipidemia after consuming 80 g (in-shell) pistachios for 12 weeks. Results showed that usual intake of pistachios not only improved glycaemic and lipid parameters but also produced improvements in vascular stiffness and endothelial function (carotid-femoral and brachial-ankle pulse wave velocity). In a recent meta-analysis of randomized, controlled-feeding clinical studies, Fogacci et al. [56] reviewed the effect of pistachio on brachial artery diameter and flow-mediated dilatation and suggested a significant effect of pistachios on endothelial reactivity, affecting brachial artery diameter but not flow-mediated dilatation. Recently, a comprehensive review [57] and several systematic reviews and meta-analyses Ghanavati et al. [58] demonstrated that pistachio consumption can elicit a beneficial effect on some cardiometabolic risk factors, and Asbaghi et al. [59] suggest the efficacy of pistachio consumption to reduce SBP. Finally, early this year, in the latest review dealing with the effect of dietary polyphenols on vascular health, Grosso and coworkers concluded that hypertension is counteracted by a plant-based dietary pattern, including pistachio, for example, since no single food is enough to control hypertension [60] (Figure 1).

4.3. *Effects of Pistachio on Glucose Metabolism*

In recent years, animal studies have unequivocally shown the positive effect of pistachio feed on glucose homeostasis in altered situations such as diabetes and metabolic syndrome (MetS). Thus, the consumption of pistachio has favorable effects on avoiding hypertriglyceridemia, hyperglycemia, hypercholesterolemia, and inflammation characteristic of a MetS induced by a fructose overload [61].

Recently, the pistachio extract administered to diabetic rats for three weeks significantly improved the lipid profile, oxidative stress, and inflammation process by reducing lipid peroxidation and increasing total antioxidant capacity [62]. In the same year, a pistachio hull hydro-alcoholic extract, along with aerobic exercise, improved passive avoidance memory in streptozotocin-induced diabetic rats [63]. More recently, a pistachio extract also reverted most parameters altered by streptozotocin-induced diabetes in rats; not only markers related to glucose homeostasis, but also those associated to ovary damage and oxidative stress [64].

Several epidemiological studies and clinical trials suggested that the regularity of nut consumption is inversely related to an increased risk of T2DM [65–70], mainly attributed to the relatively high content in dietary fibre, the presence of healthy fats and antioxidants components [69]. Specifically, the effect of consuming pistachio alone or combined with meals was evaluated on postprandial glycaemia [71,72]. Pistachios consumed alone had a minimal effect on postprandial glycaemia, and when 28, 56, or 84 g of pistachios were taken with a carbohydrate meal attenuated in a dose-dependent manner the glycaemic response [71]. These authors evaluated pistachio consumption's acute effects on postprandial glucose and insulin levels in a randomized crossover study conducted on subjects with MetS. Results showed that compared with white bread, pistachio with bread decreased postprandial glycaemia levels and increased glucagon-like peptide levels [72]. Feng et al. [73] confirmed the positive effect that 42 g of pistachio had on postprandial glycemic and gut hormone responses in women with gestational diabetes mellitus or gestational impaired glucose tolerance, by inducing significantly lower postprandial glucose, insulin, and gastric inhibitory polypeptide (GIP) but higher glucagon-like peptide-1 (GLP-1) levels compared to 100 g of whole-wheat bread.

In a controlled clinical trial, healthy young men were randomly addressed to a Mediterranean diet or a Mediterranean diet where monounsaturated fat content was replaced by pistachios (equivalent to 20% of daily caloric intake) for four weeks. Subjects who followed the Mediterranean diet supplemented with pistachios showed a significant decrease in fasting plasma glucose concentrations as compared to the control [40]. Following the Mediterranean diet-based nutritional intervention, a study carried out by Melero et al. [74] confirmed the positive influence on the pregnancy of this dietary pattern by reducing the rate of gestational diabetes mellitus in 600 Spanish women who had to intake about 25–30 g of pistachios at least 3 days a week. Besides, a nutritional clinical trial based on the Mediterranean diet during pregnancy seems to be related with a decrease in the offspring's hospital admission, especially in women with pre-gestational body mass index (BMI) < 25 kg/m2 [75]. Later, the consumption of 0, 42, and 70 g/day of pistachios by 70 subjects with MetS for 12 weeks, evidenced a downward trend in blood glucose levels in subjects who adhered to pistachio interventions compared to the control diet, but without reaching significant differences [51]. A similar study carried out with subjects with MetS randomized to either an unsalted pistachios diet (20% dietary energy) or a control diet for 24 weeks showed a significant decrease in glucose levels but not blood insulin levels [76]. While Hernandez-Alonso et al. [43] demonstrated that pistachios have glucose- and insulin-lowering effect in a RCT with 54 pre-diabetic subjects who consumed for four months a pistachio-supplemented diet in comparison with a control diet. The effect of replacing carbohydrate consumption with mixed nut consumption (75 g/d) that included pistachios in 117 T2DM subjects for three months showed a significant decrease in glycated hemoglobin (HbA1c) levels, improving glycaemic control [77]. Results were similar in a crossover trial with 48 diabetic participants after three months of pistachio consumption [78]. Interestingly, chronic pistachio consumption reduced oxidative damage to DNA and increased the gene expression of some telomere-associated genes, reversing certain deleterious metabolic consequences of prediabetes [79]. More recently, pistachio, among other Mediterranean products, has been recommended a promising source of multi-target agents in the treatment of MetS [80]. Despite the encouraging results reported for glucose metabolism in fasting conditions or postprandial status, additional studies are necessary to test pistachio consumption's long-term effects on insulin resistance and T2DM prevention and/or control. In a recent systematic review and meta-analysis of pistachio on glycemic control and insulin sensitivity in patients with T2DM, prediabetes and MetS, the authors concluded that pistachio nuts might cause a significant reduction in fasting blood glucose and HOMA-IR (homeostatic model assessment for insulin resistance), although HbA1c and fasting plasma insulin might not significantly improve in patients suffering from or at risk of T2DM [81]. Finally, a RCT published early this year has reported that a Mediterranean diet with additional extra virgin olive oil and pistachios decreases the occurrence of gestational diabetes mellitus [82] (Figure 1).

4.4. Effects of Pistachios on Satiety Regulation and Body-Weight Control

Although nuts, including pistachios, are still perceived by the general public to be fattening because of their high-fat content, there are several epidemiological studies that have provided strong evidence that nut consumption is associated with neither weight gain nor an increased risk of obesity [83,84]. Regarding pistachio, Li et al. [85] evaluated the effects of pistachio snack consumption on body weight and lipid levels in obese participants, which were randomly assigned to consume isocaloric weight reduction diets for 12 weeks with an afternoon snack of either 53 g of salted pistachios or 56 g of salted pretzels. Both groups lost weight, but the pistachio-supplemented group showed a higher BMI reduction than the pretzel-supplemented group. Regarding the already mentioned study of Wang et al., [51], results showed that the supplementation with pistachios (42 or 70 g/d) for 12-weeks did not lead to weight gain or an increase in waist-to-hip-ratio in Chinese subjects with MetS. Likewise, Gulati et al. [76] evaluated the effects of pistachios on body composition in 60 individuals with MetS randomised to either the pistachio (20% dietary energy) or control group for 24 weeks, and no significant differences were detected in body weight, although a significant decrease in waist circumference was observed. However, Parham et al. [78] found a significant reduction in BMI after 50 g pistachio consumption for 12 weeks compared to the control diet.

Other clinical studies where apart from studying the effect of pistachio consumption on biomarkers of cardiovascular health, also evaluated the impact on weight or/and BMI, did not observe differences between participants belonging to the supplemented diet with nuts and the control group [38–41]. A more recent study developed by Fantino et al. [86], concluded that the daily intake of 44 g pistachios improved quality diet without disturbing body weight in healthy women. While Rock et al. [87] evaluated the impact of consuming 42 g/d of pistachios by non-diabetic overweight/obese adults assigned to a four-month behavioural weight loss intervention against a similar group without consuming pistachios, observing a similar reduction of weight, BMI, and waist circumference (Figure 1).

Among the various explanations of why the consumption of pistachios does not induce overweight, being a very energetic food, is their high satiating power, inefficiency in the absorption of the energy they contain, a possible increment in energy expenditure at rest and an increase in fat oxidation (revised by Tan et al. [84]). Moreover, the crunchy physical structure of nuts in general, and pistachio in particular, has demonstrated its positive influence on satiety [88]. In this sense, recently, a randomized controlled pilot study was carried out to assess the effects of a daily pistachio afternoon snack on next-meal energy intake, satiety and anthropometry in 30 healthy French women were tutored to consume either 56 g of pistachios or 56 g of isoenergetic/equiprotein savoury biscuit as an afternoon snack. Results revealed that both afternoon snacks provided a similar subjective feeling of satiety, and pistachios consumption did not affect body weight or composition [89]. Lately, pistachio treatment in obese mice fed a high-fat diet showed neuroprotective effects, including decreased brain apoptosis, decreased brain lipid, and oxidative stress with the improvement of mitochondrial function [90]. A recent meta-analysis of RCTs by Xia et al. [91] reported that a diet with pistachios reduced BMI and had no significant effects on body weight and waist circumference. Indeed, recent research has focused on pistachio's applications as a plant-based snack, particularly for appetite control and healthy weight management [92]. Due to the importance of satiety regulation on body-weight control, clinical trials with pistachios need to be carried out in the future to establish this aspect fully.

4.5. Effect of Pistachios on Inflammatory State

Chronic low-grade inflammation has been related with insulin resistance, diabetes, atherosclerosis, obesity and MetS. A few research studies have evaluated the effects of nut intake on inflammation with different results. A previous study suggested that a diet supplemented with pistachio improved some inflammation biomarkers in healthy young men; by decreasing serum interleukin 6 but without changing CRP and tumour necrosis factor-alpha (TNFα) levels [40]. Recently, proanthocyanidins extracted from Sicilian pistachio was the major bioactive able to modulate the inflammatory response of human intestinal epithelial cells through the inhibition of nuclear factor kappa B (NF-κB) activation [93]. More recently, the anti-inflammatory effect in vivo of pistachio was shown in a rat model of ulcerative colitis inflammation [94]. In the same year, the results of another study showed that usual pistachio intake improved inflammation in obese mice, probably due to the positive modulation of the microbiota composition [95]. More recently, an aqueous leaf extract obtained from *Pistacia lentiscus* improved acute acetic acid-induced colitis in rats, which was associated with its ability to reduce inflammation and oxidative stress [96] (Figure 1).

4.6. Effects of Pistachios on Oxidative Stress

Pistachios contain polyphenols that act as radical scavengers, neutralising reactive oxygen species (ROS) and enhancing endogenous antioxidant defences. Many studies have demonstrated the antioxidant activity of extracts obtained from different pistachio parts in both in vitro [97–101] and in vivo models, such as in animals [64,97,99,102–108] and humans [7,15,54,55,76,109]. Therefore, antioxidants present in pistachios could have significant effects on the regulation of oxidative stress and a reduced risk of chronic diseases. Remarkably, previous research on the antioxidant capacity and chemo-preventive potential of pistachio phenolic compounds in cell culture models has been performed in cell types from thyroid, lung, skin, and monocyte/macrophage, but not in endothelial cells, which is the more suitable cell type to study endothelial dysfunction and its potential effect on cardiovascular capacity. This inattention should be repaired in the near future.

Regarding human studies, a study carried out on 44 healthy volunteers showed an increased blood antioxidant potential determined by the evaluation of biomarkers of lipid peroxidation in those participants consuming pistachios compared with those assigned to a control diet without nuts [38]. A cross-over RCT developed by Kay et al. [110] on 28 hypercholesterolaemic adults showed that the consumption of diets containing 10% and 20% of energy from pistachios increased γ-tocopherol, lutein and β-carotene concentrations in serum, and decreased oxLDL concentration compared with the un-supplemented group. Sari et al. [40] evaluated the effect of the Mediterranean diet supplemented with pistachios by replacing the monounsaturated fat content on thirty-two healthy young men for four weeks in a prospective study. Results revealed an increase in total antioxidant status and superoxide dismutase, along with a decrease in inflammation and other oxidative markers. In a more recent randomized, double-blind, placebo-controlled trial, the antioxidant efficacy of a *Pistacia lentiscus* supplement in inflammatory bowel disease was assessed [111]. 60 patients were randomly allocated to *Pistacia lentiscus* supplement (2.8 g/day) or to placebo for three months and oxLDL, oxLDL/HDL, and oxLDL/LDL decreased significantly in the intervention group, confirming its antioxidant activity. These afore-mentioned results have demonstrated the beneficial effects of pistachios on the factor risks of CVD (Figure 1).

4.7. Effects of Pistachios on Cancer

In cell culture studies, mastic gum resin has shown anticancer effect in bile duct cancer (cholangiocarcinoma) (KMBC), pancreatic carcinoma (PANC-1), gastric adenocarcinoma (CRL-1739), and colonic adenocarcinoma (COLO205) cells [112]. *P. lentiscus* extract showed moderate activity against liver cancer [113] as well as breast cancer [114,115]. Raw and roasted pistachio showed a chemo-preventive potential regarding colon cancer [116]. The essential oil from *Pistacia lentiscus* aerial parts induced oxidative stress and apoptosis in human thyroid carcinoma cells [98]. Pistachio green hull extract induced apoptosis through multiple signaling pathways by causing oxidative stress on colon cancer cells [117]; whereas the cytotoxic fraction from *Pistacia vera* red hull ethyl acetate extract showed anticancer activity on breast cancer both in cell culture and in mice [118]. Mastiha is a natural aromatic resin obtained from the trunk and branches of the mastic tree (*Pistacia lentiscus* L. var latifolius Coss or *Pistacia lentiscus* var. Chia). Similar to the results reported with green hull extracts by Koyuncu and coworkers, mastic has shown a powerful anticancer effect in colon cancer cells [119]. In the same year, mastic gum essential oils exhibited selective cytotoxicity against SK-MEL-30 (melanoma), A559 (lung), PANC-1 (pancreatic), and U87MG (glioblastoma-astrocytoma) human cells lines, whereas HeLa (cervix adenocarcinoma) cells exhibited more sensitivity to the treatment [120]. In a very recent overview of pistachio effects on human health [121], beneficial effects on inflammation and oxidative stress were generally reported for mastic consumption, while a study developed in a cancer bioassay model showed an increase in biomarkers associated with the formation of hepatic preneoplastic lesions after mastic administration [122]. In contrast, mastic exhibited an antihepatotoxic activity in carbon tetrachloride-intoxicated rats, by reducing alkaline phosphatase activity and bilirubin levels [123]. Moreover, recently, a current systematic review and meta-analysis of observational studies showed the inverse association between pistachio consumption and the cancer risk and mortality [124]. Finally, early this year, the nanoliposomal formulation of pistachio hull extract showed promising anti-cancer effects through Bax/Bcl2 modulation [125] (Figure 1).

4.8. Effect of Pistachios on Intestinal Microbiota

Dietary fibre and phytochemicals present in nuts can reach the proximal colon and modulate the microbiota composition. Effects of almond and pistachio consumption on gut microbiota composition were evaluated in a crossover RCT for 18 days with 0, 1.5, and three servings/day. Pistachio showed a higher effect than almond on gut microbiota, increasing the number of potentially beneficial butyrate-producing bacteria, whereas *bifidobacteria* were unaffected [126]. A more recent study led by Yanni et al. [127] in diabetic rats demonstrated that dietary pistachio for four weeks restored normal flora and enhanced the presence of beneficial microbes (*lactobacilli* and *bifidobacteria*, among others) in the rat model of streptozotocin-induced diabetes. Likewise, chronic intake of pistachio for 16 weeks significantly enhanced the level of healthy bacteria genera such as *Parabacteroides*, *Dorea*, *Allobaculum*, *Turicibacter*, *Lactobacillus*, and *Anaeroplasma*, and decreased bacteria related to inflammation, such as *Oscillospira*, *Desulfovibrio*, *Coprobacillus*, and *Bilophila* in mice fed a high-fat diet [95]. Finally, Creedon et al. [128] carried out a systematic review of RCT to check the impact on gut microbiota and gut function that nut consumption has in healthy people. Seven RCTs were eligible, and only one of them was developed with pistachios. Nut consumption significantly increased *Clostridium*, *Dialister*, *Lachnospira* and *Roseburia* and significantly decreased parabacteroides. Nonetheless, it is convenient to carry out new studies to expand knowledge about the effect of pistachio on intestinal microbiota and gain robust conclusions (Figure 1).

5. Consumption and Uses of Pistachios

Regarding the estimated world consumption of pistachios (in-shell basis) per capita (Kg/year), the top five countries were Turkey (2.06 kg/year), Syria (1.83 kg/year), Israel (1.79 kg/year), Spain (1.43 kg/year), and Germany (0.98 kg/year) in 2019 [8].

Nuts in general, and pistachio in particular, are consumed fresh, roasted, or as salty snacks. Furthermore, gourmet and functional oils, flours, plant-based milk substitutes, spreads, cereal bars, and bakery goods, among others, may be made with nuts, with high requirements in terms of nut quality. Thus, nut oils are obtained by mechanical pressing without refining, and due to their sensorial attributes, the pistachio oil is an appreciated product from the gastronomy point of view (as a substitute for butter or margarine, as a dressing for vegetables, and as an ingredient of haute cuisine) [129]. The composition and properties of virgin pistachio oils processed employing different cultivars were characterized by Ojeda-Amador et al. [130] in 2018; highlighting the attractive added value of this type of vegetable oil. Recently, the development of functional edible oils enriched with pistachio and walnut phenolic extracts has been described [131], which constitute an exciting strategy to increase the phenolic intake in the diet. A phenolic concentration of 340–570 mg/kg was reached in the different enriched edible oils, showing a great antioxidant activity (up to 54 mmol/kg Trolox, as DPPH). Furthermore, the functional properties and stability of these formulations have been improved employing emulsion and micro emulsion systems [132]. Nut flour has aroused interest as a substitute for wheat flour for gluten-free formulations to use in bakery products such as bread, cakes and cookies [133] and with an interesting nutritional profile. Sanchiz et al. [134] evaluated pistachio flour after combining heat and pressure processing, resulting in a higher phenolic content and excellent physical properties of pistachio flour. Non-dairy milk alternatives answer the needs of people with milk protein allergy and lactose intolerance, and its market is growing at breakneck speed. The pistachio represents an excellent raw material for obtaining vegetable-based milk, as described by Sethi et al. [135]. Nut spread has also been formulated with pistachio, which constitutes an excellent alternative to increase its added value [136]. In the same line, pistachio has been used as a predominant ingredient in the formulation of cereal bars, which is very well accepted by consumers [137].

Furthermore, pistachio oil is a valued ingredient for cosmetic care formulations, e.g., shampoos, hair conditioners, soaps, skin creams, and lip balms, [129]. Recently, distilled leaves from pistachio have been reported as a potential cosmeceutical ingredient because of their transdermal diffusion and anti-elastase and anti-tyrosinase activities [138].

Such commercial presentations enlarge the use of pistachio as value-added ingredients, motivating their cultivation and production. Pistachio constitutes an ideal ingredient for new functional foods and satisfies the needs of more limiting diets, such as vegetarianism or veganism [139].

6. Barriers and Facilitators to Pistachio Consumption

Population nut consumption does not meet the current recommendation of intake despite their widespread demonstrated health benefits. In this sense, a recent review identified some barriers and facilitators to nut consumption [140], which should be taken into account in elaborating future strategies to encourage nut intake.

The most extended feeling about nut consumption is the belief that eating nuts would cause weight gain, despite the cumulative results reporting no effect on it [51,76,78,85]. An interesting and recent study that evidence this issue was that developed by Brown et al. [141], who evaluated the nut perceptions among the general public. Results revealed that over half the respondents reported they would eat more nuts if they were advised to do it by a dietitian or doctor, and the most frequently selected deterrent to increasing nut consumption was the potential weight gain (66%). Likewise, a survey of 710 interviewees in New Zealand identified as a barrier to regular nut consumption the perception that nuts are high in fat, which could affect negatively on control weight [142]. Another common barrier to nut consumption was their high cost. The study developed by Brown et al. [141] in 710 people found the cost (67%) as the most frequently selected deterrent to increase nut intake. Recently, a survey of 124 participants from the United States [143] highlighted an inverse relationship between the price and the intention to consume nuts. Dentition issues may act as a possible barrier to regular nut consumption, as was, for example, reported in the New Zealand study where it was mentioned as a common barrier to nut intake [142]. Other presentations of nuts, such as nut butter, would constitute interesting alternative forms of nuts for those with poor dentition. Finally, allergy to nuts constitutes an important nut intake barrier, as was reported by Australian health professionals in the study developed by Tran et al. [144]. In addition, this barrier is extrapolated to people who lived with or were in close contact with someone allergic to nuts, since up to 8% of participants compatible with this situation declared not to consume nuts or nut butter.

Conversely, the healthy properties associated with nut consumption constitute a strong facilitator, but certain demographic characteristics such as a higher level of education or socio-economic status [145–147], healthier lifestyle [145,148], and higher levels of physical activity [145], are determining a higher intake of nut. Likewise, increased age appears to be related to higher nut consumption [145,148]. In contrast, intake of nut butter may be higher among younger people [147]. Health professionals have a key role in encouraging nut intake, although their knowledge about health benefits derived from nut consumption could be improved.

The aforementioned data could help to identify the most vulnerable groups to not meeting the recommendations for the nut intake, very important to design targeted dietetic strategies to reach the minimal pistachio consumption.

From the evidence given and discussed in this review, it is clear why pistachio should be a regular food in our diet, since there are several well-documented health benefits related to the consumption of this kind of nut. However, this is probably clear to food and nutrition experts, just yet to reach the general population.

To increase pistachio consumption several strategies should be used. To effectively inform the population about the health benefits of its consumption, specific promotion campaigns should eb designed focused on different target consumers, e.g., kids, teenagers, adults, sportsmen/sportswomen, elderly, and so on. As mentioned, dietitians or doctors should be involved to increase people's confidence in this message.

Moreover, campaigns should also promote the consumption of the different types of pistachio's products, several of them described above. The fusion between gastronomy and nutrition is generally a successful story; increasing the number of recipes using pistachios as an ingredient but also developing new ways of employing them, such as virgin oils, nut butter, bakery products, etc.

Well-designed educational tasks, such as MyPlate or similar food guidelines, should reinforce nut consumption, that of pistachio in particular in this case, within the fruits and vegetables food group.

Since nuts are consumed roasted because it enhances their sensory characteristics, it should also be very relevant to advance the optimization of roasting conditions with the aim of maximizing the effects of pistachio's bioactive polar and non-polar components; which must also involve a proper consumer's sensory evaluation to ensure that preferences for the new product remain still high.

Finally, as commented in this review, a complementary strategy to ensure an adequate intake of pistachio's bioactive components should involve the development of functional foods that contain phenolic extracts.

7. Conclusions

The pistachio is valued as an important source of bioactive components that, compared to other nuts, possesses a healthier nutritional profile with low-fat content composed mainly of MUFA, a high source of vegetable protein and dietary fibre, remarkable content of minerals, especially potassium, and an excellent source of vitamins, such as vitamins C and E. Its phytochemicals rich composition such as tocopherols, carotenoids and, importantly, phenolic compounds make pistachio an interesting food for its potential to prevent prevalent pathologies. The first health claim approved by the United States FDA in 2003 informed that there is scientific evidence that suggests, although does not demonstrate, that eating 1.5 oz (42.5 g) each day of nuts within a low saturated fat and cholesterol diet, may diminish the heart disease risk, was a turning point for nuts. A growing interest in confirming nuts effect on health has led to an increase in the quantity of research on nuts. Although pistachio has been less explored than other nuts (walnut, almonds, hazelnut, etc.), some studies provide evidence of its beneficial effects on CVD risk factors beyond the lipid-lowering effect. Different studies showed that diets supplemented with pistachios present a preventive effect to T2DM due to its effect of improving markers of glucose homeostasis, decreasing oxidative stress, alleviating post-prandial hyperglycemia and reducing the rate of gestational diabetes, among others. Regarding the most concern issue regarding pistachio intake, recent results showed neither weight gain nor obesity risk, but rather the opposite. A tendency to decrease and/or maintain weight as well as improve and/or maintain anthropometric parameters was observed and, in part, due to its positive effect on controlling appetite and enhancing satiety feeling, although more studies are needed to evaluate this aspect thoroughly. Likewise, the pistachio could play an important role in cancer prevention due to its ability to induce cytotoxicity and apoptosis in neoplastic cells, as well as the modulatory activity of signaling pathways involved in the regulation of cancer observed in carcinogenic cell models. The phytochemicals would be behind the demonstrated anti-inflammatory and antioxidant activities associated with pistachios, which would explain the ability to prevent the chronic diseases mentioned (CVD, T2DM, obesity, and cancer). Finally, there are more and more studies that inform about the pistachio's ability to modulate the activity of gut microbiota, increasing the abundance of beneficial bacteria, although more studies are needed to gain robust conclusions.

This review provides enough information and evidence of the usual pistachio intake's health benefits. However, specific barriers continue to make it difficult to reach the intake requirements for nuts, such as the belief that it is a fatty product, the cost, or teething problems. Except for the allergenic problems, the rest are solvable with more research and information from the population and health professionals.

Author Contributions: Conceptualization, L.G., G.F., M.D.S. and R.M.; investigation, R.M. and L.G.; methodology, R.M. and L.G.; data curation, L.G. and R.M.; project administration, L.G. and R.M.; resources, L.G.; supervision, R.M. and L.G.; writing—original draft, L.G. and R.M.; writing—review & editing, L.G., G.F., M.D.S. and R.M. All authors have read and agreed to the published version of the manuscript.

Funding: This research was funded by project 202270E021 funded by CSIC.

Conflicts of Interest: The authors declare no conflict of interest.

Abbreviations

ABTS	2,2′-Azino-bis (3-ethylbenzothiazoline-6-sulphonic acid)
BMI	Body mass index
BP	Blood pressure
CRP	C-reactive protein
CVD	Cardiovascular disease
FRAP	Ferric reducing antioxidant power
DBP	Diastolic blood pressure
DRI	Dietary reference intake
DPPH	2,2-Diphenyl-1-picrylhydrazyl
FMD	Flow-mediated dilation
GAE	gallic acid equivalent
GIP	Gastric inhibitory polypeptide
GLP-1	Glucagon-like peptide-1
HbA1c	Glycated hemoglobin
HDL	High-density lipoprotein
HOMA-IR	Homeostatic model assessment for insulin resistance
LDL	Low-density lipoprotein
MetS	Metabolic syndrome
MUFA	Monounsaturated fatty acid
NF-κB	Nuclear factor kappa B
ORAC	Oxygen radical absorbance capacity
oxLDL	oxidised-LDL
PUFA	Polyunsaturated fatty acids
RCT	Randomized controlled trials
RDA	Recommended dietary allowance
ROS	Reactive oxygen species
SBP	Systolic blood pressure
SFA	Saturated fatty acids
sVCAM-1	Soluble vascular cell adhesion molecule-1

T2DM	Type 2 diabetes mellitus
TC	Total cholesterol
TE	Trolox equivalent
TEAC	Trolox equivalent antioxidant capacity
TG	Tryglycerides
TNFα	Tumour necrosis factor-alpha
TPP	Total polar phenolics

References

1. Alasalvar, C.; Bolling, B.W. Review of nut phytochemicals, fat-soluble bioactives, antioxidant components and health effects. *Br. J. Nutr.* **2015**, *113*, S68–S78. [CrossRef] [PubMed]
2. Amarowicz, R.; Gong, Y.; Pegg, R.B. Recent advances in our knowledge of the biological properties of nuts. In *Wild Plants, Mushrooms and Nuts: Functional Food Properties and Applications*; Ferreira, I.C.F.R., Morales, P., Barros, L., Eds.; Blackwell: Hoboken, NJ, USA, 2017; ISBN 9781118944646.
3. Machado de Souza, R.G.; Schincaglia, R.M.; Pimentel, G.D.; Mota, J.F. Nuts and human health outcomes: A systematic review. *Nutrients* **2017**, *9*, 1311. [CrossRef] [PubMed]
4. FDA, Food and Drug Administration Docket No. 02P-0505. 2003. Available online: https://www.weightwatchers.com/images/1033/dynamic/articles/2010/03/FDA_Approved_Qualified_Health_Claim_on_Nuts.pdf (accessed on 4 July 2022).
5. Estruch, R.; Ros, E.; Salas-Salvado, J.; Covas, M.I.; Corella, D.; Aros, F.; Gomez-Gracia, E.; Ruiz-Gutierrez, V.; Fiol, M.; Lapetra, J.; et al. Primary prevention of cardiovascular disease with a mediterranena diet supplemented with extra-virgin olive oil or nuts. *N. Engl. J. Med.* **2018**, *378*, e34. [CrossRef] [PubMed]
6. Kim, Y.; Keogh, J.B.; Clifton, P.M. Benefits of nut consumption on insulin resistance and cardiovascular risk factors: Multiple potential mechanisms of actions. *Nutrients* **2017**, *22*, 1271. [CrossRef]
7. Terzo, S.; Baldassano, S.; Caldara, G.F.; Ferrantelli, V.; Lo Dico, G.L.; Mulè, F.; Amato, A. Health benefits of pistachios consumption. *Nat. Prod. Res.* **2019**, *33*, 715–726. [CrossRef]
8. I.N.&D. Nuts & Dried Fruits Statistical Yearbook 2020/2021. 2021. Available online: https://www.iranpistachio.org/fa/images/INC/INC_Statistical_Yearbook_2019-2020.pdf?msclkid=89309733d05511ec99551a66d22f0de1) (accessed on 4 July 2022).
9. Couceiro, J.F. (Ed.) *El Cultivo del Pistachio*; Ediciones Mundi-Prensa: Madrid, Spain, 2017; ISBN 9788484767220.
10. Nadimi, A.E.; Ahmadi, Z.; Falahati-Pour, S.K.; Mohamadi, M.; Nazari, A.; Hassanshahi, G.; Ekramzadeh, M. Physicochemical properties and health benefits of pistachio nuts. *Int. J. Vitam. Nutr. Res.* **2020**, *90*, 564–574. [CrossRef]
11. Mandalari, G.; Barreca, D.; Gervasi, T.; Roussell, M.A.; Klein, B.; Feeney, M.J.; Carughi, A. Pistachio nuts (*Pistacia vera* L.): Production, nutrients, bioactives and novel health effects. *Plants* **2022**, *11*, 18. [CrossRef]
12. Neale, E.P.; Tapsell, L.C. Nuts in Healthy Dietary Patterns and Dietary Guidelines. In *Health Benefits of Nuts and Dried Fruits*; Alasalvar, C., Salas-Salvado, J., Ros, E., Sabate, J., Eds.; CRC Press: Boca Raton, FL, USA, 2020.
13. Afshin, A.; Sur, P.J.; Fay, K.A.; Cornaby, L.; Ferrara, G.; Salama, J.S.; Mullany, E.C.; Abate, K.H.; Abbafati, C.; Abebe, Z. Health effects of dietary risks in 195 countries, 1990–2017: A systematic analysis for the Global Burden of Disease Study 2017. *Lancet* **2019**, *393*, 1958–1972. [CrossRef]
14. Tsantili, E.; Takidelli, C.; Christopoulos, M.V.; Lambrinea, E.; Rouskas, D.; Roussos, P.A. Physical, compositional and sensory differences in nuts among pistachio (*Pistachia vera* L.) varieties. *Sci. Hortic.* **2010**, *125*, 562–568. [CrossRef]
15. Bulló, M.; Juanola-Falgarona, M.; Hernández-Alonso, P.; Salas-Salvadó, J. Nutrition attributes and health effects of pistachio nuts. *Br. J. Nut.* **2015**, *113*, S79–S93. [CrossRef]
16. Ros, E.; Singh, A.; O'Keefe, J.H. Nuts: Natural pleiotropic nutraceuticals. *Nutrients* **2021**, *13*, 3269. [CrossRef] [PubMed]
17. Kris-Etherton, P.M.; Zhao, G.; Binkoski, A.E.; Coval, S.M.; Etherton, T.D. The effects of nuts on coronary heart disease risk. *Nutr. Rev.* **2001**, *59*, 103–111. [CrossRef] [PubMed]
18. Halvorsen, B.L.; Carlsen, M.H.; Phillips, K.M.; Bøhn, S.K.; Holte, K.; Jacobs, D.R., Jr.; Blomhoff, R. Content of redox-active compounds (ie, antioxidants) in foods consumed in the United States. *Am. J. Clin. Nutr.* **2006**, *84*, 95–135. [CrossRef] [PubMed]
19. Phillips, K.M.; Ruggio, D.M.; Ashraf-Khorassani, M. Phytosterol composition of nuts and seeds commonly consumed in the United States. *J. Agric. Food Chem.* **2005**, *53*, 9436–9445. [CrossRef] [PubMed]
20. Stuetz, W.; Schlörmann, W.; Glei, M. B-vitamins, carotenoids and α-/γ-tocopherol in raw and roasted nuts. *Food Chem.* **2017**, *221*, 222–227. [CrossRef]
21. Pérez-Jiménez, J.; Neveu, V.; Vos, F.; Scalbert, A. Identification of the 100 richest dietary sources of polyphenols: An application of the Phenol-Explorer database. *Eur. J. Clin. Nutr.* **2010**, *64*, S112–S120. [CrossRef]
22. USDA. Available online: http://www.ars.usda.gov/nutrientdata (accessed on 2 May 2022).
23. Chang, S.K.; Alasalvar, C.; Bolling, B.W.; Shahidi, F. Nuts and their co-products: The impact of processing (roasting) on phenolics, bioavailability, and health benefits—A comprehensive review. *J. Funct. Foods* **2016**, *26*, 88–122. [CrossRef]
24. Ojeda-Amador, R.M.; Salvador, M.D.; Fregapane, G.; Gómez-Alonso, S. Comprehensive study of the phenolic compound profile and antioxidant activity of eight pistachio cultivars and their residual cakes and virgin oils. *J. Agric. Food Chem.* **2019**, *67*, 3583–3594. [CrossRef]

25. Ojeda-Amador, R.M.; Trapani, S.; Fregapane, G.; Salvador, M.D. Phenolics, tocopherols, and volatiles changes during virgin pistachio oil processing under different technological conditions. *Eur. J. Lipid Sci. Technol.* **2018**, *120*, 180–221. [CrossRef]
26. Shahidi, F.; Ambigaipalan, P. Phenolics and polyphenolics in foods, beverages and spices: Antioxidant activity and health effects—A review. *J. Funct. Foods* **2015**, *18*, 820–897. [CrossRef]
27. Rodríguez-Bencomo, J.J.; Kelebek, H.; Sonmezdag, A.S.; Rodríguez-Alcalá, L.M.; Fontecha, J.; Serkan, S. Characterization of the aroma-active, phenolic, and lipid profiles of the pistachio (*Pistacia vera* L.) nut as affected by the single and double roasting process. *J. Agric. Food Chem.* **2015**, *63*, 7830–7839. [CrossRef] [PubMed]
28. Sonmezdag, A.S.; Kelebek, H.; Selli, S. Effect of hulling methods and roasting treatment on phenolic compounds and physicochemical properties of cultivars 'Ohadi' and 'Uzun' pistachios (*Pistacia vera* L.). *Food Chem.* **2019**, *272*, 418–426. [CrossRef] [PubMed]
29. Wu, X.; Prior, R.L. Identification and characterization of anthocyanins by high-performance liquid chromatography−electrospray ionization−tandem mass spectrometry in common foods in the United States: vegetables, nuts, and grains. *J. Agric. Food Chem.* **2005**, *53*, 3101–3113. [CrossRef] [PubMed]
30. Guo, F.; Moellering, D.R.; Garvey, W.T. The progression of cardiometabolic disease: Validation of a new cardiometabolic disease staging system applicable to obesity. *Obesity* **2014**, *22*, 110–118. [CrossRef]
31. Estruch, R.; Ros, E.; Salas-Salvadó, J.; Covas, M.-I.; Corella, D.; Arós, F. Primary prevention of cardiovascular disease with a Mediterranean diet. *N. Engl. J. Med.* **2013**, *368*, 1279–1290. [CrossRef]
32. Mayhew, A.J.; de Souza, R.J.; Meyre, D.; Anand, S.S.; Mente, A. A systematic review and meta-analysis of nut consumption and incident risk of CVD and all-cause mortality. *Br. J. Nutr.* **2016**, *115*, 212–225. [CrossRef]
33. Mohammadifard, N.; Salehi-Abargouei, A.; Salas-Salvadó, J.; Guasch-Ferré, M.; Humphries, K.; Sarrafzadegan, N. The effect of tree nut, peanut, and soy nut consumption on blood pressure: A systematic review and meta-analysis of randomized controlled clinical trials. *Am. J. Clin. Nutr.* **2015**, *101*, 966–982. [CrossRef]
34. Viguiliouk, E.; Kendall, C.W.C.; Blanco Mejia, S.; Cozma, A.I.; Ha, V.; Mirrahimi, A.; Jayalath, V.H.; Augustin, L.S.A.; Chiavaroli, L.; Leiter, L.A.; et al. Effect of tree nuts on glycemic control in diabetes: A systematic review and meta-analysis of randomized controlled dietary trials. *PLoS ONE* **2014**, *9*, e103376. [CrossRef]
35. Jackson, C.L.; Hu, F.B. Long-term associations of nut consumption with body weight and obesity. *Am. J. Clin. Nutr.* **2014**, *100*, 408–411. [CrossRef]
36. Taş, N.G.; Gökmen, V. Phenolic compounds in natural and roasted nuts and their skins: A brief review. *Curr. Opin. Food Sci.* **2017**, *14*, 103–109. [CrossRef]
37. Edwards, K.; Kwaw, I.; Matud, J.; Kurtz, I. Effect of pistachio nuts on serum lipid levels in patients with moderate hypercholesterolemia. *J. Am. Coll. Nutr.* **1999**, *18*, 229–232. [CrossRef] [PubMed]
38. Kocyigit, A.; Koylu, A.A.; Keles, H. Effects of pistachio nuts consumption on plasma lipid profile and oxidative status in healthy volunteers. *Nutr. Metab. Cardiovasc. Dis.* **2006**, *16*, 202–209. [CrossRef] [PubMed]
39. Gebauer, S.K.; West, S.G.; Kay, C.D.; Alaupovic, P.; Bagshaw, D.; Kris-Etherton, P.M. Effects of pistachios on cardiovascular disease risk factors and potential mechanisms of action: A dose-response study. *Am. J. Clin. Nutr.* **2008**, *88*, 651–659. [CrossRef] [PubMed]
40. Sari, I.; Baltaci, Y.; Bagci, C.; Davutoglu, V.; Erel, O.; Celik, H.; Ozer, O.; Aksoy, N.; Aksoy, M. Effect of pistachio diet on lipid parameters, endothelial function, inflammation, and oxidative status: A prospective study. *Nutrition* **2010**, *26*, 399–404. [CrossRef]
41. Sheridan, M.J.; Cooper, J.N.; Erario, M.; Cheifetz, C.E. Pistachio nut consumption and serum lipid levels. *J. Am. Coll. Nutr.* **2007**, *26*, 141–148. [CrossRef]
42. Aldemir, M.; Okulu, E.; Neşelioğlu, S.; Erel, O.; Kayıgil, O. Pistachio diet improves erectile function parameters and serum lipid profiles in patients with erectile dysfunction. *Int. J. Impot. Res.* **2011**, *23*, 32–38. [CrossRef]
43. Hernandez-Alonso, P.; Baldrich-Mora, M.; Juanola-Falgarona, M.; Bullo, M. Beneficial effect of pistachio consumption on glucose metabolism, insulin resistance, inflammation, and related metabolic risk markers: A randomized clinical trial. *Diabetes Care* **2014**, *37*, 3098–3105. [CrossRef]
44. Lippi, G.; Cervellin, G.; Mattiuzzi, C. More pistachio nuts for improving the blood lipid profile. Systematic review of epidemiological evidence. *Acta Biomed.* **2016**, *87*, 5–12.
45. Holligan, S.D.; West, S.G.; Gebauer, S.K.; Kay, C.D.; Kris-Etherton, P.M. A moderate-fat diet containing pistachios improves emerging markers of cardiometabolic syndrome in healthy adults with elevated LDL levels. *Br. J. Nutr.* **2014**, *112*, 744–752. [CrossRef]
46. Liu, K.; Hui, S.; Wang, B.; Kaliannan, K.; Guo, X.; Liang, L. Comparative effects of different types of tree nut consumption on blood lipids: A network meta-analysis of clinical trials. *Am. J. Clin. Nutr.* **2020**, *111*, 219–227. [CrossRef]
47. Hadi, A.; Pourmasoumi, M.; Kazemi, M.; Najafgholizadeh, A.; Marx, W. Efficacy of synbiotic interventions on blood pressure: A systematic review and meta-analysis of clinical trials. *Crit. Rev. Food Sci. Nutr.* **2022**, *62*, 5582–5591. [CrossRef] [PubMed]
48. Gunathilake, M.; Van, N.T.H.; Kim, J. Effects of nut consumption on blood lipid profile: A meta-analysis of randomized controlled trials. *Nutr. Metab. Cardiovasc. Dis.* **2022**, *32*, 537–549. [CrossRef]
49. West, S.G.; Gebauer, S.K.; Kay, C.D.; Bagshaw, D.M.; Savastano, D.M.; Diefenbach, C.; Kris-Etherton, P.M. Diets containing pistachios reduce systolic blood pressure and peripheral vascular responses to stress in adults with dyslipidemia. *Hypertension* **2012**, *60*, 58–63. [CrossRef] [PubMed]

50. Sauder, K.A.; McCrea, C.E.; Ulbrecht, J.S.; Kris-Etherton, P.M.; West, S.G. Pistachio nut consumption modifies systemic hemodynamics, increases heart rate variability, and reduces ambulatory blood pressure in well-controlled type 2 diabetes: A randomized trial. *J. Am. Heart Assoc.* **2014**, *3*, e000873. [CrossRef]
51. Wang, X.; Li, Z.; Liu, Y.; Lv, X.; Yan, W. Effects of pistachios on body weight in Chinese subjects with metabolic syndrome. *Nutr. J.* **2012**, *11*, 20. [CrossRef] [PubMed]
52. Orem, A.; Balaban Yucesan, F.; Orem, C.; Akcan, B.; Vanizor Kural, B.; Alasalvar, C.; Shahidi, F. Hazelnut-enriched diet improves cardiovascular risk biomarkers beyond a lipid-lowering effect in hypercholesterolemic subjects. *J. Clin. Lipidol.* **2013**, *7*, 123–131. [CrossRef] [PubMed]
53. Katz, D.L.; Davidhi, A.; Yanchou Njike, V. Effects of walnuts on endothelial function in overweight adults with visceral obesity: A randomized, controlled, crossover trial. *J. Am. Coll. Nutr.* **2012**, *31*, 415–423. [CrossRef] [PubMed]
54. Morgillo, S.; Hill, A.M.; Coates, A.M. The effects of nut consumption on vascular function. *Nutrients* **2019**, *11*, 116. [CrossRef]
55. Kasliwal, R.R.; Bansal, M.; Mehrotra, R.; Yeptho, K.P.; Trehan, N. Effect of pistachio nut consumption on endothelial function and arterial stiffness. *Nutrition* **2015**, *31*, 678–685. [CrossRef]
56. Fogacci, F.; Cicero, A.F.G.; Derosa, G.; Rizzo, M.; Veronesi, M.; Borghi, C. Effect of pistachio on brachial artery diameter and flow-mediated dilatation: A systematic review and meta-analysis of randomized, controlled-feeding clinical studies. *Crit. Rev. Food Sci. Nutr.* **2019**, *59*, 328–335. [CrossRef]
57. Ahmad, R.; Almubayedh, H.; Ahmad, N.; Naqvi, A.A.; Riaz, M. Ethnobotany, ethnopharmacology, phytochemistry, biological activities and toxicity of *Pistacia chinensis* subsp. integerrima: A comprehensive review. *Phytother. Res.* **2020**, *34*, 2793–2819. [CrossRef] [PubMed]
58. Ghanavati, M.; Rahmani, J.; Clark, C.C.T.; Hosseinabadi, S.M.; Rahimlou, M. Pistachios and cardiometabolic risk factors: A systematic review and meta-analysis of randomized controlled clinical trials. Complement. *Ther. Med.* **2020**, *52*, 102513. [CrossRef] [PubMed]
59. Asbaghi, O.; Hadi, A.; Campbell, M.S.; Venkatakrishnan, K.; Ghaedi, E. Effects of pistachios on anthropometric indices, inflammatory markers, endothelial function and blood pressure in adults: A systematic review and meta-analysis of randomised controlled trials. *Br. J. Nutr.* **2020**, *17*, 718–729. [CrossRef] [PubMed]
60. Grosso, G.; Godos, J.; Currenti, W.; Micek, A.; Falzone, L.; Libra, M.; Giampieri, F.; Forbes-Hernández, T.Y.; Quiles, J.L.; Battino, M.; et al. The effect of dietary polyphenols on vascular health and hypertension: Current evidence and mechanisms of action. *Nutrients* **2022**, *14*, 545. [CrossRef]
61. Jamshidi, S.; Hejazi, N.; Golmakani, M.-T.; Tanideh, N. Wild pistachio (*Pistacia atlantica mutica*) oil improve metabolic syndrome features in rats with high fructose ingestion. *Iran J. Basic Med. Sci.* **2018**, *21*, 1255–1261.
62. Hosseini, S.; Nili-Ahmadabadi, A.; Nachvak, S.M.; Dastan, D.; Moradi, S.; Abdollahzad, H.; Mostafai, R. Antihyperlipidemic and antioxidative properties of *Pistacia atlantica* subsp. kurdica in streptozotocin-induced diabetic mice. *Diabetes Metab. Syndr. Obes.* **2020**, *13*, 1231–1236. [CrossRef]
63. Gorabi, S.A.; Mohammadzadeh, H.; Rostampour, M. The effects of ripe pistachio hulls hydro-alcoholic extract and aerobic training on learning and memory in streptozotocin-induced diabetic male rats. *Basic Clin. Neurosci.* **2020**, *11*, 525–534. [CrossRef]
64. Behmanesh, M.A.; Poormoosavi, S.M.; Pareidar, Y.; Ghorbanzadeh, B.; Mahmoodi-kouhi, A.; Najafzadehvarzi, H. *Pistacia atlantica's* effect on ovary damage and oxidative stress in streptozotocin-induced diabetic rats. *JBRA Assist. Reprod.* **2021**, *25*, 28–33. [CrossRef]
65. Jenkins, D.J.A.; Hu, F.B.; Tapsell, L.C.; Josse, A.R.; Kendall, C.W.C. Possible benefit of nuts in type 2 diabetes. *J. Nutr.* **2008**, *138*, 1752S–1756S. [CrossRef]
66. Kochar, J.; Gaziano, J.M.; Djoussé, L. Nut consumption and risk of type II diabetes in the Physicians' Health Study. *Eur. J. Clin. Nutr.* **2010**, *64*, 75–79. [CrossRef]
67. Ros, E. Health Benefits of Nut Consumption. *Nutrients* **2010**, *2*, 652–682. [CrossRef]
68. Damavandi, R.D.; Eghtesadi, S.; Shidfar, F.; Heydari, I.; Foroushani, A.R. Effects of hazelnuts consumption on fasting blood sugar and lipoproteins in patients with type 2 diabetes. *J. Res. Med. Sci.* **2013**, *18*, 314–321.
69. Pan, A.; Sun, Q.; Manson, J.E.; Willett, W.C.; Hu, F.B. Walnut consumption is associated with lower risk of type 2 diabetes in women. *J. Nutr.* **2013**, *143*, 512–518. [CrossRef]
70. Pachi, V.K.; Mikropoulou, E.V.; Gkiouvetidis, P.; Siafakas, K.; Argyropoulou, A.; Angelis, A.; Mitakou, S.; Halabalaki, M. Traditional uses, phytochemistry and pharmacology of Chios mastic gum (*Pistacia lentiscus* var. Chia, Anacardiaceae): A review. *J. Ethnopharmacol.* **2020**, *254*, 112485. [CrossRef]
71. Kendall, C.W.C.; Josse, A.R.; Esfahani, A.; Jenkins, D.J.A. The impact of pistachio intake alone or in combination with high-carbohydrate foods on post-prandial glycemia. *Eur. J. Clin. Nutr.* **2011**, *65*, 696–702. [CrossRef]
72. Kendall, C.W.C.; West, S.G.; Augustin, L.S.; Esfahani, A.; Vidgen, E.; Bashyam, B.; Sauder, K.A.; Campbell, J.; Chiavaroli, L.; Jenkins, A.L.; et al. Acute effects of pistachio consumption on glucose and insulin, satiety hormones and endothelial function in the metabolic syndrome. *Eur. J. Clin. Nutr.* **2014**, *68*, 370–375. [CrossRef]
73. Feng, X.; Liu, H.; Li, Z.; Carughi, A.; Ge, S. Acute effect of pistachio intake on postprandial glycemic and gut hormone responses in women with gestational diabetes or gestational impaired glucose tolerance: A randomized, controlled, crossover study. *Front. Nutr.* **2019**, *6*, 186. [CrossRef]

74. Melero, V.; García de la Torre, N.; Assaf-Balut, C.; Jiménez, I.; del Valle, L.; Durán, A.; Bordiú, E.; Valerio, J.J.; Herraiz, M.A.; Izquierdo, N.; et al. Effect of a Mediterranean diet-based nutritional intervention on the risk of developing gestational diabetes mellitus and other maternal-fetal adverse events in hispanic women residents in Spain. *Nutrients* **2020**, *12*, 3505. [CrossRef]
75. Melero, V.; Assaf-Balut, C.; De La Torre, N.G.; Jiménez, I.; Bordiú, E.; Del Valle, L.; Valerio, J.; Familiar, C.; Durán, A.; Runkle, I.; et al. Benefits of adhering to a Mediterranean diet supplemented with extra virgin olive oil and pistachios in pregnancy on the health of offspring at 2 years of age. Results of the San Carlos gestational diabetes mellitus prevention study. *J. Clin. Med.* **2020**, *9*, 1454. [CrossRef]
76. Gulati, S.; Misra, A. Sugar Intake, Obesity, and Diabetes in India. *Nutrition* **2014**, *30*, 192–197. [CrossRef]
77. Jenkins, D.J.A.; Kendall, C.W.C.; Banach, M.S.; Srichaikul. Nuts as a replacement for carbohydrates in the diabetic diet. *Diabetes Care* **2011**, *34*, 1706–1711. [CrossRef] [PubMed]
78. Parham, M.; Heidari, S.; Khorramirad, A.; Hozoori, M.; Hosseinzadeh, F.; Bakhtyari, L.; Vafaeimanesh, J. Effects of pistachio nut supplementation on blood glucose in patients with type 2 diabetes: A randomized crossover trial. *Rev. Diabet. Stud.* **2014**, *11*, 190–196. [CrossRef] [PubMed]
79. Canudas, S.; Hernández-Alonso, P.; Galié, S.; Muralidharan, J.; Morell-Azanza, L.; Zalba, G.; García-Gavilán, J.; Martí, A.; Salas-Salvadó, J.; Bulló, M. Pistachio consumption modulates DNA oxidation and genes related to telomere maintenance: A crossover randomized clinical trial. *Am. J. Clin. Nutr.* **2019**, *109*, 1738–1745. [CrossRef]
80. Bagetta, D.; Maruca, A.; Lupia, A.; Mesiti, F.; Catalano, R.; Romeo, I.; Moraca, F.; Ambrosio, F.A.; Costa, G.; Artese, A.; et al. Mediterranean products as promising source of multi-target agents in the treatment of metabolic syndrome. *Eur. J. Med. Chem.* **2020**, *186*, 111903. [CrossRef] [PubMed]
81. Nowrouzi-Sohrabi, P.; Hassanipour, S.; Sisakht, M.; Daryabeygi-Khotbehsara, R.; Savardashtaki, A.; Fathalipour, M. The effectiveness of pistachio on glycemic control and insulin sensitivity in patients with type 2 diabetes, prediabetes and metabolic syndrome: A systematic review and meta-analysis. *Diabetes Metab. Syndr. Clin. Res. Rev.* **2020**, *14*, 1589–1595. [CrossRef] [PubMed]
82. Assaf-Balut, C.; García de la Torre, N.; Durán, A.; Fuentes, M.; Bordiú, E.; del Valle, L.; Familiar, C.; Ortolá, A.; Jiménez, I.; Herraiz, M.A.; et al. A Mediterranean diet with additional extra virgin olive oil and pistachios reduces the incidence of gestational diabetes mellitus (GDM): A randomized controlled trial: The St. Carlos GDM prevention study. *PLoS ONE* **2017**, *12*, e0185873. [CrossRef] [PubMed]
83. Vadivel, V.; Kunyanga, C.N.; Biesalski, H.K. Health benefits of nut consumption with special reference to body weight control. *Nutrition* **2012**, *28*, 1089–1097. [CrossRef]
84. Tan, S.Y.; Dhillon, J.; Mattes, R.D. A review of the effects of nuts on appetite, food intake, metabolism, and body weight. *Am. J. Clin. Nutr.* **2014**, *100* (Suppl. 1), 412S–422S. [CrossRef]
85. Li, Z.; Song, R.; Nguyen, C.; Zerlin, A.; Karp, H.; Naowamondhol, K. Pistachio nuts reduce triglycerides and body weight by comparison to refined carbohydrate snack in obese subjects on a 12-week weight loss program. *J. Am. Coll. Nutr.* **2010**, *29*, 198–203. [CrossRef]
86. Fantino, M.; Bichard, C.; Mistretta, F.; Bellisle, F. Daily consumption of pistachios over 12 weeks improves dietary profile without increasing body weight in healthy women: A randomized controlled intervention. *Appetite* **2020**, *144*, 104483. [CrossRef]
87. Rock, C.L.; Zunshine, E.; Nguyen, H.T.; Perez, A.O.; Zoumas, C.; Pakiz, B.; White, M.M. Effects of pistachio consumption in a behavioral weight loss intervention on weight change, cardiometabolic factors, and dietary intake. *Nutrients* **2020**, *12*, 2155. [CrossRef]
88. Cassady, B.A.; Hollis, J.H.; Fulford, A.D.; Considine, R.V.; Mattes, R.D. Mastication of almonds: Effects of lipid bioaccessibility, appetite, and hormone response. *Am. J. Clin. Nutr.* **2009**, *89*, 794–800. [CrossRef]
89. Carughi, A.; Bellisle, F.; Dougkas, A.; Giboreau, A.; Feeney, M.J.; Higgs, J. A randomized controlled pilot study to assess effects of a daily pistachio (*Pistacia vera*) afternoon snack on next-meal energy intake, satiety, and anthropometry in French women. *Nutrients* **2019**, *11*, 767. [CrossRef]
90. Nuzzo, D.; Galizzi, G.; Amato, A.; Terzo, S.; Picone, P.; Cristaldi, L.; Mulè, F.; Di Carlo, M. Regular intake of pistachio mitigates the deleterious effects of a high fat-diet in the brain of obese mice. *Antioxidants* **2020**, *9*, 317. [CrossRef]
91. Xia, K.; Yang, T.; An, L.-Y.; Lin, Y.-Y.; Qi, Y.-X.; Chen, X.-Z.; Sun, D.-L. The relationship between pistachio (*Pistacia vera* L) intake and adiposity. A systematic review and meta-analysis of randomized controlled trials. *Medicine* **2020**, *99*, e21136. [CrossRef]
92. Higgs, J.; Styles, K.; Carughi, A.; Roussell, M.A.; Bellisle, F.; Elsner, W.; Li, Z. Plant-based snacking: Research and practical applications of pistachios for health benefits. *J. Nutr. Sci.* **2021**, *10*, e87. [CrossRef]
93. Gentile, C.; Perrone, A.; Attanzio, A.; Tesoriere, L.; Livrea, M.A. Sicilian pistachio (*Pistacia vera* L.) nut inhibits expression and release of inflammatory mediators and reverts the increase of paracellular permeability in IL-1b-exposed human intestinal epithelial cells. *Eur. J. Nutr.* **2015**, *54*, 811–821. [CrossRef]
94. Ostovan, M.; Bagher Fazljou, S.M.; Khazraei, H.; Araj Khodaei, M.; Torbati, M. The anti-inflammatory effect of *Pistacia lentiscus* in a rat model of colitis. *J. Inflamm. Res.* **2020**, *13*, 369–376. [CrossRef]
95. Terzo, S.; Mulè, F.; Caldara, G.F.; Baldassano, S.; Puleio, R.; Vitale, M.; Cassata, G.; Ferrantelli, V.; Amato, A. Pistachio consumption alleviates inflammation and improves gut microbiota composition in mice fed a high-fat diet. *Int. J. Mol. Sci.* **2020**, *21*, 365. [CrossRef]

96. Zahouani, Y.; Rhouma, K.B.; Kacem, K.; Sebai, H.; Sakly, M. Aqueous leaf extract of *Pistacia lentiscus* improves acute acetic acid-induced colitis in rats by reducing inflammation and oxidative stress. *J. Med. Food* **2021**, *24*, 697–708. [CrossRef]
97. Khedir, S.B.; Mzid, M.; Bardaa, S.; Moalla, D.; Sahnoun, Z.; Rebai, T. In vivo evaluation of the anti-inflammatory effect of *Pistacia lentiscus* fruit oil and its effects on oxidative stress. *Evid. Based Complement. Alternat. Med.* **2016**, *2016*, 6108203. [PubMed]
98. Catalani, S.; Palma, F.; Battistelli, S.; Benedetti, S. Oxidative stress and apoptosis induction in human thyroid carcinoma cells exposed to the essential oil from *Pistacia lentiscus* aerial parts. *PLoS ONE* **2017**, *12*, e0172138. [CrossRef] [PubMed]
99. Paterniti, I.; Impellizzeri, D.; Cordaro, M.; Siracusa, R.; Bisignano, C.; Gugliandolo, E.; Carughi, A.; Esposito, E.; Mandalari, G.; Cuzzocrea, S. The anti-inflammatory and antioxidant potential of pistachios (*Pistacia vera* L.) in vitro and in vivo. *Nutrients* **2017**, *9*, 915. [CrossRef] [PubMed]
100. Mohamed, K.; Zine, K.; Fahima, K.; Abdelfattah, E.; Sharifudin, S.M.; Duduku, K. NiO nanoparticles induce cytotoxicity mediated through ROS generation and impairing the antioxidant defense in the human lung epithelial cells (A549): Preventive effect of *Pistacia lentiscus* essential oil. *Toxicol. Rep.* **2018**, *5*, 480–488. [CrossRef] [PubMed]
101. Ghzaiel, I.; Zarrouk, A.; Nury, T.; Libergoli, M.; Florio, F.; Hammouda, S.; Menetrier, F.; Avoscan, L.; Yammine, A.; Samadi, M.; et al. Antioxidant properties and cytoprotective effect of *Pistacia lentiscus* L. seed oil against 7β-hydroxycholesterol-induced toxicity in C2C12 myoblasts: Reduction in oxidative stress, mitochondrial and peroxisomal dysfunctions and attenuation of cell death. *Antioxidants* **2021**, *10*, 1772. [CrossRef] [PubMed]
102. Tolooei, M.; Mirzaei, A. Effects of Pistacia atlantica extract on erythrocyte membrane rigidity, oxidative stress, and hepatotoxicity induced by CCl4 in rats. *Glob. J. Health Sci.* **2015**, *7*, 32–38. [CrossRef]
103. Abidi, A.; Kourda, N.; Feki, M.; Khamsa, S.B. Protective Effect of Tunisian Flaxseed Oil against Bleomycin-Induced Pulmonary Fibrosis in Rats. *Nutr. Cancer* **2017**, *69*, 490–497. [CrossRef]
104. Chebab, S.; Mekircha, F.; Leghouchi, E. Potential protective effect of *Pistacia lentiscus* oil against chlorpyrifos-induced hormonal changes and oxidative damage in ovaries and thyroid of female rats. *Biomed. Pharmacother.* **2017**, *96*, 1310–1316. [CrossRef]
105. Saidi, S.A.; Ncir, M.; Chaaben, R.; Jamoussi, K.; van Pelt, J.; Elfeki, A. Liver injury following small intestinal ischemia reperfusion in rats is attenuated by *Pistacia lentiscus* oil: Antioxidant and anti-inflammatory effects. *Arch. Physiol. Biochem.* **2017**, *123*, 199–205. [CrossRef]
106. Ammari, M.; Othman, H.; Hajri, A.; Sakly, M.; Abdelmelek, H. Pistacia lentiscus oil attenuates memory dysfunctionand decreases levels of biomarkers of oxidative stress inducedby lipopolysaccharide in rats. *Brain Res. Bull.* **2018**, *140*, 140–147.
107. Hong, M.Y.; Groven, S.; Marx, A.; Rasmussen, C.; Beidler, J. Anti-inflammatory, antioxidant, and hypolipidemic effects of mixed nuts in atherogenic diet-fed rats. *Molecules* **2018**, *23*, 3126. [CrossRef] [PubMed]
108. Bagheri, S.; Sarabi, M.M.; Khosravi, P. Effects of *Pistacia atlantica* on oxidative stress markers and antioxidant enzymes expression in diabetic rats. *J. Am. Coll. Nutr.* **2019**, *38*, 267–274. [CrossRef] [PubMed]
109. Sauder, K.A.; McCrea, C.E.; Ulbrecht, J.S.; Kris-Etherton, P.M.; West, S.G. Effects of pistachios on the lipid/lipoprotein profile, glycemic control, inflammation, and endothelial function in type 2 diabetes: A randomized trial. *Metabolism* **2015**, *64*, 1521–1529. [CrossRef] [PubMed]
110. Kay, C.D.; Gebauer, S.K.; West, S.G.; Kris-Etherton, P.M. Pistachios Increase Serum Antioxidants and Lower Serum Oxidized-LDL in Hypercholesterolemic Adults. *J. Nutr.* **2010**, *140*, 1093–1098. [CrossRef]
111. Papada, E.; Forbes, A.; Amerikanou, C.; Torovic, L.; Kalogeropoulos, N.; Tzavara, C.; Triantafillidis, J.K.; Kaliora, A.C. Antioxidative efficacy of a *Pistacia lentiscus* supplement and its effect on the plasma amino acid profile in inflammatory bowel disease: A randomised, double-blind, placebo-controlled trial. *Nutrients* **2018**, *10*, 1779. [CrossRef]
112. Rahman, H.S. Phytochemical analysis and antioxidant and anticancer activities of mastic gum resin from *Pistacia atlantica* subspecies kurdica. *Onco Targets Ther.* **2018**, *11*, 4559–4572. [CrossRef]
113. Kirollos, F.N.; Elhawary, S.S.; Salama, O.M.; Elkhawas, Y.A. LC-ESI-MS/MS and cytotoxic activity of three *Pistacia* species. *Nat. Prod. Res.* **2018**, *33*, 1747–1750. [CrossRef]
114. Pasban-Aliabadi, H. Effects of baneh (*Pistacia atlantica*) gum on human breast cancer cell line (MCF-7) and its interaction with anticancer drug doxorubicin. *Iran. J. Pharm. Res.* **2019**, *18*, 1959–1966.
115. Dousti, M.; Sari, S.; Saffari, M.; Kelidari, H.; Asare-Addo, K.; Nokhodchi, A. Loading *Pistacia atlantica* essential oil in solid lipid nanoparticles and its effect on apoptosis of breast cancer cell line MDA-MB-231. *Pharm. Dev. Technol.* **2022**, *27*, 63–71. [CrossRef]
116. Glei, M.; Ludwig, D.; Lamberty, J.; Fischer, S.; Lorkowski, S.; Schlörmann, W. Chemopreventive potential of raw and roasted pistachios regarding colon carcinogenesis. *Nutrients* **2017**, *18*, 1368. [CrossRef]
117. Koyuncu, I.; Gönel, A.; Temiz, E.; Karaoğul, E.; Uyar, Z. Pistachio green hull extract induces apoptosis through multiple signaling pathways by causing oxidative stress on colon cancer cells. *Anticancer. Agents Med. Chem.* **2021**, *21*, 725–737. [CrossRef] [PubMed]
118. Seifaddinipour, M.; Farghadani, R.; Namvar, F.; Bin Mohamad, J.; Muhamad, N.A. In vitro and in vivo anticancer activity of the most cytotoxic fraction of pistachio hull extract in breast cancer. *Molecules* **2020**, *25*, 1776. [CrossRef] [PubMed]
119. Ostovan, M.; Andabar, M.H.; Khazraei, H.; Bagher Fazljou, S.M.; Khodabandeh, Z.; Shamsdin, S.A.; Araj Khodaei, M.; Torbati, M. The short-term effects of *Pistacia lentiscus* oil and sesame oil on liver and kidney pathology of rats and human cancer cell lines. *Galen Med. J.* **2020**, *9*, e2001. [CrossRef] [PubMed]
120. Tabanca, N.; Nalbantsoy, A.; Kendra, P.E.; Demirci, F.; Demirci, B. Chemical characterization and biological activity of the mastic gum essential oils of *Pistacia lentiscus* var. chia from Turkey. *Molecules* **2020**, *25*, 2136. [CrossRef] [PubMed]

121. Soulaidopoulos, S.; Tsiogka, A.; Chrysohoou, C.; Lazarou, E.; Aznaouridis, K.; Doundoulakis, I.; Tyrovola, D.; Tousoulis, D.; Tsioufis, K.; Vlachopoulos, C.; et al. Overview of chios mastic gum (*Pistacia lentiscus*) effects on human health. *Nutrients* **2022**, *14*, 590. [CrossRef]
122. Doi, K.; Wei, M.; Kitano, M.; Uematsu, N.; Inoue, M.; Wanibuchi, H. Enhancement of preneoplastic lesion yield by Chios Mastic Gum in a rat liver medium-term carcinogenesis bioassay. *Toxicol. Appl. Pharmacol.* **2009**, *234*, 135–142. [CrossRef]
123. Janakat, S.; Al-Merie, H. Evaluation of hepatoprotective effect of *Pistacia lentiscus, Phillyrea latifolia* and *Nicotiana glauca*. *J. Ethnopharmacol.* **2002**, *83*, 135–138. [CrossRef]
124. Naghshi, S.; Sadeghian, M.; Nasiri, M.; Mobarak, S.; Asadi, M.; Sadegh, O. Association of total nut, tree nut, peanut, and peanut butter consumption with cancer incidence and mortality: A comprehensive systematic review and dose-response meta-analysis of observational studies. *Adv. Nutr.* **2021**, *12*, 793–808. [CrossRef]
125. Harandi, H.; Khanamani Falahati-pour, S.K.; Mahmoodi, M.; Faramarz, S.; Maleki, H.; Nasab, F.B.; Shiri, H.; Fooladi, S.; Nematollahi, M.H. Nanoliposomal formulation of pistachio hull extract: Preparation, characterization and anti-cancer evaluation through Bax/Bcl_2 modulation. *Mol. Biol. Rep.* **2022**, *49*, 2735–2743. [CrossRef]
126. Ukhanova, M.; Wang, X.; Baer, D.J.; Novotny, J.A.; Fredborg, M.; Mai, V. Effects of almond and pistachio consumption on gut microbiota composition in a randomised cross-over human feeding study. *Br. J. Nutr.* **2014**, *111*, 2146–2152. [CrossRef]
127. Yanni, A.E.; Mitropoulou, G.; Prapa, I.; Agrogiannis, G.; Kostomitsopoulos, N.; Bezirtzoglou, E.; Kourkoutas, Y.; Karathanos, V.T. Functional modulation of gut microbiota in diabetic rats following dietary intervention with pistachio nuts (*Pistacia vera* L.). *Metabol. Open* **2020**, *7*, 100040. [CrossRef] [PubMed]
128. Creedon, A.S.; Hung, E.S.; Berry, S.E.; Whelan, K. Nuts and their effect on gut microbiota, gut function and symptoms in adults: A systematic review and meta-analysis of randomised controlled trials. *Nutrients* **2020**, *12*, 2347. [CrossRef] [PubMed]
129. Catalán, L.; Alvarez-Orti, M.; Pardo-Giménez, A.; Gómez, R.; Rabadán, A.; Pardo, J.E. Pistachio oil: A review on its Chemical composition, extraction systems, and uses. *Eur. J. Lipid Sci. Technol.* **2017**, *119*, 1600126. [CrossRef]
130. Ojeda-Amador, R.M.; Fregapane, G.; Salvador, M.D. Composition and properties of virgin pistachio oils and their by-products from different cultivars. *Food Chem.* **2018**, *240*, 123–130. [CrossRef] [PubMed]
131. Fregapane, G.; Guisantes-Batan, E.; Ojeda-Amador, R.M.; Salvador, M.D. Development of functional edible oils enriched with pistachio and walnut phenolic extracts. *Food Chem.* **2020**, *310*, 125917. [CrossRef] [PubMed]
132. Fregapane, G.; Cabezas Fernández, C.; Salvador, M.D. Emulsion and Microemulsion Systems to Improve Functional Edible Oils Enriched with Walnut and Pistachio Phenolic Extracts. *Foods* **2022**, *11*, 1210. [CrossRef]
133. Arribas, C.; Cabellos, B.; Sánchez, C.; Cuadrado, C.; Guillamón, E.; Pedrosa, M.M. The impact of extrusion on the nutritional composition, dietary fiber and in vitro digestibility of gluten-free snacks based on rice, pea and carob flour blends. *Food Funct.* **2017**, *18*, 3654–3663. [CrossRef]
134. Sanchiz, A.; Pedrosa, M.; Guillamon, E.; Arribas, C.; Cabellos, B.; Linacero, R.; Cuadrado, C. Influence of boiling and autoclave processing on the phenolic content, antioxidant activity and functional properties of pistachio, cashew and chestnut flours. *LWT* **2019**, *105*, 250–256. [CrossRef]
135. Sethi, S.; Tyagi, S.K.; Anurag, R.K. Plant-Based Milk Alternatives an Emerging Segment of Functional Beverages: A Review. *J. Food Sci. Technol.* **2016**, *53*, 3408–3423. [CrossRef]
136. Shakerardekani, A.; Karim, R.; Ghazali, H.M.; Chin, N.L. Oxidative Stability of Pistachio (*Pistacia vera* L.) Paste and Spreads. *J. Am. Oil Chem. Soc.* **2015**, *92*, 1015–1021. [CrossRef]
137. Polmann, G.; Badia, V.; Danielski, R.; Salvador-Ferreira, S.R.; Block, J.M. Nuts and nut-based products: A meta-analysis from intake health benefits and functional characteristics from recovered constituents. *Food Rev. Int.* **2022**, *29*, 197–210. [CrossRef]
138. Elloumi, W.; Maalej, A.; Ortiz, S.; Michel, S.; Chamkha, M.; Boutefnouchet, S.; Sayadi, S. *Pistacia lentiscus* L. Distilled Leaves as a Potential Cosmeceutical Ingredient: Phytochemical Characterization, Transdermal Diffusion, and Anti-Elastase and Anti-Tyrosinase Activities. *Molecules* **2022**, *27*, 855. [CrossRef] [PubMed]
139. Kahleova, H.; Levin, S.; Barnard, N.D. Vegetarian Dietary Patterns and Cardiovascular Disease. *Prog. Cardiovasc. Dis.* **2018**, *61*, 54–61. [CrossRef] [PubMed]
140. Neale, E.P.; Tran, G.; Brown, R.C. Barriers and facilitators to nut consumption: A narrative review. *Int. J. Environ. Res. Public Health* **2020**, *17*, 9127. [CrossRef]
141. Brown, R.C.; Gray, A.R.; Yong, L.C.; Chisholm, A.; Leong, S.L.; Tey, S.L. A comparison of perceptions of nuts between the general public, dietitians, general practitioners, and nurses. *Peer J.* **2018**, *6*, e5500. [CrossRef]
142. Yong, L.C.; Gray, A.R.; Chisholm, A.; Leong, S.L.; Tey, S.L.; Brown, R.C. Barriers to and facilitators and perceptions of nut consumption among the general population in New Zealand. *Public Health Nutr.* **2017**, *20*, 3166–3182. [CrossRef]
143. Hong, L.; Yao, L.; Xie, P.; Li, W. An empirical study on consumer purchase intention for nuts and influencing factors—Survey based on consumers from Zhejiang. *Food Control* **2020**, *117*, 107343. [CrossRef]
144. Tran, G. Investigating the Role of Nutrition Information and Misinformation in Dietetic Practice: A Case Study Exploring the Perceptions of Nut Consumption in Health Professionals and Consumers. Bachelor's Thesis, University of Wollongong, Wollongong, Australia, 2020.
145. O'Neil, C.E.; Nicklas, T.A. Tree Nut Consumption Is Associated with Better Nutrient Adequacy and Diet Quality in Adults: National Health and Nutrition Examination Survey 2005–2010. *Nutrients* **2015**, *7*, 595–607. [CrossRef]

146. Nikodijevic, C.J.; Probst, Y.C.; Batterham, M.J.; Tapsell, L.C.; Neale, E.P. Nut consumption in a representative survey of Australians: A secondary analysis of the 2011–2012 National Nutrition and Physical Activity Survey. *Public Health Nutr.* **2020**, *23*, 3368–3378. [CrossRef]
147. Brown, R.C.; Tey, S.L.; Gray, A.R.; Chisholm, A.; Smith, C.; Fleming, E.; Blakey, C.; Parnell, W. Patterns and predictors of nut consumption: Results from the 2008/09 New Zealand Adult Nutrition Survey. *Br. J. Nutr.* **2014**, *112*, 2028–2040. [CrossRef]
148. Dikariyanto, V.; Berry, S.E.; Pot, G.K.; Francis, L.; Smith, L.; Hall, W.L. Tree nut snack consumption is associated with better diet quality and CVD risk in the UK adult population: National Diet and Nutrition Survey (NDNS) 2008–2014. *Public Health Nutr.* **2020**, *23*, 3160–3169. [CrossRef] [PubMed]

nutrients

Article

Sustained Consumption of a Decaffeinated Green Coffee Nutraceutical Has Limited Effects on Phenolic Metabolism and Bioavailability in Overweight/Obese Subjects

Miguel Ángel Seguido, Rosa Maria Tarradas, Susana González-Rámila, Joaquín García-Cordero, Beatriz Sarriá, Laura Bravo-Clemente and Raquel Mateos *

Department of Metabolism and Nutrition, Institute of Food Science, Technology and Nutrition (ICTAN-CSIC), Spanish National Research Council (CSIC), José Antonio Nováis 10, 28040 Madrid, Spain; m.seguido@ictan.csic.es (M.Á.S.); rosa.tarradas.valero@gmail.com (R.M.T.); s.gonzalez@ictan.csic.es (S.G.-R.); j.garcia@ictan.csic.es (J.G.-C.); beasarria@ictan.csic.es (B.S.); lbravo@ictan.csic.es (L.B.-C.)
* Correspondence: raquel.mateos@ictan.csic.es

Abstract: Knowledge on the bioavailability of coffee (poly)phenols mostly come from single dose postprandial studies. This study aimed at investigating the effects of regularly consuming a green coffee phenolic extract (GCPE) on the bioavailability and metabolism of (poly)phenols. Volunteers with overweight/obesity consumed a decaffeinated GCPE nutraceutical containing 300 mg hydroxycinnamates twice daily for two months. Plasma and urinary pharmacokinetics, and fecal excretion of phenolic metabolites were characterized by LC-MS-QToF at weeks 0 and 8. Fifty-four metabolites were identified in biological fluids. Regular consumption of the nutraceutical produced certain changes: reduced forms of caffeic, ferulic and coumaric acids in urine or 3-(3′-hydroxypenyl)propanoic, and 3,4-dihydroxybenzoic acids in feces significantly increased ($p < 0.05$) after 8 weeks; in contrast, coumaroylquinic and dihydrocoumaroylquinic acids in urine decreased ($p < 0.05$) compared to baseline excretion. The sum of intestinal and colonic metabolites increased after sustained consumption of GCPE, without reaching statistical significance, suggesting a small overall effect on (poly)phenols' bioavailability.

Keywords: green coffee; hydroxycinnamates; bioavailability; pharmacokinetics; microbial catabolites; biotransformation pathways

Citation: Seguido, M.Á.; Tarradas, R.M.; González-Rámila, S.; García-Cordero, J.; Sarriá, B.; Bravo-Clemente, L.; Mateos, R. Sustained Consumption of a Decaffeinated Green Coffee Nutraceutical Has Limited Effects on Phenolic Metabolism and Bioavailability in Overweight/Obese Subjects. *Nutrients* **2022**, *14*, 2445. https://doi.org/10.3390/nu14122445

Academic Editor: Cristian Del Bo

Received: 18 May 2022
Accepted: 9 June 2022
Published: 13 June 2022

Publisher's Note: MDPI stays neutral with regard to jurisdictional claims in published maps and institutional affiliations.

1. Introduction

Green coffee beans are rich in phenolic compounds, mainly hydroxycinnamate esters, collectively known as hydroxycinnamic or chlorogenic acids [1], as opposed to roasted coffee where most phenolic compounds are lost during roasting or transformed and incorporated into melanoidins [2]. Phenol-rich extracts obtained from unroasted coffee beans (green coffee phenolic extracts, GCPE) are commonly used as dietary supplements or nutraceuticals. Consumption of GCPE is increasing due to the reported health benefits of green coffee (poly)phenols, with special emphasis on their effects on diabetes and obesity in humans [3], preventing insulin resistance, reducing appetite [4], regulating body weight [5], and favoring weight loss [6]. Green coffee phenolic compounds have also shown cardioprotective properties due to their antihypertensive, antihyperlipidemic, antifibrotic, or anti-inflammatory activities [7,8], even showing antiproliferative and cytotoxic effects in human cancer cell lines [9].

The metabolism, pharmacokinetics, and bioavailability of coffee phenols have been reported in several studies [10–12], in particular, green coffee hydroxycinnamic acids have proven to be extensively metabolized and partially absorbed in healthy subjects after acute consumption of a green/roasted coffee brew, with long permanence in the host, thus favoring their bioactivity [13]. In fact, some of these phenols can be directly absorbed in the small intestine, although most reach the large intestine where they are catabolized by the intestinal microbiota. These catabolites can exert specific functions, although the physiological role of phenols may be altered by the biotransformation of parent compounds [14]. It is important to bear in mind that the interaction between (poly)phenols and microbiota is bidirectional. The microbiota may act on (poly)phenols to enhance their conversion into active and bioavailable microbial catabolites or "postbiotics" [15–17]. Reciprocally, (poly)phenols may also modulate the gut microbiota ecology, exerting a prebiotic-like action [15,16], which might potentially affect the microbial catabolism of phenolic compounds.

Up to date, most studies on the bioavailability and metabolism of phenolic compounds are single dose, acute, postprandial assays with plasma and urine sample collection up to 12–24 h after intake of the test food or supplement. However, the long-term effects of regular consumption on the bioavailability of (poly)phenols have been scarcely studied so far. Only in Mena et al., (2021) [18] has the absorption, metabolism, and pharmacokinetic profile of roasted coffee phenolics after one month of repeated consumption of one or three cups of coffee or a confectionary containing coffee and cocoa been assessed. However, in this work, the authors did not study the bioavailability of coffee phenolics at the beginning of the intervention, and thus the final results could not be compared with the initial metabolism to assess potential modifications over prolonged consumption. In a previous study, these authors did study the effect of the sustained consumption of tablets containing green tea and green coffee phenolic extracts for 8 weeks, finding important differences in the urinary excretion of flavanol-derived catabolites, but less remarkable changes in the excretion of metabolites derived from green coffee hydroxycinnamates [19].

Bearing in mind the above points, the initial hypothesis of the present study was that the sustained intake of green coffee (poly)phenols may modify the overall bioavailability and metabolism of these compounds, likely due to the prolonged exposure of the intestinal microbiota, which would lead to a higher and more uniform concentration of the associated catabolites in the gastrointestinal tract and in plasma after absorption. In addition, considering the increasing consumption of GCPE as dietary supplements or nutraceuticals, usually aimed at weight management, it is important to understand the bioavailability and metabolism of phenols in the GCPE nutraceutical in the target population (i.e., subjects with obesity/overweight), which might differ from that in a beverage brewed from the green coffee beans. Therefore, the present work aimed at studying the possible changes in the absorption and metabolism profiles of hydroxycinnamates in plasma, urine, and feces after regular consumption of a GCPE nutraceutical by subjects with overweight/obesity, comparing the absorption, metabolism, and pharmacokinetic profile at baseline and after eight weeks of daily intake.

2. Materials and Methods

2.1. Chemicals

Ascorbic, 5-caffeoylquinic acid, 3′,4′-dihydroxycinnamic acid, 4′-hydroxy-3′-methoxycinnamic acid, 4′-hydroxycinnamic acid; hippuric acid, 3-(3′,4′-dihydroxyphenyl)propanoic acid, 3-(4′-hydroxy-3′-methoxyphenyl)propanoic acid, 3-(4′-hydroxyphenyl)propanoic acid, 3-(3′-hydroxyphenyl)propanoic acid, 3′,4′-dihydroxyphenylacetic acid, 4′-hydroxy-3′-methoxyphenylacetic acid, 4′-hydroxyphenylacetic acid, 3′-hydroxyphenylacetic acid, 3,4-dihydroxybenzoic acid, 4-hydroxy-3-methoxybenzoic acid, 4-hydroxybenzoic acid, and 3-hydroxybenzoic acid were purchased from Sigma-Aldrich (Madrid, Spain). 3,5-dicaffeoylquinic acid was purchased from PhytoLab (Vestenbergsgreuth, Germany). HPLC-grade solvents were acquired from Panreac (Madrid, Spain). Ultrapure water from MilliQ system (Millipore, Bedford, MA, USA) was used throughout the experiment.

2.2. Phenolic Composition of the GCPE

The green coffee phenolic extract (GCPE) was provided by Quimifarma Laboratorios S.L. (Toledo, Spain), which was a decaffeinated, phenol-rich (45%) commercial extract. The phenolic content and composition of the GCPE was verified by in-house HPLC-

DAD analysis, as in previous studies [1,13]. Briefly, one gram of GCPE was dissolved in 100 mL of MilliQ water and filtered through a PVDF 0.45 μm filter. Phenols were separated on a Superspher 100 RP18 column (4.6 mm × 250 mm i.d., 4 μm; Agilent Technologies, Santa Clara, CA, USA) preceded by an ODS RP18 guard column kept in a thermostatic oven at 30 °C using an Agilent 1200 series liquid chromatographic system equipped with an autosampler, quaternary pump, and diode-array detector (DAD). An aliquot (20 μL) was eluted at a flow rate of 1 mL/min using a mobile phase of 1% formic acid in deionized water (solvent A) and acetonitrile (solvent B). The solvent gradient changed from 10% to 20% solvent B over 5 min, 20% to 25% solvent B over 30 min, 25% to 35% solvent B over 10 min, and then was maintained isocratically for 5 min, returning to the initial conditions over 10 min. Chromatograms were recorded at 320 nm, and 5-caffeoylquinic and 3,5-dicaffeoylquinic acids were used to calculate the mono- and di-acylcinnamate esters content, respectively, by external standard method. Samples were analyzed in triplicate.

The phenolic composition of the GCPE is detailed in Supplementary Table S1. Caffeoylquinic and dicaffeoylquinic acids were the main phenolic compounds detected in the extract, accounting for 85% of the quantified hydroxycinnamates, followed by 7.6% feruloylquinic and 4.6% of caffeoylferuloylquinic acids. The nutraceutical was prepared considering the total phenolic content of the GCPE (458 mg/g) so that each sachet contained 0.66 g of GCPE, providing 300 mg of phenolic compounds. Excipient (whey protein) and aroma compounds (two flavors were available, vanilla and chocolate) were added to the formulation to facilitate dispersion in water and increase acceptability.

2.3. Participants, Study Design, and Sample Collection

This study is part of a larger intervention carried out in 29 subjects with overweight or obesity designed to assess the effect of sustained consumption of GCPE on weight control, blood pressure, lipid metabolism, and glucose homeostasis. Of these participants, nine volunteers (8 men and 1 woman), aged 44 ± 9 years and with a body mass index (BMI) of 31 ± 4 agreed to be included in the present bioavailability study. Details of the recruitment process, inclusion and exclusion criteria, and ethical clearance are reported elsewhere [20]. Subjects signed an additional specific written informed consent to participate in the present bioavailability study. The main biochemical and vital characteristics of the volunteers at baseline (week 0) and at the end of the intervention (week 8) are shown in Supplementary Table S2. No adverse events were reported.

Participants attended the Human Nutrition Unit (HNU) at the Institute of Food Science, Technology and Nutrition (ICTAN-CSIC) on two consecutive days at the beginning of the intervention (week 0) and also on two consecutive days at the end of the study, after consuming daily two doses of the GCPE nutraceutical, providing 600 mg/d phenols, during two months (week 8). On each occasion, on the first visit day (Figure 1), subjects arrived at the HNU after an overnight (12 h) fast and provided a urine sample. A licensed health care professional placed an intravenous catheter in the subjects' nondominant arm and collected a fasting blood sample (baseline, 0 h). Then, subjects drank the soluble GCPE nutraceutical dissolved in 200 mL of water. Subsequently, blood samples were collected in EDTA-coated tubes at 0.5, 1, 1.5, 2, 3, 4, 6, 8, 10, and 11 h. On the sampling days, volunteers remained at the HNU, where a (poly)phenol-free diet was provided. Half an hour after consuming the nutraceutical, volunteers ate a breakfast consisting of ham, cheese, white bread, and a yogurt. Lunch (paella, bread, and yogurt) and an afternoon snack (muffin and yogurt) were eaten 5 and 9 h, respectively, after consuming the nutraceutical. Water and isotonic beverages were freely available. Volunteers were instructed to eat a (poly)phenol-free dinner that night and to attend the HNU on the next day after a 12 h fast. Then, a final blood sample was obtained from the antecubital vein 24 h after consuming the nutraceutical. The same protocol was followed during visits at the end of the 8-week intervention. On each occasion, two days before each visit to the HNU, subjects consumed a low (poly)phenol diet free of foods containing hydroxycinnamates such as coffee, herbal teas, red wine, artichokes, whole grain cereals, vegetables, fruits and fruit juices, and dried fruits. Only banana, watermelon, cantaloupes, and potatoes were allowed.

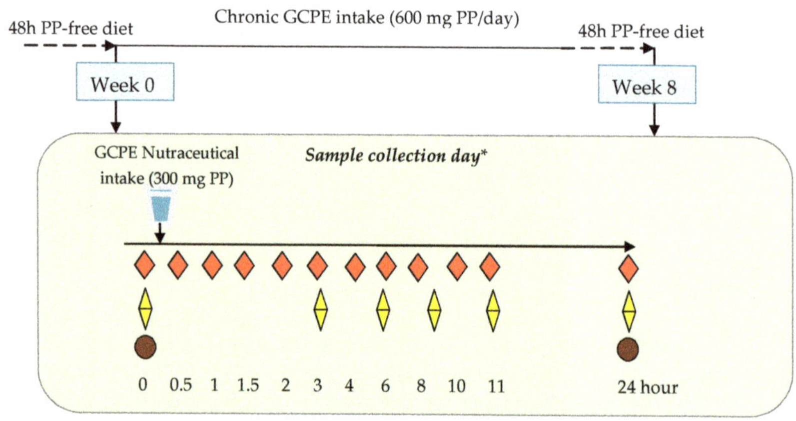

Figure 1. Schematic view of the study and collection of blood, urine, and fecal samples. * On each sampling day, volunteers consumed a single dose of the GCPE nutraceutical dissolved in water. GCPE: green coffee phenolic extract; PP: (poly)phenols.

Blood samples were immediately centrifuged (10 min, 3000 rpm, 4 °C) to obtain plasma. Urine was collected in plastic flasks containing preservative (0.5 g ascorbic acid) and sampled at different intervals: 0–3 h, 3–6 h, 6–9 h, 9–11 h, and 11–24 h after GCPE consumption, measuring total urine volume at each interval. Fecal samples were provided at baseline (−2 to 0 h) and 24 h after ingestion of the nutraceutical at the beginning and end of the intervention. All biological samples were stored at −80 °C until analysis.

2.4. Plasma, Urine and Fecal Samples Processing and Analysis by HPLC-ESI-QToF

To extract plasma metabolites, 10 µL of 50% (*v*/*v*) aqueous formic acid, and 900 µL cold acetonitrile containing 50 µL of 10% (*w*/*v*) ascorbic acid were added to 400 µL defrosted plasma. Samples were vortexed and centrifuged (14,000 rpm, 10 min, 4 °C), repeating the extraction and combining the two supernatants, which were reduced to dryness under nitrogen [13]. Dried samples were resuspended in 150 µL of 0.1% aqueous formic acid (containing 10% acetonitrile acidified with 0.1% formic acid), centrifuged (15 min, 14,000 rpm, 4 °C), and the supernatant collected. Urine samples were diluted 1:1 with Milli-Q water and centrifuged (14,000 rpm, 20 min, 4 °C) prior to analysis. Phosphate-buffered saline (3 mL) was added to a representative 300 mg sample of thawed feces, homogenized in an ultrasound bath during 10 min, centrifuged (14,000 rpm, 20 min, 4 °C), and the supernatants collected.

An amount of 30 µL of plasma and 5 µL of urine and fecal extracts, all previously filtered through 0.45 µm cellulose acetate membrane filters, were injected into an Agilent 1200 series LC system coupled to an Agilent 6530A Accurate-Mass Quadrupole Time-Of-Flight (Q-ToF) with ESI-Jet Stream Technology (Agilent Technologies, Santa Clara, CA, USA). A reverse-phase Ascentis Express C18 (15 cm × 3 mm, 2.7 µm) column (Sigma-Aldrich Quimica, Madrid, Spain), preceded by a Supelco 55215-U guard column (3 mm × 5 mm, 2.7 µm), was used for separation. Mobile phase A was 0.1% formic acid in Milli-Q water, and mobile phase B was acetonitrile containing 0.1% formic acid at a 0.3 mL/min flow rate. Details of the solvent gradient and Q-ToF acquisition conditions are given in the Supplementary Information. Metabolites were identified based on their retention time using authentic standards when possible. Those metabolites for which there were no available standards were tentatively quantified using the calibration curves of their corresponding phenolic precursors, as specified in the Supplementary Information, which contains additional information on the preparation and validation of calibration curves. Urine concentration of excreted metabolites was normalized to the total volume excreted in each studied interval.

2.5. Pharmacokinetic and Statistical Analysis

The PKsolver add-on program was used to perform pharmacokinetic analyses in Microsoft Excel, including maximum concentration (C_{max}), area under curve between 0–24 h ($AUC_{0–24}$), and time to reach maximum concentration (T_{max}) of metabolites in plasma and urine. All the results were statistically analyzed with SPSS software (version 27.0; SPSS, Inc., IBM Company, Armonk, NY, USA). The Shapiro–Wilk test was used to assess data normality. In view of the lack of normality, and considering the small sample size, comparisons between week 0 and week 8 were performed by the nonparametric Wilcoxon test for paired comparisons. The level of significance was $p < 0.05$. All data are expressed as mean ± standard error of the mean (SEM) unless specified otherwise.

3. Results

3.1. Identification and Quantification of Plasma Metabolites

Twenty-eight compounds derived from GCPE hydroxycinnamate intake were identified and quantified in the 0–24 h plasma samples. The identification of native compounds in the administered nutraceutical facilitated the targeting of their potential metabolites by searching for their exact mass and confirmation by fragmentation patterns (MS/MS). Supplementary Table S3 shows the retention time (RT), molecular formula, accurate mass of the quasimolecular ion $[M-H]^-$ after negative ionization, and the MS^2 fragments of the main compounds identified in plasma using LC-QToF. Peak plasma concentrations (C_{max}), time to reach C_{max} (T_{max}), and area under the curve ($AUC_{0–24h}$) values of the 28 metabolites detected in plasma at baseline and after consuming the GCPE product for 8 weeks are detailed in Table 1. Figure 2 presents the plasmatic pharmacokinetic profiles for some representative metabolites.

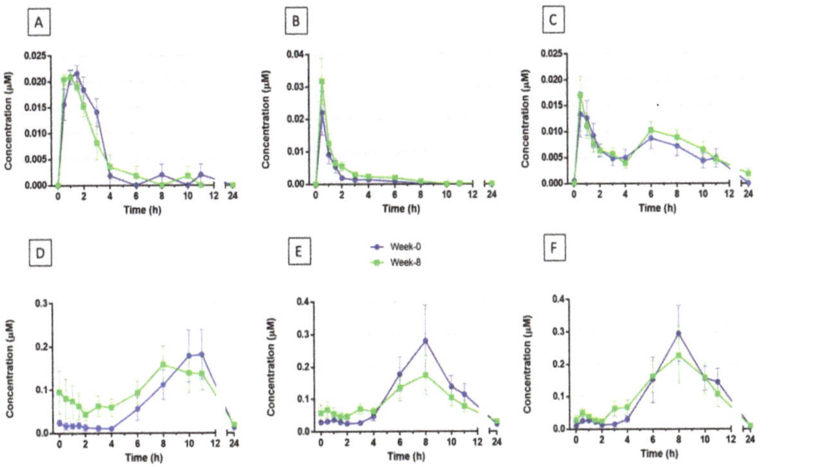

Figure 2. Plasma kinetic profile of (**A**) 5-caffeoylquinic acid, (**B**) isoferulic acid, (**C**) 3′-methoxycinnamic acid-4-sulfate, (**D**) 3-(3′,4′-dihydroxyphenyl)propanoic acid, (**E**) 3-(4′-hydroxy-3′-methoxyphenyl)propanoic acid, and (**F**) 3-(3′-hydroxyphenyl)propanoic acid-4′-sulfate after consumption of a nutraceutical containing 300 mg of hydroxycinnamic acids. Values are means ± SEMs (n = 9). Blue lines: week 0 of the intervention. Green lines: week 8 of the intervention.

Table 1. Pharmacokinetics of plasma metabolites after consumption of the GCPE nutraceutical containing 300 mg of hydroxycinnamic acids at the beginning (week 0) and the end of the intervention (week 8).

	C_{max} (μM)		T_{max} (h) or Range [a]		AUC_{0-24h} (μM min^{-1})	
Metabolite	**Week 0**	**Week 8**	**Week 0**	**Week 8**	**Week 0**	**Week 8**
Intestinal absorption						
5-Caffeoylquinic acid	0.023 ± 0.001	0.022 ± 0.001	1.3 ± 0.2 **	0.7 ± 0.1 **	0.08 ± 0.02	0.062 ± 0.009
5-Feruloylquinic acid	0.032 ± 0.004	0.034 ± 0.005	0.9 ± 0.1	0.9 ± 0.1	0.09 ± 0.02	0.08 ± 0.02
4-Feruloylquinic acid	0.029 ± 0.004	0.031 ± 0.004	0.9 ± 0.1	0.9 ± 0.1	0.06 ± 0.01	0.07 ± 0.01
4′-Hydroxy-3′-methoxycinnamic acid (*Ferulic acid, FA*)	Traces [b,**]	0.008 ± 0.001 **	1.8 ± 0.6 *	0.6 ± 0.1 *	0.034 ± 0.006 **	0.09 ± 0.02 **
3′-Hydroxy-4′-methoxycinnamic acid (*isoFerulic acid, iFA*)	0.023 ± 0.006	0.032 ± 0.007	0.5 ± 0.1	0.5 ± 0	0.025 ± 0.007 **	0.043 ± 0.008 **
3′,4′-Dimethoxycinnamic acid	Traces [b]	Traces [b]	(2–4) [a]	(2–4) [a]	0.007 ± 0.003	0.01 ± 0.006
3′-Methoxycinnamic acid-4′-glucuronide (*FA-4′-glucuronide*)	0.016 ± 0.009	0.009 ± 0.002	1.2 ± 0.3 and 4.7 ± 0.3	1.2 ± 0.3 and 6.0 ± 0.7	0.06 ± 0.01	0.063 ± 0.02
3′-Methoxycinnamic acid-4′-sulfate (*FA-4′-sulfate*)	0.017 ± 0.004	0.017 ± 0.004	0.78 ± 0.09 and 6.8 ± 0.7	0.56 ± 0.06 and 7.2 ± 0.6	0.11 ± 0.03	0.13 ± 0.02
Cinnamic acid-4′-glucuronide (*CoA-4′-glucuronide*)	0.039 ± 0.002	0.038 ± 0.002	1.4 ± 0.3 and 7.0 ± 0.9	2.1 ± 0.2 and 9 ± 2	0.72 ± 0.08	0.71 ± 0.07
Cinnamic acid-4′-sulfate (*CoA-4′-sulfate*)	0.02 ± 0.02	0.03 ± 0.01	1.7 ± 0.9 and 6 ± 3	0 ± 0 and 11 ± 3	0.2 ± 0.2	0.4 ± 0.2
Colonic absorption						
3-(3′,4′-Dihydroxyphenyl)propanoic acid (*Dihydrocaffeic acid, DHCA*)	0.22 ± 0.06	0.22 ± 0.05	10.2 ± 0.3	9.4 ± 0.5	2.0 ± 0.6	2.1 ± 0.5
3-(4′-Hydroxy-3′-methoxyphenyl)propanoic acid (*Dihydroferulic acid, DHFA*)	0.3 ± 0.1	0.25 ± 0.06	7.6 ± 0.4	8.4 ± 0.6	2.2 ± 0.6	1.8 ± 0.3
3-(3′-Hydroxy-4′-methoxyphenyl) propanoic acid (*Dihydroisoferulic acid, DHiFA*)	0.07 ± 0.02	0.07 ± 0.02	7 ± 1	7 ± 1	0.5 ± 0.2	0.6 ± 0.2
3-(3′,4′-Dimethoxyphenyl)propanoic acid (*Dihydrodimethoxycinnamic acid*)	0.11 ± 0.02	0.12 ± 0.01	4 ± 1	10 ± 4	1.2 ± 0.1	1.3 ± 0.1
3-(3′-Hydroxyphenyl)propanoic acid-4′-sulfate (*DHCA-4′-sulfate*)	0.34 ± 0.09	0.33 ± 0.07	7.9 ± 0.5	8.4 ± 0.6	2.3 ± 0.5	2.1 ± 0.4
3-(3′-Methoxyphenyl)propanoic acid-4′-glucuronide (*DHFA-4′-glucuronide*)	0.13 ± 0.03	0.12 ± 0.02	8.7 ± 0.6	8.7 ± 0.6	1.1 ± 0.3	0.8 ± 0.1
3-(4′-Methoxyphenyl)propanoic acid-3′-glucuronide (*DHiFA-3′-glucuronide*)	0.06 ± 0.02	0.07 ± 0.03	9.0 ± 0.6	8.1 ± 0.8	0.6 ± 0.4	0.7 ± 0.4
3-(3′-Methoxyphenyl)propanoic acid-4′-sulfate(*DHFA-4′-sulfate*)	0.19 ± 0.08	0.17 ± 0.05	8.1 ± 0.4	8.4 ± 0.6	1.5 ± 0.7	1.1 ± 0.3
3-(4′-Methoxyphenyl)propanoic acid-3′-sulfate(*DHiFA-3′-sulfate*)	0.04 ± 0.01	0.11 ± 0.06	7 ± 1	11.6 ± 3.2	0.3 ± 0.1	0.8 ± 0.5
Feruloylglycine	Traces [b]	Traces [b]	7 ± 1	8.4 ± 0.6	0.029 ± 0.007	0.026 ± 0.004
Other microbial metabolites						
4′-Hydroxy-3′-methoxyphenylacetic acid	0.14 ± 0.06	0.12 ± 0.05	7 ± 4	4 ± 3	1.4 ± 0.9	1.5 ± 0.8
4′-Hydroxyphenylacetic acid	2 ± 1	7 ± 3	2 ± 2	0.6 ± 0.1	2 ± 1	36 ± 23
3′-Hydroxyphenylacetic acid	0.20 ± 0.07	0.15 ± 0.05	8 ± 2	5 ± 1	2.6 ± 0.9	1.6 ± 0.6
4-Hydroxy-3-methoxybenzoic acid	1.1 ± 0.2	1.6 ± 0.4	7 ± 3	7 ± 2	9 ± 2	14 ± 4
4-Hydroxybenzoic acid	0.09 ± 0.01	0.08 ± 0.01	7 ± 2	5 ± 3	0.9 ± 0.2	0.5 ± 0.2
3-Hydroxybenzoic acid	0.08 ± 0.02	0.10 ± 0.03	8 ± 1 *	4 ± 1 *	0.7 ± 0.3	0.8 ± 0.3
4′-Hydroxyhippuric acid	0.08 ± 0.01	0.09 ± 0.01	12 ± 3	8 ± 2	1.2 ± 0.2	1.3 ± 0.2
3′-Hydroxyhippuric acid	1.1 ± 0.3	1.3 ± 0.2	10.0 ± 0.4	9.2 ± 0.5	10 ± 3	13 ± 3

Values are means ± SEM (n = 9). * $p < 0.05$ ** $p < 0.01$ week 0 vs. week 8. C_{max}, maximum plasma concentration; T_{max}, time to reach the maximum plasma concentration; AUC_{0-24h}, area under the curve. [a] Range where the metabolite showed the highest value. [b] At trace levels, pharmacokinetic parameters were not determined.

Some unmetabolized compounds originally contained in the nutraceutical (5-caffeoylquinic acid (Figure 2A), 4- and 5-feruloylquinic acids) were also detected in plasma, as well as hydroxycinnamic acids (3′,4′-dimethoxycinnamic acid, 4′-hydroxy-3′-methoxycinnamic acid (*Ferulic acid, FA*), and 3′-hydroxy-4′-methoxycinnamic acid (*isoFerulic acid, iFA*) (Figure 2B)) resulting from the hydrolysis of their respective monoacylquinic esters and/or methylation of 3′,4′-dihydroxycinnamic acid (*Caffeic acid*). Some of these compounds were extensively metabolized into phase II sulfated and glucuronidated derivatives (3′-methoxycinnamic acid-4′-glucuronide (*FA-4′-glucuronide*) and 3′-methoxycinnamic acid-4′-sulfate (*FA-4′-sulfate*) (Figure 2C)). No free 4′-hydroxycinnamic acid (*Coumaric acid, CoA*) was detected, but its conjugated forms were present in plasma. According to the time of appearance in plasma, this group of metabolites showed early absorption in the small intestine, with T_{max} values between 0.5 and 2.1 h after GCPE intake (Table 1). Nevertheless, despite having a similar pharmacokinetic profile to these earlier-absorbed metabolites,

FA-4′-sulfate and FA-4′-glucuronide showed a second C_{max} peak in plasma around 5–8 h after ingestion, displaying a biphasic plasma profile (Figure 2C), as well as sulfated and glucuronidated forms of CoA. Overall, these compounds had low concentrations in plasma, showing C_{max} values from 8 to 39 nM (Table 1). Comparing C_{max}, T_{max} and $AUC_{0–24h}$ values at the beginning of the trial (week 0) with those obtained after 8-week supplementation with the GCPE nutraceutical, FA values were significantly different between the two time points. Thus, a shortening of T_{max} from 1.8 h to 0.6 h was observed at week 0 vs. week 8, while C_{max} values increased from traces to 8 nM after supplementation, as well as its $AUC_{0–24h}$ values from 34 nM min^{-1} at week 0 to 90 nM min^{-1} at the end of the intervention (Table 1). Similarly, $AUC_{0–24h}$ also increased for *i*FA from 25 to 43 nM min^{-1}. Moreover, T_{max} values for 5-caffeoylquinic showed a significant reduction after 8 weeks of GCPE supplementation (from 1.3 to 0.7 h).

Another important group of metabolites found in plasma corresponded to the reduced forms of hydroxycinnamic acids, such as 3-(3′,4′-dihydroxyphenyl)propanoic acid (*Dihydrocaffeic acid, DHCA*) (Figure 2D), 3-(4′-hydroxy-3′-methoxyphenyl)propanoic acid (*Dihydroferulic acid, DHFA*) (Figure 2E), 3-(3′-hydroxy-4′-methoxyphenyl)propanoic acid (*Dihydroisoferulic acid, DHiFA*), and 3-(3′,4′-dimethoxyphenyl)propanoic acid (*Dihydrodimethoxycinnamic acid*). All these metabolites were extensively transformed into their phase II derivatives: 3-(3′-hydroxyphenyl)propanoic acid-4′-sulfate (*DHCA-4′-sulfate*) (Figure 2F), 3-(3′-methoxyphenyl)propanoic acid-4′-glucuronide (*DHFA-4′-glucuronide*), 3-(3′-methoxyphenyl) propanoic acid-4′-sulfate (*DHFA-4′-sulfate*), 3-(4′-methoxyphenyl)propanoic acid-3′-glucuronide (*DHiFA-3′-glucuronide*), and 3-(4′-methoxyphenyl)propanoic acid-3′-sulfate (*DHiFA-3′-sulfate*). Feruloylglicine was also present in plasma at trace levels. These compounds appeared in plasma for longer than their precursors after GCPE consumption, reaching their C_{max} between 4 and 11 h postintake (Table 1), thus with kinetics compatible with colonic absorption.

In addition, this group of metabolites showed higher C_{max} values than their precursors, being DHCA-4′-sulfate, DHFA, and DHCA the predominant metabolites (C_{max} values ranged from 220 nM to 340 nM, Table 1). No significant differences were observed for C_{max}, T_{max}, or $AUC_{0–24h}$ values of each metabolite between baseline (week 0) and the end of the intervention (week 8).

Finally, several microbial metabolites identified as hydroxyphenylacetic, hydroxybenzoic, and hydroxyhippuric acid derivatives were also detected in plasma, with 4′-hydroxyphenylacetic and 4-hydroxy-3-methoxybenzoic acids being the most abundant catabolites in this matrix. Some changes were observed between week 0 and 8 with slightly higher C_{max} values at the end of the intervention, but these were not statistically significant (Table 1). Similarly, with the exception of 3-hydroxybenzoic acid, which T_{max} decreased from 8 h to 4 h, no significant differences were observed in pharmacokinetic parameters after 8 weeks of nutritional intervention in this group of microbial catabolites.

3.2. Identification and Quantification of Urinary Metabolites

A total of 46 metabolites derived from GCPE consumption were identified in urine, 22 of which were also detected in plasma. Compound identification features are shown in Supplementary Table S3. Pharmacokinetic parameters (C_{max}, T_{max} and $AUC_{0–24h}$) at the beginning and at the end of the 8-week supplementation period are shown in Supplementary Table S4. The amounts excreted at the different collection intervals, before and after 8 weeks of regular nutraceutical consumption, are given in Supplementary Tables S5 and S6, respectively. The total urinary recovery from 0 to 24 h is shown in Table 2 and the excretion percentages, together with some significant changes in representative metabolites, are summarized in Figure 3.

Table 2. Cumulative excretion (0–24 h) of urinary metabolites after the intake of GCPE nutraceutical at the beginning (week 0) and the end of the intervention (week 8).

	Total 0–24 h (μmol)	
Metabolite	**Week 0**	**Week 8**
Intestinal absorption		
3-Caffeoylquinic acid	0.21 ± 0.05	0.18 ± 0.03
5-Caffeoylquinic acid	0.33 ± 0.06	0.33 ± 0.04
4-Caffeoylquinic acid	0.13 ± 0.07	0.15 ± 0.07
3-Feruloylquinic acid	0.29 ± 0.07	0.34 ± 0.07
5-Feruloylquinic acid	1.5 ± 0.2	1.5 ± 0.2
4-Feruloylquinic acid	0.4 ± 0.2	0.4 ± 0.1
Coumaroylquinic acid	0.14 ± 0.03	0.10 ± 0.02 *
3′,4′-Dihydroxycinnamic acid (*Caffeic acid, CA*)	0.42 ± 0.06	0.45 ± 0.09
4′Hydroxycinnamic acid-3′-sulfate (*CA-3′-sulfate*)	3.7 ± 0.5	4.4 ± 0.8
3′-Hydroxy-4′-methoxycinnamic acid (*isoFerulic acid, iFA*)	1.7 ± 0.2	2.1 ± 0.2
3′-Methoxycinnamic acid-4′-glucuronide (*FA-4′-glucuronide*)	3.4 ± 0.4	3.7 ± 0.5
4′-Methoxycinnamic acid-3′-glucuronide (*iFA-3′-glucuronide*)	4.4 ± 0.5	6 ± 1
3′-Methoxycinnamic acid-4′-sulfate (*FA-4′-sulfate*)	32 ± 5	33 ± 5
4′-Methoxycinnamic acid-3′-sulfate (*iFA-3′-sulfate*)	1.1 ± 0.3	2.1 ± 0.9
TOTAL—Intestinal metabolites	50 ± 6	59 ± 8
Colonic absorption		
3-(3′,4′-Dihydroxyphenyl)propanoic acid (*Dihydrocaffeic acid, DHCA*)	16 ± 2	31 ± 5 **
3-(4′-Hydroxy-3′-methoxyphenyl) propanoic acid (*Dihydroferulic, DHFA*)	0.5 ± 0.2	1.1 ± 0.4
3-(3′-Hydroxy-4′-methoxyphenyl) propanoic acid (*Dihydroisoferulic, DHiFA*)	3.7 ± 0.4	3.2 ± 0.2
3-(4′-Hydroxyphenyl)propanoic acid (*Dihydrocoumaric acid, DHCoA*)	4 ± 1	6 ± 2 **
3-(3′,4′-Dimethoxyphenyl)propanoic acid (*Dihydrodimethoxycinnamic acid*)	0.7 ± 0.2	0.63 ± 0.08
3-(4′-Hydroxyphenyl)propanoic acid-3′-glucuronide (*DHCA-3′-glucuronide*)	0.6 ± 0.2	0.5 ± 0.1
3-(3′-Hydroxyphenyl)propanoic acid-4′-sulfate (*DHCA-4′-sulfate*)	8 ± 2	10 ± 2*
3-(4′-Hydroxyphenyl)propanoic acid-3′-sulfate (*DHCA-3′-sulfate*)	9 ± 2	9 ± 3
3-(3′-Methoxyphenyl)propanoic acid-4′-glucuronide (*DHFA- 4′-glucuronide*)	5 ± 1	7 ± 1*
3-(4′-Methoxyphenyl)propanoic acid-3′-glucuronide (*DHiFA- 3′-glucuronide*)	2.9 ± 0.5	3.1 ± 0.7
3-(3′-Methoxyphenyl)propanoic acid-4′-sulfate (*DHFA- 4′-sulfate*)	10 ± 2	9 ± 2
3-(4′-Methoxyphenyl)propanoic acid-3′-sulfate (*DHiFA- 3′-sulfate*)	3 ± 1	2.6 ± 0.8
3-(Phenyl)propanoic acid-4′-glucuronide (*DHCoA-4′-glucuronide*)	2.1 ± 0.3	1.9 ± 0.3
3-(Phenyl)propanoic acid-4′-sulfate (*DHCoA-4′-sulfate*)	21 ± 6	21 ± 4
3-Dihydrocaffeoylquinic acid	0.35 ± 0.07	0.27 ± 0.08 **
5-Dihydrocaffeoylquinic acid	0.03 ± 0.02	0.03 ± 0.01

Table 2. *Cont.*

Metabolite	Total 0–24 h (μmol) Week 0	Week 8
Intestinal absorption		
4-Dihydrocaffeoylquinic acid	0.06 ± 0.02	0.04 ± 0.02
3-Dihydroferuloylquinic acid	0.3 ± 0.1	0.5 ± 0.07
5-Dihydroferuloylquinic acid	0.3 ± 0.1	0.18 ± 0.07
4-Dihydroferuloylquinic acid	0.06 ± 0.02	0.05 ± 0.02
Dihydrocoumaroylquinic acid	0.5 ± 0.1	0.4 ± 0.1
Dihydrocoumaroylquinic acid	0.22 ± 0.08	0.17 ± 0.07 *
Feruloylglycine	19 ± 5	25 ± 5
*Iso*Feruloylglicine	0.44 ± 0.04	0.6 ± 0.1
TOTAL—Colonic metabolites	106 ± 16	132 ± 17
Other microbial metabolites		
3′,4′-Dihydroxyphenylacetic acid	1.2 ± 0.2	1.3 ± 0.2 *
4′-Hydroxy-3′-methoxyphenylacetic acid	13.2 ± 0.9	12 ± 1
3′-Hydroxyphenylacetic acid	12 ± 3	9 ± 2
3,4-Dihydroxybenzoic acid	0.13 ± 0.02	0.18 ± 0.04
4-Hydroxybenzoic acid	1.3 ± 0.2	1.01 ± 0.09 *
3-Hydroxybenzoic acid	0.91 ± 0.07	1.0 ± 0.1
4′-Hydroxyhippuric acid	14 ± 2	14 ± 2
3′-Hydroxyhippuric acid	32 ± 5	46 ± 8 **
TOTAL—Other microbial metabolites	75 ± 7	85 ± 8
TOTAL Colonic + other microbial met.	181 ± 21	217 ± 21
TOTAL INTESTINAL + COLONIC + OTHERS	231 ± 26	274 ± 30

Values are means ± SEM (n = 9). * $p < 0.05$ ** $p < 0.01$ week 0 vs. week 8.

Unmetabolized hydroxycinnamoylquinic acids (3-, 4-, 5-caffeoylquinic, 3-, 4-, 5-feruloylquinic and coumaroylquinic acids) represented a minor part (1.3%) of the total polyphenols excreted in urine at baseline, decreasing to 1.1% at week 8 (Figure 3). In addition, this group of compounds showed the earliest absorption, since their C_{max} appeared between 0–6 h postintake (Supplementary Table S4). Nevertheless, the only statistically significant difference between week 0 and week 8 in this group of metabolites corresponded to the total urinary recovery (collected from 0 to 24 h after product intake) of coumaroylquinic acid, which decreased from 0.14 μmoles at week 0 to 0.10 μmoles after nutraceutical intervention (week 8) (Table 2).

Lower amounts (0.9%) of free hydroxycinnamic acids, such as 3′,4′-dihydroxycinnamic acid (*Caffeic acid, CA*) or *iFA* were also detected at baseline (week 0) and at week 8 (Figure 3). In turn, their phase II derivatives (4′-hydroxycinnamic acid-3′-sulfate (*CA-3′-sulfate*), FA-4′-glucuronide, FA-4′-sulfate, 4′-methoxycinnamic acid-3′-glucuronide (*iFA-3′-glucuronide*) and 4′-methoxycinnamic acid-3′-sulfate (*iFA-3′-sulfate*)) accounted for 19.4% and 18.1% of phenolic excretion at week 0 and 8, respectively (Figure 3A). Actually, FA-4′-sulfate was the most abundant compound excreted both at baseline (32 μmol/24 h) and after the intervention (33 μmol/24 h) (Table 2). All these metabolites (hydroxycinnamates, hydroxycinnamic acids, and their phase II derivatives) were mainly excreted in the interval from 0 to 3 h, pointing to their early absorption in the small intestine. However, glucuronidated and sulfated FA derivatives were also excreted in great amounts between 11 and 24 h, emphasizing the biphasic profile observed in plasma (Figure 2C). C_{max} and T_{max} values showed no significant changes in these compounds between week 0 and 8. As for the $AUC_{0\text{–}24h}$, there was a statistically significant increase of the $AUC_{0\text{–}24h}$ values of *i*FA-3-sulfate and *iFA*-3-glucuronide after sustained consumption of the nutraceutical (Supplementary Table S4). Total amounts excreted of these early-absorption metabolites were 50 μmol (21.6% of the phenolic excretion) and 59 μmol (20.1%) before and after sustained consumption of GCPE, respectively (Table 2).

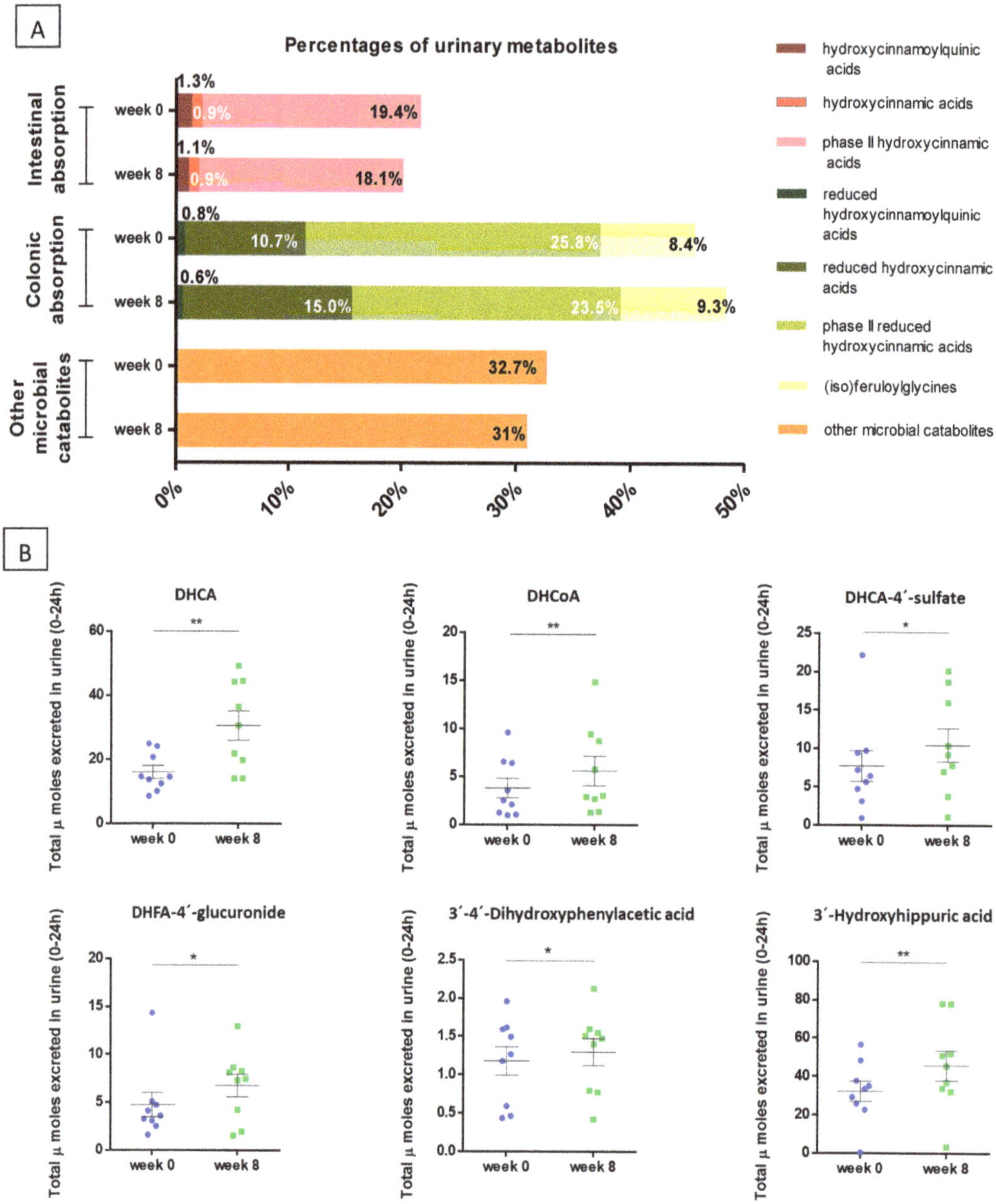

Figure 3. 24 h cumulative urinary excretion represented as (**A**) percentages of the main groups of phenolic metabolites identified, and (**B**) selected metabolites with significant increases in their excretion at the beginning (week 0) and after 8 weeks consuming the GCPE nutraceutical. DHCA: Dihydrocaffeic acid; DHCoA: Dihydrocoumaric acid; DHCA-4′-sulfate: 3-(3′-Hydroxyphenyl)propanoic acid-4′-sulfate; DHFA-4′-glucuronide: 3-(3′-Methoxyphenyl)propanoic acid-4′-glucuronide. Values are means ± SEMs (n = 9). * $p < 0.05$ ** $p < 0.01$ week 0 vs. week 8.

Reduced forms of hydroxycinnamic acids (DHCA, DHFA, DH*i*FA, 3-(4′-hydroxyphenyl) propanoic acid (*Dihydrocoumaric acid, DHCoA*), and dihydrodimethoxycinnamic acid), together with their sulfated and glucuronidated phase II derivatives, constituted the main group of phenolics excreted in urine (Figure 3A). Feruloylglicine and *iso*feruloylglicine were also detected, showing delayed kinetics compatible with colonic absorption (Supplementary Tables S5 and S6), with feruloylglycine being one of the most abundant metabolites in this group (with excretions of 19 and 25 μmol in 24 h at baseline and after 8 weeks of supplementation, respectively). Lastly, reduced forms of hydroxycinnamoylquinic acids (3-, 4- and 5-dihydrocaffeoylquinic acids, 3-, 4-, and 5-dihydroferuloylquinic acids and two isomers of dihydrocoumaroylquinic acid) were also identified in urine, although with lower abundance (0.8% of total phenolics at week 0 and 0.6% at week 8, Figure 3A). Taken together, all these metabolites showed delayed kinetics compatible with colonic absorption (high excretion between 6 and 24 h postintake, Supplementary Tables S5 and S6) and amounted up to 45.7% and 48.4% (week 0 and week 8, respectively) of the total compounds quantified in urine. DHCA stands out among the most abundant catabolites in urine, which excretion significantly increased from 16 to 31 μmol/24 h after the nutritional intervention (Table 2, Figure 3B). The concentration of other metabolites also increased after 8 weeks, such as DHCA-4′-sulfate (which increased from 8 to 10 μmol/24 h), DHFA-4′-glucuronide (from 5 to 7 μmol/24 h), and DHCoA (from 4 to 6 μmol/24 h) (Figure 3B). In turn, urinary concentration of dihydrocaffeoylquinic acid and dihydrocoumaroylquinic acid significantly decreased from 0.35 to 0.27 and from 0.22 to 0.17 μmol/24 h after the intervention, respectively (Table 2). Therefore, quantitative changes between baseline and week 8 did not follow a clear pattern, probably due to the high interindividual variability observed. However, there was a trend towards an increase in urinary elimination of colonic-absorption metabolites (from 106 μmoles/24 h at week 0 to 132 at week 8), as well as an apparent prevalence of sulfated forms over glucuronidated metabolites (Table 2). It is worth pointing out that some metabolites, such as DHCA, DHCA-4′-sulfate, DHCoA, or DHFA-4′-glucuronide, were present in the basal urine samples obtained before the nutraceutical intake (Supplementary Tables S5 and S6), with excretion at 0 h significantly higher at the end of the intervention (week 8) compared to baseline (week 0, $p < 0.01$). In terms of pharmacokinetic parameters (Supplementary Table S4), all the colonic compounds showed T_{max} values mainly between 9 and 24 h. However, at the end of the intervention, T_{max} values tended to decrease, with these changes being statistically significant ($p < 0.05$) for sulfate and glucuronide conjugates of DHCoA. In addition, when the $AUC_{0\text{–}24h}$ values from week 0 and week 8 were compared, statistically significant changes ($p < 0.05$) were also found for DHCA (increasing from 96 to 125 μM min^{-1}), 3-(phenyl)propanoic acid-4′-sulfate (*DHCoA-4′-sulfate*) (from 161 to 251 μM min^{-1}), and 3-dihydroferuloylquinic acid (from 1.6 to 4 μM min^{-1}). Regarding C_{max}, only DHCA values significantly increased at the end of the intervention (from 7.7 to 10 μM).

Finally, some microbial metabolites, such as derivatives of hydroxyphenylacetic and hydroxbenzoic acids, along with hydroxyhippuric acid, could also be detected in urine in high concentrations (Table 2). These metabolites accounted up to 32.7% and 31% of the total phenolics at week 0 and 8, respectively (Figure 3A), showing extensive excretion in the last collection period (11–24 h) in accordance with their microbial origin. Among them, only 3′,4′-dihydroxyphenylacetic acid and 3′-hydroxyhippuric acid showed a statistically significant increase after the intervention (Table 2, Figure 3B). In turn, 4-hydroxybenzoic acid decreased from 1.3 to 1.01 μmol/24 h at week 0 and 8, respectively. No significant differences were observed in the pharmacokinetic parameters or in the total excretion of these microbial metabolites.

In summary, the total amount of hydroxycinnamate metabolites excreted in 24 h urine reached 231 μmol at week 0 and 274 μmol after 8 weeks of daily intake of the nutraceutical (Table 2), which represents, respectively, 27.3% and 32.3% of the 847 μmol (300 mg) of phenols consumed. Although the total amount of urinary metabolites was higher at week 8, it was not statistically different compared to baseline (week 0).

3.3. Identification and Quantification of Fecal Metabolites

As shown in Table 3, eighteen hydroxycinnamic acid derivatives were quantified in feces collected at 0 and 24 h after consuming the GCPE nutraceutical at week 0 and after 8 weeks of daily supplementation. These compounds included minor amounts of hydroxycinnamates and hydroxycinnamic acids, like 3-, 4-, and 5-feruloylquinic acids, CA, FA, and CoA, in which the total concentrations after 8 weeks of intervention were higher both at 0 and 24 h, but without reaching statistical significance.

Table 3. Amount of fecal metabolites excreted at 0 h and 24 h after consumption of the GCPE nutraceutical at the baseline (week 0) and the end of the intervention (week 8).

	0 h (μmol/g)		24 h (μmol/g)	
Metabolite	**Week 0**	**Week 8**	**Week 0**	**Week 8**
Intestinal absorption				
5-Feruloylquinic acid	0.0009 ± 0.0007	0.002 ± 0.002	0.002 ± 0.001	0.0003 ± 0.0003
3-Feruloylquinic acid	0.003 ± 0.003	0.0010 ± 0.0007	0.003 ± 0.003	0.003 ± 0.002
4-Feruloylquinic acid	N.D.	0.001 ± 0.001	N.D.	0.0004 ± 0.0004
3′,4′-Dihydroxycinnamic acid (*Caffeic acid, CA*)	N.D.	0.003 ± 0.002	0.0003 ± 0.0003	0.003 ± 0.003
4′-Hydroxy-3′-methoxycinnamic acid (*Ferulic acid, FA*)	0.005 ± 0.002	0.009 ± 0.003	0.011 ± 0.007	0.019 ± 0.007
4′-Hydroxycinnamic acid (*Coumaric acid, CoA*)	N.D.	0.006 ± 0.004	0.002 ± 0.002	0.003 ± 0.001
TOTAL—Intestinal metabolites	0.009 ± 0.004	0.022 ± 0.007	0.017 ± 0.008	0.028 ± 0.009
Colonic absorption				
3-(3′,4′-Dihydroxyphenyl)propanoic acid (*Dihydrocaffeic acid, DHCA*)	0.011 ± 0.007	0.005 ± 0.002	0.008 ± 0.006	0.0001 ± 0.0001
3-(4′-Hydroxy-3′-methoxyphenyl)propanoic acid (*Dihydroferulic, DHFA*)	0.013 ± 0.004	0.018 ± 0.003	0.019 ± 0.007	0.014 ± 0.003
3-(4′-Hydroxyphenyl)propanoic acid (*Dihydrocoumaric acid, DHCoA*)	0.005 ± 0.003	0.006 ± 0.006	0.005 ± 0.005	0.014 ± 0.005
3-(3′-Hydroxyphenyl)propanoic acid (*Dihydroisocoumaric acid, DHiCoA*)	0.21 ± 0.07	0.4 ± 0.1	0.24 ± 0.07	0.25 ± 0.07
5-Dihydrocaffeoylquinic acid	0.04 ± 0.01	0.06 ± 0.02	0.05 ± 0.01	0.08 ± 0.02
TOTAL—Colonic metabolites	0.28 ± 0.07	0.5 ± 0.1	0.33 ± 0.08	0.36 ± 0.08
Other microbial metabolites				
3′,4′-Dihydroxyphenylacetic acid	N.D.	N.D.	N.D.	0.010 ± 0.008
4′-Hydroxy-3′-methoxyphenylacetic acid	0.007 ± 0.005	0.0006 ± 0.0006	0.02 ± 0.01	0.003 ± 0.003
3′-Hydroxyphenylacetic acid	0.31 ± 0.09	0.18 ± 0.08	0.21 ± 0.08	0.11 ± 0.04
3,4-Dihydroxybenzoic acid	0.016 ± 0.006 *	0.026 ± 0.008 *	0.018 ± 0.005	0.04 ± 0.01
4-Hydroxy-3-methoxybenzoic acid	0.06 ± 0.03	0.006 ± 0.006	0.09 ± 0.05	N.D.
4-Hydroxybenzoic acid	0.004 ± 0.004	0.006 ± 0.004	0.0006 ± 0.0004	0.01 ± 0.01
3-Hydroxybenzoic acid	0.03± 0.02	0.020 ± 0.008	0.03 ± 0.01	0.020 ± 0.007
TOTAL—Other microbial metabolites	0.4 ± 0.1	0.24 ± 0.09	0.4 ± 0.1	0.19 ± 0.07
TOTAL Colonic + other microbial met.	0.7 ± 0.2	0.8 ± 0.2	0.7 ± 0.1	0.6 ± 0.1
TOTAL INTESTINAL + COLONIC + OTHERS	0.7 ± 0.2	0.8 ± 0.2	0.7 ± 0.1	0.6 ± 0.1

Values are means ± SEM (n = 9). * $p < 0.05$ week 0 vs. week 8. N.D.: Not detected.

Regarding colonic metabolites, some compounds already quantified in plasma and urine were also present in feces, such as, DHFA, DHCoA, and 5-dihydrocaffeoylquinic acids. In addition, unconjugated 3-(3′-hydroxyphenyl)propanoic acid (*Dihydroisocoumaric acid, DHiCoA*) was also found in fecal samples, which corresponded to one of the main metabolites excreted in feces. Of note, the concentration of this metabolite in samples collected at 0 h increased from 0.21 μmol/g at week 0 to 0.4 μmol/g at week 8, without reaching statistical significance ($p > 0.05$). This group of colonic metabolites accounted for 39.7% of the total metabolites excreted in the samples collected at week 0 and increased up to 66.7% after 8 weeks of intervention, respectively, which might suggest an accumulative effect derived from the sustained daily intake of the nutraceutical.

Finally, another large group of phenolic metabolites present in feces were the microbial metabolites, including 3′,4′-dihydroxyphenylacetic acid, 4′-hydroxy-3′-methoxyphenylacetic acid, 3′-hydroxyphenylacetic acid, 3,4-dihydroxybenzoic acid, 4-hydroxy-3-methoxybenzoic acid, and 3- and 4- hydroxybenzoic acids (Table 3). This group constituted 59.1% and 51.5% of the total metabolites found at week 0 in the samples collected before and 24 h after consuming the nutraceutical, respectively. Contrary to what was observed for the colonic metabolites, the excretion of these catabolites was lower at the end of the intervention, amounting to only 30.5% and 32.9% of the total phenolics in 0 and 24 h feces, respectively, at week 8. In this group, 3′-hydroxyphenylacetic acid and 3,4-dihydroxybenzoic acids were the most abundant compounds, and particularly the latter showed a significant increase at 0 h, from 0.016 μmol/g to 0.026 μmol/g at week 0 and after the 8-week intervention, respectively ($p < 0.05$). In turn, excretion of 3′-hydroxyphenylacetic acid decreased from 0.31 μmol/g at week 0 to 0.18 μmol/g at week 8 (samples collected at 0 h; concentration in samples collected at 24 h decreased from 0.21 to 0.11 μmol/g, at week 0 and week 8, respectively).

4. Discussion

Nutraceuticals based on (poly)phenols-rich vegetable extracts, such as green coffee, are widely used to combat overweight, obesity, and associated diseases [6]. In this context, understanding the bioavailability and metabolic fate of dietary phenols are key to better know about their potential effects on human health [21]. However, it is still poorly understood how sustained consumption of phenolic compounds may affect their absorption and metabolism, as most bioavailability studies have been acute assays that determined plasma and/or urinary metabolites up to 12–24 h after consuming a single dose of the nutraceutical or food containing the bioactive compound(s). Therefore, the present work aimed to assess the bioavailability of green coffee hydroxycinnamates at the beginning and after 8 weeks of sustained consumption of a GCPE nutraceutical in order to explore any potential adaptive effect in the metabolic profile of this type of phenolic compounds after long-term consumption.

Quantitatively, results are in agreement with those previously obtained in our research group with a green/roasted coffee blend [13], showing that hydroxycinnamate esters were partially absorbed and extensively metabolized in the intestinal tract, with the gut microbiota playing an important role in catabolism. Total urinary excretion of absorbed metabolites accounted for 27.3% of the 847 μmol ingested at week 0, increasing up to 32.3% at the end of the 8-week intervention, with percentages in line with values reported by other authors after coffee consumption [13,22,23]. Qualitatively, no different metabolites were identified before and after regular consumption of the nutraceutical, suggesting no major changes on the biotransformation of phenolic compounds. This small increase in total urinary excretion of metabolites (only 5% between week 0 and week 8) was not statistically significant. However, considering the low bioavailability of green coffee hydroxycinnamates this increment is of interest, and it could be attributed to the sustained consumption of the nutraceutical. The question remains whether prolonging the intake of phenolic-rich foods might have a more relevant effect on the overall bioavailability or, on the contrary, saturation might occur, as suggested by Mena et al. in their repeated-dose study, where they observed a reduced bioavailability of coffee phenolic acids as the daily amount consumed during 4 weeks increased [18].

Unmetabolized parent compounds (caffeoyl-, feruloyl-, and *p*-coumaroylquinic acids), which were the major components of the nutraceutical (Supplementary Table S1), were detected in minor amounts in plasma (Table 1), urine (Table 2), and feces (Table 3), pointing to an extensive metabolization. Lower C_{max} values of 5-caffeoylquinic and 4-,5-feruloylquinic acids (from 22 to 34 nM) were observed in plasma (Table 1), peaking at short times (T_{max} between 0.9 and 1.3 h); they were also excreted in the urine, along with other monoacylquinic acids like coumaroylquinic acid, mainly within 0 to 3 h after consumption of the nutraceutical (Supplementary Table S4). These kinetics are compatible with that observed in previous studies [13,22], indicating a rapid absorption of these compounds in the small intestine. At this stage, the hydrolysis of monoacylquinic and diacylquinic acids were hydrolyzed by mammalian esterases, giving rise to free hydroxycinnamic acids (CA, FA and CoA), which are transformed into sulfated, glucuronidated, and methylated phase II derivatives by catechol-*O*-methyltransferase (COMT), sulfotransferases (SULT), or UDP-glucuronosyltransferases (UGT). All the aforementioned metabolites (hydroxycinnamates, hydroxycinnamic acids, and their phase II derivatives) reached the bloodstream and were excreted mainly at short times, between 0 and 3 h after intake, amounting up to 21.6% and 20.1% of all urinary metabolites excreted in 0–24 h at week 0 and at the end of the 8-week intervention, respectively. It is noteworthy that glucuronidated and sulfated forms showed a second excretion peak, from 6 up to 9 h, in agreement with the observed behavior in plasma (Figure 2C), which may be related to biphasic kinetics due to the enterohepatic circulation and/or colonic metabolism of the hydroxycinnamates [24]. Furthermore, the prevalence of feruloylquinic acid derivatives over caffeoylquinic forms is notorious; in fact, FA-4′-sulfate was one of the most abundant urinary metabolites, up to 32 μmol at week 0 and 33 μmol after the nutraceutical intervention. This suggests extensive methylation of caffeoylquinic acids via COMT [22], considering that these compounds were the most abundant in the GCPE nutraceutical (Supplementary Table S1). Although some significant changes were found in the pharmacokinetic parameters of FA, iFA, and 5-caffeoylquinic acid in plasma (Table 1), along with a reduction in coumaroylquinic acid in urine (Table 2), there were no qualitative differences or a clear pattern in this group of metabolites at the beginning vs. the end of the intervention, probably due to the large interindividual variability.

Most of the hydroxycinnamic acid derivatives reached the colon and were substrates for microbial reductases before their absorption and conjugation into phase II metabolites. It is well known that microbiota esterases are able to hydrolyze the phenolic–quinic acid linkage, and then the released phenolic acids are converted into reduced forms (namely dihydroxycinnamic acids), which are absorbed through the colonic epithelium and transported via portal circulation to the liver [13,24]. These reduced forms can be conjugated by phase II enzymes, which results in a wide range of sulfated and glucuronidated microbial derivatives that reach the systemic circulation and, at the end, are excreted in the urine. It should be noted that among the colonic catabolites, sulfation was the predominant phase II transformation and, to a lower extent, glucuronidation, in agreement with Sanchez-Bridge et al. [25]. These colonic metabolites, together with other colonic ferulic acid derivatives such as feruloylglycine, and a minor group constituted by the dihydromonoacylquinic acids, formed the predominant group of metabolites in plasma and urine, accounting for 45.7% and 48.4% of total phenolics excreted at week 0 and after 8 weeks of intervention, respectively (Figure 3A). These results highlight once again the important role of the microbiota in the metabolism of hydroxycinnamates. Plasma concentration of reduced catabolites was higher than that of their parent compounds, with C_{max} values ranging from 60 to 340 nM, although no significant differences were found in any pharmacokinetic parameter when comparing week 0 vs. 8 (Table 1). One of the few statistically significant differences obtained in the present study was DHCA C_{max} in urine samples, which increased from week 0 to 8, along with DHCoA-4′-sulfate and 3-dihydroferuloylquinic acid, in which $AUC_{0–24h}$ also increased significantly from week 0 to week 8 (Supplementary Table S4). In general, the pharmacokinetic parameters showed an upward trend after the 8-week green coffee nutraceutical intake. These results were in line with the higher amount of colonic metabolites quantified from 0–24 h at week 8 (136 μmoles) versus 106 μmol at week 0, although this difference was not statistically significant, in contrast to the individual amounts of DHCA, DHCoA, DHCA-4′-sulfate, and DHFA-4′-glucuronide, which increased significantly (Table 2, Figure 3B). It is worth noting that these metabolites, along with DHCA-3′-sulfate, DHCoA-4′-sulfate, and feruloylglycine, were the most abundant metabolites in urine. Interestingly, they were present at baseline (time 0 h before GCPE consumption), which could derive either from hydroxycinnamate consumption before the 48 h restriction, reinforcing the idea of their delayed elimination, or from the biotransformation of other phenolic compounds present in nonrestricted foods. Interestingly, DHCoA-4′-sulfate and feruloylglycine, along with unconjugated 3-(3′-hydroxyphenyl)propanoic acid (DH*i*CoA), were the only coffee-derived metabolites in which urinary excretion was increased in the study by Mena et al. after 8 weeks of daily consumption of tablets containing green tea and green coffee phenolic extracts [19]. In our study, only feruloylglycine excretion in urine was higher after 8 weeks of daily intake of the GCPE nutraceutical, yet without reaching the level of significance ($p > 0.05$) (Table 2).

Considering these results, future studies should extend the urine collection period beyond 24 h. Lastly, the higher basal excretion of this group of colonic metabolites at week 8 compared to week 0 (79 vs. 43 μmoles, respectively, Supplementary Tables S5 and S6) might be affected by the previous dose of the nutraceutical taken at week 8, which was consumed only 17 h before sample collection. However, this group of compounds was present in low amounts in fecal samples obtained at week 0 and 8 (Table 3), probably due to extensive microbial catabolism to hydroxyphenylacetic acid and derived metabolites. Remarkably, 5-dihydrocaffeoylquinic acid, a high molecular weight compound and hallmark of hydroxycinnamate metabolism, was found in fecal samples too (Table 3).

On the other hand, high amounts of low molecular weight compounds, termed as 'other microbial metabolites' (hydroxyphenylacetic and hydroxybenzoic acid derivatives), have been quantified in plasma, urine, and feces, resulting from extensive colonic biotransformation. These compounds were present in the biological samples before nutraceutical consumption, since they are not exclusively derived from the catabolism of hydroxycinnamic acids and can participate in other biotransformation pathways. Among the significant increase of some of microbial metabolites in urine outstands 3′-hydroxyhippuric acid, which raised from 32 μmol at week 0 up to 46 μmol at week 8, becoming the most abundant metabolite quantified in urine (Table 2, Figure 3B). However, it cannot be considered a biomarker of coffee intake since it can derive from multiple biotransformation pathways, as just mentioned. Instead, the high levels of FA-4′-sulfate, DHCA, feruloylglycine, or DHCoA-4′-sulfate in urine could be positively correlated with the intake of chlorogenic acids. Therefore, these compounds might be proposed as biomarkers of coffee intake or compliance, and thus be used to assess adherence in intervention trials with coffee as well as a marker of habitual dietary intake of hydroxycinnamates in observational studies.

Overall, the concentration of phenolic metabolites excreted in urine increased from week 0 (231 μmol) to week 8 (274 μmol), without reaching the level of statistical significance (Table 2). This increase was slightly higher for colonic metabolites (which excretion was 24.5% higher in week 8 vs. week 0) compared to intestinal or other low molecular weight catabolites (that increased 18–20% in week 8 vs. week 0). However, this apparently higher absorption of colonic metabolites was not reflected in the circulating concentrations of these compounds in plasma (Table 1).

The amounts of plasmatic, urinary, and fecal metabolites after consumption of the GCPE nutraceutical varied noticeably between participants (detailed in Supplementary Information), showing no clear patterns in the excretion of phenolic metabolites among volunteers. On the contrary, each subject may show a different response depending on the biological sample analyzed, which increases the complexity of interindividual variability. This is in agreement with previous studies that highlight interindividual variability as an important factor affecting the bioavailability of (poly)phenols, pointing to differences in the intestinal microbiota of each person as a key factor, although understanding the underlying causes on interindividual variability is still incomplete and challenging [26]. Certainly, the microbial bioconversion capacity of each person is related to their specific microbiota composition, which influences the final metabolites produced, their bioavailability and biotransformation, and thus, the final impact on the health of the host [27,28]. For example, DHCA and DHFA, in addition to their precursors and other minor catabolites, are present in the upper regions of the large intestine, where they can act as antioxidants and prebiotics [29]. Nevertheless, the influence of other factors, such as age, sex, physiological status, diet, and dose, also need to be taken into account [19]. Therefore, more studies are needed to better understand the influence of these factors on the bioavailability of phenolic compounds.

This study has several limitations: it would have been interesting to extend urinary and fecal sample collection up to 48 h to recover delayed excreted metabolites; also, changes in the intestinal microbiota have not been addressed. In addition, the sample size was estimated following similar previous studies on the urinary excretion of phenolic compounds from coffee [13,19], but we do not perform power calculation due to the small number of studies available as well as the high interindividual variability observed in them. Strengths of the study: pharmacokinetic analysis of the long-term effects of coffee consumption on (poly)phenols bioavailability has been studied. Moreover, the collection of fecal samples enabled a more complete view of the bioavailability of hydroxycinnamic acids. To end, background features were controlled, such as volunteers' physical activity and dietary habits. Lifestyle characteristics and socioeconomic status of the volunteers was homogenous. Intake of the nutraceutical at each visit was also monitored through the indicated markers.

In conclusion, this study confirms that phenolic compounds contained in the GCPE nutraceutical are highly metabolized throughout the gastrointestinal tract in a population of overweight and obese subjects, being differentially absorbed in the upper intestine compared to the colon. The colonic microbiota has played a key role in the metabolism of coffee hydroxycinnamates, since most of the metabolites characterized were formed at the colon. In addition, when the bioavailability before and after 8 weeks of daily consumption of the green coffee nutraceutical were compared, a higher trend in the absorption of GCPE was observed after regular consumption, but the metabolic profiles in plasma, urine, and feces did not statistically change. The present study contributes to better understand the effect of sustained consumption of a phenol-rich extract such as GCPE, since there are limited data on repeated exposure to phenols in bioavailability studies. This could help to develop refined dietary strategies and recommendations to optimize the beneficial effects of phenol-rich foods. Furthermore, FA-4′-sulfate, DHCA, feruloylglycine, or DHCoA-sulfate may be proposed as biomarkers of hydroxycinnamate intake, due to the high levels of these compounds in the biological fluids analyzed.

Supplementary Materials: The following supporting information can be downloaded at: https://www.mdpi.com/article/10.3390/nu14122445/s1, Table S1. Quantities of polyphenols in the green coffee phenolic extract (GCPE) identified by HPLC-DAD; Table S2. Subject characteristics at baseline (week 0) and after 8 weeks of GCPE intervention. Table S3. Chromatographic and spectrometric properties of the identified phenolic metabolites; Table S4. Urine kinetic profile of phenolic metabolites after the ingestion of the GCPE nutraceutical at the beginning (week 0) and at the end of the intervention (week 8); Table S5. Amount of urine metabolites excreted 0–24 h after consumption of the GCPE nutraceutical at the beginning of the intervention (week 0); Table S6. Amount of urine metabolites excreted 0–24 h after consumption of the GCPE nutraceutical at the end of the intervention (week 8).

Author Contributions: Conceptualization L.B.-C., R.M. and B.S.; Investigation B.S., R.M. and L.B.-C.; Methodology M.Á.S., R.M.T., S.G.-R., J.G.-C., B.S., L.B.-C. and R.M.; Data curation M.Á.S., R.M.T. and R.M.; Project administration L.B.-C., B.S. and R.M.; Resources L.B.-C., R.M., and B.S.; Supervision R.M.; Writing—original draft M.Á.S. and R.M.; Writing—review and editing R.M., B.S. and L.B.-C. All authors have read and agreed to the published version of the manuscript.

Funding: This research was funded by project AGL2015-69986-R from the Spanish Ministry of Science and Innovation and project PIE 202070E184 funded by CSIC. MAS and JGC had predoctoral grants funded by Comunidad de Madrid (PEJD-2018-PRE/SAL-9104) and (PEJD-2017-PRE_BIO-4225), respectively.

Institutional Review Board Statement: The study was approved by the Bioethics Committee of Consejo Superior de Investigaciones Científicas (CSIC) and the Clinical Research Ethics Committee of Hospital Universitario Puerta del Hierro (Madrid, Spain) (approval number PI-162-18). It was retrospectively registered in ClinicalTrials.gov (NCT05009615).

Informed Consent Statement: Informed consent was obtained from all subjects involved in the study.

Data Availability Statement: Data have not been deposited in any open repository, but they are available upon request to the authors.

Acknowledgments: We are grateful to volunteers participating in the study and to the Analysis Service Unit (USTA) of ICTAN for the LC-MS analysis.

Conflicts of Interest: The authors declare no conflict of interest.

References

1. Baeza, G.; Sarriá, B.; Bravo, L.; Mateos, R. Exhaustive qualitative LC-DAD-MS^n analysis of Arabica green coffee beans: Cinnamoyl-glycosides and cinnamoylshikimic acids as new polyphenols in green coffee. *J. Agric. Food Chem.* **2016**, *64*, 9663–9674. [CrossRef] [PubMed]
2. Bastian, F.; Hutabarat, O.S.; Dirpan, A.; Nainu, F.; Harapan, H.; Emran, T.B.; Simal-Gandara, J. From plantation to cup: Changes in bioactive compounds during coffee processing. *Foods* **2021**, *10*, 2827. [CrossRef] [PubMed]
3. Pimpley, V.; Patil, S.; Srinivasan, K.; Desai, N.; Murthy, P.S. The chemistry of chlorogenic acid from green coffee and its role in attenuation of obesity and diabetes. *Prep. Biochem. Biotechnol.* **2020**, *50*, 969–978. [CrossRef] [PubMed]
4. Roshan, H.; Nikpayam, O.; Sedaghat, M.; Sohrab, G. Effects of green coffee extract supplementation on anthropometric indices, glycaemic control, blood pressure, lipid profile, insulin resistance and appetite in patients with the metabolic syndrome: A randomised clinical trial. *Br. J. Nutr.* **2018**, *119*, 250–258. [CrossRef] [PubMed]
5. Haidari, F.; Samadi, M.; Mohammadshahi, M.; Jalali, M.T.; Engali, K.A. Energy restriction combined with green coffee bean extract affects serum adipocytokines and the body composition in obese women. *Asia Pac. J. Clin. Nutr.* **2017**, *26*, 1048–1054. [PubMed]
6. Gorji, Z.; Varkaneh, H.K.; Talaei, S.; Nazary-Vannani, A.; Clark, C.; Fatahi, S.; Rahmani, J.; Salamat, S.; Zhang, Y. The effect of green-coffee extract supplementation on obesity: A systematic review and dose-response meta-analysis of randomized controlled trials. *Phytomedicine* **2019**, *63*, 153018. [CrossRef] [PubMed]
7. Ali, S.S.; Ahmad, W.; Budin, S.B.; Zainalabidin, S. Implication of dietary phenolic acids on inflammation in cardiovascular disease. *Rev. Cardiovasc. Med.* **2020**, *21*, 225–240. [CrossRef]
8. Chen, H.; Huang, W.; Huang, X.; Liang, S.; Gecceh, E.; O Santos, H.; Khani, V.; Jiang, X. Effects of green coffee bean extract on C-reactive protein levels: A systematic review and meta-analysis of randomized controlled trials. *Complementary Ther. Med.* **2020**, *52*, 102498. [CrossRef]
9. Amigo-Benavent, M.; Wang, S.; Mateos, R.; Sarriá, B.; Bravo, L. Antiproliferative and cytotoxic effects of green coffee and yerba mate extracts, their main hydroxycinnamic acids, methylxanthine and metabolites in different human cell lines. *Food Chem. Toxicol.* **2017**, *106*, 125–138. [CrossRef]
10. Lang, R.; Dieminger, N.; Beusch, A.; Lee, Y.M.; Dunkel, A.; Suess, B.; Skurk, T.; Wahl, A.; Hauner, H.; Hofmann, T. Bioappearance and pharmacokinetics of bioactives upon coffee consumption. *Anal. Bioanal. Chem.* **2013**, *405*, 8487–8503. [CrossRef]
11. Marmet, C.; Actis-Goretta, L.; Renouf, M.; Giuffrida, F. Quantification of phenolic acids and their methylates, glucuronides, sulfates and lactones metabolites in human plasma by LC-MS/MS after oral ingestion of soluble coffee. *J. Pharm. Biomed. Anal.* **2014**, *88*, 617–625. [CrossRef]
12. Clifford, M.N.; Kerimi, A.; Williamson, G. Bioavailability and metabolism of chlorogenic acids (acyl-quinic acids) in humans. *Compr. Rev. Food Sci. Food Saf.* **2020**, *19*, 1299–1352. [CrossRef]
13. Gómez-Juaristi, M.; Martínez-López, S.; Sarria, B.; Bravo, L.; Mateos, R. Bioavailability of hydroxycinnamates in an instant green/roasted coffee blend in humans. Identification of novel colonic metabolites. *Food Funct.* **2018**, *9*, 331–343. [CrossRef]
14. Kawabata, K.; Yoshioka, Y.; Terao, J. Role of intestinal microbiota in the bioavailability and physiological functions of dietary polyphenols. *Molecules* **2019**, *24*, 370. [CrossRef]
15. Cortés-Martín, A.; Selma, M.V.; Tomás-Barberán, F.A.; González-Sarrías, A.; Espín, J.C. Where to look into the puzzle of polyphenols and health? The postbiotics and gut microbiota associated with human metabotypes. *Mol. Nutr. Food Res.* **2020**, *64*, e1900952. [CrossRef]
16. Faraldo, T.A.; Macedo, M.; Aymoto, N.M.; Maria, F. The two-way polyphenols-microbiota interactions and their effects on obesity and related metabolic diseases. *Front. Nutr.* **2019**, *6*, 188. [CrossRef]
17. Lee, H.C.; Jenner, A.M.; Low, C.S.; Lee, Y.K. Effect of tea phenolics and their aromatic fecal bacterial metabolites on intestinal microbiota. *Res. Microbiol.* **2006**, *157*, 876–884. [CrossRef]
18. Mena, P.; Bresciani, L.; Tassotti, M.; Rosi, A.; Martini, D.; Antonini, M.; Cas, A.D.; Bonadonna, R.; Brighenti, F.; Del Rio, D. Effect of different patterns of consumption of coffee and a cocoa-based product containing coffee on the nutrikinetics and urinary excretion of phenolic compounds. *Am. J. Clin. Nutr.* **2021**, *114*, 2107–2118. [CrossRef]
19. Mena, P.; Ludwig, I.A.; Tomatis, V.B.; Acharjee, A.; Calani, L.; Rosi, A.; Brighenti, F.; Ray, S.; Griffin, J.L.; Bluck, L.J.; et al. Inter-individual variability in the production of flavan-3-ol colonic metabolites: Preliminary elucidation of urinary metabotypes. *Eur. J. Nutr.* **2019**, *58*, 1529–1543. [CrossRef]
20. García-Cordero, J.; Sierra-Cinos, J.L.; Seguido, M.A.; González-Rámila, S.; Mateos, R.; Bravo-Clemente, L.; Sarriá, B. Regular consumption of green coffee phenol, oat β-glucan and green coffee phenol/oat β-glucan supplements does not change body composition in subjects with overweight and obesity. *Foods* **2022**, *11*, 679. [CrossRef]
21. Dima, C.; Assadpour, E.; Dima, S.; Jafari, S.M. Bioavailability and bioaccessibility of food bioactive compounds; overview and assessment by in vitro methods. *Compr. Rev. Food Sci. Food Saf.* **2020**, *19*, 2862–2884. [CrossRef] [PubMed]
22. Stalmach, A.; Williamson, G.; Crozier, A. Impact of dose on the bioavailability of coffee chlorogenic acids in humans. *Food Funct.* **2014**, *5*, 1727–1737. [CrossRef] [PubMed]
23. Farah, A.; Monteiro, M.; Donangelo, C.M.; Lafay, S. Chlorogenic acids from green coffee extract are highly bioavailable in humans. *J. Nutr.* **2008**, *138*, 2309–2315. [CrossRef] [PubMed]

24. Marín, L.; Miguélez, E.M.; Villar, C.J.; Lombó, F. Bioavailability of dietary polyphenols and gut microbiota metabolism: Antimicrobial properties. *BioMed Res. Int.* **2015**, *2015*, 905215. [CrossRef] [PubMed]
25. Sanchez-Bridge, B.; Renouf, M.; Sauser, J.; Beaumont, M.; Actis-Goretta, L. The roasting process does not influence the extent of conjugation of coffee chlorogenic and phenolic acids. *BioFactors* **2016**, *42*, 259–267. [CrossRef]
26. Manach, C.; Milenkovic, D.; Van de Wiele, T.; Rodriguez-Mateos, A.; de Roos, B.; Garcia-Conesa, M.T.; Landberg, R.; Gibney, E.R.; Heinonen, M.; Tomás-Barberán, F.; et al. Addressing the inter-individual variation in response to consumption of plant food bioactives: Towards a better understanding of their role in healthy aging and cardiometabolic risk reduction. *Mol. Nutr. Food Res.* **2017**, *61*, 1600557. [CrossRef]
27. Selma, M.V.; Espín, J.C.; Tomás-Barberán, F.A. Interaction between phenolics and gut microbiota: Role in human health. *J. Agric. Food Chem.* **2009**, *57*, 6485–6501. [CrossRef]
28. Etxeberria, U.; Fernández-Quintela, A.; Milagro, F.I.; Aguirre, L.; Martínez, J.A.; Portillo, M.P. Impact of polyphenols and polyphenol-rich dietary sources on gut microbiota composition. *J. Agric. Food Chem.* **2013**, *61*, 9517–9533. [CrossRef]
29. Ludwig, I.A.; de Peña, M.P.; Cid, C.; Crozier, A. Catabolism of coffee chlorogenic acids by human colonic microbiota. *BioFactors* **2013**, *39*, 623–632. [CrossRef]

MDPI
St. Alban-Anlage 66
4052 Basel
Switzerland
www.mdpi.com

Nutrients Editorial Office
E-mail: nutrients@mdpi.com
www.mdpi.com/journal/nutrients

www.ingramcontent.com/pod-product-compliance
Lightning Source LLC
Chambersburg PA
CBHW041016270926
51850CB00024B/405
* 9 7 8 3 0 3 6 5 9 2 6 2 6 *